April 1998

Theory of Transport Properties of Semiconductor Nanostructures

ELECTRONIC MATERIALS SERIES

Electronic Materials Series:
This series, devoted to electronic materials subjects of active research interest, provides coverage of basic scientific concepts as well as relating the subjects to the elctronic applications and providing details of the electronic systems, circuits or devices in which the materials are used. The series will be a useful reference source at senior undergarduate and graduate level as well as for research workers in industrial laboratories who wish to broaden their knowledge into a new field.

Series editors:

Professor A.F.W. Willoughby,
Department of Engineering Materials,
University of Southampton,
UK.

Professor R. Hull
Department of Material
Science and Engineering,
University of Virginia,
USA.

Series adviser:

Dr Peter Capper,
GEC–Marconi Infra-Red Ltd,
Southampton,
UK

Titles available:

1. *Widegap II-VI Compounds for Opto-electronic Applications.*
   Edited by E. Ruda

2. *High Temperature Electronics.*
   Edited by M. Willander and H.L. Hartnagel

3. *Narrow-gap II-VI Compounds for Optoelectronic and Electromagnetic Applications.*
   Edited by Peter Capper

4. *Theory of Transport Porperties of Semiconductor Nanostructures.*
   Edited by Eckehard Schöll

Theory of Transport Properties of Semiconductor Nanostructures

Edited by
Eckehard Schöll
Institut für Theoretische Physik
Technische Universität Berlin
Germany

CHAPMAN & HALL
London · Weinheim · New York · Tokyo · Melbourne · Madras

Published by Chapman & Hall, an imprint of Thomson Science, 2–6 Boundary Row, London SE1 8HN, UK

Thomson Science, 2–6 Boundary Row, London SE1 8HN, UK

Thomson Science, 115 Fifth Avenue, New York, NY 10003, USA

Thomson Science, Suite 750, 400 Market Street, Philadelphia, PA 19106, USA

Thomson Science, Pappelallee 3, 69469 Weinheim, Germany

First edition 1998

Thomson Science is a division of International Thomson Publishing I(T)P®

Typeset in LaTeX 2E by Focal Image Ltd, London

Printed in Great Britain by St Edmundsbury Press, Bury St Edmunds

ISBN 0 412 73100 2

A catalogue record for this book is available from the British Library

∞ Printed on acid-free text paper, manufactured in accordance with ANSI/NISO Z39.48-1992 (Permanence of Paper).

Contents

Preface

Recent advances in the fabrication of semiconductors have created almost unlimited possibilities to design structures on a nanometre scale with extraordinary electronic and optoelectronic properties. The theoretical understanding of electrical transport in such nanostructures is of utmost importance for future device applications. This represents a challenging issue of today's basic research since it requires advanced theoretical techniques to cope with the quantum limit of charge transport, ultrafast carrier dynamics and strongly nonlinear high-field effects.

This book, which appears in the electronic materials series, presents an overview of the theoretical background and recent developments in the theory of electrical transport in semiconductor nanostructures.

It contains 11 chapters which are written by experts in their fields. Starting with a tutorial introduction to the subject in Chapter 1, it proceeds to present different approaches to transport theory. The semiclassical Boltzmann transport equation is in the centre of the next three chapters. Hydrodynamic moment equations (Chapter 2), Monte Carlo techniques (Chapter 3) and the cellular automaton approach (Chapter 4) are introduced and illustrated with applications to nanometre structures and device simulation. A full quantum-transport theory covering the Kubo formalism and nonequilibrium Green's functions (Chapter 5) as well as the density matrix theory (Chapter 6) is then presented. The next four chapters are devoted to more specific topics which are of great current interest, such as phase-coherent transport in mesoscopic structures including quantized Hall samples, quantum point contacts and resonant tunnelling structures (Chapter 7), chaotic dynamics in lateral antidot superlattices (Chapter 8), Bloch oscillations and Wannier–Stark localization in vertical superlattices (Chapter 9) and high-field transport and field domain formation in superlattices (Chapter 10). Finally, Chapter 11 provides an overview of the different scattering processes which are of relevance in semiconductor nanostructures.

The book is aimed at physicists, electronic engineers, materials scientists and applied mathematicians. It may be used in research, as professional reference in microelectronics, optoelectronics and graduate teaching. It covers issues associated with recent advances in device simulation, quantum transport and nonlinear dynamics in semiconductors. It should be useful to graduate students as well as to established researchers by providing state-of-the-art overviews of theoretical methods, results and applications in this field. Wherever possible, it is attempted not only to introduce the general framework of the theoretical approaches, but also to apply these to the simulation of specific semiconductor devices such as, for instance, EEPROM memory cells, field-effect and high-electron mobility transistors and resonant tunnelling diodes. At the same time, it is hoped that this book will stimulate further developments by discussing novel effects which have not yet been used in commercial devices, such as the Coulomb drag effect, mesoscopic conductors and quantum-point contacts, terahertz emission by Bloch oscillations and nonlinear self-sustained space charge oscillations in superlattices, to mention but a few examples.

I wish to thank many colleagues for their stimulating influence and, in particular, my collaborators S. Bose, D. Merbach, F. Prengel, G. Schwarz and A. Wacker for their cooperation. Special thanks are due to the contributors of the different chapters whose excellent and timely collaboration only has made possible this volume and to Chapman and Hall Publishers for their efficient cooperation. Partial support by Deutsche Forschungsgemeinschaft is gratefully acknowledged.

Berlin, August 1997
Eckehard Schöll

CHAPTER 1

Introduction

Eckehard Schöll

Technische Universität Berlin, Institut für Theoretische Physik, Hardenbergstr. 36, 10623 Berlin, Germany

1.1 INTRODUCTION

This book deals with charge transport in semiconductor structures whose spatial dimensions are on a nanometre scale. With the advent of modern semiconductor growth technologies such as molecular beam epitaxy (MBE) or metal-organic chemical vapour deposition (MOCVD), artificial structures composed of different materials with layer widths of only a few nanometres have been grown and additional lateral patterning by electron-beam lithography or other lithographic or etching techniques (ion beam, x-ray, scanning probe microscopies) can impose lateral dimensions of quantum confinement in the 10 nm regime. Thus, it has become possible to design and fabricate semiconductor structures whose vertical and lateral dimensions are controlled on an atomic length scale. This has given us the unprecedented capability to tailor devices with extraordinary electrical and optical properties. If the geometrical dimensions of the semicondutur structures reach the order of the characteristic physical length scales of transport which will be discussed below, the transport and optical properties are no longer determined by the bare material constants but will heavily depend upon the size and geometry of the device. This has opened up a vast field of research activity which is, on one hand, of fundamental interest since it pushes the border of physically accessible phenomena in semiconductors to the ultimate quantum limit, but which is also, on the other hand, of utmost importance with respect to applications since the miniaturization and ultralarge-scale integration of electronic components is still in progress. For example, channel lengths of field-effect transistors (FETs) of 100 nm are nowadays standard in mass-production Megabit chips, and memory chips with 10^9 transistors on a single chip are anticipated in a few years. While these are based on silicon technology, the class of GaAs–AlGaAs materials is essential for high-speed electronic and optoelectronic nanometre devices such as for example the high electron mobility transistor (HEMT), the modulation doped field effect transistor (MODFET) and injection laser diodes. Even single-electron tunnelling has been realized in

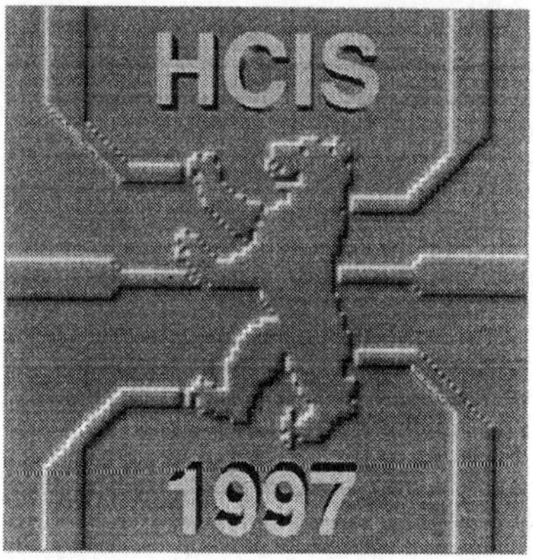

Figure 1.1 *Logo of the International Conference on Nonequilibrium Carrier Dynamics in Semiconductors (HCIS-10) in Berlin, 1997, showing a fictitious semiconductor nanostructure patterned as the Berlin bear. Logo design: F. Prengel (Technical University of Berlin).*

GaAs nanostructures. A whole new class of promising optoelectronic devices are quantum dot lasers based on InAs or InGaAs grown on strained GaAs substrates. Under appropriate growth conditions the InAs or InGaAs spontaneously forms self-organized regular arrays of islands ('quantum dots') with a diameter on the nanometre scale, leading to highly efficient laser emission at low threshold currents with high temperature stability. Those few examples may serve to demonstrate the applicative and technological relevance of the subject.

There is a need for a thorough theoretical understanding of the transport properties of such semiconductor nanostructures. Over the years various different approaches and concepts have been developed [1–7]. Workshops, schools and international conference series such as those on hot carriers in semiconductors (HCIS) [8], electronic properties of two-dimensional systems (EP2DS) [9], modulated semiconductor structures (MSS) [10] and semiconductor microstructures and microdevices (ICSMM) [11] have focused on both experimental and theoretical aspects of these topics (Fig. 1.1). It is an aim of the current research activity not only to tailor and optimize structures with respect to desired transport properties, but conversely also to draw conclusions on the quality and perfection of the growth processes themselves from transport measurements.

In this book basic state-of-the-art theoretical methods, concepts and results will be presented and applied to a variety of specific transport phenomena which are of broad current interest.

Sections 1.1.1–1.1.3 give three issues which should be noted as they are associated with transport in nanostructures.

1.1.1 Device simulation

First, there is a fruitful mutual interaction of transport theory and device simulation. The limitations of classical drift-diffusion device simulators [12] have been long surpassed by nowadays' nanostructure devices and new means of device simulation must be explored. The need for new simulation strategies has stimulated research on novel approaches to transport theory. Conversely, state-of-the-art concepts of charge transport are applied to simulate realistic semiconductor device structures and real electronic materials. For instance, the concept of hydrodynamic simulation using not only the carrier densities but also the mean carrier momenta and energies as dynamic variables has been successfully applied to logic and memory components with complex three-dimensional structures and realistic geometries [13]. Also, the Monte Carlo method has been extensively employed to a realistic simulation of nanostructure devices. Hereby not only the overall carrier densities but also the distribution of carriers over different energy states can be correctly described. A statistical ensemble of carriers with different energies is simulated, taking into account their acceleration in the electric field as well as detailed scattering processes [14]. While the inclusion of complicated realistic boundary conditions is generally difficult within this approach, other methods relying on a discretization of the time as well as the real space and state space in cells ('cellular automata') have been devised as an efficient alternative scheme which can cope with arbitrarily complex geometries, boundary conditions and pronounced spatial inhomogeneities [15].

1.1.2 Quantum transport

Secondly, with smaller and smaller dimensions, the conventional semiclassical picture of charge transport based on the Boltzmann transport equation must be replaced by fully quantum-mechanical concepts. Quantum-mechanical effects become important on length and timescales given by the uncertainty relations. On timescales τ shorter than that given by the energy-time uncertainty $\tau \Delta E \geq \hbar$, energies ΔE cannot be resolved. If τ is the mean time between collisions, this effect leads to a collisional broadening of the energy levels by ΔE, i.e. the energy is not strictly conserved. If we require, for instance, ΔE to be much smaller than an optical phonon energy of 50 meV, τ should be much larger than 10^{-14} s. Similarly, the Heisenberg uncertainty relation requires $\Delta x \Delta p \geq \hbar/2$, where x and p are position and momentum, respectively. In the classical picture electrons are considered as particles with well-defined position and momenta in between two collisions. This approximation requires that the quantum-mechanical wave packet describing the electron has a momentum uncertainty Δp much less than the average momentum p and a position uncertainty Δx much less than the mean-free path. Further, in the classical Boltzmann transport theory collisions are assumed to be instantaneous in time and point-like in space. However, if a strong electric field is applied, the electron momentum and energy will change appreciably during the collision and the transition rates will be modified. This is called the intracollisional field effect. In devices of length 0.1 μm and fields of the order of 10^5 V cm^{-1}, for example, electron energies of the order of 1 eV and average times between collisions of 10^{-14} s can be reached, and a noticeable effect will occur. Thus, the conditions of classical transport can easily be violated in semiconductor nanostructures and quantum transport theories must be applied [16].

1.1.3 Nonlinear dynamics

Thirdly, if a bias of typically a few volts is applied across a nanostructure, electric fields up to 10^6 V cm^{-1} can easily arise. Therefore we are dealing with the high-field regime where strong nonlinearities and considerable electron heating have to be considered. In the electric field the charge carriers may acquire average kinetic energies many orders of magnitude above the value of $(3/2)k_B T_L$, k_B being the Boltzmann constant, which characterizes thermal equilibrium with the crystal lattice at temperature T_L. An 'electron temperature' T_e may be associated with that kinetic energy which can reach 10^4 K in real devices [17]. Far from thermodynamic equilibrium the carrier dynamics is described by nonlinear equations and instabilities or nonequilibrium phase transitions may arise [18–22]. The viewpoint of a semiconductor as a nonlinear dynamic system has provided a fruitful concept in the understanding and predicting of complex transport behaviour such as the self-organized formation of inhomogeneous current density and field distributions in the form of current filaments or high-field domains, threshold switching, bifurcations of different solution branches leading to multistability and hysteresis, spontaneous current or voltage oscillations or chaotic dynamics. In fact, the nonlinear spatio-temporal dynamics of carriers has become an active and still growing field of research, in particular in the realm of semiconductor nanostructures.

All these three issues will be touched in this book.

1.2 WHAT ARE NANOSTRUCTURES?

A typical feature of semiconductor structures fabricated by epitaxial growth and lateral patterning is carrier confinement. It means that the charge carriers cannot move freely in all directions of space but are confined by potential barriers which form at interfaces between different materials. If this confinement occurs on a nanometre scale the semiconductor behaves as a system of reduced dimensionality. Depending on whether the confinement occurs in one, two or even all three spatial directions, the carriers can move only in the remaining two, one or zero directions, respectively and hence the electronic system is called two-, one- or zero-dimensional, respectively. Figure 1.2 schematically shows three prototypes of nanostructures with reduced dimensionality: (a) a **quantum well** formed of a GaAs epitaxial layer sandwiched between two $Al_xGa_{1-x}As$ layers

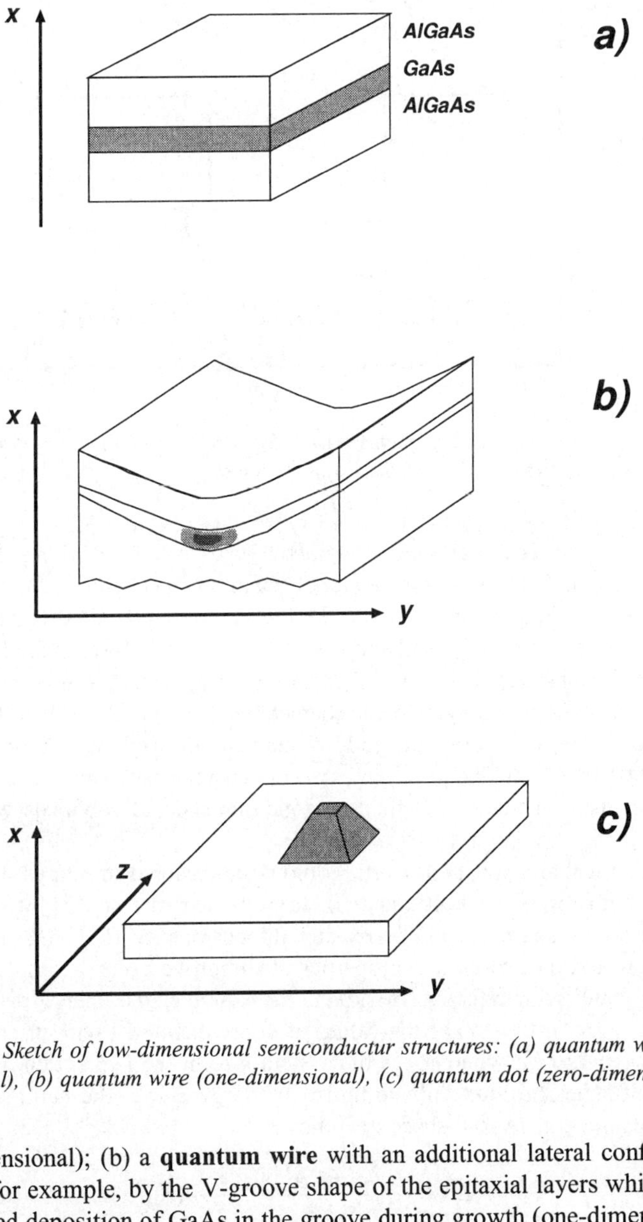

Figure 1.2 *Sketch of low-dimensional semiconductur structures: (a) quantum well (two-dimensional), (b) quantum wire (one-dimensional), (c) quantum dot (zero-dimensional).*

(two-dimensional); (b) a **quantum wire** with an additional lateral confinement effected, for example, by the V-groove shape of the epitaxial layers which leads to enhanced deposition of GaAs in the groove during growth (one-dimensional); (c) a **quantum dot** grown, for example, by the Stranski–Krastanov growth mode forming an island on a strained substrate (zero-dimensional).

Figure 1.3 shows the schematic energy diagram of an AlGaAs/GaAs quantum well heterostructure. Because the energy gap of AlGaAs is larger than that of GaAs, there is an energy offset in the conduction band E_C at the interfaces

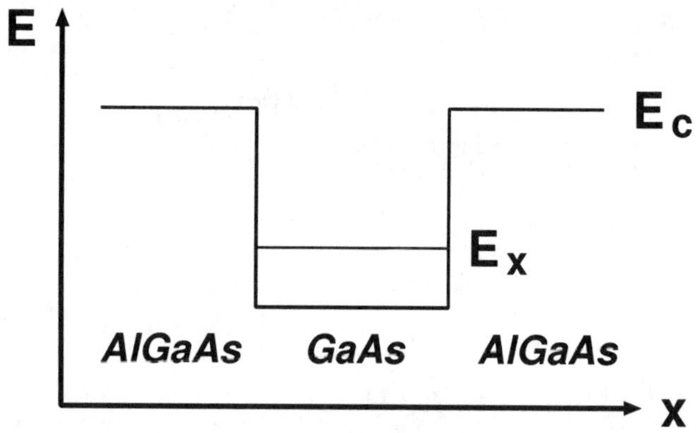

Figure 1.3 *Energy diagram of a quantum well.* E_c *denotes the bulk conduction band and* E_x *denotes the ground level in the GaAs quantum well.*

between the two materials and a quantum well is formed along the confining x-direction. It follows from textbook quantum mechanics [23] that bound states exist in this quantum well and the lowest energy level is denoted by E_x. However, propagation of the electron is still possible in the remaining two (y- and z-) directions and therefore the energy E_x represents the bottom of a two-dimensional conduction (sub-)band characterized by Bloch vectors k_y and k_z. Depending upon the width and depth of the quantum well, several bound levels may exist, corresponding to several two-dimensional subbands. Similarly, the energy levels in a quantum wire form one-dimensional subbands with Bloch vector k_z.

The electrical and optical properties of low-dimensional structures are severely changed with respect to bulk material. In particular, the density of states generally becomes sharper with decreasing dimensionality (Fig. 1.4) and hence the optical and electro-optical properties are improved. For a simple isotropic parabolic band with effective mass m^*, for example, the energy dispersion is $E = \hbar^2k^2/(2m^*)$. In k-space the values of k_i are quantized with discrete differences $\Delta k_i = 2\pi/L_i$, where L_i is the system size in the i-direction. Hence, the number of states per unit volume up to an energy E, i.e. the number of states contained in a sphere in k-space of radius $k(E) = (2m^*E/\hbar^2)^{1/2}$, is given by

$$N(E) = \frac{2}{(2\pi)^3}\frac{4\pi k^3}{3} = \frac{1}{3\pi^2}k(E)^3, \tag{1.2.1}$$

the factor of 2 is present to allow for spin degeneracy. The density of states is obtained as $D(E) = \mathrm{d}N(E)/\mathrm{d}E$, i.e. for the three-dimensional case

$$D_{3D}(E) = \frac{1}{2\pi^2}\left(\frac{2m^*}{\hbar^2}\right)^{3/2} E^{1/2}. \tag{1.2.2}$$

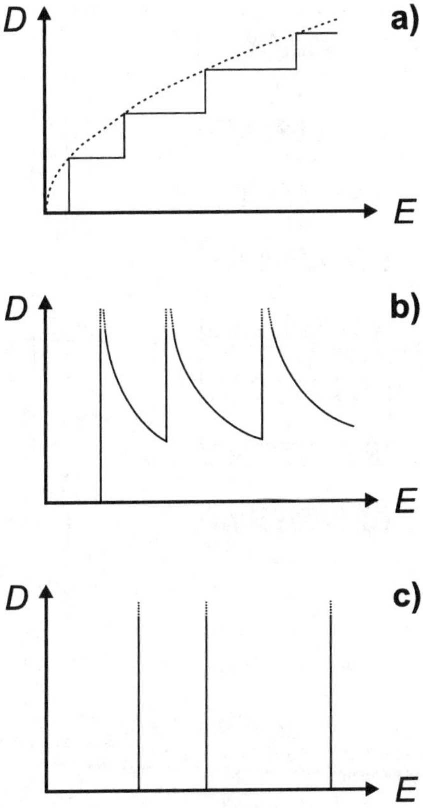

Figure 1.4 *Density of states D(E) of low-dimensional semiconductor structures: (a) quantum well (two-dimensional); the broken curve represents the three-dimensional density of states, (b) quantum wire (one-dimensional), (c) quantum dot (zero-dimensional).*

For the two-dimensional case with $E = E_\nu + \hbar^2 k^2/(2m^*)$, $k^2 = k_y^2 + k_z^2$, $\nu = 1, 2, \ldots$ an analogous argument yields for the number of states per unit area in each individual subband

$$N(E) = \frac{2}{(2\pi)^2}\pi k^2 = \frac{1}{2\pi}k(E)^2 = \frac{m^*}{\pi\hbar^2}E \qquad (1.2.3)$$

and hence

$$D_{2D}(E) = \frac{m^*}{\pi\hbar^2}\sum_\nu \Theta(E - E_\nu) \qquad (1.2.4)$$

where $\Theta(E)$ is the Heaviside function.

For the one-dimensional case with confinement in two directions $E = E_{\nu,\mu} + \hbar^2 k^2/(2m^*)$, $k^2 = k_z^2$, $\nu, \mu = 1, 2, \ldots$ we obtain for the number of states per

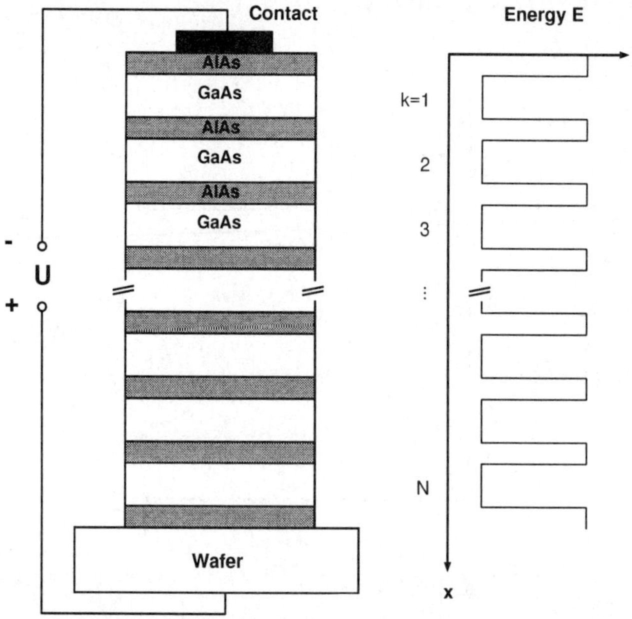

Figure 1.5 *Superlattice structure of alternating GaAs and AlAs layers and corresponding energy diagram.*

unit length in each subband

$$N(E) = \frac{2}{2\pi} 2k = \frac{2}{\pi}\left(\frac{2m^* E}{\hbar^2}\right)^{1/2} \tag{1.2.5}$$

and hence

$$D_{1D}(E) = \frac{1}{\pi}\left(\frac{2m^*}{\hbar^2}\right)^{1/2} \sum_{\nu\mu} (E - E_{\nu\mu})^{-1/2} \Theta(E - E_{\nu\mu}). \tag{1.2.6}$$

Finally, for the zero-dimensional case the spectrum is completely discrete and the density of states becomes a series of δ-peaks

$$D_{0D}(E) = 2 \sum_{\nu\mu\lambda} \delta(E - E_{\nu\mu\lambda}). \tag{1.2.7}$$

Another important nanostructure arises if several alternating layers of two materials with different bandgaps are grown as shown in Fig. 1.5 for the GaAs/AlAs system. The energy diagram shows a periodic modulation of the conduction band with a period given by the sum of the GaAs quantum well width and the AlAs barrier width. Therefore this artificial structure is called a **superlattice** [24]. For sufficiently thin barriers the different quantum wells are so strongly coupled that a one-dimensional energy band is formed in the growth direction. However,

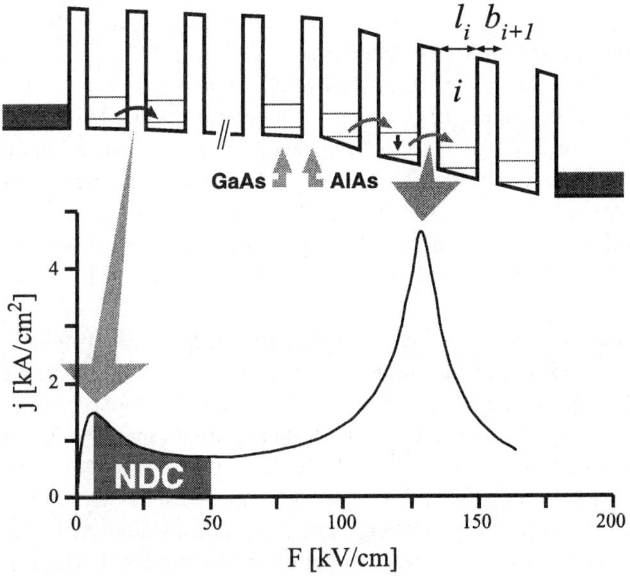

Figure 1.6 *Superlattice exhibiting domain formation. The associated current density (j) versus field (F) characteristic shows NDC. The low-field domain corresponds to sequential tunnelling between equivalent levels of adjacent quantum wells (low-field peak of the j(F) characteristic), while the high-field domain corresponds to resonant tunnelling between different levels of adjacent wells (high-field peak).*

because the superlattice period is much larger than the atomic lattice constant, the resulting Brillouin zones and bandwidths are much smaller than those of atomic lattices and the obtained band structure is called a **miniband**. For larger barrier widths the coupling is weaker and sequential tunnelling of electrons between different wells plays a dominant role. A number of interesting transport phenomena occur in such superlattices if a voltage U is applied as shown in Fig. 1.5; these will be discussed in detail in forthcoming chapters.

For example, if a strong bias is applied to a weakly coupled superlattice, the current density j as a function of the electric field F displays a strongly nonlinear relation as sketched in Fig. 1.6. With increasing field the current density first rises and then drops again as the overlap between the energy levels (two-dimensional subbands) in adjacent wells decreases, thereby displaying negative differential conductivity (NDC). Upon a further increase of the field the current density rises again up to a sharp peak which occurs when the ground level in one well is aligned with the second level in the neighbouring well, i.e. when the field F satisfies the relation $eFd = \Delta E$ where ΔE is the intersub-band spacing and d is the superlattice period. Then resonant tunnelling between adjacent wells produces a large current. Such an N-shaped $j(E)$ characteristic may give rise to instabilities. When the superlattice is biased into the regime of NDC, the homogeneous field distribution may break up into a low-field domain where the field is near the first peak of the $j(E)$ characteristic and a high-field domain where the field is close to the second, resonant-tunnelling peak. This is schematically indicated by the different slopes of the potential drop in the left and right part of the superlattice structure depicted in Fig. 1.6. The field distribution must adjust such that the total applied voltage drops between the two contacts. If the space charge available (by doping or optical generation of carries) is not sufficient to form a stable domain boundary, self-generated current oscillations may occur instead.

Other features which are associated with transport in strongly coupled superlattices are **Bloch oscillations**. These result from coherent motion of wave packets through the periodic miniband structure of the superlattice, which leads to THz emission of radiation with a frequency given by $\nu = eFd/h$. This can be understood by observing that a Bloch electron in a potential with period d is accelerated in an applied uniform field F until it reaches the edge of the Brillouin zone at $k = \pi/d$. After Bragg reflection its quasimomentum is changed to $k = -\pi/d$ and it starts its next cycle through the miniband. The time t it takes to complete one cycle is given by $\hbar\Delta k = eFt$.

A different type of superlattice which has recently attracted much attention is obtained if the electronic system is modulated laterally (i.e. perpendicular to the growth direction) in a periodic array of quantum wires or dots. Such lateral superlattices can be generated by lithographic techniques in combination with etching or adding of metallic gates. The latter structures are designed to change the strength of the electrostatic potential modulation in the plane of the two-dimensional electron gas. This leads to one- or two-dimensional lateral superlattices described by one- or two-dimensional minibands. As a special two-dimensional lateral nanostructure the so-called antidot superlattices have been studied extensively. Here the lateral potential modulation is so strong that it is seen by the carriers as strongly repulsive barriers arranged in a two-dimensional periodic array. Therefore the electrons can move everywhere except inside these potential barriers (antidots), complementary to the case of normal quantum dots. This system is a realization of the Sinai billiard which in classical nonlinear dynamics has been shown to exhibit chaos.

The potential profile of quantum wells and barriers is intimately connected with the charge transport properties of the nanostructure. Tunnelling through barriers provides one transport mechanism and thermionic emission of hot-electrons which have enough kinetic energy to overcome the barrier is another. Both mechanisms depend strongly upon the applied bias. The double barrier resonant tunnelling diode consists just of two barriers but the principle which produces N-shaped negative differential conductivity of the $j(F)$ characteristic is similar to that of the supperlattice shown in Fig. 1.6. The heterostructure hot-electron diode is even simpler in that it requires only one heterojunction. It possesses an S-shaped $j(F)$ characteristic with two stable current states at a given voltage in

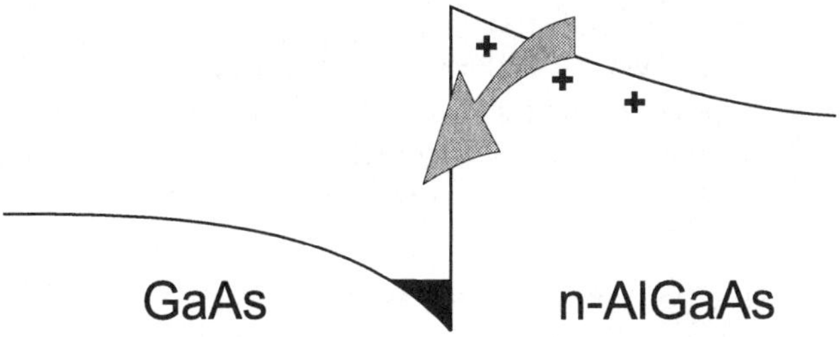

Figure 1.7 *Conduction band energy diagram of an n-AlGaAs/GaAs heterojunction. Note that the Fermi level E_F is aligned in thermodynamic equilibrium. A two-dimensional electron gas is formed in the triangular potential well.*

a certain range, the lower one corresponding to tunnelling through the barrier and the upper one corresponding to thermionic emission over the barrier.

An important feature connected with barriers and wells is the ubiquitous presence of space charges. This, according to Poisson's equation, induces a nonlinear feedback between the charge carrier distribution and the electric potential distribution governing the transport. This mutual interdependence is particularly pronounced in the case of semiconductor heterostructures and low-dimensional structures where abrupt junctions between different materials on an atomic length scale cause conduction band discontinuities resulting in potential barriers and wells. The local charge accumulation in these potential wells, together with nonlinear transport processes across the barriers have been found to provide a number of mechanisms for negative differential conductivity, bistability of the current at a given voltage and nonlinear dynamics [25]. The effect of space charges can be illustrated most simply by the example of a single heterojunction (Fig. 1.7). If n-doped AlGaAs and intrinsic or p-doped GaAs are joined together, electrons spill over from the higher potential in the AlGaAs into the GaAs, leaving behind positively charged donors. This space charge gives rise to an electrostatic potential that causes the bands to bend near the interface such that a narrow triangular potential well is formed in the GaAs layer at the interface. In this well a thin accumulation layer of electrons constitutes a two-dimensional electron gas. Since the electrons are spatially separated from the host impurities in the AlGaAs barrier, they experience very little scattering and their mobility is extremely high in such modulation-doped structures, reaching values of up to 10^7 cm^2 (Vs)$^{-1}$. Real space transfer of electrons back into the AlGaAs barrier with increasing voltage, i.e. by carrier heating, can reduce the mobility and lead to N-shaped negative differential conductivity. If the GaAs is p-doped, the type of majority carriers is even reversed in this so-called inversion layer. Such inversion layers were first demonstrated in Si metal-oxide-semiconductor (MOS) devices, where the oxide plays the role of the wide-gap AlGaAs and they are essential in the operation of all field effect transistors.

In conclusion, by controlled layer-by-layer epitaxial growth of heterostructures in combination with lateral patterning intricate artificial nanostructures with arbitrary shapes of barriers and wells can be designed. Such band-structure engineering can produce novel semiconductor devices with desired transport properties.

1.3 PHYSICAL LENGTH SCALES IN TRANSPORT

The transport properties of semiconductor nanostructures depend on a number of characteristic length scales and their relation with the system size. A conductor usually only shows classical ohmic behaviour if its dimension is much larger than all of these lengths. The length scales vary widely from one material to the other and also depend upon temperature, electric and magnetic field, impurity concentration, etc.

1.3.1 Mean-free path

The mean-free path l_m is the average distance that an electron travels before it experiences elastic scattering which destroys its initial momentum. The dominant elastic scattering mechanism is impurity scattering. The mean-free path is related to the momentum relaxation time τ_m by $l_m = v\tau_m$ with the average carrier velocity v. In high mobility semiconductors at $T < 4$ K, l_m is typically in the 10–100 μm regime.

1.3.2 Phase-relaxation length

The phase-relaxation length l_ϕ is the average distance that an electron travels before it experiences inelastic scattering which destroys its initial coherent state. Typical scattering events, such as electron–phonon or electron–electron collisions, change the energy of the electron and randomize its quantum-mechanical phase. Impurity scattering may also contribute to phase relaxation if the impurity has an internal degree of freedom so that it can change its state. For example, magnetic impurities have an internal spin that fluctuates with time. In high-mobility degenerate semiconductors, phase relaxation often occurs on a timescale τ_ϕ which is of the same order or shorter than the momentum-relaxation time τ_m. Then $l_\phi = v_F\tau_\phi$ holds with the Fermi velocity v_F. In low-mobility semiconductors the momentum-relaxation time τ_m can be considerably shorter than the phase-relaxation time τ_ϕ and diffusive motion may occur over a phase-coherent region; then $l_\phi^2 = D\tau_\phi$ with a diffusion constant $D = v_F^2\tau_m/2$.

1.3.3 de Broglie wavelength

The de Broglie wavelength $\lambda = 2\pi/k = h/(2m^*E)^{1/2}$ is related to the kinetic energy of an electron. It defines the length scale on which quantum-mechanical effects, i.e. the wave-like nature of the electron, become important. According to Pauli's exclusion principle the electrons in a metal or degenerate semiconductor at $T = 0$ fill up all states up to the Fermi energy E_F corresponding to the Fermi velocity $v_F = (2E_F/m^*)^{1/2}$ or the Fermi wave vector $k_F(E) = (2m^*E_F)^{1/2}/\hbar$. It is determined by the electron density per unit volume $n_{3D} = k_F^3/(3\pi^2)$ or per unit area $n_{2D} = k_F^2/(2\pi)$ in the three- or two-dimensional case, respectively. For a two-dimensional electron density of 10^{11} cm^{-2}, for example, the Fermi wavelength is about 75 nm.

1.3.4 Magnetic length

In the presence of a magnetic field (inductance B) the electron energy is quantized in Landau levels $E_N = (N + \frac{1}{2})\hbar\omega_c$ where $\omega_c = eB/m^*$ is the cyclotron frequency. The magnetic length $l_B = (\hbar/eB)^{1/2}$ characterizes the extension of the cyclotron orbit. The importance of the magnetic length lies in the fact that it can be tuned over a large range by changing the magnetic field. Thus, a magnetic field provides additional means of reducing the effective dimensionality of the system.

1.3.5 Thermal length

The thermal length $l_T = \hbar v_F/(k_B T)$ is connected with the average excess energy of thermal electrons $k_B T$. The phase of an electron travelling at the Fermi velocity is undetermined within l_T, due to thermal fluctuations of the electron energy.

Depending upon the values of those characteristic lengths in comparison with the system size L, different transport regimes can be distinguished.

- *Classical diffusive transport.* For macroscopic dimensions $L \gg l_m, l_\phi$ the carrier experiences many elastic and inelastic collisions so that the energy and the momentum is relaxed and the average velocity is given by the electron drift velocity $v = -\mu F$ with the mobility $\mu = e\tau_m/m^*$ following in the simplest case from Drude theory. Diffusion occurs due to gradients in the carrier density and the diffusion current density is given by $eD\nabla n$ where the diffusion constant D is related to the mobility by the Einstein relation $eD = \mu k_B T$.
- *Coherent transport.* For system sizes L smaller than the phase-relaxation length l_ϕ the quantum-mechanical wave function of the charge carriers has a well-defined phase throughout the system. Quantum-interference phenomena such as Aharonov–Bohm oscillations or universal conductance fluctuations can be observed in transport.
- *Ballistic transport.* When the system size L becomes smaller than the mean-free path l_m, a carrier can cross the device without any scattering. The carrier momentum grows due to the accelerating force of the electric field.

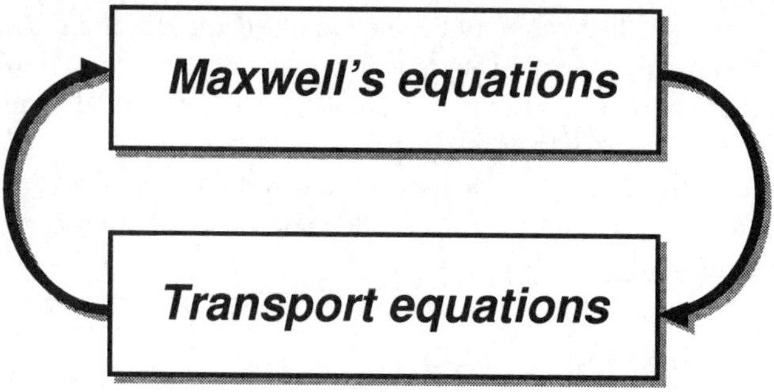

Figure 1.8 *Scheme of the self-consistently coupled dynamic equations for the field and carrier distributions.*

- *Quantum-size effects.* If the system size L in one or several directions is of the order of the de Broglie wavelength λ, size quantization occurs. Propagation in those directions is no longer possible due to quantum confinement and the density of states is modified as discussed in section 1.1.

1.4 HIERARCHY OF MODELLING APPROACHES

In this book various theoretical approaches to charge transport will be used. From a conceptual point of view, these may be grouped into a hierarchy of transport models where different approximations are used at different levels. In this section the levels of that hierarchy will be surveyed and the principal approximations which lead from one level to the next will be pointed out. Starting from a microscopic quantum-kinetic description we shall thus eventually arrive at the macroscopic classical drift-diffusion theory.

A full theoretical description of charge transport in semiconductor nanostructures requires a self-consistent solution of the coupled problem of Maxwell's equations for the fields and a suitable set of transport equations for the carriers taken from this hierarchy (Fig. 1.8). This is important because the presence of space charges leads to a mutual nonlinear interdependence of both. For example, the potential profile including band-bending effects due to space charges can be determined in thermodynamic equilibrium by the simultaneous solution of Schrödinger and Poisson's equation, where the electrostatic potential calculated from Poisson's equation for a given carrier density distribution enters the Schrödinger equation and, conversely, the free-carrrier density is determined by the electron wave function calculated from the Schrödinger equation.

1.4.1 Quantum kinetics

At the most fundamental level, the dynamics of electrons should be described by quantum mechanics. The Schrödinger picture or the Heisenberg picture may

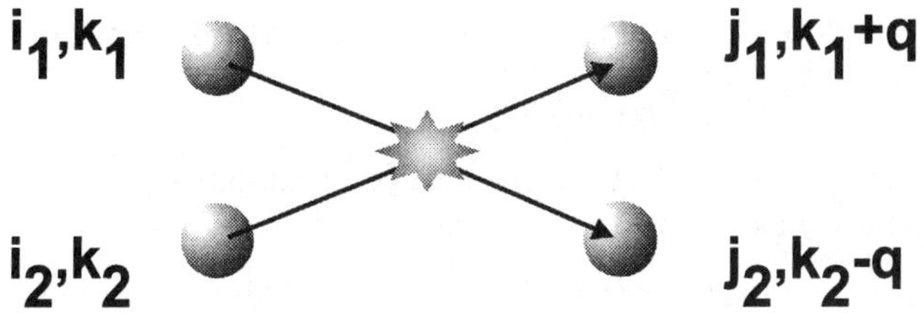

Figure 1.9 *Electron–electron scattering process (schematic).*

be used. In the Heisenberg picture, the dynamics of an operator A is given by

$$\mathrm{i}\hbar \frac{\mathrm{d}}{\mathrm{d}t} A = [A, H] \tag{1.4.1}$$

where H is the Hamiltonian. In the framework of second quantization, which is appropriate to the many-body theory, one introduces the creation and annihilation operators of an electron with wave vector k in the jth conduction subband, e.g. of a quantum well or quantum wire, $c^{\dagger}_{j,k}$ and $c_{j,k}$, respectively. Likewise, $d^{\dagger}_{j,k}$ and $d_{j,k}$ are creation and annihilation operators of holes with wave vector k in the jth valence subband.

The Hamiltonian H can be expressed as a sum of single-particle contributions of the form

$$H_0 = \sum_{j,k} \epsilon_{j,k} c^{\dagger}_{j,k} c_{j,k}, \tag{1.4.2}$$

where $\epsilon_{j,k}$ is the single-electron energy of an electron with wave vector k in the jth subband and $c^{\dagger}_{j,k} c_{j,k}$ is the corresponding number operator (and analogously for holes) and two-particle contributions describing the Coulomb interaction

$$H_{cc} = \frac{1}{2} \sum_{j_1,j_2,i_2,i_1} \sum_{k_1,k_2,q} V_{j_1 j_2 i_2 i_1}(q) c^{\dagger}_{j_1,k_1+q} c^{\dagger}_{j_2,k_2-q} c_{i_2,k_2} c_{i_1,k_1} \tag{1.4.3}$$

where $V_{j_1 j_2 i_2 i_1}(q)$ is the Coulomb matrix element, and additional coupling terms with external fields, phonons, etc.

In the language of second quantization, equation (1.4.3) describes elementary processes where two electrons from subbands i_1 and i_2 with wave vectors k_1 and k_2 are scattered into subbands j_1 and j_2 with wave vectors $k_1 + q$ and $k_2 - q$, respectively (Fig. 1.9).

In the density matrix formalism the quantum-statistical ensemble of carriers is described by the single-particle density matrices which are given in the k-representation by the ensemble average $\langle\ \rangle$ of the operators $c^{\dagger}_{j,k} c_{i,k}$, $d^{\dagger}_{j,k} d_{i,k}$, $d^{\dagger}_{j,-k} c_{i,k}$, etc. For simplicity, we shall restrict ourselves to electrons in the following.

All single-particle density matrices can be summarized in the form

$$f_{ji}(k) = \langle c^{\dagger}_{j,k} c_{i,k} \rangle \qquad i, j = 1, \ldots, n. \tag{1.4.4}$$

Note that the two operators are to be taken at equal times.

In particular, this comprises the following.

- The distribution functions of electrons in the different subbands of the conduction band, given by the intrasubband density matrices:

$$f_{ii}(k) = \langle c^{\dagger}_{i,k} c_{i,k} \rangle. \tag{1.4.5}$$

- The intersubband polarizations, given by the intersubband density matrices:

$$f_{ji}(k) = f_{ij}{}^{*}(k) = \langle c^{\dagger}_{j,k} c_{i,k} \rangle \qquad j \neq i. \tag{1.4.6}$$

The dynamics of these density matrices can be derived from the Heisenberg equation of motion (1.4.1) for the single-particle operator $A = c^{\dagger}_{j,k} c_{i,k}$ by taking the ensemble average. The temporal change of the expectation values is composed of two contributions corresponding to the commutator with H_0 and H_{cc}, respectively. The dynamics induced by the interaction with the crystal lattice (H_0)

$$\mathrm{i}\hbar \frac{\mathrm{d}}{\mathrm{d}t} f_{j_1 i_1}(k_1) \bigg|_0 = (\epsilon_{i_1,k_1} - \epsilon_{j_1,k_1}) f_{j_1 i_1}(k_1) \tag{1.4.7}$$

describes merely a rotation of the polarizations ($j_1 \neq i_1$) in the complex plane with a frequency corresponding to the intersubband splitting energy at the respective k-value.

The Coulomb interaction of the electrons (H_{cc}) leads to a coupling of the single-particle density matrices to two-particle density matrices $\langle c^{\dagger}_{j_1',k_1'} c^{\dagger}_{j_2',k_2'} c_{j_2,k_2} c_{j_1,k_1} \rangle$. A set of dynamic equations for the time evolution of these two-particle density matrices can be set up in the same way using equation (1.4.1). However, those equations of motion contain the three-particle density matrices, etc. In order to obtain a closed set of equations, this infinite hierarchy of many-particle correlations must be truncated and this is obtained by replacing the expectation value of a product of field-operator pairs with the product of their expectation values, i.e. by introducing a mean-field (or Hartree–Fock) approximation at some stage. If the two-particle density matrices are retained as independent dynamic variables, the resulting quantum kinetics—with respect to the one-particle densities alone—is not Markovian, i.e. it contains memory effects. Moreover, it includes renormalizations of the single-particle energies and internal fields due to many-body effects, and electronic coherence properties represented by the polarizations and higher-order correlations.

In the preceding survey spatial homogeneity has been assumed, so that the single-particle density matrices do not contain off-diagonal elements with respect to k, because those would not correspond to a momentum-conserving process.

Spatially inhomogeneous systems can be described by Wigner functions which are obtained by performing a Fourier transform with respect to the relative momentum. The Wigner function can be conceived as a quantum extension of the semiclassical distribution function $f(r, k, t)$. However, it is not possible to attribute a direct probabilistic interpretation to the Wigner function since it is in general not positive definite.

An alternative approach to quantum transport is via Green's functions. The general concept of a Green's function involves a quantity which describes how a system in a state q' at the initial time t' evolves into a state q at a later time t. With respect to quantum dynamics the Green's function $G(q, t, q', t')$ may be defined by the time evolution of a quantum state

$$\Phi(q, t) = \mathrm{i}\hbar \int \mathrm{d}q'\, G(q, t, q', t')\Phi(q', t') \tag{1.4.8}$$

where q stands for the set of system variables. The two solutions G propagating the system forward or backward in time, respectively, are denoted by a superscript r (retarded) or a (advanced), respectively. In the framework of second quantization the single-particle retarded Green's functions (with $\hbar = 1$) can be written as:

$$G^r_{m,n}(t, t', \boldsymbol{k}) = -\mathrm{i}\Theta(t - t')\langle\{c_{m,\boldsymbol{k}}(t), c^\dagger_{n,\boldsymbol{k}}(t')\}\rangle, \tag{1.4.9}$$

where $\{A, B\}$ denotes the anticommutator between the Fermi operators in the Heisenberg picture. Note that the two times are usually not equal, in contrast to the density matrix theory. While in thermodynamic equilibrium the occupation of states is given by the Fermi distribution, in the theory of nonequilibrium Green's functions the lesser functions

$$G^<_{m,n}(t, t', \boldsymbol{k}) = \mathrm{i}\langle c^\dagger_{n,\boldsymbol{k}}(t')c_{m,\boldsymbol{k}}(t)\rangle \tag{1.4.10}$$

have to be considered as well. When evaluated at $t = t'$, they equal the density matrix and determine the electron densities as well as the polarizations. While the density matrix formalism is particularly useful for systems with long phase-relaxation times since the coherent temporal evolution of the energetically resolved electron distribution can be directly obtained from a coupled set of differential equations, the Green's function approach can efficiently deal with strong scattering processes. The advantage of the Green's function approach is that nonperturbational techniques allow for a partial summation of some interactions up to infinite order. This technique results in integrodifferential equations which display also quantum correlations and memory effects but are generally difficult to solve.

1.4.2 Semiclassical Boltzmann equation

At a semiclassical level the carrier kinetics under the influence of electric and magnetic fields is described by the Boltzmann transport equation for the semiclassical carrier distribution function $f(\boldsymbol{r}, \boldsymbol{k}, t)$, where $\boldsymbol{r}$ is the spatial coordinate and $\boldsymbol{k}$ is the wave vector:

$$\frac{\partial f}{\partial t} + \boldsymbol{v}_g \boldsymbol{\nabla}_{\boldsymbol{r}} f + \frac{q}{\hbar}(\boldsymbol{F} + \boldsymbol{v}_g \times \boldsymbol{B})\boldsymbol{\nabla}_{\boldsymbol{k}} f = \left(\frac{\partial f}{\partial t}\right)_{\text{coll}} . \tag{1.4.11}$$

Here $\boldsymbol{v}_g = \hbar^{-1}\boldsymbol{\nabla}_{\boldsymbol{k}}\epsilon(k)$ is the group velocity. The collision integral $(\frac{\partial f}{\partial t})_{\text{coll}}$ includes in principle all scattering processes such as phonon, impurity and electron–electron scattering, generation and recombination processes. For intraband single-electron processes, e.g.

$$\left(\frac{\partial f}{\partial t}\right)_{\text{coll}} = \int (W(\boldsymbol{k}', \boldsymbol{k}) f(\boldsymbol{r}, \boldsymbol{k}', t) - W(\boldsymbol{k}, \boldsymbol{k}') f(\boldsymbol{r}, \boldsymbol{k}, t)) z \, \mathrm{d}^3 k' \tag{1.4.12}$$

holds, where $W(\boldsymbol{k}, \boldsymbol{k}')$ is the probability per unit time for a transition from state $\boldsymbol{k}$ to $\boldsymbol{k}'$, z is the density of states in k-space. The carrier charge is $q = \pm e$ for holes or electrons, respectively.

The Boltzmann equation relies on the assumption that:

- the distribution function varies little over the de Broglie wavelength;
- the carrier density is sufficiently low so that only binary collisions occur;
- the time between successive collisions is much longer that the duration of a collision;
- the density gradients are small over the range of the interparticle potential.

The set of quantum kinetic equations of the density matrix theory can be reduced to the semiclassical Boltzmann transport equation under the following approximations. The general procedure consists of the adiabatic elimination of variables involving quantum-mechanical correlations by means of a Markov approximation under the assumption that initially the system was uncorrelated. In particular, interband and intersubband polarizations and two-particle density matrices can be eliminated by formally integrating the respective dynamic equations. This yields non-Markovian convolution integrals in time whose kernel contains oscillating terms of the form $\exp(-\mathrm{i}\Omega t)$ where $\hbar\Omega = E_{\text{in}} - E_{\text{out}}$ represents the total energy balance of the interaction process. Assuming that the distribution functions and field amplitudes occurring also in the integral are sufficiently slowly varying functions of time, the Markov limit can be performed by taking these functions out of the integral and approximating the remaining oscillating time integral by the energy-conserving delta function $\delta(\Omega)$ in the infinite-time limit. Thus, the semiclassical rates of the interaction processes are regained in the familiar form of Fermi's golden rule in first-order quantum-mechanical perturbation theory:

$$W(\boldsymbol{k}', \boldsymbol{k}) = \frac{2\pi}{\hbar}|M|^2 \delta(E_{\text{in}} - E_{\text{out}}) \tag{1.4.13}$$

where M is the matrix element of the interaction process. Applying this procedure to electron–photon, electron–phonon or electron–electron interactions yields the respective collision integrals of the Boltzmann equation.

1.4.3 Hydrodynamic balance equations

At the hydrodynamic level of description the detailed kinetics in k-space is averaged over, and only slow, macroscopic quantities such as the carrier density $n(\boldsymbol{r},t) = \int f(\boldsymbol{r},\boldsymbol{k},t) z\, \mathrm{d}^3k$, the mean momentum per carrier $p(\boldsymbol{r},t) = \langle \hbar \boldsymbol{k} \rangle$ and the mean energy per carrier $w_n(\boldsymbol{r},t) = \langle \epsilon(\boldsymbol{k}) \rangle$ are considered as dynamic variables. Here the brackets denote the semiclassical ensemble average $\langle A \rangle = n^{-1} \int A(\boldsymbol{k}) f(\boldsymbol{r},\boldsymbol{k},t) z\, \mathrm{d}^3k$. The resulting hydrodynamic balance equations have the form of continuity equations for the carrier density, the mean momentum density and the mean energy density. For simplicity we confine our attention to a nondegenerate, isotropic parabolic band structure $\epsilon(\boldsymbol{k}) = \hbar^2k^2/(2m^*)$. The mean group velocity $\boldsymbol{v}(\boldsymbol{r},t) = \langle \boldsymbol{v}_g \rangle$ is then related to the carrier temperature $T_e = (m^*/3k_B)\langle (\boldsymbol{v}_g - \langle \boldsymbol{v}_g \rangle)^2 \rangle$ and the mean momentum and energy by:

$$\boldsymbol{p} = m^*\boldsymbol{v}, \qquad w_n = \frac{m^*}{2}\boldsymbol{v}^2 + \frac{3}{2}k_B T_e. \tag{1.4.14}$$

Hydrodynamic balance equations for $n, \boldsymbol{p}$ and w_n (moments of the Boltzmann equation) are obtained by multiplying (1.4.11) by appropriate powers of $\boldsymbol{k}$ and integrating over the first Brillouin zone. Since each balance equation is coupled to the next higher moment, we obtain an infinite hierarchy of moment equations, which can be truncated by the following approximations.

- The electron temperature is a scalar, i.e. the momentum flux tensor reduces to the scalar electron pressure nk_BT_e.

- The heat flux $\boldsymbol{j}_Q = \frac{1}{2}nm^*\langle (\boldsymbol{v}_g - \langle \boldsymbol{v}_g \rangle)^2 (\boldsymbol{v}_g - \langle \boldsymbol{v}_g \rangle) \rangle$ is approximated by Fourier's law $\boldsymbol{j}_Q = -\kappa \boldsymbol{\nabla}_{\boldsymbol{r}} T_e$ with thermal conductivity κ.

- The collision integrals are expressed in terms of generation–recombination rates, momentum relaxation rates and energy relaxation rate, respectively, via

$$\int \left(\frac{\partial f}{\partial t}\right)_{\text{coll}} z\, \mathrm{d}^3k = \phi(n, w_n) \tag{1.4.15}$$

$$\int \hbar \boldsymbol{k} \left(\frac{\partial f}{\partial t}\right)_{\text{coll}} z\, \mathrm{d}^3k = -n\frac{\boldsymbol{p}}{\tau_m} \tag{1.4.16}$$

$$\int \epsilon(\boldsymbol{k}) \left(\frac{\partial f}{\partial t}\right)_{\text{coll}} z\, \mathrm{d}^3k = -n\frac{w_n - w_{n0}}{\tau_e(w_n)} \tag{1.4.17}$$

with mean-energy-dependent energy relaxation time τ_e, momentum relaxation time τ_m and $w_{n0} = \frac{3}{2}k_BT_L$.

The following closed set of hydrodynamic equations is obtained:

$$\frac{\partial n}{\partial t} + \boldsymbol{\nabla}_{\boldsymbol{r}}(n\boldsymbol{v}) = \phi(n, w_n) \tag{1.4.18}$$

$$\frac{\partial \boldsymbol{p}}{\partial t} + (\boldsymbol{v}\boldsymbol{\nabla}_{\boldsymbol{r}})\boldsymbol{p} + \frac{1}{n}\boldsymbol{\nabla}_{\boldsymbol{r}}(nk_B T_e) - q(\boldsymbol{F} + \boldsymbol{v} \times \boldsymbol{B}) = -\frac{\boldsymbol{p}}{\tau_m} \tag{1.4.19}$$

$$\frac{\partial w_n}{\partial t} + (\boldsymbol{v}\boldsymbol{\nabla}_{\boldsymbol{r}})w_n + \frac{1}{n}\boldsymbol{\nabla}_{\boldsymbol{r}}(nk_B T_e \boldsymbol{v}) - \frac{\kappa}{n}\Delta T_e - q\boldsymbol{v}\boldsymbol{F} = -\frac{(w_n - w_{n0})}{\tau_e(w_n)}. \tag{1.4.20}$$

1.4.4 Classical drift-diffusion theory

If momentum and energy relaxation occur faster than all other processes, $\boldsymbol{p}$ and w_n can be eliminated adiabatically and the carrier densities remain as the only dynamic variables on this slow timescale. Transport may then be described within classical drift-diffusion theory. Specifically, from the momentum balance equation (1.4.19), $\boldsymbol{p}$ can be eliminated adiabatically by setting $d\boldsymbol{p}/dt \equiv \partial \boldsymbol{p}/\partial t + (\boldsymbol{v}\boldsymbol{\nabla}_{\boldsymbol{r}})\boldsymbol{p} = 0$. Thus (1.4.19) leads to a current density as a function of field

$$\boldsymbol{j} = -en\boldsymbol{v} = en\mu_B \boldsymbol{F}_{\text{tot}} + en\mu\mu_B(\boldsymbol{B} \times \boldsymbol{F}_{\text{tot}}) + en\mu^2\mu_B \boldsymbol{B}(\boldsymbol{B}\boldsymbol{F}_{\text{tot}}) \tag{1.4.21}$$

with the mobility $\mu = (e/m^*)\tau_m$, $\mu_B = \mu/(1 + (\mu B)^2)$ and

$$\boldsymbol{F}_{\text{tot}} = \boldsymbol{F} + \frac{1}{en}\boldsymbol{\nabla}_{\boldsymbol{r}}(nk_B T_e), \tag{1.4.22}$$

where we have restricted ourselves to electrons and a constant momentum relaxation time. More sophisticated treatments distinguish between drift and Hall mobility.

Neglecting the spatial variations of w_n and hence of T_e and setting $B = 0$, the usual drift-diffusion expression for the current density

$$\boldsymbol{j} = en\mu\boldsymbol{F} + eD\boldsymbol{\nabla}_{\boldsymbol{r}} n \tag{1.4.23}$$

is recovered, if the Einstein relation $\mu k_B T_e = eD$ is used.

Furthermore, if energy relaxation occurs fast, temporal and spatial derivatives in (1.4.20) can be neglected, yielding a local energy-field relation:

$$w_n = w_{n0} + e\tau_e\mu F^2. \tag{1.4.24}$$

The hierarchy of transport equations is summarized in Fig. 1.10.

1.5 SCOPE OF THIS BOOK

This book is organized as follows. After this introduction, Chapters 2–6 present a hierarchy of different approaches to transport theory, descending from a more

Drift-diffusion equations $\qquad n(\underline{r}, t)$

$$\dot{n} = f(n) + D\Delta n + \underline{\nabla} \cdot (n\mu\underline{F})$$

hydrodynamic: $\qquad n(\underline{r}, t)$ **density**

$$\dot{n} + \underline{\nabla} \cdot (n\underline{v}) = \varphi(n, w_n)$$
$$\dot{\underline{p}} + \underline{\nabla} \ldots$$
$$\dot{w}_n + \underline{\nabla} \ldots$$

kinetic: $\qquad f(\underline{r}, \underline{k}, t)$ **distribution fct.**
semiclassical Boltzmann equation

$$\dot{f} + \underline{v}_g \cdot \underline{\nabla}_r f - \frac{e}{\hbar}\underline{F} \cdot \underline{\nabla}_k f = \left(\frac{\partial f}{\partial t}\right)_{\text{coll.}}$$

Quantum kinetics: $\qquad f_{ij}(k) = \langle c_{i,k}^{\dagger} c_{j,k}\rangle$
density matrix equations

$$\frac{d}{dt}\langle c_{i,k}^{\dagger} c_{j,k}\rangle = \frac{i}{\hbar}\langle\left[H, c_{i,k}^{\dagger} c_{j,k}\right]\rangle$$

...

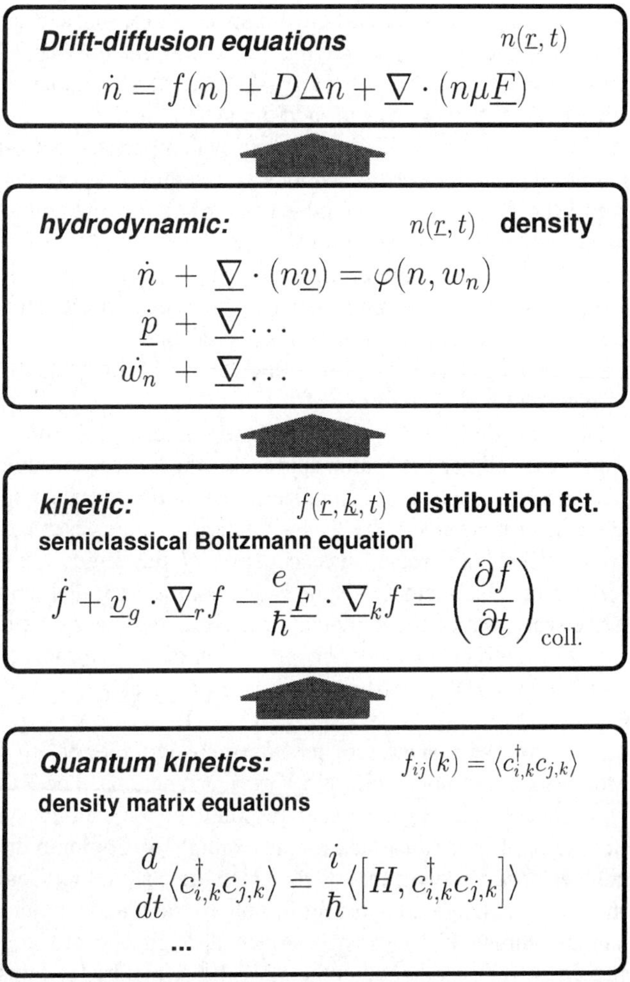

Figure 1.10 *Hierarchy of transport equations.*

macroscopic level (hydrodynamic simulation) via semiclassical Monte Carlo techniques and cellular automaton approaches to a full microscopic quantum-transport theory covering both nonequilibrium Green's functions and density matrix theory. In Chapters 7–10 the formalism is applied to more specific topics such as phase-coherent transport in mesoscopic structures, chaotic dynamics in lateral superlattices, Bloch oscillations and Wannier–Stark localization, field domain formation in superlattices. Chapter 11 provides an overview of the different scattering processes which are of relevance in semiconductor nanostructures.

In Chapter 2 the hydrodynamic model is derived and applied to device simulation. Hot-electron effects, in particular impact ionization and carrier injection, are discussed and as an application a nonvolatile memory cell (electrically erasable programmable read only memory—EEPROM) is simulated. It consists of a silicon MOS transistor with an additional floating gate and is based on hot-electron injection from the channel region, which permits us to store charge for a long time, typically 10 years, without the need of a power supply.

In Chapter 3 the Monte Carlo method is used to solve the semiclassical Boltzmann transport equation. It consists of a simulation in $(\boldsymbol{r}, \boldsymbol{k})$ space of the carrier motion, subject to scattering effects selected stochastically. The underlying physical models as well as the fundamentals of the method are introduced. The method is illustrated by applications to bulk semiconductors, low-dimensional structures and devices. The extension of the Monte Carlo method to quantum transport is also outlined.

An alternative method for the solution of the semiclassical Boltzmann transport equation is the cellular automaton approach, which is reviewed in Chapter 4. Here the Boltzmann equation is discretized on a lattice in $(\boldsymbol{r}, \boldsymbol{k})$ space and in time and the carrier motion and the scattering events are represented by discrete automata rules which determine the occupation of the lattice sites after each time step. Therefore this method is suited for massively parallel computers and can very effectively reduce the computational cost. The macroscopic dynamics is recovered as ensemble averages. The simulation of an ultrashort channel vertical MOSFET and a HEMT is presented as a practical application to device modelling.

In Chapter 5 two approaches to quantum-transport theory are introduced: the Kubo formalism and nonequilibrium Green's functions. The Kubo formula describes the conductivity as the linear response of a quantum system to an external electric field. It is illustrated by the example of Coulomb drag, i.e. momentum transfer between two subsystems which are coupled by Coulomb interaction without direct charge transfer (for instance, a double quantum well or two nearby quantum wires). If a current is driven through one of the subsystems, then an induced current is dragged in the other, described by the transconductivity which is evaluated by diagrammatic techniques. The nonequilibrium Green's function technique is required for nonlinear, far-from-equilibrium conditions. It is applied to a resonant tunnelling device where a small, possibly strongly interacting mesoscopic region is coupled to noninteracting leads.

Chapter 6 presents another approach to quantum transport: the density matrix theory. It is particularly suited for the study of ultrafast carrier dynamics in the coherent, nonlinear transport regime far from thermodynamic equilibrium, which cannot be described by a semiclassical theory. The general framework of the density matrix theory including the correlation expansion of the many-particle density matrices for various types of interactions is reviewed and its semiclassical limit is discussed. Examples include the carrier generation by an ultrashort laser pulse, four-wave mixing, quantum beats and THz emission from quantum wells, the dynamics of coherent phonons in quantum wells and electron–phonon quantum kinetics in quantum wires.

Chapter 7 focuses on dynamic and nonlinear transport in mesoscopic structures. Conductors are called mesoscopic if their dimensions are larger than microscopic objects such as atoms but smaller than macroscopic samples. The transmission theory (Landauer–Büttiker formalism) is used to describe mesoscopic transport. The capacitance, the linear low-frequency admittance (AC conductance) and the weakly nonlinear DC transport of low-dimensional phase-coherent conductors is investigated. The theory applied is an extension of the scattering approach to conduction which now includes the long-range Coulomb interaction in a self-consistent way. As examples, quantized Hall samples, the quantum point contact, the resonant-tunnelling barrier and the metallic diffusive conductor are discussed.

In Chapter 8 magnetotransport in lateral antidot superlattices is considered. The electrons move ballistically in the two-dimensional plane of the lateral lattice, subjected to the periodic potential of the antidots and to a high magnetic field (up to 10 T) applied perpendicular to that plane. The classical dynamics of these electrons shows chaotic behaviour. Its quantum-mechanical signature is displayed in the energy spectrum and in the magnetoresistivity of this system. The magnetic field-dependent band structure is similar to Hofstadter's butterfly. The classical as well as the quantum transport regime (where the Fermi wavelength is as large as the superlattice period) are discussed.

In Chapter 9 vertical superlattices composed of strongly coupled multiple quantum wells are studied. The ultrafast carrier dynamics including tunnelling and scattering processes is described by the density matrix theory. Two equivalent pictures are discussed: the Bloch oscillation and the Wannier–Stark representation. If a moderate electric field F is applied to a superlattice of period d, the miniband splits into a series of levels spaced by eFd, the Wannier–Stark ladder and the associated eigenfunctions are localized at different quantum wells. Bloch oscillations, i.e. coherent motion of wave packets through the periodic miniband structure, leads to coherent THz emission at the frequency $\omega = eFd/\hbar$ if the scattering rate is much smaller than ω. A comparison with recent experiments which have detected Bloch oscillations is drawn.

In Chapter 10 high-field effects related to the vertical transport in weakly coupled multiple quantum-well structures are reviewed. A self-contained microscopic quantum-transport model for the calculation of the well-to-well currents without fit parameters is presented. The model yields the well-known peaks in the current-field relation in quantitative agreement with experiments. Moreover, the formation of stationary field domains as well as self-sustained current oscillations due to moving charge accumulations are described. The underlying physics of these nonlinear phenomena is discussed in detail.

Finally, in Chapter 11, the different scattering processes in low-dimensional structures are surveyed. Optical phonons, acoustic phonons, ionized impurities, interface roughness and alloy scattering are treated, based on quantum-mechanical time-dependent perturbation theory as given by Fermi's golden rule.

The various processes contribute to the scattering rate, the momentum relaxation rate and the energy relaxation rate. Those quantities enter the transport description which was discussed in previous chapters; in particular, the scattering rate is needed for the semiclassical Monte Carlo approach, while the momentum and energy relaxation rates are required, for example in the hydrodynamic model.

REFERENCES

[1] Mahan, G. D. (1990) *Many-Particle Physics*, Plenum, New York.

[2] Beenakker, C. W. J. and van Houten, H. (1991) Quantum transport in semiconductor nanostructures *Solid-State Physics* (eds H. Ehrenreich and D. Turnbull), Academic Press, New York, Vol. 44.

[3] Landsberg, P. T. (ed.) (1992) *Handbook on Semiconductors*, 2nd edn, Elsevier, Amsterdam, Vol. 1.

[4] Haug, H. and Koch, S. W. (1993) *Quantum Theory of the Optical and Electronic Properties of Semiconductors*, 2nd edn, World Scientific, Singapore.

[5] Ferry, D. K., Grubin, H. L., Jacoboni, C. and Jauho, A.-P. (eds) (1995) *Quantum Transport in Ultrasmall Devices*, Plenum Press, New York.

[6] Datta, S. (1995) *Electronic Transport in Mesoscopic Systems*, Cambridge University Press, Cambridge.

[7] Haug, H. and Jauho, A.-P. (1996) *Quantum Kinetics in Transport and Optics of Semiconductors*, Springer, Berlin.

[8] *Proc. Int. Conf. Nonequilibrium Carrier Dynamics in Semiconductors (HCIS-10, Berlin (1997)) (Special Issue of Phys. Stat. Solidi b).*

[9] *Proc. Int. Conf. Electronic Properties of Two-Dimensional Systems (EP2DS-12, Tokyo (1997)) (Special Issue of Physica B).*

[10] *Proc. Int. Conf. Modulated Semiconductor Structures (MSS-8, Santa Barbara (1997)) (Special Issue of Physica B).*

[11] *Int. Conf. Semiconductor Microstructures and Microdevices (ICSMM, Liege (1996)) (Special Issue of Superlatt. Microstructures).*

[12] Markovich, P. A. (1986) *The Stationary Semiconductor Device Equations*, Springer, Wien.

[13] Ferry, D. K., Gardner, C. and Ringhofer, C. (eds) (1997) *Proc. 4th Internat. Workshop on Computational Electronics, Tempe, Az.* VLSI Design, to be published.

[14] Jacoboni, C. and Lugli, P. (1989) *The Monte Carlo Method for Semiconductor Device Simulation*, Springer, Wien.

[15] Kometer, K., Zandler, G. and Vogl, P. (1992) *Phys. Rev.* B **46** 1382.

[16] Rossi, F., Brunetti, R. and Jacoboni, C. (1992) *Hot Carriers in Semiconductor Nanostructures: Physics and Applications* (ed. J. Shah), Academic Press, Boston, p. 153.

[17] Balkan, N. (ed.) (1997) *Hot Electrons in Semiconductors: Physics and Devices*, Oxford University Press, Oxford.

[18] Schöll, E. (1987) *Nonequilibrium Phase Transitions in Semiconductors*, Springer, Berlin.

[19] Shaw, M. P., Mitin, V. V., Schöll, E. and Grubin, H. L. (1992) *The Physics of Instabilities in Solid State Electron Devices*, Plenum Press, New York.

[20] Thomas, H. (ed.) (1992) *Nonlinear Dynamics in Solids*, Springer, Berlin.

[21] Niedernostheide, F. J. (ed.) (1995) *Nonlinear Dynamics and Pattern Formation in Semiconductors and Devices*, Springer, Berlin.

[22] Engel, H., Niedernostheide, F. J., Purwins, H. G. and Schöll, E. (eds) (1996) *Self-organization in Activator-inhibitor-systems: Semiconductors, Gas-discharge and Chemical Active Media*, Wissenschaft und Technik Verlag, Berlin.

[23] Bastard, G. (1988) *Wave Mechanics Applied to Semiconductor Heterostructures*, Les Editions de Physique, Les Ulis Cedex, France.

[24] Grahn, H. T. (ed.) (1995) *Semiconductor Superlattices, Growth and Electronic Properties*, World Scientific, Singapore.

[25] Balkan, N., Ridley, B. K. and Vickers, A. J. (eds) (1993) *Negative Differential Resistance and Instabilities in two-dimensional Semiconductors*, Plenum Press, New York.

CHAPTER 2

Hydrodynamic simulation of semiconductor devices

Massimo Rudan†, Martino Lorenzini† and Rossella Brunetti‡

† Dipartimento di Elettronica, Informatica e Sistemistica, Università di Bologna, Viale Risorgimento 2, 40136 Bologna, Italy
‡ Istituto Nazionale di Fisica della Materia, and Dipartimento di Fisica, Università di Modena, Via Campi 213/A, 41100 Modena, Italy

2.1 INTRODUCTION

Recently the hydrodynamic model has become popular in the field of analysis and simulation of semiconductor devices*. The model has the merit of providing, along with the concentration and current density of the carriers, also their average energy and average energy flux. At the same time, its equations basically retain the same structure of the simpler drift-diffusion model; because of this, it has been possible to incorporate the hydrodynamic equations into existing device-analysis codes, thus exploiting a number of robust solution schemes already available there.

The hydrodynamic model is derived by applying the moment technique to the Boltzmann transport equation (BTE) and truncating the series of moments at a suitable order. This yields a number of partial differential equations in the $\mathbf{r}, t$ space, specifically, the continuity equations for the carrier number, momentum, average energy and average energy flux. In this procedure, due to the integration over the wave vector space and to the series truncation, some of the information originally carried by the distribution function is lost. However, the information provided by the continuity equations indicated above is in most cases sufficient to describe the electrical behaviour of realistic semiconductor devices.

It is worth noting that the term **hydrodynamic model** is not always given exactly the same meaning in the existing literature. What is meant here by this term is a model derived through the following steps:

* Part of this work was carried out within ADEQUAT (JESSI BT11, ESPRIT 8002) and DESSIS (ESPRIT 6075).

1. The BTE is taken in its general form, without preliminary manipulations or simplifications of the collision terms.
2. The functions used to calculate the moments account for the full shape of the semiconductor bands.
3. The moments are calculated without preliminarily multiplying the BTE by the inverse scattering rate.

The simplifications and the closure condition are introduced after the completion of the above steps which, in themselves, do not introduce approximations.

As mentioned above, the equations of the hydrodynamic model basically have the same structure as in the simpler drift-diffusion model. Thanks to this, the numerical solution of the hydrodynamic equations can efficiently be carried out by an exponential-fitting scheme [1], a generalization of the well-known Scharfetter–Gummel method [2]. Since the number of equations is larger, the computational cost of the solution is higher than in the drift-diffusion case; however, it is still of the same order of magnitude and is largely compensated by a better accuracy. In contrast to the drift-diffusion model, where the carrier temperature is kept at the same value as the lattice temperature, the hydrodynamic model provides the carrier temperature as an independent variable. In the typical operating regime of semiconductor devices, the latter may be found substantially larger than the lattice temperature, particularly near the drain edge of the channel and around the drain junction in MOS transistors. In such regions the carriers tend to spread apart because of the higher energy, hence, the concentration and velocity rendered by the hydrodynamic equations are different from those of the drift-diffusion ones.

An intrinsic property of the hydrodynamic equations, as of any model based upon moments of the BTE, is that the model coefficients are built up by means of integrals involving the distribution function. Because of this, it is necessary to obtain the coefficients independently and express them in terms of the problem's unknown variables and other parameters. This is done either experimentally or by a full solution of the BTE; the result is a table model or, sometimes, a set of analytic functions.

The more accurate calculation of the carrier concentration and velocity, along with the information about the carrier temperature, is important for describing a number of phenomena indicated by the general term **hot-carrier effects**, which play a relevant role in modern devices. Carriers are called 'hot' when their average kinetic energy exceeds the equilibrium value, or when their temperature exceeds that of the lattice. The two definitions are not fully equivalent because average energy and temperature are equivalent concepts only near equilibrium, however, they are qualitatively similar. In the regions where carriers are hot, the high-energy tail of the distribution function becomes more populated and, as a consequence, a substantial number of carriers have enough energy to overcome barriers which would otherwise be unsurmountable. Examples of this are impact ionization, whose energy threshold in silicon is of the order of 1 eV, and injection from the semiconductor into the gate insulator in MOS devices, whose energy threshold for the $Si–SiO_2$ system is of the order of 3 eV.

In principle, for the description of phenomena such as impact ionization or carrier injection, it is necessary to determine the distribution function. Although the latter may be calculated by a direct solution of the BTE, such a costly procedure is often avoided and the problem is tackled by starting from the observation that the distribution function decays rapidly as the microscopic energy increases. Because of this, for phenomena characterized by an energy threshold it is reasonable to base the calculations upon an approximation of the distribution function's tail which only depends on the microscopic energy and is sufficiently accurate in the energy region near the threshold. Following this concept, a number of models have been introduced in which the form of the tail is prescribed as far as its dependence on the energy is concerned, but is parametrized by functions that are available in the hydrodynamic model, typically, carrier concentration and temperature. Such models have been applied to the analysis of impact ionization [3, 4] and carrier injection (e.g., [5, 6]). In this way, it became possible to couple the advantage of standard methods, applicable to solving the hydrodynamic equations on realistic structures, with a finer analysis involving the use of the distribution function over specific energy ranges. Such an approach proved very efficient and has been used in the analysis and design of complicated structures [7, 8].

This chapter is organized as follows. In section 2.2 the application of the moment scheme to the BTE is shown, including the effect of the magnetic induction. Section 2.3 depicts the approximations that lead from the general expression of the moments to the hydrodynamic equations. Section 2.4 reports a number of results about the analysis of the model's coefficients, while section 2.5 shows applications of the model to the case of memory cells, where carrier injection into the gate oxide occurs.

2.2 STATISTICAL AVERAGES AND MOMENTS OF THE BTE

The transport of carriers in semiconductor devices can be profitably described in terms of suitable averages using the distribution function of the carriers, $f = f(\boldsymbol{r}, \boldsymbol{k}, t)$, as a weighing function. The vector $\boldsymbol{r} \equiv (x_1, x_2, x_3)$ is the central position vector of the wave packet representing the carrier and belongs to the physical domain of the semiconductor. The vector $\boldsymbol{k} \equiv (k_1, k_2, k_3)$, in turn, is the central wave vector of the wave packet and belongs to the first Brillouin zone B. The explicit dependence on t accounts for a possible time dependence of the boundary conditions. The distribution function is positive and bounded and its meaning is that the product $f(\boldsymbol{r}, \boldsymbol{k}, t)\mathrm{d}^3r\mathrm{d}^3k$ equals the number of wave packets that, at time t, have the position vector belonging to the elementary volume d^3r centred on $\boldsymbol{r}$ and the wave vector belonging to the elementary volume d^3k centred on $\boldsymbol{k}$.

2.2.1 General definitions

The above-mentioned averages provide the macroscopic quantities of interest, such as the concentration, average velocity and average energy. For a general scalar quantity $\alpha(\boldsymbol{k})$ the average is defined

$$\overline{\alpha}(\boldsymbol{r}, t) = \frac{\int_B \alpha(\boldsymbol{k}) f(\boldsymbol{r}, \boldsymbol{k}, t)\, \mathrm{d}^3 k}{\int_B f(\boldsymbol{r}, \boldsymbol{k}, t)\, \mathrm{d}^3 k}; \tag{2.2.1}$$

by way of example, the average energy of the carriers is defined by letting $\alpha(\boldsymbol{k}) = \epsilon(\boldsymbol{k})$, where $\epsilon(\boldsymbol{k})$ is the microscopic kinetic energy, i.e. the eigenvalue of Schrödinger's equation in the periodic crystal potential, referred to the extremum of the band. In turn, each component of the average velocity of the carriers is defined by letting $\alpha(\boldsymbol{k}) = u_i(\boldsymbol{k})$, where $u_i(\boldsymbol{k})$ is the corresponding component of the group velocity

$$\boldsymbol{u}(\boldsymbol{k}) = \frac{1}{\hbar} \mathrm{grad}_k \epsilon(\boldsymbol{k}). \tag{2.2.2}$$

It is worth remembering that ϵ is an even function of $\boldsymbol{k}$, whence $\boldsymbol{u}(\boldsymbol{k})$ is odd. The procedure by which the macroscopic quantities of interest are derived from the distribution function is essentially the same for each band of the semiconductor. Here the calculations will deal with the electrons of the conduction band, hence f will indicate the distribution function of such carriers; in particular, the concentration n and average energy $w_n = \overline{\epsilon}$ of the electrons are given by

$$n(\boldsymbol{r}, t) = \int_B f(\boldsymbol{r}, \boldsymbol{k}, t)\, \mathrm{d}^3 k \tag{2.2.3}$$

and

$$n(\boldsymbol{r}, t) w_n(\boldsymbol{r}, t) = \int_B \epsilon(\boldsymbol{k}) f(\boldsymbol{r}, \boldsymbol{k}, t)\, \mathrm{d}^3 k, \tag{2.2.4}$$

where the expression of n stems from the definition of the distribution function and that of w_n derives from (2.2.1). Similarly, letting $\alpha(\boldsymbol{k}) = u_i(\boldsymbol{k})$ one finds the components of the average velocity $\boldsymbol{v}_n = \overline{\boldsymbol{u}}$. Another quantity of interest that is related to the average velocity is the random velocity $\boldsymbol{c}_n$; the definitions of $\boldsymbol{v}_n$ and $\boldsymbol{c}_n$ read

$$n(\boldsymbol{r}, t) \boldsymbol{v}_n(\boldsymbol{r}, t) = \int_B \boldsymbol{u} f \mathrm{d}^3 k \tag{2.2.5}$$

and

$$\boldsymbol{c}_n(\boldsymbol{r}, \boldsymbol{k}, t) = \boldsymbol{u}(\boldsymbol{k}) - \boldsymbol{v}_n(\boldsymbol{r}, t). \tag{2.2.6}$$

Finally, the average energy flux $\boldsymbol{P}_n = \overline{\epsilon \boldsymbol{u}}$ is derived from

$$n(\boldsymbol{r}, t) \boldsymbol{P}_n(\boldsymbol{r}, t) = \int_B \epsilon(\boldsymbol{k}) \boldsymbol{u}(\boldsymbol{k}) f(\boldsymbol{r}, \boldsymbol{k}, t)\, \mathrm{d}^3 k. \tag{2.2.7}$$

2.2.2 The moment scheme

The evolution equation for the distribution function is the BTE

$$\frac{\partial f}{\partial t} + \boldsymbol{u}\cdot \mathrm{grad}_r f - \frac{q}{\hbar}(\boldsymbol{E} + \boldsymbol{u} \wedge \boldsymbol{B}) \cdot \mathrm{grad}_k f = C, \tag{2.2.8}$$

where $\boldsymbol{E}$ and $\boldsymbol{B}$ are the electric field and magnetic induction (independent of $\boldsymbol{u}$), q is the elementary charge, $\hbar$ is the reduced Planck constant and C is the collision term [9]. In semiconductors, the existence of more than one band makes it convenient to express the collision term as

$$C = C' - \frac{f - \tilde{f}}{\tau}, \tag{2.2.9}$$

where the first and second term on the right-hand side refer to the interband and intraband transitions, respectively [10, 11]. In (2.2.9) it is

$$\frac{1}{\tau} = \int_B H(\boldsymbol{k} \to \boldsymbol{k}')\, \mathrm{d}^3k' \tag{2.2.10}$$

$$\tilde{f} = \tau(\boldsymbol{k}) \int_B f(\boldsymbol{r}, \boldsymbol{k}', t) H(\boldsymbol{k}' \to \boldsymbol{k})\, \mathrm{d}^3k', \tag{2.2.11}$$

where $H(\boldsymbol{k} \to \boldsymbol{k}')$ represents the probability per unit time of a transition from an initial state $\boldsymbol{k}$ to a final state $\boldsymbol{k}'$. The inverse of time τ thus represents the total scattering rate. In equilibrium, the interband and intraband terms vanish independently from each other, whence $(C')^{\mathrm{eq}} = 0$, $\tilde{f}^{\mathrm{eq}} = f^{\mathrm{eq}}$. From (2.2.10) and (2.2.11) it follows

$$\begin{aligned}\int_B \frac{f - \tilde{f}}{\tau}\, \mathrm{d}^3k &= \int_B \int_B f(\boldsymbol{r}, \boldsymbol{k}, t) H(\boldsymbol{k} \to \boldsymbol{k}')\, \mathrm{d}^3k'\, \mathrm{d}^3k \\ &\quad - \int_B \int_B f(\boldsymbol{r}, \boldsymbol{k}', t) H(\boldsymbol{k}' \to \boldsymbol{k})\, \mathrm{d}^3k' \mathrm{d}^3k = 0.\end{aligned} \tag{2.2.12}$$

This result is easily understood by remembering that f refers to one band and that $(f - \tilde{f})/\tau$ in (2.2.9) refers to the intraband collisions, namely those where the initial and final state of the electron belong to the same band; hence, the corresponding terms cancel each other in the double integral of (2.2.12). Defining $C_n = \int_B C \mathrm{d}^3k$ and using (2.2.12) one finds that the only collisions of importance in C_n are the interband ones, namely, those by which the electrons: 1. enter the conduction band originating from a different band or from a trap, or 2. leave the conduction band towards a different band or trap.

It follows

$$C_n = \int_B C \mathrm{d}^3k = \int_B C' \mathrm{d}^3k = \left(\frac{\partial n}{\partial t}\right)_{\mathrm{coll}}, \tag{2.2.13}$$

where the last expression on the right reminds us that the integral of C provides the time variation of n due to collisions.

The BTE is a differential equation in the $\boldsymbol{r}, \boldsymbol{k}, t$ space. On the other hand, for the description of carrier transport in semiconductor devices it is often sufficient to rely on functions defined over the $\boldsymbol{r}, t$ space only; this is typically achieved by multiplying both sides of (2.2.8) by suitable functions $\alpha(\boldsymbol{k})$ and integrating the result over the $\boldsymbol{k}$ space—more specifically, over the Brillouin zone. In this procedure the coordinates of the $\boldsymbol{k}$ space are saturated and a set of differential equations in the $\boldsymbol{r}, t$ space only is left, which are called **moments** of the BTE. Although the term 'moment' is more specifically used when α is a polynomial in $\boldsymbol{k}$, it will be used here with a broader meaning; in particular, α will be taken as one of the following:

$$1, \qquad u_i, \qquad \epsilon, \qquad \epsilon u_i, \tag{2.2.14}$$

$i = 1, 2, 3$. The resulting equation in the r, t space takes the general form

$$\int_B \alpha \left(\frac{\partial f}{\partial t} + \boldsymbol{u} \cdot \mathrm{grad}_r f - \frac{q}{\hbar} (\boldsymbol{E} + \boldsymbol{u} \wedge \boldsymbol{B}) \cdot \mathrm{grad}_k f \right) \mathrm{d}^3 k = \int_B \alpha C \mathrm{d}^3 k \tag{2.2.15}$$

and, as will be shown in section 3, constitutes a continuity equation for $n\overline{\alpha}$.

2.2.3 *Evaluation of the moments*

The explicit form of the terms of (2.2.15) will be evaluated here. For the time derivative one finds, recalling (2.2.1) and (2.2.3),

$$\int_B \alpha \frac{\partial f}{\partial t} \mathrm{d}^3 k = \int_B \frac{\partial (\alpha f)}{\partial t} \mathrm{d}^3 k = \frac{\partial}{\partial t} \int_B \alpha f \, \mathrm{d}^3 k = \frac{\partial}{\partial t} (n \overline{\alpha}). \tag{2.2.16}$$

The second term on the left-hand side of (2.2.15) is found by means of the identity $\alpha \boldsymbol{u} \cdot \mathrm{grad}_r f = \mathrm{div}_r (\alpha f \boldsymbol{u})$, which holds because α and the group velocity (2.2.2) are independent of $\boldsymbol{r}$; one finds

$$\begin{aligned} \int_B \alpha \boldsymbol{u} \cdot \mathrm{grad}_r f \mathrm{d}^3 k &= \int_B \mathrm{div}_r (\alpha f \boldsymbol{u}) \mathrm{d}^3 k \\ &= \mathrm{div}_r \int_B \alpha f \boldsymbol{u} \, \mathrm{d}^3 k = \mathrm{div}_r (n \overline{\alpha \boldsymbol{u}}). \end{aligned} \tag{2.2.17}$$

To proceed, it will be assumed that the behaviour of the distribution function at the boundary Γ of the Brillouin zone is such that $\alpha(\boldsymbol{k}) f(\boldsymbol{r}, \boldsymbol{k}, t) \to 0$ as $\boldsymbol{k} \to \boldsymbol{k}_\Gamma$. This amounts to assuming that there are no carriers at the boundary of B. From a practical standpoint, this hypothesis has the same effect as that of [10, 11], where the calculation at hand was carried out by replacing the Brillouin zone with an infinite domain and by assuming that the distribution function in a nonequilibrium condition vanishes at infinity in the same exponential-like fashion as it does at equilibrium, where it becomes proportional to the Fermi statistics. Using the identity $\alpha \mathrm{grad}_k f = \mathrm{grad}_k (\alpha f) - f \mathrm{grad}_k \alpha$, the term of (2.2.15) containing the electric field becomes

$$-\frac{q}{\hbar}\int_B \alpha \boldsymbol{E}\cdot \mathrm{grad}_k f \mathrm{d}^3 k = -\frac{q}{\hbar}\sum_{j=1}^{3} E_j \int_B \frac{\partial(\alpha f)}{\partial k_j}\mathrm{d}^3 k + \frac{q}{\hbar}\boldsymbol{E}\cdot\int_B f \mathrm{grad}_k \alpha \, \mathrm{d}^3 k, \tag{2.2.18}$$

where the sum on the right-hand side vanishes thanks to the hypothesis above. Then,

$$-\frac{q}{\hbar}\int_B \alpha \boldsymbol{E}\cdot \mathrm{grad}_k f \mathrm{d}^3 k = \frac{q}{\hbar} n \overline{\mathrm{grad}_k \alpha} \cdot \boldsymbol{E} = \frac{q}{\hbar} n \overline{\mathrm{grad}_k \alpha \cdot \boldsymbol{E}}, \tag{2.2.19}$$

where the last equality holds because $\boldsymbol{E}$ is independent of $\boldsymbol{k}$. The term of (2.2.15) containing the magnetic induction is treated more easily by rewriting the mixed product as $\alpha \boldsymbol{u}\wedge\boldsymbol{B}\cdot\mathrm{grad}_k f = \mathrm{grad}_k f \wedge \alpha\boldsymbol{u}\cdot\boldsymbol{B}$ and by using the identity $\mathrm{grad}_k f \wedge \alpha\boldsymbol{u} = \mathrm{rot}_k(f\alpha\boldsymbol{u}) - f\mathrm{rot}_k(\alpha\boldsymbol{u})$. The integral of $\mathrm{rot}_k(f\alpha\boldsymbol{u})\cdot\boldsymbol{B}$ vanishes because of the hypothesis on the behaviour of f at the boundary, whence

$$-\frac{q}{\hbar}\int_B \alpha\boldsymbol{u}\wedge\boldsymbol{B}\cdot \mathrm{grad}_k f \mathrm{d}^3 k = \frac{q}{\hbar}\boldsymbol{B}\cdot\int_B f\mathrm{rot}_k(\alpha\boldsymbol{u})\,\mathrm{d}^3 k = \frac{q}{\hbar} n \overline{\mathrm{rot}_k(\alpha\boldsymbol{u})\cdot\boldsymbol{B}}. \tag{2.2.20}$$

The above can be given a different form thanks to the identity $\mathrm{rot}_k(\alpha\boldsymbol{u}) = \alpha\mathrm{rot}_k\boldsymbol{u} + \mathrm{grad}_k\alpha\wedge\boldsymbol{u}$, where the first term on the right-hand side does not contribute since, recalling (2.2.2), it is $\mathrm{rot}_k\boldsymbol{u} = (1/\hbar)\mathrm{rot}\,\mathrm{grad}_k\epsilon = 0$. It follows

$$-\frac{q}{\hbar}\int_B \alpha\boldsymbol{u}\wedge\boldsymbol{B}\cdot \mathrm{grad}_k f \mathrm{d}^3 k = \frac{q}{\hbar}\int_B f\mathrm{grad}_k\alpha\cdot\boldsymbol{u}\wedge\boldsymbol{B}\,\mathrm{d}^3 k = \frac{q}{\hbar} n \overline{\mathrm{grad}_k\alpha\cdot\boldsymbol{u}\wedge\boldsymbol{B}}. \tag{2.2.21}$$

From this result it is easily found that the term with the magnetic induction vanishes when α depends on $\boldsymbol{k}$ through the microscopic energy alone, $\alpha = \alpha(\epsilon)$; in this case it is $\mathrm{grad}_k\alpha = (\mathrm{d}\alpha/\mathrm{d}\epsilon)\mathrm{grad}_k\epsilon$ whence, from (2.2.2), $\mathrm{grad}_k\alpha\cdot\boldsymbol{u}\wedge\boldsymbol{B} = (\mathrm{d}\alpha/\mathrm{d}\epsilon)\hbar\boldsymbol{u}\cdot\boldsymbol{u}\wedge\boldsymbol{B} = 0$. To summarize, letting $\boldsymbol{\beta} = \overline{\mathrm{grad}_k\alpha}$, $\boldsymbol{\gamma} = \overline{\mathrm{rot}_k(\alpha\boldsymbol{u})}$ and using (2.2.19)–(2.2.21), the term of (2.2.15) containing the electric field and the magnetic induction can be written as

$$-\frac{q}{\hbar}\int_B \alpha(\boldsymbol{E}+\boldsymbol{u}\wedge\boldsymbol{B})\cdot\mathrm{grad}_k f\,\mathrm{d}^3 k = \frac{q}{\hbar} n(\boldsymbol{\beta}\cdot\boldsymbol{E}+\boldsymbol{\gamma}\cdot\boldsymbol{B}) \tag{2.2.22}$$

or, equivalently,

$$-\frac{q}{\hbar}\int_B \alpha(\boldsymbol{E}+\boldsymbol{u}\wedge\boldsymbol{B})\cdot\mathrm{grad}_k f\,\mathrm{d}^3 k = \frac{q}{\hbar} n \overline{\mathrm{grad}_k\alpha\cdot(\boldsymbol{E}+\boldsymbol{u}\wedge\boldsymbol{B})}. \tag{2.2.23}$$

As for the right-hand side of (2.2.15), recalling (2.2.9) and defining $C'_\alpha(\boldsymbol{r}, t)$, $\tau_\alpha(\boldsymbol{r}, t)$ as

$$C'_\alpha(\boldsymbol{r}, t) = \frac{1}{\overline{\alpha}} \int_B \alpha C' \, \mathrm{d}^3 k, \qquad \frac{1}{\tau_\alpha} = \frac{\int_B (f\alpha/\tau) \, \mathrm{d}^3 k}{\int_B f\alpha \, \mathrm{d}^3 k}, \tag{2.2.24}$$

the collision term of (2.2.15) yields

$$\int_B \alpha C \, \mathrm{d}^3 k = \overline{\alpha} C'_\alpha(\boldsymbol{r}, t) - \frac{n\overline{\alpha}}{\tau_\alpha} + \int_B \tilde{f} \frac{\alpha}{\tau} \, \mathrm{d}^3 k. \tag{2.2.25}$$

The collision term involving $\tilde{f}$ in (2.2.25) is now rewritten by defining the dimensionless function

$$\eta_\alpha(\boldsymbol{r}, t) = \frac{\int_B (\tilde{f}\alpha/\tau) \, \mathrm{d}^3 k}{\int_B (f\alpha/\tau) \, \mathrm{d}^3 k} \tag{2.2.26}$$

which, in combination with the second of (2.2.24), yields

$$-\frac{n\overline{\alpha}}{\tau_\alpha} + \int_B \tilde{f} \frac{\alpha}{\tau} \, \mathrm{d}^3 k = -n \frac{\overline{\alpha} - \eta_\alpha \overline{\alpha}}{\tau_\alpha}. \tag{2.2.27}$$

It is worth observing that (2.2.27) does not introduce a simplification in the collisions terms, but merely expresses them in a more convenient form. In equilibrium it is $\tilde{f}^{\mathrm{eq}} = f^{\mathrm{eq}}$ whence $\eta_\alpha^{\mathrm{eq}} = 1$. From (2.2.16), (2.2.17), (2.2.22) and (2.2.27) one finally finds

$$\frac{\partial}{\partial t}(n\overline{\alpha}) + \mathrm{div}_r(n\overline{\alpha \boldsymbol{u}}) + \frac{q}{\hbar} n(\boldsymbol{\beta} \cdot \boldsymbol{E} + \boldsymbol{\gamma} \cdot \boldsymbol{B}) = \overline{\alpha} C'_\alpha - n \frac{\overline{\alpha} - \eta_\alpha \overline{\alpha}}{\tau_\alpha}. \tag{2.2.28}$$

2.3 THE HYDRODYNAMIC MODEL

In section 2.2 the derivation of the BTE's moment was shown, leading to the general form (2.2.28). In equilibrium the distribution function f^{eq} is independent of t and depends on $\boldsymbol{k}$ through ϵ only; since $\epsilon(\boldsymbol{k})$ is even, f^{eq} is an even function of $\boldsymbol{k}$ as well. The equilibrium distribution function retains a dependence on $\boldsymbol{r}$ if the semiconductor is nonuniform; this happens, for instance, when the dopant concentration depends on position. In equilibrium the transitions balance each other; the right-hand side of (2.2.28) thus vanishes. The term with the magnetic induction also vanishes, as is seen from (2.2.21); in this case, in fact, f is to be replaced with f^{eq}, which depends on ϵ only. In conclusion, (2.2.28) reduces to

$$\mathrm{div}_r(n\overline{\alpha \boldsymbol{u}})^{\mathrm{eq}} + \frac{q}{\hbar}(n\boldsymbol{\beta} \cdot \boldsymbol{E})^{\mathrm{eq}} = 0. \tag{2.3.1}$$

It is easily found that (2.3.1) yields the identity $0 = 0$ when $\alpha(\boldsymbol{k})$ is even. This, in contrast, is not true when α is odd; in such case the equilibrium condition consists of the balance between the first term, due to the spatial nonuniformity of $n\overline{\alpha \boldsymbol{u}}$ and the second term, proportional to the carrier concentration and linearly dependent on the electric field. In a nonequilibrium condition the two terms do not balance each other and contribute to the transport in the semiconductor; in particular, $\mathrm{div}_r(n\overline{\alpha \boldsymbol{u}})$ is called the **diffusive term** while $(q/\hbar)n(\boldsymbol{\beta}\cdot\boldsymbol{E}+\boldsymbol{\gamma}\cdot\boldsymbol{B})$, due to the existence of the external fields, is called the **drift term**. Strictly speaking, the diffusive term and drift term are used to indicate contributions proportional to $\mathrm{grad}_r n$ and $n\boldsymbol{E}$, which are derived from a simplified form of (2.2.15) where $\boldsymbol{B} = 0$ and α is equal to u_i and proportional to k_i. Here the definition is used in a broader meaning.

2.3.1 Continuity equations for the carrier number and average energy

Remembering that the BTE (2.2.8) is the continuity equation for f in the $\boldsymbol{r}, \boldsymbol{k}$ space, one sees from the derivation of section 2.2 and from the form of (2.2.28) that the latter is the continuity equation for $n\overline{\alpha}$ in the $\boldsymbol{r}$ space. As mentioned in section 2.2, here α is chosen as one of the quantities listed in (2.2.14); in particular, letting $\alpha = 1$ yields the continuity equation for n. In this case one finds from (2.2.23) that the electric field and magnetic induction do not contribute while, from (2.2.13) and (2.2.24), it is $C'_\alpha = C_n$; in turn, the contribution of the intraband collisions vanishes because of (2.2.12). In conclusion, the continuity equation for the electron number takes the usual form

$$\frac{\partial n}{\partial t} + \mathrm{div}_r(n\boldsymbol{v}_n) = C_n, \qquad (2.3.2)$$

where the definition of the average velocity $\boldsymbol{v}_n = \overline{\boldsymbol{u}}$ has been used. It is worth observing that the contribution of the intraband collisions vanishes for $\alpha = 1$, whereas it becomes dominant for the other choices of α listed in (2.2.14); in fact, in such cases the intraband collisions no longer cancel out and their scattering rates turn out to be much higher than those of the interband collisions. This allows one to adopt an approximation for C'_α, namely that of using the equality $C'_\alpha = C_n$, which is exact for $\alpha = 1$, also for any other α; in other words it is assumed that, since the contribution of C'_α to the right-hand side of (2.2.28) is small when $\alpha \neq 1$, the error introduced by letting $C'_\alpha = C_n$ is negligible. Expanding the time derivative in (2.2.28) and using the approximation above yields, thanks to (2.3.2),

$$n\frac{\partial \overline{\alpha}}{\partial t} + \mathrm{div}_r(n\overline{\alpha \boldsymbol{u}}) - \bar{\alpha}\,\mathrm{div}_r(n\boldsymbol{v}_n) + \frac{q}{\hbar}n[\boldsymbol{\beta}\cdot\boldsymbol{E} + \boldsymbol{\gamma}\cdot\boldsymbol{B}] = -n\frac{\overline{\alpha} - \eta_\alpha\overline{\alpha}}{\tau_\alpha}, \qquad (2.3.3)$$

where only the intraband collisions appear. Due to its simpler form, equation (2.3.3) will be used in the following to derive the moments with $\alpha \neq 1$; it will be referred to as the continuity equation for $\overline{\alpha}$ (although such a definition would only properly apply after dividing (2.3.3) by n). In particular, the continuity equation for the average energy of the electrons, w_n, is found by letting $\alpha = \epsilon$ and using (2.2.4). Also in this case the term with the magnetic induction does not contribute because α depends on $\boldsymbol{k}$ through ϵ alone. In contrast, the term with the electric field does contribute because $\boldsymbol{\beta} = \text{grad}_k\epsilon = \hbar\overline{\boldsymbol{u}} = \hbar\boldsymbol{v}_n$. In conclusion,

$$n\frac{\partial w_n}{\partial t} + \text{div}_r(n\boldsymbol{P}_n) - w_n\text{div}_r(n\boldsymbol{v}_n) + qn\boldsymbol{v}_n \cdot \boldsymbol{E} = -n\frac{w_n - \eta_{wn}w_n}{\tau_{wn}}, \tag{2.3.4}$$

where the definition (2.2.7) of the average energy flux $\boldsymbol{P}_n$ has been used and, recalling (2.2.24)–(2.2.26), it is

$$\frac{1}{\tau_{wn}} = \frac{\int_B (f\epsilon/\tau)\,\mathrm{d}^3k}{\int_B f\epsilon\,\mathrm{d}^3k}, \qquad \eta_{wn} = \frac{\int_B (\tilde{f}\epsilon/\tau)\,\mathrm{d}^3k}{\int_B (f\epsilon/\tau)\,\mathrm{d}^3k}. \tag{2.3.5}$$

Time τ_{wn} is called the **energy-relaxation time** of the electrons.

2.3.2 Continuity equations for the momentum and average energy flux

In the same manner as above, letting $\alpha = u_i$ one finds from (2.3.3) the continuity equation for the ith component of the average velocity of the electrons, $\overline{u}_i = v_{ni}$. For the calculations it is useful to introduce a rank-2 tensor $\mathcal{A}$ whose components are defined as $\mathcal{A}_{ij} = \hbar^{-2}\partial^2\epsilon/\partial k_i\partial k_j = \hbar^{-1}\partial u_i/\partial k_j$; and a vector $\boldsymbol{A}_i \equiv (\mathcal{A}_{i1}, \mathcal{A}_{i2}, \mathcal{A}_{i3}) = \hbar^{-1}\text{grad}_k u_i$. The elements of $\mathcal{A}$ and $\boldsymbol{A}_i$ are dimensionally the inverse of a mass. Thanks to the definitions above one finds for the term with the electric field $(q/\hbar)n\overline{\boldsymbol{\beta}} \cdot \boldsymbol{E} = qn\overline{\boldsymbol{A}}_i \cdot \boldsymbol{E} = qn\overline{\boldsymbol{A}_i \cdot \boldsymbol{E}}$. As for the term with the magnetic induction one has, remembering the derivation of (2.2.20), (2.2.21), $(q/\hbar)n\overline{\boldsymbol{\gamma}} \cdot \boldsymbol{B} = qn\overline{\boldsymbol{A}_i \wedge \boldsymbol{u}} \cdot \boldsymbol{B} = qn\overline{\boldsymbol{A}_i \cdot \boldsymbol{u} \wedge \boldsymbol{B}}$. In conclusion,

$$n\frac{\partial \overline{u}_i}{\partial t} + \text{div}_r(n\overline{u_i\boldsymbol{u}}) - \overline{u}_i\text{div}_r(n\overline{\boldsymbol{u}}) + qn\overline{\boldsymbol{A}}_i \cdot \boldsymbol{E}$$
$$+ qn\overline{\boldsymbol{A}_i \wedge \boldsymbol{u}} \cdot \boldsymbol{B} = -n\frac{\overline{u}_i - \eta_{pni}\overline{u}_i}{\tau_{pni}}, \tag{2.3.6}$$

with

$$\frac{1}{\tau_{pni}} = \frac{\int_B (fu_i/\tau)\,\mathrm{d}^3k}{\int_B fu_i\,\mathrm{d}^3k}, \qquad \eta_{pni} = \frac{\int_B (\tilde{f}u_i/\tau)\,\mathrm{d}^3k}{\int_B (fu_i/\tau)\,\mathrm{d}^3k}. \tag{2.3.7}$$

If the crystal potential was absent it would be $\epsilon = \hbar^2k^2/(2m)$, $\boldsymbol{u} = \hbar\boldsymbol{k}/m$ and equation (2.3.6) multiplied by m would provide the continuity equation for the electron momentum $\boldsymbol{p} = \hbar\boldsymbol{k}$; for this reason it is usually referred to, also in the present case, as the momentum-continuity equation of the electrons. Letting $i = 1, 2, 3$, the first of (2.3.7) defines a diagonal tensor $\hat{\tau}_{pn}$ which is called the **momentum-relaxation time** of the electrons. Finally, letting $\alpha = \epsilon u_i$ one finds from (2.3.3) the continuity equation for the ith component of the average energy flux, $\overline{\epsilon u}_i = P_{ni}$. The definition $\boldsymbol{e}_i = \epsilon\boldsymbol{A}_i + u_i\boldsymbol{u}$ will be useful in the calculation; in fact, observing that $\text{grad}_k(\epsilon u_i) = \hbar\boldsymbol{e}_i$ and that $\text{rot}_k(\epsilon u_i\boldsymbol{u}) = \hbar\boldsymbol{e}_i \wedge \boldsymbol{u} = \hbar\epsilon\boldsymbol{A}_i \wedge \boldsymbol{u}$, one finds here $(q/\hbar)n\overline{\boldsymbol{\beta}} \cdot \boldsymbol{E} = qn\overline{\boldsymbol{e}}_i \cdot \boldsymbol{E} = qn\overline{\boldsymbol{e}_i \cdot \boldsymbol{E}}$ and

$(q/\hbar)n\gamma \cdot \boldsymbol{B} = qn\overline{\epsilon \boldsymbol{A}_i \wedge \boldsymbol{u}} \cdot \boldsymbol{B} = qn\overline{\epsilon \boldsymbol{A}_i \cdot \boldsymbol{u} \wedge \boldsymbol{B}}$. In conclusion,

$$n\frac{\partial \overline{\epsilon u_i}}{\partial t} + \mathrm{div}_r(n\overline{\epsilon u_i \boldsymbol{u}}) - \overline{\epsilon u_i}\,\mathrm{div}_r(n\overline{\boldsymbol{u}}) + qn\overline{\boldsymbol{e}}_i \cdot \boldsymbol{E}$$
$$+qn\overline{\epsilon \boldsymbol{A}_i \wedge \boldsymbol{u}} \cdot \boldsymbol{B} = -n\frac{\overline{\epsilon u_i} - \eta_{qni}\overline{\epsilon u_i}}{\tau_{qni}}, \tag{2.3.8}$$

with

$$\frac{1}{\tau_{qni}} = \frac{\int_B (f\epsilon u_i/\tau)\,\mathrm{d}^3k}{\int_B f\epsilon u_i\,\mathrm{d}^3k}, \qquad \eta_{qni} = \frac{\int_B (\tilde{f}\epsilon u_i/\tau)\,\mathrm{d}^3k}{\int_B (f\epsilon u_i/\tau)\,\mathrm{d}^3k}. \tag{2.3.9}$$

Equation (2.3.8) is referred to as the continuity equation for the energy flux of the electrons. Again, letting $i = 1, 2, 3$, the first of (2.3.9) defines a diagonal tensor $\hat{\tau}_{qn}$, which is called the **energy-flux relaxation time** or **heat-relaxation time** of the electrons.

2.3.3 The convective terms and time derivatives

The continuity equations derived so far, namely (2.3.2), (2.3.4), (2.3.6) and (2.3.8) are still rather cumbersome in view of the application to the analysis of realistic semiconductor devices. In the following, a number of simplifications will be illustrated which lead to the hydrodynamic model [1, 10, 12–16]. To begin with, it is possible to simplify the diffusive terms in (2.3.6) and (2.3.8); remembering the definition (2.2.6) of the random velocity $\boldsymbol{c}_n$ and observing that $\overline{\boldsymbol{c}}_n = 0$, one finds in fact $\overline{u_i \boldsymbol{u}} = \overline{c_{ni}\boldsymbol{c}_n} + \overline{u}_i\overline{\boldsymbol{u}}$ and, similarly, $\overline{\epsilon u_i \boldsymbol{u}} = \overline{\epsilon u_i \boldsymbol{c}_n} + \overline{\epsilon u_i}\,\overline{\boldsymbol{u}}$. It follows

$$\mathrm{div}_r(n\overline{u_i\boldsymbol{u}}) = \mathrm{div}_r(n\overline{c_{ni}\boldsymbol{c}_n}) + \overline{u}_i\mathrm{div}_r(n\overline{\boldsymbol{u}}) + n\overline{\boldsymbol{u}}\cdot\mathrm{grad}_r\overline{u}_i, \tag{2.3.10}$$
$$\mathrm{div}_r(n\overline{\epsilon u_i\boldsymbol{u}}) = \mathrm{div}_r(n\overline{\epsilon u_i\boldsymbol{c}_n}) + \overline{\epsilon u_i}\mathrm{div}_r(n\overline{\boldsymbol{u}}) + n\overline{\boldsymbol{u}}\cdot\mathrm{grad}_r(\overline{\epsilon u_i}). \tag{2.3.11}$$

The last term on the right-hand side of (2.3.10) and (2.3.11) (**convective term**) is neglected because in typical applications the motion of the carriers is limited to the subsonic regime [11]. A second approximation involves the time derivatives on the left-hand side of the continuity equations. Such derivatives only differ from zero if the distribution function depends explicitly on time, i.e. when time-dependent boundary conditions are imposed. In the practical cases, the maximum frequency of the boundary conditions is lower by many orders of magnitude than the inverse of the times τ_{wn}, τ_{pni}, and τ_{qni}, namely $|\partial w_n/\partial t| \ll |w_n/\tau_{wn}|$, $|\partial \overline{u}_i/\partial t| \ll |\overline{u}_i/\tau_{pni}|$, and $|\partial \overline{\epsilon u_i}/\partial t| \ll |\overline{\epsilon u_i}/\tau_{qni}|$, which makes it possible to neglect the time derivatives of w_n, $\overline{u}_i$, $\overline{\epsilon u_i}$ anyhow. A steady-state approximation is thus assumed in the continuity equations (2.3.4), (2.3.6) and (2.3.8). The above argument does not apply to the case of (2.3.2) because only the interband transitions take place there, whose characteristic times are much longer that those of the intraband transitions. As a consequence, the term $\partial n/\partial t$ in (2.3.2) must be retained when the boundary conditions depend on time. Introducing the approximations indicated above, (2.3.4), (2.3.6) and

(2.3.8) become

$$\mathrm{div}_r(n\boldsymbol{P}_n) - w_n \mathrm{div}_r(n\boldsymbol{v}_n) + qn\boldsymbol{v}_n \cdot \boldsymbol{E} = -n\frac{w_n - \eta_{wn} w_n}{\tau_{wn}}, \qquad (2.3.12)$$

$$\mathrm{div}_r(n\overline{c_{ni}\boldsymbol{c}_n}) + qn(\overline{\boldsymbol{A}_i} \cdot \boldsymbol{E} + \overline{\boldsymbol{A}_i \wedge \boldsymbol{u}} \cdot \boldsymbol{B}) = -n\frac{\overline{u}_i - \eta_{pni}\overline{u}_i}{\tau_{pni}}, \qquad (2.3.13)$$

$$\mathrm{div}_r(n\overline{\epsilon u_i \boldsymbol{c}_n}) + qn(\overline{\boldsymbol{e}_i} \cdot \boldsymbol{E} + \overline{\epsilon \boldsymbol{A}_i \wedge \boldsymbol{u}} \cdot \boldsymbol{B}) = -n\frac{\overline{\epsilon u_i} - \eta_{qni}\overline{\epsilon u_i}}{\tau_{qni}} \qquad (2.3.14)$$

which, along with (2.3.2), constitute a system of four first-order differential equations. The latter is coupled with the Maxwell equations whose solution provides $\boldsymbol{E}$ and $\boldsymbol{B}$. The unknowns of the system are n, $\boldsymbol{v}_n = \overline{\boldsymbol{u}}$, w_n, $\boldsymbol{P}_n = \overline{\epsilon \boldsymbol{u}}$, $\overline{c_{ni}\boldsymbol{c}_n}$, $\overline{\epsilon u_i \boldsymbol{c}_n}$ and $\boldsymbol{e}_i$, $i = 1, 2, 3$. The coefficients are C_n, τ_{wn}, τ_{pni}, τ_{qni}, η_{wn}, η_{pni}, η_{qni} and $\overline{\boldsymbol{A}}_i$, $i = 1, 2, 3$. One sees that some of the unknowns and coefficients of the continuity equations (2.3.13), (2.3.14) have a tensor form, hence the transport properties are expected to exhibit some degree of anisotropy. Investigations about the coefficients have been carried out by different techniques, specifically the spherical-harmonics expansion method to determine the dependence on the average energy [12] and the Monte Carlo method to also study the anisotropy [17–19]; some results of more recent investigations are shown in section 2.4.

2.3.4 Derivation of the hydrodynamic model

Since the number of unknowns in (2.3.2), (2.3.12)–(2.3.14) exceeds the number of equations, further simplifications are necessary in order to solve the system. For this, a perturbative approach is typically assumed, which simplifies the intraband-collision terms and reduces some of the tensors in (2.3.12) and (2.3.14) to scalars. First, remembering (2.3.3), the terms of the form $\eta_\alpha \overline{\alpha}$ are replaced with the equilibrium value $\overline{\alpha}^{\mathrm{eq}}$; the latter, in turn, is zero when α is an odd function of $\boldsymbol{k}$. Then, tensors $\mathcal{A}$ and $\epsilon\mathcal{A}$ are replaced with $\frac{1}{3}\mathrm{tr}(\mathcal{A})\mathcal{I}$ and $\frac{1}{3}\mathrm{tr}(\epsilon\mathcal{A})\mathcal{I}$, respectively, where $\mathcal{I}$ is the identity, this leading to the definitions

$$\frac{1}{m_{pn}} = \frac{1}{3}\mathrm{tr}(\overline{\mathcal{A}}), \qquad \frac{1}{m_{qn}} = \frac{1}{3}\mathrm{tr}\left(\frac{\overline{\epsilon\mathcal{A}}}{\overline{\epsilon}}\right) \qquad (2.3.15)$$

and, similarly,

$$\frac{1}{m_{bni}} = \frac{1}{3}\mathrm{tr}\left(\frac{\overline{u_i\mathcal{A}}}{\overline{u}_i}\right), \qquad \frac{1}{m_{sni}} = \frac{1}{3}\mathrm{tr}\left(\frac{\overline{u_i\epsilon\mathcal{A}}}{\overline{u_i\epsilon}}\right). \qquad (2.3.16)$$

In the same way, the first term on the left-hand side of (2.3.13), (2.3.14) yields

$$\mathrm{div}_r(n\overline{c_{ni}\boldsymbol{c}_n}) = \frac{1}{3}\frac{\partial}{\partial x_i}(n\overline{c_n^2}) \qquad (2.3.17)$$

and

$$\mathrm{div}_r(n\overline{\epsilon u_i c_n}) = \frac{1}{3}\frac{\partial}{\partial x_i}(n\overline{\epsilon c_n^2}), \qquad (2.3.18)$$

respectively. In particular, (2.3.18) is obtained by neglecting $\overline{u_i \epsilon c_n}$ with respect to $\overline{\epsilon c_{ni} c_n}$. Similarly, neglecting $v_n v_n$ with respect to $\overline{c_n c_n}$, one finds $\overline{uu} = (\overline{c_n^2}/3)\mathcal{I}$ and finally, using the second of (2.3.15) and remembering that $\overline{\epsilon} = w_n$, one finds $\overline{\epsilon \mathcal{A}} = (w_n/m_{qn})\mathcal{I}$. Adding the last two expressions yields

$$\overline{\epsilon \mathcal{A}} + \overline{uu} = \frac{1}{m_{qn}} s_n \mathcal{I}, \qquad s_n = w_n + \tfrac{1}{3} m_{qn} \overline{c_n^2}. \qquad (2.3.19)$$

To proceed, two electron temperatures T_n and $\tilde{T}_n$ must be defined through

$$\tfrac{1}{2} m_{pn}^* \overline{c_n^2} = \tfrac{3}{2} k_B T_n, \qquad \tfrac{1}{2} m_{qn}^* \overline{\epsilon c_n^2} = \tfrac{3}{2} k_B \tilde{T}_n s_n, \qquad (2.3.20)$$

where k_B is the Boltzmann constant. For a general shape of the band it is necessary to specify the value of the masses in (2.3.20); a possible choice is $m_{pn}^* = m_{pn}^{\mathrm{eq}}$, $m_{qn}^* = m_{qn}^{\mathrm{eq}}$, which are calculated from (2.3.15) using the equilibrium distribution function f^{eq}. However, m_{pn}^{eq} and m_{qn}^{eq} are independent of position only for a Boltzmann distribution, because in such case the part of the distribution function containing $E_C - E_F(\boldsymbol{r})$ cancels out in the calculation of the average (here, E_C is the lower limit of the conduction band and E_F is the Fermi energy which depends on position when the doping in the semiconductor is not uniform); thus it is preferable to specify m_{pn}^* and m_{qn}^* as the masses obtained from (2.3.15) using the equilibrium distribution function of the nondegenerate semiconductor. Using (2.3.15)–(2.3.20) in (2.3.13) and (2.3.14) and expressing the result in vector form one finds

$$-n\boldsymbol{v}_n = \frac{\hat{\tau}_{pn}}{m_{pn}^*}\mathrm{grad}(nk_B T_n) + \frac{\hat{\tau}_{pn}}{m_{pn}} qn\boldsymbol{E} + \frac{\hat{\tau}_{pn}}{m_{bn}} qn(\boldsymbol{v}_n \wedge \boldsymbol{B}), \qquad (2.3.21)$$

with $1/m_{bn} = \frac{1}{3}\sum_i 1/m_{bni}$ and

$$-n\boldsymbol{P}_n = \frac{\hat{\tau}_{qn}}{m_{qn}^*}\mathrm{grad}(nk_B \tilde{T}_n s_n) + \frac{\hat{\tau}_{qn}}{m_{qn}} qns_n\boldsymbol{E} + \frac{\hat{\tau}_{qn}}{m_{sn}} qn(\boldsymbol{P}_n \wedge \boldsymbol{B}), \qquad (2.3.22)$$

with $1/m_{sn} = \frac{1}{3}\sum_i 1/m_{sni}$. The vector products involving the unknowns in (2.3.21), (2.3.22) can be eliminated by means of vector identities; the details are given in appendix A.1. Equations (2.3.21) and (2.3.22) are also called **transport equations**, respectively, for the electrons and electron energy. In view of the continuity equation for the average energy (2.3.12), the simplification occurs on the right-hand side only yielding

$$\mathrm{div}_r(n\boldsymbol{P}_n) - w_n \mathrm{div}_r(n\boldsymbol{v}_n) + qn\boldsymbol{v}_n \cdot \boldsymbol{E} = -n\frac{w_n - w_n^{\mathrm{eq}}}{\tau_{wn}}. \qquad (2.3.23)$$

It is worth noting that for a parabolic band, the masses m_{pn}, m_{qn}, m_{bn}, and

m_{sn} reduce to the same constant m^*, which can be used in (2.3.19)–(2.3.22). As a consequence, definitions (2.3.19) and (2.3.20) yield $w_n = \frac{1}{2}m^*v_n^2 + \frac{3}{2}k_B T_n$ and, respectively, $s_n = \frac{1}{2}m^*v_n^2 + \frac{5}{2}k_B T_n$. In particular, $\frac{3}{2}k_B T_n$ and $\frac{5}{2}k_B T_n$ are called the **thermal part** of w_n and s_n, and $\frac{1}{2}m^*v_n^2$ is called the **convective part**. The above expressions of w_n and s_n can be assumed to hold at low energies and are further simplified as $w_n \simeq \frac{3}{2}k_B T_n$ and $s_n \simeq \frac{5}{2}k_B T_n$ because in the practical cases the thermal part is dominant over the convective part. For a general shape of the band the relations $w_n(T_n), s_n(T_n)$ deviate from linearity but can still be recovered at the expense of introducing further parameters [16]. In conclusion the system to be solved is made of two scalar equations (2.3.2), (2.3.23) and two vector equations (2.3.21), (2.3.22); it is assumed that the relaxation times, masses and fields are known, that the expressions of w_n, s_n in terms of T_n are given and that the expression of the interband term C_n is also known. Hence, the unknowns of the model are n, T_n for the first group and $\boldsymbol{v}_n, \boldsymbol{P}_n$ for the second group; the structure of the model's equation is discussed in detail, for example in [10, 11]. Here it is important to notice that $\tilde{T}_n$ is an additional unknown—essentially deriving from a moment of an order higher than that of $\boldsymbol{P}_n$, which must be related to the other unknowns to make the system solvable. The relation expressing $\tilde{T}_n$ is called the **closure condition**; following [16], $\tilde{T}_n$ is expressed here by giving it the form that holds in the case of a parabolic band and a Maxwellian distribution function, namely $\tilde{T}_n = T_n$. The derivation is given in appendix A.2. Because of the analogy with the equations of fluid mechanics, the system made of equations (2.3.2), (2.3.23) and (2.3.21), (2.3.22), supplemented with the closure condition, is called the **hydrodynamic model**. It is easily seen that, if the simpler closure condition $T_n = T_L$ is chosen, where T_L is the lattice temperature, the two-equation system made of (2.3.2) and (2.3.21) becomes closed and, in particular, it provides the anisotropic generalization of the well-known drift-diffusion model.

2.4 MODEL COEFFICIENTS

In the derivation of the continuity equation for the average energy (2.3.23) and transport equations (2.3.21), (2.3.22) a number of approximations have been introduced; as discussed in section 2.3, they consist of letting $C'_\alpha = C_n$ with $\alpha \neq 1$ (section 2.3.1), neglecting the convective term and time derivative in the higher-order moments (section 2.3.3), adopting the perturbation approach for the collision terms and introducing the scalar effective masses (section 2.3.4). In order to solve the equations it is necessary to determine the masses and relaxation times appearing in the coefficients; for such an analysis, in turn, a full solution of the BTE is needed, incorporating those features of the band structure and scattering mechanisms that are relevant for a correct description of the problem. The outcome of this is a set of coefficients expressed in terms of problem unknowns (e.g. the average energy w_n [12]) or other parameters, such as the electric field; typically, the coefficients are given in table form and become part of the input data for the solution of the hydrodynamic model. The basic hypothesis underlying this procedure is that, although some approximations are involved in the calculation, the tables are applicable in a realistic range of device operation. In general, the tables are calculated for bulk materials, whence a possible effect of the gradients of the unknowns on the coefficients is lost [20]; in contrast, the surface-scattering effect, which is paramount in MOS devices, has recently been incorporated in the full solution of the BTE based on the spherical-harmonics expansion [21].

In the following, results of relaxation-time calculations are shown, obtained by means of a Monte Carlo simulator for electron transport in Si accounting for the full three-dimensional electron dynamics in the $\boldsymbol{k}$ space for a homogeneous system and including six ellipsoidal, nonparabolic valleys associated to the minima of the conduction band [22, 23]. The relaxation times have been determined through the Monte Carlo calculation of the distribution function, the collision integral on the right-hand side of the BTE and the averages in (2.3.5), (2.3.7) and (2.3.9). This approach allows one to investigate another approximation which is typically adopted, namely that of replacing the relaxation-time tensors $\hat{\tau}_{pn}$, $\hat{\tau}_{qn}$ of the transport equations (2.3.21) and (2.3.22) with scalars. This assumption is in principle not justified under nonequilibrium conditions. In fact, two types of anisotropy can be relevant. The electric field has a disturbing effect on the cubic symmetry of the crystal; this is expected to reflect into some degree of anisotropy the distribution function, whose moments over $\boldsymbol{k}$ then acquire an anisotropic nature. Furthermore, the electron populations around different minima of the band, equivalent by crystal symmetry, show nonequivalent dynamical and scattering properties due to the different effect of the electric field; the results of the calculation are thus expected to depend on the field direction. To obtain results in a number of physical conditions that are most relevant for the applications, the range of the electric field $0.1 \leq E \leq 100$ kV cm^{-1} along the $\langle 100 \rangle$ and $\langle 111 \rangle$ directions at 77 and 300 K has been investigated, where anisotropy effects are known to be relevant [23]. When transport is analysed at much higher electric fields and electron energies (typically above 100 kV cm^{-1} and 1 eV, respectively) the anisotropy effects become progressively less relevant because the carriers spread more uniformly over larger and larger portions of the Brillouin zone [24]. It is worth observing that, since u_i and ϵu_i are odd functions of $\boldsymbol{k}$ and τ is essentially even, a direct numerical evaluation of (2.3.7) and (2.3.9) provides statistically sound information only in the direction parallel to the electric field, that is, where the distribution function is drifted with respect to the equilibrium one. As a consequence, only one component, namely, τ_{pn3}, τ_{qn3} was calculated directly from (2.3.7) and (2.3.9), since the average quantities in that case were different from zero. However, the transverse momentum and energy-flux relaxation times can be obtained from the two transverse microscopic autocorrelation functions of velocity and energy flux through an alternative procedure extensively described in [25]. In fact, it is observed that transverse to

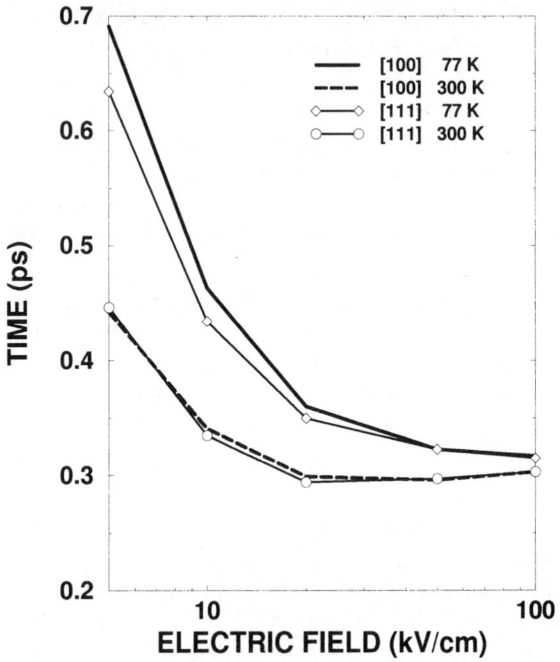

Figure 2.1 *Energy-relaxation time τ_w as a function of the electric-field strength for Si at 77 and 300 K, and at different directions of the electric field.*

the field direction, the decay of the two autocorrelation functions above is well approximated by an exponential decay, at least in the range of time where they are sensibly different from zero. Thus, the corresponding relaxation times can easily be extracted by a fitting procedure.

Figure 2.1 shows the energy-relaxation time τ_w of (2.3.23) as a function of the electric-field strength at 77 and 300 K and at different directions of the electric field. Figure 2.2 shows the longitudinal and transverse component of the momentum-relaxation time $\hat{\tau}_p$ at 300 K and at different directions of the electric field; next, Fig. 2.3 shows the longitudinal and transverse component of the energy-flux relaxation time $\hat{\tau}_q$ at 77 K and at different directions of the electric field. A general comment valid for the three relaxation times is that the anisotropy effects become larger as the temperature falls, as happens for other nonequilibrium physical properties of the electron gas. Furthermore, at both temperatures the relaxation times seem to depend very weakly on the electric-field direction. Only τ_w at 77 K for $\boldsymbol{E}\|\langle 100\rangle$ is higher than the corresponding value for the case $\boldsymbol{E}\|\langle 111\rangle$. A significant difference between the longitudinal and transverse components has been detected for both $\hat{\tau}_p$ and $\hat{\tau}_q$ in all the situations considered. This seems to suggest that the anisotropy with respect to the electric

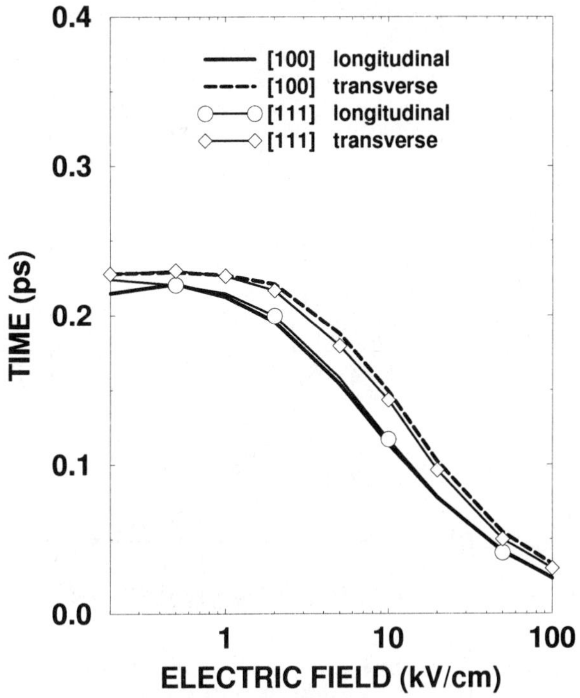

Figure 2.2 *Longitudinal and transverse component of the momentum-relaxation time* $\hat{\tau}_p$ *as a function of the electric-field strength for Si at 300 K and at different directions of the electric field.*

field direction observed in many microscopic properties of the system (e.g. the drift velocity [23]) is mainly related to the different effective masses along the field in the different valleys and not to different relaxation processes.

Finally, the expression for a number of macroscopic coefficients typically used in transport theory is derived from (2.3.2) and (2.3.22). For instance, the mobility, conductivity and diffusivity tensors for the carrier momentum are given by $\hat{\mu}_{pn} = q\hat{\tau}_{pn}/m_{pn}$, $\hat{\sigma}_{pn} = q\hat{\mu}_{pn}n$ and $\hat{D}_{pn} = k_B T_L \hat{\tau}_{pn}/m^*_{pn}$. The remaining coefficient of (2.3.21) is brought to the same form by letting $\hat{\mu}_{bn} = q\hat{\tau}_{pn}/m_{bn}$ and $\hat{\sigma}_{bn} = q\hat{\mu}_{bn}n$. Multiplying both sides of (2.3.21) by q and using the above definitions yields the electron current density

$$\boldsymbol{J}_n = -qn\boldsymbol{v}_n = q\hat{D}_{pn}\mathrm{grad}_r(nT_n/T_L) + \hat{\sigma}_{pn}\boldsymbol{E} + \hat{\sigma}_{bn}(\boldsymbol{v}_n \wedge \boldsymbol{B}). \tag{2.4.1}$$

For $\boldsymbol{J}_n = 0$ and n = constant (2.4.1) reduces to $\boldsymbol{E} = Q\mathrm{grad}_r T_n$, where $Q = -k_B m_{pn}/(qm^*_{pn})$ is the thermopower [26]. To proceed it is useful to define the dimensionless functions $\lambda = \mathrm{d}s_n/\mathrm{d}(k_B T_n)$ and $\delta = m_{pn}m^*_{qn}/(m_{qn}m^*_{pn})$; remembering the discussion in section 2.3.4, at low energies it is $\lambda \to \frac{5}{2}$ and

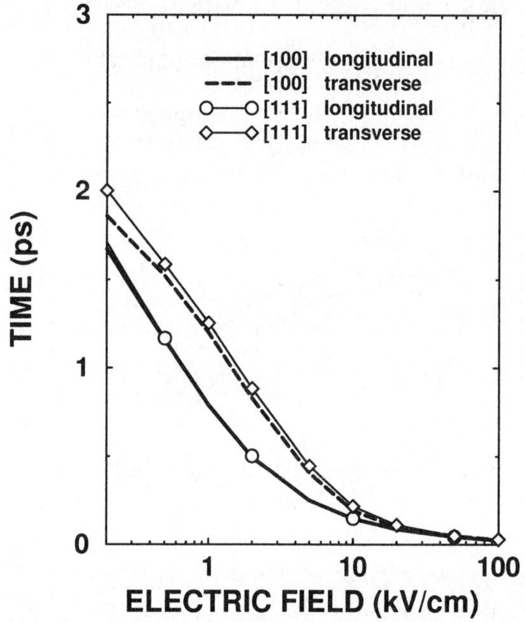

Figure 2.3 *Longitudinal and transverse component of the energy-flux relaxation time* $\hat{\tau}_q$ *as a function of the electric-field strength for Si at 77 K and at different directions of the electric field.*

$\delta \to 1$. Letting $\boldsymbol{B} = 0$ and eliminating $qn\boldsymbol{E}$ from (2.3.21) and (2.3.22) yields

$$-n\boldsymbol{P}_n = \hat{\tau}_{pn}^{-1}\hat{\tau}_{qn}\frac{m_{pn}}{m_{qn}}\frac{s_n}{q}\boldsymbol{J}_n + \hat{\tau}_{qn}\lambda\frac{k_B^2}{m_{qn}^*}nT_n\mathrm{grad}_r T_n$$
$$+\hat{\tau}_{qn}(1-\delta)\frac{s_n}{m_{qn}^*}\mathrm{grad}_r(nk_BT_n). \qquad (2.4.2)$$

On account of the magnitude of $1 - \delta$, the energy flux is essentially the sum of two terms, one proportional to the current density and the other proportional to the gradient of the carrier temperature. The coefficients

$$\hat{\Pi}_n = \hat{\tau}_{pn}^{-1}\hat{\tau}_{qn}\frac{m_{pn}}{m_{qn}}\frac{s_n}{q}, \qquad \hat{\kappa}_n = \hat{\tau}_{qn}\lambda\frac{k_B^2}{m_{qn}^*}nT_n \qquad (2.4.3)$$

are the tensor-form generalization of the Peltier coefficient and thermal conductivity of the carriers, respectively.

2.5 EXAMPLES OF APPLICATION TO HOT-CARRIER EFFECTS

The analysis of a functional device will be illustrated here, using the hydrodynamic equations. As already mentioned in section 2.1, the information provided

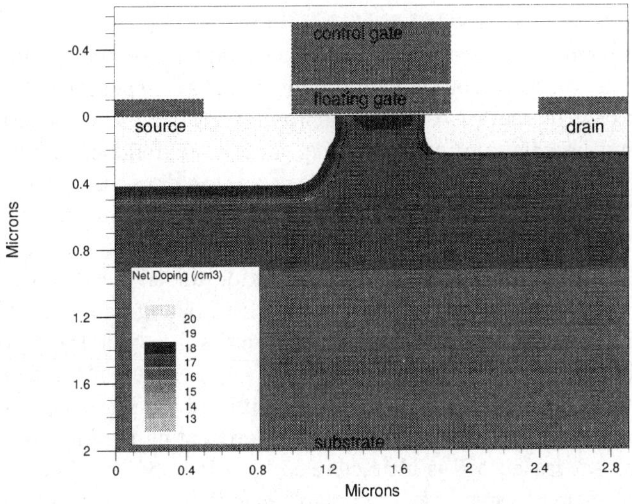

Figure 2.4 *Cross section of a flash-EEPROM cell along the channel length.*

by the hydrodynamic model makes it possible to describe the behaviour of devices in which hot-carrier effects play a significant role. An example of this is given by the nonvolatile memories, where carrier heating is a basic aspect because it is exploited to perform the programming operation. This class of device is extremely important in industrial applications. On the other hand, the physical mechanisms involved in the programming operation are still subject to investigation, which makes this topic also an interesting research issue. For this reason, the modelling and simulation of a type of nonvolatile memory has been chosen as an example, namely, a flash-EEPROM cell.

2.5.1 The flash-EEPROM cell

The flash-EEPROM cell is a derivative of the standard EPROM technology. Fundamentally, it consists of an MOS transistor with an additional layer of polysilicon within the oxide between the gate and silicon substrate. The additional layer, being completely surrounded by an insulating oxide, is electrically isolated and is referred to as a **floating gate**, whereas the top gate is referred to as a **control gate**. A cross section of a flash-EEPROM cell along the channel length is shown in Fig. 2.4, which refers to an n-channel device. The threshold voltage of the device depends on the amount of charge stored within the floating gate, where it is trapped by the potential barrier at the polysilicon-oxide interface. In normal operating conditions the confinement can be maintained for a long time, typically about 10 years, without the need of a power supply. The charge is brought into the floating gate, i.e. the cell is programmed, by injection of hot electrons from the channel region. This requires a high lateral electric field in silicon ($\sim 10^5$ V cm^{-1}), so that strong carrier heating occurs. As already mentioned in section 1, this means that the average kinetic energy of the electrons is larger than the equilibrium value; as a consequence, a number of electrons overcome the barrier of approximately 3.1 eV at the silicon-oxide interface and are able to reach the floating gate. A programmed cell is characterized by a high-threshold voltage, since the negative stored charge tends to screen the channel from the effect of the positive voltage applied to the control gate. The charge is removed from the floating gate, i.e. the cell is erased, by Fowler–Nordheim electron tunnelling through the oxide under the floating gate, that occurs in a region defined by the overlap of the floating gate and source junction. The requirement of a high electric field across the oxide ($\sim$ 10 MV cm^{-1}) imposes the adoption of a thinner oxide ($\sim$ 10 nm) with respect to the standard EPROM technology and of a double-diffused graded source junction to maintain a high voltage during erasure while preventing the junction avalanche breakdown. An erased cell is characterized by a low-threshold voltage so that, in a normal memory operation, the control-gate voltage is high enough to turn the cell on.

Due to the complicated geometry, an accurate simulation of a flash-EEPROM cell demands a three-dimensional approach. In particular, a correct description of the transition region between the active device and the thick oxide is crucial to determine the capacitive coupling between the control gate and floating gate, a critical parameter for the cell performance. Figure 2.5 represents a two-dimensional section of the mesh in the direction orthogonal to the current flow. The shape of the floating gate has been derived from a TEM micrograph of the cell; the two-dimensional grid has then been replicated in the third dimension to provide a complete discretization mesh featuring about 50 000 nodes, which is a practical limit for present-day workstations in order to keep the computational cost acceptable.

2.5.2 Hot-electron injection and related mechanisms

A typical programming condition for a flash-EEPROM cell is $V_{cg} = 12$ V and $V_d = 5$ V, where V_{cg} is the control-gate voltage and V_d is the drain voltage. All voltages are referred to the substrate contact. Due to the capacitive coupling between the control and floating gates, the floating-gate voltage V_{fg} of an unprogrammed cell increases, giving rise to an inverted channel. The drain voltage builds up the lateral channel field responsible for hot-electron generation. Depending on V_d, impact ionization may occur and some of the generated electrons add to the hot-electron population. The electrons reach their maximum average energy after crossing the position of the peak of the lateral electric field and, as previously mentioned, a fraction of them are injected into the gate oxide. There, the positive voltage of the floating gate pulls the electrons towards the floating gate. The gate current consists of the electrons that cross the oxide, i.e. of only a part of the injected carriers. In fact, an electron in the oxide experiences a number of scattering events and may eventually be scattered back. The build-up of a negative charge Q_{fg} within the floating gate progressively increases the threshold voltage V_t of the cell; this is well described by the relation

$$\Delta V_t = -\frac{Q_{fg}}{C_{pp}}, \tag{2.5.1}$$

where C_{pp} is the capacitance between the control and floating gates. At the same time, the floating-gate voltage decreases. When the latter reaches the drain voltage, the electric field in the oxide reverses in proximity of the drain junction and the point of field inversion is further shifted towards the source as V_{fg} decreases. As a result, the electric-field distribution within the thin oxide cannot be considered as one-dimensional even as a first approximation. The importance to accurately model this situation is illustrated by the experimental curves. When the floating-gate voltage reaches the drain voltage, the charge injection slows down as is revealed by the sudden change of slope visible in the programming characteristics.

Apart from electron injection, other phenomena occur during the programming cycle, which are briefly described below; since their influence on the programming cycle is less relevant, they have not been considered in the simulations shown in the following. Among these are electron trapping within the oxide and field-dependent trap generation [27]. Also, although most of the holes created by impact ionization are collected by the substrate contact, a fraction of them accelerate towards the source and gain energy; in analogy with electron injection they may be injected into the gate oxide. However, for this to occur, their energy must exceed the Si–SiO_2 valence-band offset, which is approximately 4.8 eV. Since the barrier height for hole injection is much larger than that for electrons, the hole-injection mechanism is much less efficient. Furthermore, an appreciable hole gate current can only be detected when V_{fg} is sensibly lower than V_d, so that the transverse electric field strongly favours the injection of holes. As the floating-gate voltage increases, electron injection dominates by orders of magnitude. Another phenomenon worth mentioning is that some carriers, which do not acquire a sufficient energy to surmount the energy barrier, can nevertheless reach the oxide by tunnelling. A number of authors have taken into account this contribution to the gate current by including an additional barrier-lowering term [28] or by explicitly considering a tunnelling current [29]. Recent Monte Carlo results seem to indicate that the tunnelling contribution to the gate current can be neglected with respect to the hot-current one [30].

2.5.3 Physical model

Early formulations of the gate current were based on the **lucky-electron** model [31, 32], originally introduced for calculating impact ionization. In this approach, an electron is supposed to be 'lucky' enough to travel ballistically in the elec-

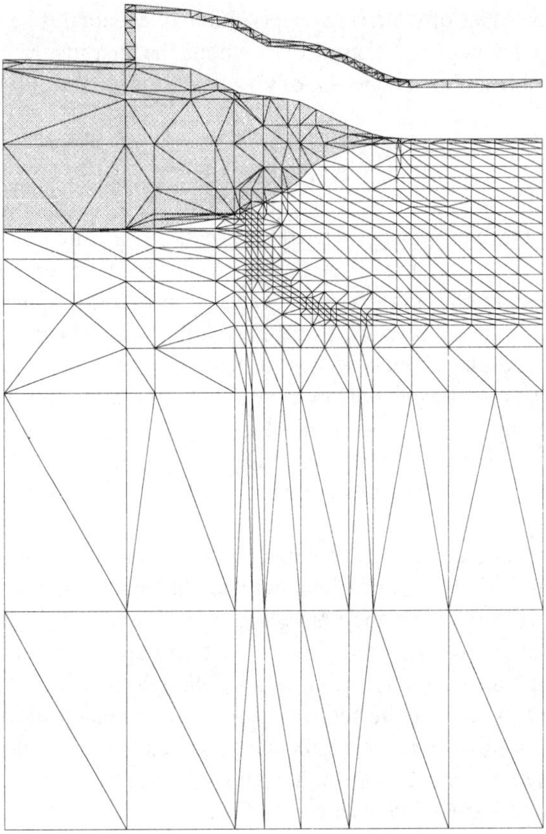

Figure 2.5 *Flash-EEPROM cell: cross section of the mesh orthogonal to the current flow.*

tric field without scattering, thus acquiring the energy necessary for injection. Once the fitting parameters of the model have properly been adjusted, the lucky-electron approach accounts for the gate current rather well, but the fitting procedure has to be repeated when the doping profile and, thereby, the electric field is changed. The electron heating is in fact a nonlocal effect, determined at each point by the complete profile of the electric field rather than by its local value. This explains the extreme sensitivity of the lucky-electron model to the doping profile. Alternatively, the hot-electron injection has been modelled as a thermionic emission from the heated electron gas over the barrier at the Si–SiO_2 interface. In this approach, the electron gas is characterized by an effective temperature, higher than the lattice one and derived analytically from an energy-conservation equation, to be used together with a Maxwellian [33, 34] or shifted-Maxwellian [35] energy distribution. One fundamental advantage of this approach over the lucky-electron model is that it establishes a nonlocal relationship between the electron temperature and the field; on the other hand, the gate current predicted by this method is larger than that actually observed. Investigations based on the Monte Carlo technique pointed out that the electrons in the high-field domain have a strong non-Maxwellian distribution, i.e. a single-electron temperature is not adequate to describe the whole electron-distribution function. As a consequence, different non-Maxwellian expressions for the energy distribution function have been proposed, usually derived from Monte Carlo results by a fitting procedure [36], or obtained directly from the Boltzmann equation under simplifying assumptions [37]. Furthermore, gate-current models have been incorporated into hydrodynamic simulators [38, 39], which are less demanding in terms of computation time than the Monte Carlo ones and are consequently commonly used in the technology development.

The expression of the current injected from Si into SiO_2 has the general form

$$J_{\text{inj}} = -q \int_D \boldsymbol{u} f \, \mathrm{d}^3 k, \tag{2.5.2}$$

where the space coordinates $\boldsymbol{r}$ in the distribution function are calculated at the Si–SiO_2 interface and D is the subdomain of the Brillouin zone B such that the electrons with $\boldsymbol{k} \in D$ overcome the barrier. Introducing the transformation $\boldsymbol{k} \to (\epsilon, \zeta, \eta)$, the energy range turns out to be $[\epsilon_B, \infty)$, where ϵ_B denotes the height of the barrier at the Si–SiO_2 interface. Denoting by H the Jacobian determinant of the transformation, (2.5.2) becomes

$$J_{\text{inj}} = -q \int_{\epsilon_B}^{+\infty} \iint_{\zeta_D \eta_D} \boldsymbol{u} H f \, \mathrm{d}\zeta \, \mathrm{d}\eta \, \mathrm{d}\epsilon. \tag{2.5.3}$$

Letting Q be the density of states in $\boldsymbol{r}$, $\boldsymbol{k}$, defining

$$g_D = \iint_{\zeta_D \eta_D} H Q \, \mathrm{d}\zeta \, \mathrm{d}\eta, \qquad f_D = \frac{1}{g_D} \iint_{\zeta_D \eta_D} H f \, \mathrm{d}\zeta \, \mathrm{d}\eta \tag{2.5.4}$$

and introducing a velocity $\boldsymbol{v}_D$ such that

$$g_D f_D \boldsymbol{v}_D = \iint_{\zeta_D \eta_D} \boldsymbol{u} H f \, \mathrm{d}\zeta \, \mathrm{d}\eta, \tag{2.5.5}$$

the expression of the injected current transforms into

$$J_{\text{inj}} = -q \int_{\epsilon_B}^{+\infty} g_D f_D \boldsymbol{v}_D \, \mathrm{d}\epsilon. \tag{2.5.6}$$

Note that, repeating the above calculation after replacing $\boldsymbol{u}$ with 1, the number of electrons whose $\boldsymbol{k}$ belongs to D is found to be

$$\int_D f \, \mathrm{d}^3 k = \int_{\epsilon_B}^{+\infty} g_D f_D \, \mathrm{d}\epsilon. \tag{2.5.7}$$

As a consequence, f_D and g_D constitute, respectively, the energy-distribution function and the density of states in energy associated to D. In the following examples the approach outlined in section 1 has been adopted, namely that of using an approximated description of the distribution function in the energy region near the threshold. Several approximate models have been proposed in the literature for the integrand of (2.5.6) [5, 6, 40]; due to its simpler form, that of [40] has been tested here, which consists of replacing $\boldsymbol{v}_D$ with $v_\perp$ and letting $g_D f_D v_\perp = nA\epsilon^{3/2}\exp[-\chi(\epsilon/\epsilon_0)^3]$. In this model, A, χ are fitting parameters and ϵ_0^2 is a linear function of T_n. It should be stressed that, even if the integrand of (2.5.6) was calculated exactly, equation (2.5.6) is in itself insufficient to calculate the gate current. This is easily understood by observing that the Si–SiO_2 interface is not an external boundary of the device, hence the physical model must be completed with some description of the current transport within the oxide. In principle, such a description should lead to an expression similar to (2.5.2) accounting for the injection from SiO_2 to Si. In the simulations shown in the following, this problem has been faced in a simplified manner; in particular, the current loss due to electron scattering within the oxide has been accounted for by multiplying $\boldsymbol{J}_{\text{inj}}$ by the factor $\exp(-l/\lambda_{\text{ox}})$, where λ_{ox} is the electron mean-free path in SiO_2 ($\lambda_{\text{ox}} = 32$ Å from [41]) and $l = x_0/\sin\theta$. In turn, θ is the angle between vector $-q\boldsymbol{E}_{\text{ox}}$ and the interface and, letting t_{ox} be the oxide thickness, x_0 is taken equal to

$$x_B = \left(\frac{q}{16\pi\kappa_{\text{ox}}\varepsilon_0 E_{\text{ox}\perp}}\right)^{1/2} \qquad \text{if } x_B < t_{\text{ox}}, \tag{2.5.8}$$

or $x_0 = t_{\text{ox}}$ otherwise. In this way, the barrier lowering due to the image force is taken into account. In (2.5.8), ε_0 is the vacuum permittivity and $\kappa_{\text{ox}} = 2.15$ is the optical dielectric constant of SiO_2. The expressions above have been used adopting the 'critical injection-angle' technique [42] and a critical angle $\theta_c = 70°$ was taken from [36] with no additional adjustment. So doing, an electron reaches the floating gate if $\theta > \theta_c$, otherwise it is repelled back to the substate after crossing a part of the oxide region. Consistently with this approach, the injection of hot electrons was neglected in the region where the electric field in the oxide became repulsive and, thereby, the barrier height increased; on the other hand, preliminary investigations accounting for the barrier increase showed that the contribution of such region is negligible within the range of currents considered here. In view of the above considerations, it is sufficient to specify the height of the interfacial energy barrier in (2.5.6) only when the force pulls the electrons from the bulk through the oxide. Following [28], ϵ_B is expressed as

$$\epsilon_B = \epsilon_{B0} - \beta(E_{\text{ox}\perp})^{1/2}, \qquad \beta = [q^3/(4\pi\kappa_{\text{ox}}\varepsilon_0)]^{1/2}, \tag{2.5.9}$$

with $\epsilon_{B0} = 3.1$ eV. For simplicity, the height of the barrier is calculated in the direction normal to the interface. As mentioned before, an additional barrier-lowering term introduced in [28] to account for the hot-electron tunnelling probability is neglected.

2.5.4 Simulation results

The simulations have been carried out using a version of HFIELDS-3D [43, 44] incorporating the hydrodynamic model. The latter was made of equations (2.3.2), (2.3.23) and (2.3.21), (2.3.22) with $\boldsymbol{B} = 0$. Remembering the conclusions of section 4, $\hat{\tau}_p$ and $\hat{\tau}_q$ were replaced by scalars. Other simplifications relative to the transport model for the semiconductor consisted in using the parabolic-band approximation to determine the masses and in letting $w_n \simeq \frac{3}{2}k_B T_n$ and $s_n \simeq \frac{5}{2}k_B T_n$. The best fit between experiments and simulations has been obtained by using only one fitting parameter, namely, χ; all simulations shown below refer to the same set of parameters' values. Figures 2.6 and 2.7 show a comparison between experiment and simulation; the device is a flash-EEPROM cell designed for the SGS-Thomson 4 Megabit flash chip. In agreement with [45], to obtain the experimental current level it was necessary to reduce the value of χ by about a factor of 4 with respect to that proposed in [40]. The simulation is able to reproduce to an acceptable degree the sudden change in the slope of the programming characteristics, that occurs when $V_{fg} \sim V_d$. The simulated voltage-threshold shift agrees much better with the experimental data than in previous simulations using the model of [40] alone, without the critical-angle technique (e.g. [46, Fig. 9]). The overestimation in the voltage-threshold shift observed in [46] has been eliminated by including the critical-injection angle method described above. It should be noted that the shift of the experimental curves with V_d is predicted only qualitatively. Such a discrepancy cannot entirely be ascribed to difficulty in reproducing the complicated three-dimensional structure of the device; the lack of agreement also indicates the necessity of further investigations about the form of the electron-distribution function. This is in fact the object of current research activity using higher-level transport models [12, 19].

Figure 2.8 shows the injection angle θ and gate-current density as a function of the position along the channel. The largest floating-gate voltage $V_{fg} = 7.51$ V corresponds to the onset of the programming cycle. In this situation, the current-density peak is the highest and is located inside the drain region. As the injection proceeds, the floating-gate voltage decreases and the peak is reduced; at the same time, due to the decrease of the injection angle at the drain junction, the peak shifts towards the channel (with the exception of the $V_{fg} = 6.15$ V case, which will be commented on below). Eventually, at the lowest floating-gate voltage $V_{fg} = 4.53$ V it is $\theta < \theta_c$ in the whole drain region: the gate-current density occurs in the channel only and is strongly reduced. Figure 2.9 shows again the gate-current density, along with the electron temperature, as a function of the position. The increase in the electron temperature for a decreasing V_{fg} is due to the increase of the field component parallel to the interface; as a consequence, the peak shifts deeper into the drain region where the hot electrons eventually thermalize. It is interesting to note that the gate-current peak corresponding to $V_{fg} = 6.15$ V is shifted to the right with respect to the $V_{fg} = 7.51$ V case;

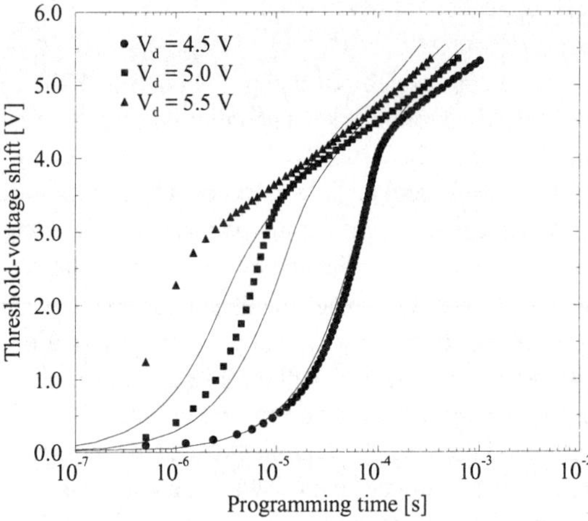

Figure 2.6 *Programming characteristics of the flash-EEPROM cell at $V_{cg} = 11$ V. Symbols refer to experiments and curves refer to simulations.*

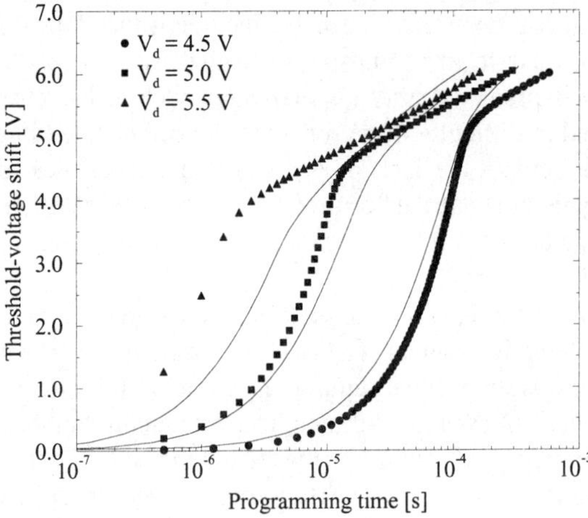

Figure 2.7 *Programming characteristics of the flash-EEPROM cell at $V_{cg} = 12$ V. Symbols refer to experiments and curves refer to simulations.*

in fact, in this case it is still $\theta > \theta_c$, hence the injection peak tends to follow the temperature peak. The lowering in the former is due to the increase in the barrier height and x_B due to the smaller $E_{\text{ox}\perp}$. As V_{fg} is further decreased, the effect of the injection angle takes place as described before.

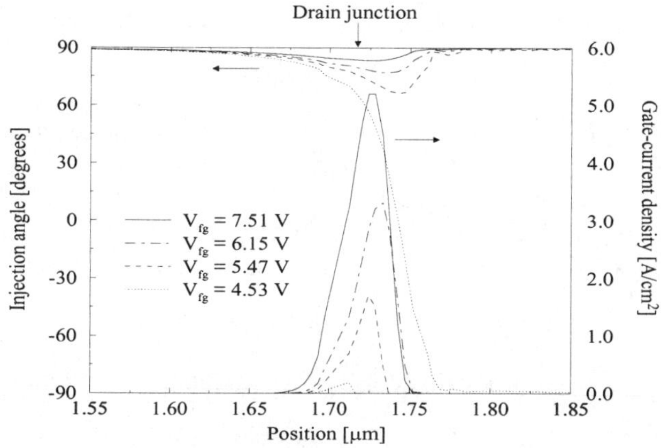

Figure 2.8 *Gate-current density and injection angle as a function of the position along the channel, with* $V_{cg} = 11$ *V and* $V_d = 5$ *V.*

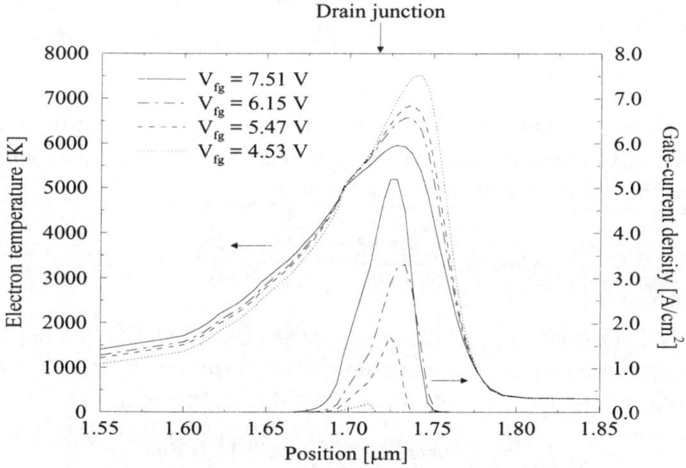

Figure 2.9 *Gate-current density and electron temperature as a function of the position along the channel, with* $V_{cg} = 11$ *V and* $V_d = 5$ *V.*

APPENDIX A

A.1 Vector products in the transport equations

The vector products on the right-hand side of (2.3.21) and (2.3.22) can be manipulated in order to eliminate the unknown $\boldsymbol{v}_n$ or $\boldsymbol{P}_n$ (e.g. [47, section 37]). To simplify the procedure, (2.3.21), (2.3.22) are first rewritten in a shorter form as

$$\boldsymbol{s} = \boldsymbol{g} - \hat{\mu}(\boldsymbol{s} \wedge \boldsymbol{B}), \tag{2.A.1}$$

where $\boldsymbol{s}$ stands for $-n\boldsymbol{v}_n$ or $-n\boldsymbol{P}_n$, $\hat{\mu}$ stands for $q\hat{\tau}_{pn}/m_{bn}$ or $q\hat{\tau}_{qn}/m_{sn}$ and $\boldsymbol{g}$ indicates the sum of the first two terms on the right-hand side of (2.3.21) or (2.3.22). Remembering the definition of the relaxation-time tensors, one sees that $\hat{\mu}$ is diagonal. The last term of (2.A.1) is written as $\hat{\mu}(\boldsymbol{s} \wedge \boldsymbol{B}) = \hat{\mu}(\boldsymbol{g} \wedge \boldsymbol{B}) + \hat{\mu}[(\boldsymbol{s} - \boldsymbol{g}) \wedge \boldsymbol{B}]$. Taking $\boldsymbol{s} - \boldsymbol{g}$ again from (2.A.1), such equation becomes

$$\boldsymbol{s} = \boldsymbol{g} + \hat{\mu}(\boldsymbol{B} \wedge \boldsymbol{g}) - \hat{\mu}\{[\hat{\mu}(\boldsymbol{s} \wedge \boldsymbol{B})] \wedge \boldsymbol{B}\}. \tag{2.A.2}$$

Letting $\boldsymbol{M} = \mu_{11}\mu_{22}\mu_{33}\hat{\mu}^{-1}\boldsymbol{B}$, the following is found:

$$\hat{\mu}\{[\hat{\mu}(\boldsymbol{s} \wedge \boldsymbol{B})] \wedge \boldsymbol{B}\} = (\boldsymbol{M} \cdot \boldsymbol{s})\boldsymbol{B} - (\boldsymbol{M} \cdot \boldsymbol{B})\boldsymbol{s}, \tag{2.A.3}$$

which generalizes a known vector identity. From (2.A.1) one also finds $\boldsymbol{M} \cdot \boldsymbol{s} = \boldsymbol{M} \cdot \boldsymbol{g}$; using this in (2.A.3) and carrying the result in (2.A.2) finally yields

$$\boldsymbol{s} = \frac{\boldsymbol{g} + \hat{\mu}(\boldsymbol{B} \wedge \boldsymbol{g}) + (\boldsymbol{M} \cdot \boldsymbol{g})\boldsymbol{B}}{1 + \boldsymbol{M} \cdot \boldsymbol{B}}. \tag{2.A.4}$$

A.2 Closure condition

As indicated in section 2.3, the approximation leading to the closure condition assumes a distribution function of the form $f = f_0 \exp[-\epsilon/(k_B T)]$, with f_0, T independent of $\boldsymbol{k}$. Equation (2.1) then becomes

$$\overline{\alpha} = \frac{\int_B \alpha \exp[-\epsilon/(k_B T)]\, \mathrm{d}^3 k}{\int_B \exp[-\epsilon/(k_B T)]\, \mathrm{d}^3 k}. \tag{2.A.5}$$

In the parabolic-band approximation $\epsilon = m^* u^2/2 = \hbar^2 k^2/(2m^*)$ whence, turning to polar coordinates, $\mathrm{d}^3 k = k^2\, \mathrm{d}k \sin\theta\, \mathrm{d}\theta\, \mathrm{d}\phi = (m^*/\hbar^2)^{3/2}\sqrt{2\epsilon}\, \mathrm{d}\epsilon\, \sin\theta\, \mathrm{d}\theta\, \mathrm{d}\phi$. If α depends on ϵ only, the integration over the angles yields

$$\overline{\alpha} = \frac{\int_0^\infty \alpha(\epsilon) \exp[-\epsilon/(k_B T)] \epsilon^{1/2}\, \mathrm{d}\epsilon}{\int_0^\infty \exp[-\epsilon/(k_B T)] \epsilon^{1/2}\, \mathrm{d}\epsilon}. \tag{2.A.6}$$

In particular, letting $\alpha = \epsilon$ and remembering that for $n = 0, 1, 2, \ldots$ it is $I_{2n} = \int_0^\infty \zeta^{2n} \exp(-\zeta^2)\, \mathrm{d}\zeta = \sqrt{\pi}(2n-1)!!/2^{n+1}$, one finds

$$\overline{\epsilon} = \frac{I_4}{I_2} k_B T = \frac{3}{2} k_B T. \tag{2.A.7}$$

Similarly, letting $\alpha = \epsilon^2$ one finds

$$\overline{\epsilon^2} = \frac{I_6}{I_2}(k_B T)^2 = \frac{15}{4}(k_B T)^2 = \frac{5}{3}(\overline{\epsilon})^2. \tag{2.A.8}$$

It is seen that the assumption on the distribution function implies $\boldsymbol{u} = \boldsymbol{c}_n$, whence the first of (2.3.20) and the second of (2.3.19) become, respectively,

$$\frac{1}{2}m^*\overline{c_n^2} = \overline{\epsilon} = \frac{3}{2}k_B T, \qquad s_n = \frac{5}{3}\overline{\epsilon} = \frac{5}{2}k_B T, \tag{2.A.9}$$

and the second of (2.3.20) becomes

$$\frac{1}{2}m^*\overline{\epsilon c_n^2} = \overline{\epsilon^2} = \frac{3}{2}k_B \tilde{T}_n s_n = \frac{15}{4}k_B^2 T \tilde{T}_n. \tag{2.A.10}$$

A comparison with (2.A.8) yields $\tilde{T}_n = T$, which is then taken as the closure condition of the general case.

REFERENCES

[1] Rudan, M. and Odeh, F. (1986) Multi-dimensional discretization scheme for the hydrodynamic model of semiconductor devices. *COMPEL*, **5** (3), 149–83.

[2] Scharfetter, D. L. and Gummel, H. K. (1969) Large-signal analysis of a silicon Read diode oscillator. *IEEE Trans. on Electron Devices* ED-**16**, 64–77.

[3] Quade, W., Schöll E. and Rudan, M. (1993) Impact-ionization within the hydrodynamic approach to semiconductor transport. *Solid-State Electronics,* **26** (10), 1493–505.

[4] Rahmat, K., White, J. and Antoniadis, D. A. (1993) Computation of drain and substrate currents in ultra-short-channels *n*MOSFETS using the hydrodynamic model. *IEEE Trans. on CAD of ICAS, CAD*-**12** (6), 817–24.

[5] Concannon, A., Mathewson, A., Piccinini, F. *et al.* (1994) Application of a novel hot carrier injection model in flash EEPROM design. In *Proc. of the 1994 ESSDERC Conference* (eds C. Hill and P. Ashburn), Edition Frontiers, France, pp. 503–6.

[6] Concannon, A., Piccinini, F., Mathewson, A. *et al.* (1995) The numerical simulation of substrate and gate currents in MOS and EPROMS, in IEDM *95 Technical Digest*, pp. 289–92.

[7] Lorenzini, M., Rudan, M. and Baccarani, G. (1996) A dual gate flash EEPROM cell with two-bit storage capacity. In *Proc. of International Nonvolatile Memory Technology Conference,* Albuquerque, pp. 84–90.

[8] Baccarani, G., Rudan, M., Lorenzini, M. *et al.* (1996) Device simulation for smart integrated systems—DESSIS. In *Proc.* ICECS *'96*, Rhodos, pp 752–55.

[9] Baccarani, G., Rudan, M., Guerrieri, R. *et al.* (1986) Physical models for numerical device simulation, volume 1 — Process and device modeling, *Advances in* CAD *for* VLSI (ed. W. L. Engl), North-Holland, Amsterdam, chapter 4, pp. 107–58

[10] Rudan, M., Gnudi, A. and Quade, W. (1993) A generalized approach to the hydrodynamic model of semiconductor equations, volume *Process and Device Modeling for Microelectronics* (ed. G. Baccarani), Elsevier, Amsterdam, chapter 2, pp. 109–54.

[11] Rudan, M. and Baccarani, G. (1995) On the structure and closure-condition of the hydrodynamic model. VLSI Design, *Special Issue* (ed. J. Jerome).

[12] Rudan, M., Vecchi, M. C. and Ventura, D. (1995) The hydrodynamic model in semiconductors — coefficient calculation for the conduction band of silicon, volume *mathematical problems in semiconductor physics* (eds P. Marcati, P. A. Markowich

and R. Natalini) *Pitman Research Notes in Mathematical Series no. 340*, Longman, London, pp. 186–214.

[13] Rudan, M., Odeh, F. and White, J. (1987) Numerical solution of the hydrodynamic model for a one-dimensional semiconductor device. *COMPEL*, **6** (3), 151–70.

[14] Forghieri, A., Guerrieri, R., Ciampolini, P. *et al.* (1988) A new discretization strategy of the semiconductor equations comprising momentum and energy balance. *IEEE Trans. on CAD of ICAS, CAD*-**7** (2), 231–42.

[15] Gnudi, A., Odeh, F. and Rudan, M. (1990) Investigation of non-local transport phenomena is small semiconductor devices. *European Trans. on Telecommunications and Related Technologies*, **1** (3), 307–12 (77–82).

[16] Thoma, R., Emunds, A., Meinerzhagen, B. *et al.* (1991) Hydrodynamic equations for semiconductors with non-parabolic band structure. *IEEE Trans. on Electron Devices*, ED-**38** (6), 1343–353

[17] Golinelli, P., Brunetti, R., Varani, L. *et al.* (1996) Monte Carlo calculation of hot-carrier thermal conductivity in semiconductors. In *Proc. of the Ninth Intl. Conf. on Hot Carriers in Semiconductors* (HCIS-IX), Chicago. (eds K. Hess, J. P. Leburton and U. Ravaioli), Plenum, New York.

[18] Brunetti, R., Vecchi, M.C. and Rudan, M. (1996) Monte Carlo analysis of anisotropy in the transport relaxation times for the hydrodynamic model, *Fourth Int. Workshop on Computational Electronics* (IWCE) (eds D. K. Ferry, C. Gardner and C. Ringhafer) VLSI Design.

[19] Brunetti, R., Golinelli, P., Reggiani, L. *et al.* (1996) Hot-carrier thermal conductivity for hydrodynamic analyses, *Proc. of the* 1996 ESSDERC *Conference* (eds G. Baccarani and M. Rudan) Edition Frontiers, pp. 829–32.

[20] Tang, T.-W., Ramaswamy, S. and Nam, J. (1993) An improved hydrodynamic transport model for silicon. IEEE *Trans. on Electron Devices*, ED-**40** (8).

[21] Vecchi, M. C., Greiner, A. and Rudan, M. (1996) Modeling surface scattering effects in the solution of the BTE based on a spherical harmonics expansion, *Proc. of the* 1996 ESSDERC *Conference* (eds G. Baccarani and M. Rudan), Edition Frontiers, France, pp. 825–8.

[22] Brunetti, R., Jacoboni, C., Nava, F. *et al.* (1981) Diffusion coefficient of hot electrons in Si. *Journal of Applied Physics*, **52** (11), 6713.

[23] Jacoboni, C. and Reggiani, L. (1983) The Monte Carlo method for the solution of charge transport in semiconductors with application to covalent materials. *Rev. Mod. Phys.*, **55** (3), 645.

[24] Laux, S. E., Fischetti, M. V. and Franck, D. J. (1990) Monte Carlo analysis of semiconductor devices: the DAMOCLES program. *IBM Journal of Research and Development*, **34**, 1663.

[25] Brunetti, R. and Jacoboni, C. (1984) Analysis of stationary and transient autocorrelation function in semiconductors. *Physical Review B15*, **29** (10), 5739.

[26] Ziman, J. M. (1960) *Electron and Phonons.* Clarendon Press, Oxford.

[27] v. Schwerin, A., Bergner, W. and Jacobs, H. (1995) Efficient and accurate simulation of EEPROM write time and its degradation using MINIMOS, *Proceedings of the Fifth* SISDEP *Conference* (eds E. Strasser, S. Selberherr and H. Stippel), Vienna, pp. 69–72.

[28] Ning, T. H., (1978) Hot-electron emission from silicon into silicon dioxide. *Solid-St. Electronics*, **21**, 273–82.

[29] Chen, Y.-Z. and Tang, T.-W. (1988) Numerical simulation of avalanche hot-carrier injection in short-channel MOSFETS *IEEE Trans. Electron. Devices*, ED-**35**, 2180–7.

[30] Fischetti, M. V., Laux, S. E. and Crabbé, E. (1995) Understanding hot-electron transport in silicon devices: is there a shortcut? *Journal of Applied Physics*, **78**, 1058–87.

[31] Tam, S., Ko, P.-K. and Hu, C. (1984) Lucky-electron model of channel hot-electron injection in MOSFETS. IEEE *Trans. Electron. Devices*, ED-**31**, 1116–25.

[32] Hu, C., Tam, S. C., Hsu, F.-C. *et al.* (1985) Hot-electron-induced MOSFET degradation—model, monitor, and improvement. IEEE *Trans. Electron. Devices*, ED-**32**, 375–85.

[33] Takeda, E., Kume, H., Toyabe, T. *et al.* (1982) Submicrometer MOSFET structure for minimizing hot-carrier generation. IEEE *Trans. Electron. Devices*, ED-**29**, 611–18.

[34] Hofmann, K. R., Werner, C., Weber, W. *et al.* (1985) Hot-electron and hole-emission effects in short *n*-channel MOSFETs. IEEE *Trans. Electron. Devices*. ED-**32**, 691–9.

[35] Wang, C T. (1988) An improved hot-electron-emission model for simulating the gate-current characteristic of MOSFETS. *Solid-State Electonics*, **31**, 229–31.

[36] Peng, J. Z., Longcor, S. and Frey, J. (1994) An integrated efficient method for deep-submicron EPROM/flash device simulation using energy transport model. IEICE *Trans. on Electronics*, **E77**–C, 166–73.

[37] Goldsman, N. and Frey, J. (1988) Electron energy distribution for calculation of gate leakage current in MOSFETs. *Solid-St. Electronics*, **31**, 1089–92.

[38] Katayama, K. and Toyabe, T. (1989) A new hot carrier simulation method based on full 3D hydrodynamic equations *in IEDM 89 Technical Digest*, Washington D.C., pp. 135–8.

[39] Keeney, S., Bez, R., Cantarelli, D. *et al.* (1992) Complete transient simulation of flash EEPROM devices. *IEEE Trans. Electron. Devices*, ED-**39**, 2750–7.

[40] Fiegna, C., Venturi, F., Melanotte, M. *et al.* (1991) Simple and efficient modeling of EPROM writing. IEEE *Trans. Electron. Devices*. ED-**38**, 603–10.

[41] Young, D. R. (1976) Electron current injected into SiO_2 from *p*-type Si depletion regions. *Journal of Applied Physics*, **47**, 2098–102.

[42] Wada, M., Shibata, R., Konaka, M. *et al.* (1981) A two-dimensional computer simulation of hot carrier effects in MOSFETS. In *IEDM 81 Technical Digest*, pp. 223–6.

[43] Ciampolini, P., Pierantoni, A. and Baccarani, G. (1991) Efficient 3-D simulation of complex structures. IEEE *Trans. on* CAD *of* ICAS, CAD-**10** (9), 1141–9.

[44] Ciampolini, P., Pierantoni, A., Liuzzo, A. *et al.* (1993) 3D simulation of silicon devices: physical models and numerical algorithms, volume *Process and Device Modeling for Microelectronics* (ed. G. Baccarani), Elsevier, Amsterdam, chapter 2, pp. 53–107.

[45] Keeney, S. N. (1992) *The Numerical Simulation of Floating Gate Non-Volatile Memory Devices*. PhD thesis, National University of Ireland.

[46] Baccarani, G., Rudan M., Lorenzini M. *et al.* (1997) Recent advances in device simulation using standard transport models. In *Fourth Int. Workshop on Computational Electronics*. (IWCE) (eds D. K. Ferry, C. Gardner and C. Ringhafer) VLSI Design.

[47] Kireev P. (1975) *La Physique des Semiconducteurs*. MIR, Moscou.

CHAPTER 3

Monte Carlo simulation of semiconductor transport

C. Jacoboni, R. Brunetti and P. Bordone

Istituto Nazionale per la Fisica della Materia and Dipartimento di Fisica, Università di Modena, Via Campi 213/A, 41100 Modena, Italy

3.1 INTRODUCTION

The introduction of the Monte Carlo (MC) method [1] for the analysis of non-linear charge transport in semiconductors is due to Kurosawa [2] who presented a study of high-field transport of holes in Ge at the semiconductor conference held in Kyoto in 1966.

Since then the method has been greatly improved and widely used to obtain the solution of the Boltzmann transport equation (BTE) for a large number of physical systems. Tremendous effort was dedicated in the 1970s and 1980s to the study of nonequilibrium transport properties of bulk systems [3]. Researchers soon realized that the MC method could be used to simulate experiments or particular physical situations of interest, even when they cannot be realized in practice. The system is idealized in the model and it can be changed almost at will in such a way as to give greater emphasis to any special effect under investigation. As in real experiments, the final results of the simulation must be interpreted in order to obtain a better understanding of the problem and they can possibly suggest new physical conditions for further experiments.

The application of MC techniques to the simulation of semiconductor devices started soon after the introduction of the method [4, 5], but it has only received great attention in the last decade [6, 7] due to the availability of MC algorithms able to handle phenomena and systems of great complexity, much closer to real devices than in the past.

In the last two decades the realization of low-dimensional structures, such as quantum wells, superlattices and quantum wires, opened a new exciting chapter of semiconductor physics and technology. Consequently, the MC method has been extended to reproduce the peculiar physical features of these new structures and again, as it happened for bulk materials, it revealed to be a very powerful tool, sometimes the only available tool, for the theoretical analysis of nonequilibrium transport and relaxation processes in less than three dimensions.

Furthermore, very advanced growth techniques, such as molecular-beam epitaxy (MBE) and metal-organic chemical-vapour deposition (MOCVD), can produce very pure samples, where the motion of the electrons may be ballistic over the sample, rather than diffusive as in macroscopic solids. Quantum effects neglected in the semiclassical transport scheme, such as intracollisional field effect and collision broadening, can influence the response of the structure under investigation. Under these conditions the semiclassical approach to charge transport based on the validity of the BTE equation is no more justified and, as a consequence, the traditional MC procedure based on this scheme cannot be used. New MC alogorithms are being developed in order to solve the new quantum equations.

Due to the amount of literature related to the subject of the present review, we will mainly concentrate on the advanced problems related to MC simulation, giving reference to older reviews for well established matter. Section 3.2 of this chapter is concerned with the physical models; the fundamentals of the method are discussed in section 3.3, while section 3.4 deals with applications and results for bulk semiconductor systems, low-dimensional structures and devices. A final section, before the conclusion, is devoted to a reformulation of the MC method for its extension to the study of quantum transport.

3.2 SEMICLASSICAL TRANSPORT IN SEMICONDUCTORS

3.2.1 Band-structure models

If transport processes are considered as involving carrier energies below 1 eV, a general and simplified model for the band structure of the whole class of cubic semiconductors with both diamond and zinc-blende symmetries can be used to interpret the microscopic properties of the semiconductors of interest. This model basically consists of one conduction band, with three sets of minima located at the Γ point, at the L points and along the Δ lines and three valence-band maxima located at the Γ point. For this band-structure model the energy–wave vector relationship $\varepsilon(\boldsymbol{k})$ is expressed through more or less simple analytical formulae, often using the concept of effective mass. The use of few simple bands is justified by the fact that the electron states occupied during the transport process lie close to the minima of the conduction band or to the top of the valence band and the whole complexity of the band structure is not involved.

When 'high-energy' electron transport is considered, with electron energies above 1 eV, more details of the band structure must be included in the physical model. This is quite often the case in microelectronic devices.

The analytical band model referred to above has been generalized to include structures of the band at higher energies [8]. The free parameters are fixed in such a way as to fit the density of states and the group velocity as functions of energy from band-structure calculations [9]. The model is particularly suited for applications in the field of MC device simulations, since it allows us to account for important physical aspects related to high-energy transport without requiring too large computer memory and CPU time [10].

When detailed microscopic effects at high energies are analysed, the full-band structure must be used in the simulation. Usually the band is calculated using a pseudopotential approach on a grid in the irreducible wedge of the Brillouin zone at the beginning of the simulation and stored in the computer memory. In order to associate the band energy to a given wave vector it is necessary to determine, inside the wedge, the wave vector equivalent by symmetry to the given one and then interpolate the function $\varepsilon(\boldsymbol{k})$ on the points of the grid closest to the point of interest. Particular care must be taken in this interpolation technique in order to avoid discontinuities in the band and in the group velocity from one cell to the neighbour cells. More details about full-band structure calculations for MC simulations can be found in [11].

When low-dimensional systems are simulated, such as quantum wells or quantum wires, in order to determine the electron energy levels the dynamics along the quantized dimensions is usually supposed to be decoupled from the dynamics along the unbounded dimensions, thus assuming translational symmetry in the free direction(s). The confining potential must be given as a simulation input and the eigenvalues and eigenfunctions of the associated Hamiltonian must be preliminarly determined from a numerical solution of Poisson and Schrödinger equations along the quantized dimensions. Sometimes self-consistency is imposed between the MC method and the solution of the Schrödinger equation, which can provide a more realistic well shape for the transport process under investigation [12]. Usually the energy–wave vector relationship associated to the nonquantized dimensions is treated assuming Bloch eigenfunctions and using simple analytical band models. A set of minibands is then formed, each of them with a band offset equal to one of the eigenvalues associated to the bounded dimensions. The peculiar energy dependence of the associated density of states (DOS) depends on the dimensionality of the system [13].

Recently, the validity of the conventional effective-mass theorem for the theoretical analysis of transport in quantum wells has been critically reviewed [14], with particular reference to the hypothesis of continuity of the slowly varying part of the electron wave function (envelope function) at an abrupt interface.

3.2.2 Scattering mechanisms: General remarks

The electronic transitions of interest for charge transport in semiconductors can be classified as intravalley when the initial and final states lie close to the same minimum, or intervalley when they lie close to different minima. In the case of holes, transitions can also be intraband or interband. When many subbands are present, for example in quantum wells, intrasubband and intersubband scattering can take place.

The most important sources of scattering that determine electron transitions in homogeneous crystals or in low-dimensional semiconductor structures are phonons, impurities, surface/interface scattering and other carriers.

Usually from the Fermi golden rule the differential scattering probability $P(\boldsymbol{k}, C; \boldsymbol{k}', C')$ for an electronic transition from an unperturbed electronic state $\boldsymbol{k}$ to a state $\boldsymbol{k}'$ and from the crystal state $|C\rangle$ to a state $|C'\rangle$ is obtained [15]:

$$P(\boldsymbol{k}, C; \boldsymbol{k}', C') = \frac{(2\pi)^4}{\hbar V^2}\left|\sum_{\boldsymbol{q}}\langle C'|\mathcal{H}'(\boldsymbol{q})|C\rangle\right|^2 \mathcal{G}\delta[\varepsilon(\boldsymbol{k}', C') - \varepsilon(\boldsymbol{k}, C)]$$

where $\mathcal{G}$ is the overlap factor and $\mathcal{H}'(\boldsymbol{q})$ is the Fourier transform of the interaction Hamiltonian. Models for such a Hamiltonian are given for each scattering agent and they have been extensively discussed in the literature as it regards both the approximations introduced in treating the specific interaction and the determination of the coupling constants.

In the following we will limit ourselves to a brief classification and physical discussion of the most important scattering agents for charge transport in semiconductors. The reader interested in technical details is referred to previous review papers [15]. Here mainly recent advances in the field will be reported.

3.2.3 Phonon scattering

(a) Bulk systems

The interaction of phonons with charge carriers is due to the deformation of the otherwise perfect crystal produced by phonon oscillations through the deformation potential mechanism or through the electrostatic forces produced by the polarization waves that accompany the phonons. The first kind of interaction is typical of covalent semiconductors; the electrostatic interaction, typical of polar materials, is called piezoelectric interaction for acoustic phonons and polar interaction for optical phonons.

The maximum energy transfer for an electronic interaction with acoustic phonons is, in general, much smaller than the electron energy. Thus, the energy dissipation of this type of scattering is important only when carriers cannot exchange energy through optical and intervalley phonons (for example in low-field and low-temperature conditions). When high temperatures and/or high fields are considered the electron energy is usually larger than the optical-phonon energy and this kind of interaction can assume the task of exchanging energy between the electrons and the crystal.

The established theory of the electron–phonon interaction has recently been revisited in connection with the inclusion of the full-band structure into the physical model [11]. More realistic models for the electron–phonon interaction have been developed for electron transport in the ultrahigh field regime. In particular they must account for the band and wave vector dependence of the scattering rates. An approach discussed extensively in the literature is based on the use

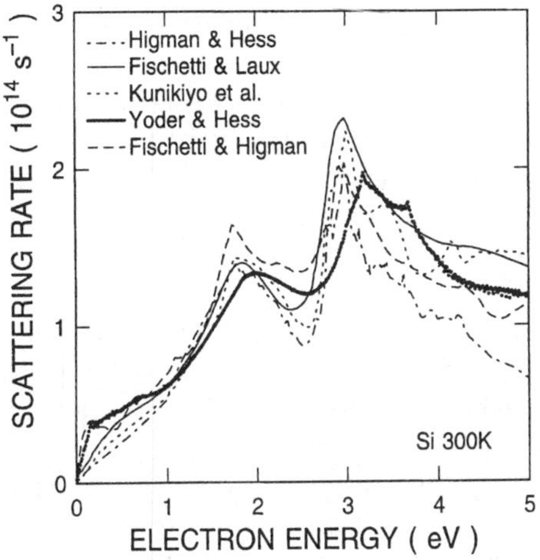

Figure 3.1 *Comparison of electron–phonon scattering rates in Si at $T = 300$ K employed in different Monte Carlo simulators [21].*

of empirical (either local or nonlocal) pseudopotentials [16] and the rigid-ion approximation [17–19]. Recently an *ab initio* calculation based on the density-functional approach has been presented [20]. However, it has been recognized that the general behaviour of the electron–phonon scattering rate is mainly determined by the DOS associated to the band structure and is less sensitive to the theoretical improvements in the interaction potential [21]. Figure 3.1 shows the total electron scattering probability (including impact ionization) as a function of energy and summarizes the results of the different approaches mentioned above. The larger discrepancies are found at high energies (above 3 eV), where empirical rates can be fitted to few and indirect data and where the particular model adopted for impact ionization is of relevance in the interpretation of the experimental results (see also section 3.2.9).

(b) Low-dimensional systems

The layered structure of low-dimensional systems has fundamental consequences on the vibrational properties of these materials, which are strongly modified with respect to the bulk case [22, 23].

Raman measurements and microscopic calculations have clearly indicated that the optical modes are confined in either one or the other constituent; i.e. the atoms of each layer tend to vibrate at the frequency of the appropriate material. In addition, in polar materials, the presence of the heterointerfaces between media with different dielectric properties leads to vibrations, called interface (IF) modes, decaying into neighbouring layers. Acoustic modes are instead much less affected by the layered structure. The dispersion relations of acoustic phonons usually prevent their confinement: for any system of two given materials there are propagating acoustic modes in both of them within the range of frequencies from zero to the maximum frequency of the elastically softer of the two of them. Elastic waves propagate in the heterostructures with a velocity given by the average of the sounds velocity of the two media.

Owing to its importance in practical cases at room temperature we will focus on optical phonons in the prototype GaAs/AlAs system. The Fröhlich Hamiltonian for electron–LO-phonon interaction can be expressed essentially in terms of a scalar electrostatic potential multiplied by the electron charge. In principle, this potential can be calculated using microscopic lattice dynamical models with suitable boundary conditions. In practice these calculations are very complex and provide results specific to a particular well width. For this reason most of the studies which have appeared in the literature for the long wavelength limit, are based on macroscopic models [24–26]. These models are called 'macroscopic' because at their starting point they assume the sample to be a continuum. They differ in their treatment of the boundary conditions imposed on the optical phonons at interfaces. Among them the so-called 'dielectric continuum model', [24, 27], which applies purely electrostatic boundary conditions at the interfaces turned out to be more appropriate. Two sets of modes are found: 1. confined modes, in the GaAs-like frequency range, whose displacements are represented by sine or cosine functions; 2. IF modes, whose potential and frequency strongly depend on the in-plane phonon wave vector. For a GaAs/AlAs structure IF modes can be both GaAs-like and AlAs-like.

An improved macroscopic phonon model that fulfils both the electrostatic boundary conditions and the continuity of the displacements at the interface has been proposed by Huang and Zhu [25].

The validity of these two macroscopic models has been confirmed by accurate studies of the electron–optical-phonon interaction that rely on fully microscopic *ab initio* calculations of the phonon spectra in two-dimensional (2D) systems [28]. A comparison among the total 2D phonon emission rate (confined plus IF modes) from microscopic calculation, the results predicted by the macroscopic models and bulk-GaAs phonon emission rate is shown in Fig. 3.2 [28]. As it can be seen, the predictions of macroscopic models are in good agreement with the microscopic calculations. As it regards the inclusion of the two-dimensional (2D) phonon rate into a MC simulator, the use of bulk GaAs phonons is in general considered to be an acceptable approximation for the total scattering rate. As an example the dependence of the total emission rate of a 50 meV electron in the first subband on the well thickness d, as obtained from the dielectric continuum model, is compared in Fig. 3.3 with the curves relative to the rates for bulk GaAs and bulk AlAs phonons evaluated at the same energy [28]. The total 2D scattering rate always falls in the range defined by the rates associated to GaAs and AlAs bulk phonons. For well thicknesses above 40 Å the rough assumption of unmodified GaAs bulk phonons may provide reasonable results.

In view of one-dimensional (1D) systems, it can be noticed that, for the

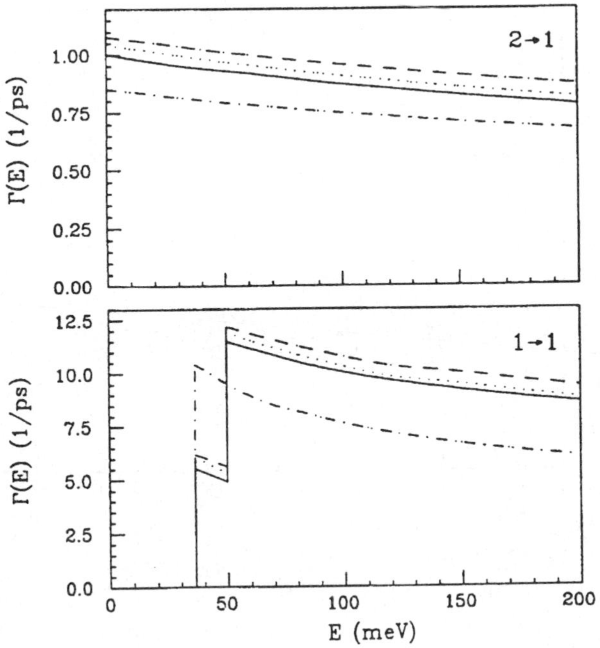

Figure 3.2 *Comparison of the total electron-phonon scattering rates (GaAs- and AlAs-like modes) obtained from the microscopic model in [28] (solid curve) with the results of macroscopic models (dielectric continuum model: dotted curve, Huang–Zhu model: dashed curve). The scattering rates for interaction with bulk GaAs phonons are also shown for comparison (dot-dashed curve) [28]. Cases (a) and (b) refer to intraband transitions relative to the lowest subband and interband transitions from the lowest to the second subband, respectively.*

relatively large wires made available by the current technology, the dielectric continuum model gives again results in good agreement with those obtained with microscopic calculations [29, 30].

Moreover, it is worth mentioning that also in the case of quantum wires the total electron–phonon scattering rates turn out to be rather similar to those obtained from bulk phonons [31, 32]. As in the 2D case, this is due to the fact that the contributions of confined and IF modes sum to roughly reproduce the bulk-phonon effect.

The total emission scattering rates for a 100 Å quantum well and a 100×300 Å quantum wire are shown in comparison with the GaAs bulk curve in Fig. 3.4 [31].

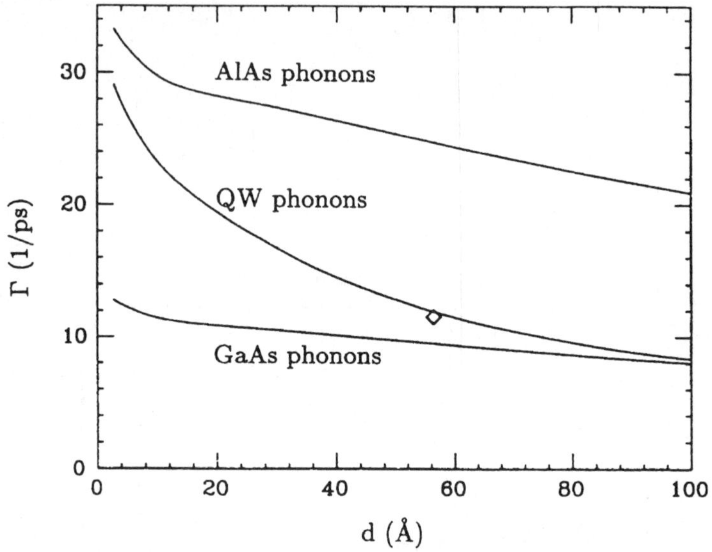

Figure 3.3 *Room-temperature phonon emission rate of a 50 meV electron in the first subband as a function of the well thickness. Curves for interaction with bulk GaAs and bulk AlAs phonons are shown together with the results of the dielectric continuum model (confined plus IF). The diamond indicates the results of the microscopic calculation for the 56 Å well [28].*

3.2.4 Impurity scattering

Ionized-impurity scattering is elastic and, therefore, it cannot control the transport process alone in the presence of an external field. Owing to the electrostatic nature of the interaction, its efficiency decreases with increasing electron energies (i.e. at high fields and/or high crystal temperatures).

The standard theory of impurity scattering is illustrated in [15]. Repeated efforts have been made in the past to improve the theory of impurity scattering beyond the traditional Born approximation. In particular, phase-shift theories have been developed [33, 34] and the screening problem has been tackled more rigorously. In our opinion, however, impurity scattering suffers from a limited knowledge of elements (for example impurity space distribution or clustering) that may overwhelm the refinement of the cross section details.

Neutral-impurity scattering has a very small cross section at normal concentrations and it only influences carrier transport at extremely low temperatures [35].

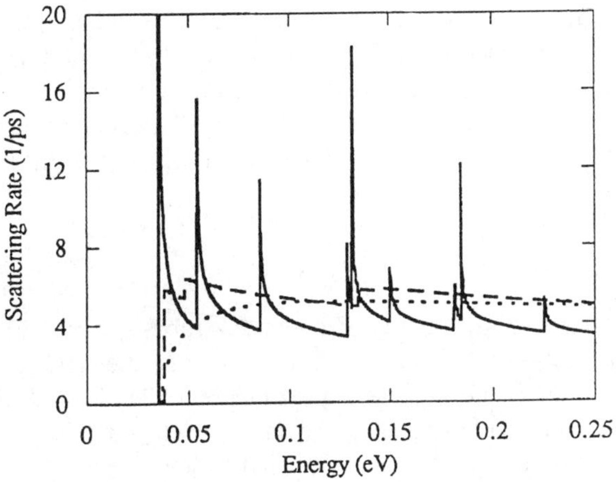

Figure 3.4 *Comparison of the room temperature electron-phonon scattering rate for a GaAs quantum wire (*100 × 300 Å*) (solid curve), a GaAs/AlGaAs quantum well (dashed curve) and bulk GaAs (dotted curve) [31].*

3.2.5 Coulomb interaction among carriers

(a) Bulk systems

Coulomb interaction among charge carriers can be decoupled into the sum of a short-range screened Coulomb interaction and a long-range contribution associated to electron–plasmon scattering [36]. The latter has been treated as the other scattering mechanisms (i.e. phonons and impurities) in the form of electron–plasmon interaction [37], or by means of the self-consistent MC–Poisson procedure (see also section 3.3.8). The main difficulty related to the treatment of the short-range scattering is the dependence of the scattering rate on the carrier distribution function, which is the unknown function of the transport problem. Thus, some sort of self-consistent calculation must be performed. Convergency is achieved when the input $f(\boldsymbol{k})$ used to evaluate scattering probabilities coincides with the $f(\boldsymbol{k})$ resulting from the simulation [38, 39]. The use of the ensemble Monte Carlo (EMC) technique greatly simplifies the treatment of carrier–carrier scattering, since the distribution function is built in the simulated ensemble of carriers [40, 41] (section 3.3.5).

In the first approach to short-range screened Coulomb interaction, the screening parameter β is treated by using the Debye approximation in the static limit (i.e. $\omega = 0$). Recently, the effect of dynamic screening has been discussed in connection to its possible influence on a number of physical conditions of relevance for devices [21]. Both static and dynamic q-dependent screening decrease as the relative momentum transfer q of the interacting particles increases.

The direct influence of the Coulomb interaction on transport effects is, in general, quite small, since these collisions do not change the total momentum and total energy of the colliding particles. It may be indirectly effective on those particular quantities which strongly depend on the shape of the distribution function (e.g. the anisotropy of the drift velocity, the efficiency of impact ionization, etc.).

(b) Low-dimensional systems

The inclusion of screening in the evaluation of electron–electron (EE) interaction for multisubband 2D systems is a difficult task. The dielectric matrix is usually described in the random-phase approximation (RPA) [42]. From a detailed study of the problem it results that the inverse screening length is independent of temperature and carrier concentration at low temperatures (degenerate conditions, Thomas–Fermi screening), while at high temperatures (nondegenerate conditions, Debye screening) it is proportional to n/T. When temperature dependence and multisubbands are included into the theoretical model for the screening length in 2D, the EE scattering rate to be used in MC simulations is enhanced due to a decrease in the screening factor.

The analysis of EE scattering rate for 2D quantum-well systems shows that the dominant contribution comes from pure intrasubband scattering where the initial and final states for both electrons belong to the same subband, due to the large value of the form factor accounting for the overlap of the carrier wave function [43]. The form factor for intersubband scattering in which the initial subbands of the two electrons are different, but neither carrier change subband index after scattering, has many analogies with the previous case, even though it is much lower in magnitude and vanishes for $q = 0$ due to the orthogonality of the eigenfunctions. This scattering between subbands however is important for transport processes since it allows an energy transfer from hot to cold subbands favouring the thermalization of electron populations in the two subbands.

The 1D case presents some peculiarities with respect to the 2D case and it deserves an independent analysis [32, 44, 43]. The fact that the carrier motion is only allowed in one direction implies that for every pair of interacting electrons there are only two available final states. If, at the end of the scattering, both carriers do not change subband, the only possible final state is that in which the eletrons exchange their crystal momentum. EE scattering is thus ineffective if both particles originate from the same subband, since they are indistinguishable. If the two electrons are in different subbands, this interaction may produce a significant energy exchange between the subbands. The interaction of two electrons that lie in different subbands, as long as they remain in the same subband after scattering, contributes to intrasubband scattering. Only intersubband scattering generates new k states in 1D systems and, therefore, it contributes to momentum redistribution. In this case, the energy difference between the initial and final subbands enters the conservation law and allows the creation of two completely new $\boldsymbol{k}$ states. However, intersubband scattering is greatly reduced due to the small value of the associated form factor and so energy exchange due to EE scattering is significantly reduced compared with that in higher-dimensional systems.

Using a full many-body approach [45] which also accounts for the long-range Coulomb interaction, the restrictions on the energy and momentum of the interacting particles disappear because energy and momentum can be redistributed over all the carriers. It is found however that the effect of the intrasubband many-body interaction is very weak and only has a significant effect on a timescale of 100 ps [45], while the intersubband scattering is already effective on a sub-picoseconds timescale.

Screening effects in 1D are usually included using the Coulomb potential 'dressed' by the dielectric function of the carrier system in the RPA [44, 43].

3.2.6 Degeneracy effects

The exclusion principle in degenerate semiconductors can, in some way, be considered as a sort of carrier–carrier interaction. Its effect has to be included when high doping, hence high carrier concentrations, are present (e.g. in the source and drain regions of a MOSFET). A method for including such a phenomenon inside a MC code was applied first for bulk materials [46] and then extended to semiconductor devices [40].

3.2.7 Surface/interface scattering

Surface and interface scattering are among the most important scattering sources in confined transport. In particular, interface scattering is essential for a correct theoretical description of electron transport in real devices. The characteristics of the interface strongly determine the mobility properties of the electron gas, sometimes becoming the limiting factor.

From a theoretical point of view in a semiclassical description an ideal surface introduces just specular reflections of the electron wave vector and it does not affect the mobility properties of a three-dimensional (3D) distribution function.

However, experiments show [47] that the interface usually exhibits a certain amount of roughness, whose effect on mobility has to be included in the theoretical description to be used in MC simulations. This effect has been modelled by the introduction of a correcting diffusive term to the totally specular reflection [48–50], representing the fraction of carriers being randomly diffused instead of being reflected by the surface. This correcting factor is then phenomenologically determined from a fitting procedure of effective mobility data for surface transport [51].

In a more rigorous theory of surface effects in confined systems a scattering rate can be evaluated that accounts for the amplitude and correlation length of the roughness [52] and the steepness of the wave function close to the surface. This model has been introduced in various MC simulators that were able to reproduce available experimental data for transport in confined systems [53, 54].

3.2.8 Generation–recombination process

Generation and recombination (GR) change the number of active carriers in the conduction process and are stimulated by photons or phonons present in the crystal. Usually GR processes take place with transitions from an impurity state in the energy gap to a state in the conduction band and vice versa, respectively. In very pure materials, when GR occurs between valence and conduction bands the necessary energy exchange is much larger and usually the process is negligible. This mechanism has been included in a MC simulator and treated following the standard theory of scattering [55]. The electron trapping times are generated according to the probability distribution specific of the particular trap considered. Since these times are usually quite long with respect to typical MC time scales (for shallow impurities typical times are of the order of 1 ns), simulations including GR are usually quite long.

3.2.9 Impact ionization

In the impact-ionization (II) process a conducting carrier in a high-energy state transfers energy to an electron in the valence band and promotes it to a conduction-band low-energy state, losing most of its energy in the process. Two low-energy carriers (a hole and an electron) are generated by this process. II is thus very efficient in relaxing carrier energy when high-energy electron transport is considered. Under these conditions it is necessary to account for the secondary electrons generated by the ionization process. Typical threshold energies for the onset of II are of the order of 1 eV. The analysis of this mechanism has been extensively developed in the literature. Theoretical models consistent with the adopted band structure have appeared in the literature [56, 11, 57–59]. They have been validated on a large amount of experimental data relative to the ionization coefficient as a function of the electric field, quantum yield as a function of the electron energy and soft X-ray photo-emission spectroscopy. In particular, the soft nature of the threshold energy has been established through comparisons of theoretical predictions with experimental results. Figure 3.5 shows a comparison among the II scattering rates obtained from the most advanced models: the agreement is good, particularly near threshold.

3.3 THE MONTE CARLO METHOD FOR BULK TRANSPORT

3.3.1 Fundamentals

The application of the MC method to charge transport in bulk semiconductors consists of a simulation in $(\boldsymbol{r}, \boldsymbol{k})$ space of the motion of one or more carriers

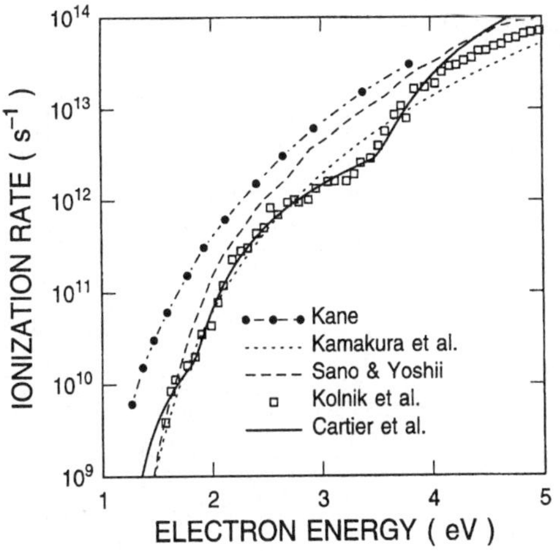

Figure 3.5 *Comparison of impact-ionization rates obtained using different impact-ionization models [21].*

inside the crystal, subject to the action of external forces due to applied electric and magnetic fields and of given scattering mechanisms. The time between two successive collisions (usually referred to as the duration of the carrier 'free flight') and the scattering events involved in the simulation are selected stochastically in accordance with some given probability distributions describing the active microscopic processes. The principles of the MC method are shown for a simple physical case in Fig. 3.6.

When a steady-state homogeneous phenomenon is studied, the motion of one single carrier simulated for a sufficiently long time will give information on the behaviour of the entire carrier gas, due to the ergodic property of the system. When nonhomogeneous or nonstationary situations are investigated a large number of carriers must be simulated for a sufficiently long time in order to obtain the desired information on the process of interest (see section 3.3.4).

The flow chart of a typical single-particle MC procedure suited to the simulation of a stationary, homogeneous transport process is presented in Fig. 3.7. For the sake of simplicity we shall refer to the case of electrons. The starting point of the program is the definition of the physical system, including basically the band structure and the parameters related to the relevant scattering processes. Physical parameters such as the lattice temperature T, the doping concentration and the external electric and/or magnetic field intensity and orientation are also specified in this segment together with some simulation parameters, such as the desired precision on the results, the duration of the simulation.

Initial conditions of motion for the electron energy ε and wave vector $\boldsymbol{k}$ must be properly assigned. Highly improbable initial situations, in fact, may require

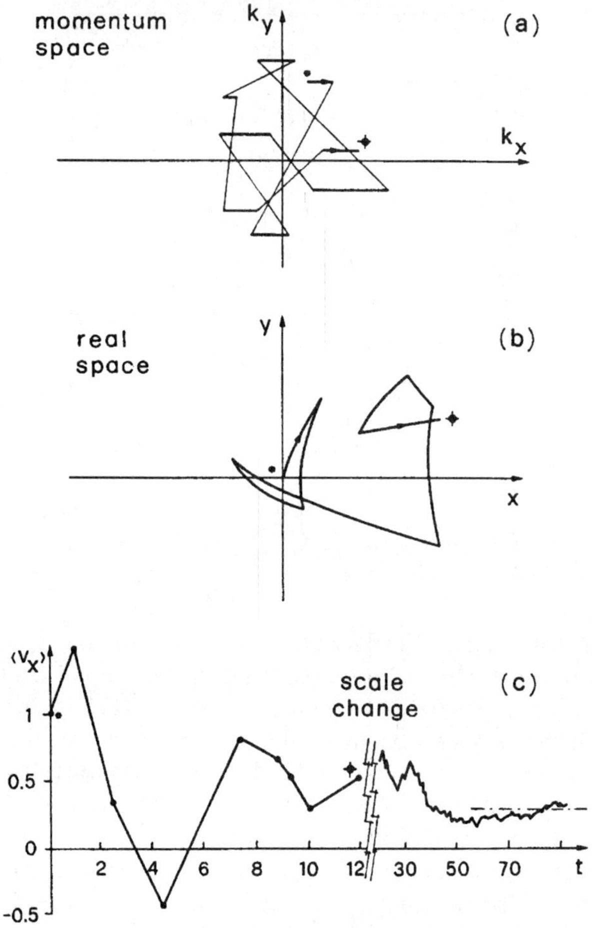

Figure 3.6 *The principles of the Monte Carlo method. For simplicity a 2D model is considered here. (a) The simulation of the sampling particle, in the wave vector space, subject to an accelerating force (field) oriented along the positive x direction. The heavy segments are due to the effect of the field during free flights; curves represent discontinuous variations of k due to scattering processes. (b) The path of the particle in real space. It is composed of eight fragments of parabolas corresponding to the eight free flights in part (a). (c) Average velocity of the particle obtained as a function of simulation time. The left section of the curve ($t < 12$) is obtained by the simulation illustrated in parts (a) and (b). The horizontal dot-dashed curve represents the 'exact' drift velocity obtained with a very long simulation time. Special symbols indicate corresponding points in the three parts of the figure (* is the starting point). All units are arbitrary [15].*

longer simulation times, since the first part of the simulation can be strongly influenced by this inappropriate choice.

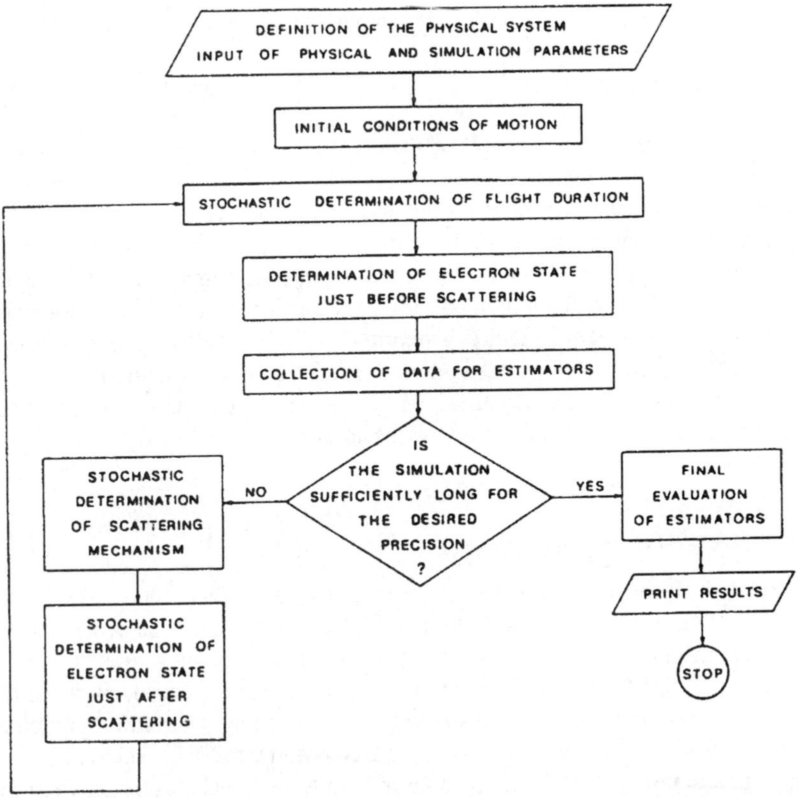

Figure 3.7 *Flow chart of a typical Monte Carlo program for a homogeneous system.*

The main loop of the simulation consists of a sequence of free flights and scattering events. The flight duration, the type of scattering and the electron state after the scattering event are chosen with distributions determined by the scattering probabilities. The estimators for the quantities of interest (e.g. drift velocity mean energy, distribution function) are updated before scattering and the results of the calculation become more and more accurate as the simulation goes on. A check on the estimator at this point can be made at each cycle or every given number of cycles in order to determine if the quantities of interest are known with the desired accuracy.

3.3.2 Determination of the flight

The electron wave vector $\boldsymbol{k}$ changes continuously between two scattering events due to the action of the external fields. The probability $P(t)\,\mathrm{d}t$ that the electron will suffer its next collision during $\mathrm{d}t$ around t is given by

$$P(t)\,\mathrm{d}t = \exp\left\{-\int_0^t P[\boldsymbol{k}(t')]\,\mathrm{d}t'\right\} P[\boldsymbol{k}(t)\,\mathrm{d}t] \tag{3.3.1}$$

where $P[\boldsymbol{k}(t)]\,\mathrm{d}t$ is the probability that an electron in the state $\boldsymbol{k}$ suffers a collision during $\mathrm{d}t$ around t.

Because of the complexity of the integral at the exponent in equation (3.3.1) this expression is in practice never used to generate stochastic free flights starting from evenly distributed random numbers applying the conventional techniques [15]. The well-known Rees method [15] provides a very simple numerical escamotage to bypass the difficulty: a fictitious scattering called 'self-scattering', is introduced such that the total scattering probability is constant and equal to Γ. If the carrier undergoes a self-scattering, its state is unchanged. A smart application of the Rees technique allows us to achieve a fast generation of free flights [60].

3.3.3 Choice of the scattering events

At the end of any free flight the electron energy and wave vector are known and all scattering probabilities $P_i(\boldsymbol{k})$ can be evaluated, where i indicates the ith scattering mechanism. The probability of self-scattering is the complement to Γ of the sum of the P_i's. A random number r between 0 and 1 is generated and the product $r\Gamma$ is compared with the successive sums of the P_i's, until a particular mechanism is selected which terminates the free flight under consideration [15]. The detailed physical information about the scattering mechanisms active in the transport process under investigation is a crucial point for any MC simulation. For two-body interactions the standard procedure used to obtain an integrated scattering probability $P(\boldsymbol{k})$ to be included in the MC simulation for each process is the Fermi golden rule, once the interaction potential is given (section 3.5.1). This approach is justified when the limit of single, separated collisions is realized and when the intracollisional field effect is negligible and it fails when high-energy transport is considered. More complex interactions involving three or more particles, such as impact ionization or phonon-assisted photon emission or absorption, require the use of the time-dependent perturbation theory at higher orders [61].

Once the scattering mechanism that caused the end of the carrier free flight has been determined, the new state of the carrier after scattering must be chosen as the final state of the scattering event. This state is generated stochastically according to the differential cross section of that particular mechanism.

3.3.4 Statistical estimators for stationary systems

The data collected at each free flight provide a determination of any quantity of interest at the end of the simulation. A procedure often used to obtain any transport quantity $A(\boldsymbol{k})$ from a MC simulation for a homogeneous ergodic system is to perform a time average as follows

$$\langle A\rangle_T = \frac{1}{T}\int_0^T \mathrm{d}t\ A[\boldsymbol{k}(t)] = \frac{1}{T}\sum_i \int_0^{t_i} \mathrm{d}t'\ A[\boldsymbol{k}(t')] \qquad (3.3.2)$$

where the integral over the simulation time T has been separated into the sum of integrals over all flights of duration t_i. When a steady state is investigated, T must be long enough to ensure an unbiased estimate of the average in equation (3.3.2) The electronic distribution function is obtained by recording the time spent by the sampling carrier in each cell of a given mesh in $\boldsymbol{k}$ space [15].

An alternative method to obtain an average quantity is the 'synchronous ensemble' method [62], which, in the case of a constant total scattering rate Γ, gives:

$$\langle A\rangle = \frac{1}{N}\sum_i A_{bi}$$

where the sum covers all N electron free flights and A_{bi} indicates the value of the quantity A evaluated at the end of the free flight immediately before the ith scattering event (self-scattering included). For a variable Γ the estimator must be properly generalized [15]. In order to estimate the statistical uncertainty of a time-average result due to the finite value of the simulation time T, the whole simulation can be divided into M fragments of duration T/M. For each of them an average value is obtained and the standard deviation associated to the average value represents the statistical uncertainty on $\langle A\rangle_T$ [15].

3.3.5 The ensemble Monte Carlo for transients and nonhomogeneous systems

When time and/or space-dependent problems are considered we cannot rely on the ergodicity of the system and an ensemble of particles must be explicitely simulated [63] (EMC).

(a) Transients

If a transient situation for a homogeneous electron gas must be studied, then many particles must be independently simulated with appropriate distributions of initial conditions. Provided the number of simulated particles is sufficiently large, histograms obtained by sampling the electron wave vectors or energies at regular intervals of time provide the distribution function $f(\boldsymbol{k}, t)$ or $f(\varepsilon, t)$, respectively. If a single quantity A is of interest, its time-dependent mean value can be obtained as the ensemble average. The standard deviation of the quantity can also be computed over the ensemble.

(b) Nonhomogeneous systems

The simulation of a steady-state phenomenon in a physical system where electron transport depends upon the position in space is of great relevance for semiconductor-device modelling and it will be discussed in section 3.3.8. To study this system an EMC is required and averages must be taken over particles at given positions.

(c) Molecular-dynamics Monte Carlo method

An alternative method for the treatment of Coulomb interaction among carriers [6] couples the usual EMC algorithm for the $\boldsymbol{k}$-space dynamics to the simulation of the real-space trajectories of an ensemble of electrons interacting via a bare Coulomb potential (molecular-dynamics Monte Carlo method). The strength of the method lies in the fact that it does not require any assumptions on the screening between carriers. Furthermore, it naturally accounts for a fundamental effect associated to a classical electron gas, namely the current-density fluctuations even when these are affected by carrier–carrier interactions.

3.3.6 Hot phonons

Under the effect of sufficiently high electric fields or in the presence of photoexcitation of charge carriers in semiconductors, nonequilibrium phonon distributions are expected to be established for the modes strongly coupled to the carrier ensemble and with nonelectronic dissipation rates lower or comparable with the carrier-interaction rates [64].

The presence of phonon disturbances brings about a modification of the high-field electron transport and relaxation of laser-generated electrons. Possible mobility effects of acoustic-phonon disturbances are restricted to lattice temperatures of at most a few Kelvin because of the weak carrier–phonon couplings and of the rapid increase of the thermalization rate of acoustic modes with temperature. In contrast to the acoustic case, the dependence of the thermalization rates of optical phonons on temperature is weak. Although the rates are very fast (of the order of inverse picoseconds), the strong coupling between carriers and LO-phonons in polar materials can lead to even faster phonon-emission rates and therefore to substantial LO-phonon amplification even at room temperature.

The interaction between carriers and nonequilibrium phonon population can be included in an EMC code [65–68]. The simulation is divided into time intervals Δt much shorter than the average scattering time for the LO-phonon scattering. At the beginning of the simulation the LO-phonon population is considered at thermal equilibrium. Then the time evolution of the phonon distribution is calculated as a function of the phonon wave vector $\boldsymbol{q}$ by setting up a histogram defined over a grid in $\boldsymbol{q}$ space. After each scattering event involving a LO-phonon the histogram is updated and at every Δt phonon–phonon interactions are accounted for by using a relaxation time approximation. To include the modification induced by the phonon disturbance on the electron phonon scattering rates, the integrated scattering probabilities for LO-phonons are calculated and tabulated at the beginning of the simulation using an artificial high value for the phonon distribution. The choice of the final state of each scattering process involving LO-phonons is made by using a rejection technique which compares the actual value of the differential scattering rate with the maximized one.

The numerical procedure described above has been used to analyse both relaxation processes of photo-excited carriers and nonohmic DC transport phenomena. In particular, the study of the relaxation process of electrons photo-excited by ultrafast laser pulses reveals that the presence of a perturbed LO-phonon population produces a reduction of the carrier cooling rate. Such a reduction, which is a function of the excitation conditions and of the lattice temperature, is due to a sizeable reabsorption of LO-phonons [69, 67, 43]. When a DC electric field is applied, corrections to the room-temperature steady-state carrier mean velocity are found, typically going from an increase (drag effect) up to 20% at low and intermediate fields to a reduction (heating effect) at higher fields up to 10% [66, 68, 70].

3.3.7 Variance-reducing techniques

A major problem related to the MC technique is the computational burden required to provide results with sufficient statistical accuracy, especially when dealing with device simulation where the phase space is at least five-dimensional (in stationary conditions: two space and three momentum coordinates). In addition, some relevant physical phenomena are very rare, sometimes due to their threshold nature (such as impact ionization, injection into SiO_2, etc.). Therefore special numerical techniques devoted to the artificial enhancement of the probability of such events are required to maintain the simulation time inside acceptable limits.

Several variance reducing techniques have been proposed and adopted which are based on two different approaches, namely 'split-and-gather' and 'weighted' MC.

In the first one, originally developed in [71] and then generalized and used by several groups (e.g. [72] and references therein), statistical weights are attributed to simulative particles according to the probability of the $(\boldsymbol{r}, \boldsymbol{k})$ state they belong to. At the same time, their number is kept inversely proportional to the same probability throughout the $(\boldsymbol{r}, \boldsymbol{k})$ domain of interest. In this way, if few 'heavy' particles in the highly probable domains are enough to reach reasonably smooth results, many 'light' particles sample the rare regions of phase space, providing many samples of rare events. When a 'heavy' particle enters a region with a lower average statistical weight, it is split into several particles with weight equal to the average of the region. Conversely, an appropriate number of light particles that have entered a region with higher average statistical weight are gathered into a single particle of such average weight. In this way, a reasonably flat distribution of samples can be obtained inside the simulation domain, while the correctness of the numerical result is recovered by the proper use of the weight in the statistical estimation of the quantities of interest. This allows us, for instance, to explore regions of the phase space otherwise several order of magnitude depressed by carrier statistics [73].

The second approach ([74] and references therein) acts on the probabilities of a rare event to enhance their relative relevance. It has been shown that the natural probability of each event can be altered at will maintaining the correctness of the final result if the statistical weight of the particle undergoing that event is modified by the ratio between the real to the altered probability. If event A has probability $P_T(\mathrm{A})$ to occur but such a probability has been artificially modified to $P_F(\mathrm{A})$, the particle weight must be corrected by the factor $P_T(\mathrm{A})/P_F(\mathrm{A})$ each time the event occurs.

In this way, specific phenomena can be investigated [75], or rare regions of the phase space can be explored.

The above technique is based on an analytical formulation that allows a MC sampling of electron paths generated with an arbitrary probability. We shall return to this point in section 3.5.2.

3.3.8 The Monte Carlo method for device simulation

As mentioned in the introduction, the MC method has also shown its usefulness in the field of device simulation, although several difficulties had to be solved.

The particular semiconductor band structure adopted in a MC device simulator has relevant effects on the accuracy of the results. In fact, since electron devices usually feature electric-field profiles with very high values, carriers are forced to visit high-energy regions of the phase space where simple approximations are no longer realistic. Consequently, the use of the full-band structure has often been adopted although simplified models have reached a satisfactory level of accuracy [76]. However, the improvement of modern computers and numerical algorithms [77] is in many cases overcoming the advantages related to the use of simplified models, no longer more efficient than their counterparts.

The problem of boundary conditions (BC) is of crucial importance when simulating spatially inhomogeneous devices. Different treatments of the BC have been reported [78–80]. The basic requirement is to impose a known distribution function at appropriate device boundaries, usually a Maxwellian distribution at the device extreme sections.

This must be done without perturbing the system, i.e. without introducing artificial effects into the simulation results. The extreme regions of a device are usually regions of very low electric field, thus a natural place to put the device BC. However, since carriers move very slowly from these regions towards the high-field ones, which are usually those of interest, one could be tempted to move the BC as close as possible to these. This can be done with some caution. In fact, the dynamics of carriers inside the high-field regions may influence the carrier distribution inside the low-field ones and vice versa and this contribution would be neglected by the introduction of a BC too close, irreversibly altering the results. Therefore, a commonly adopted rule of thumb is to include several mean-free paths of low-field region into the simulation domain to enforce the natural carrier thermalization, correctly accounting for the microscopic transport phenomena.

Another important subject in MC simulation of semiconductor devices is the consistency between the carrier densities, computed through the MC transport step and the potential that has generated the carrier motion.

The problem is less relevant where a MC simulation is performed starting from an initial guess for the potential profile obtained by a sufficiently accurate method, such as a hydrodynamic or energy-balance solvers. In this case, a non-selfconsistent MC simulation can be acceptable. Nevertheless, a self-consistency loop may be suggested anyway for most of the applications.

Two procedures are usually adopted. In the first one, the Poisson equation is solved after a MC step whose time duration is much shorter than the plasma frequency of the system under investigation ($\Delta t \approx 1$ fs; [81]). Plasma oscillations are automatically accounted for in this case. In the second scheme, the Poisson equation is solved after a fairly long MC simulation (several ps; [82]). In this case plasmon scattering must be explicitly included to account for the coupling of the collective oscillations of the electron gas to carrier transport [37]. In both cases the two steps are iterated until convergency is reached. The first scheme, as mentioned, maintains a link with the physical oscillations of the carrier gas and is inherently noisy. The second one, instead, damps the convergency error through the use of a nonlinear Poisson scheme, but stability problems may arise anyway. Stability issues have been investigated in recent studies [83, 84], showing that the statistical sampling method can have dramatic effects on the convergency of the self-consistent procedure.

3.3.9 Alternative numerical methods for semiclassical transport

We would like to conclude this section by giving the correct credit to alternative methods for the solution of the BTE. In fact, the MC method, as we know it, is not, in principle, the only method for solving this equation, although it can be considered, as said above, as the reference one. In particular, from the point of view of computational efficiency, some alternative techniques have proven their superiority and easy handling, still maintaining excellent accuracy in a wide range of physical situations. The first of them is the cellular automata method [85]. In such a technique the BTE is discretized on a fairly loose mesh of phase space. This elaboration results in a symmetrical form of transport and scattering parts of the BTE that allows for a great improvement of the numerical efficiency. A second method, in its various forms, is that of the expansion of the distribution function in spherical harmonics [86]. With this expansion the solution of the carrier transport becomes more efficient, still allowing the introduction of a variety of scattering mechanisms (impact-ionization, carrier–carrier) and physical features (band structure, Poisson self-consistency, quantization) that make this simplified technique a good trade-off between accuracy and computational efficiency [87]. The final techniques that we want to mention are those based on the use of scattering matrices [88–90]. In these techniques the carrier population is collectively moved in phase space according to the second law of classical mechanics during the deterministic parts of the simulation, while scattering takes place stochastically on different portions of the population according to precomputed scattering matrices, redistributing particles in phase space. Also in this final case, efficiency and accuracy are traded-off with some success.

3.4 RESULTS

3.4.1 Bulk systems

The study of hot-electron transport on the basis of fundamental physics and the MC method has involved solid-state physicists from all over the world for almost 30 years. A large amount of experimental data was collected in the 1970s for the most important semiconductors and applied electric fields ranging from the ohmic condition up to 10^5 V cm^{-1}. Proper physical models have been produced where the existing adjustable parameters have been fixed on the basis of the fitting of a variety of transport properties (drift velocity, diffusion coefficient, ohmic mobility, etc.). As an example, Fig. 3.8 shows theoretical and experimental data of drift velocity and diffusion coefficient of electrons in silicon as functions of the electric field strength and direction at 77 K. For an exhaustive review of the above data the reader is referred to the specific literature [15, 6].

The well-established state of the art was overcome in the late 1980s when a new impulse towards the study of bulk results with the use of more accurate models and simulators was stimulated by new experimental data at higher fields ($E \geq 100$ kV cm^{-1}), typically obtained for transport processes in devices. In order to understand and reproduce the new results some physical elements whose importance has been previously overlooked must be accounted for, such as, for example the full-band structure, an accurate impact-ionization model and realistic screening functions [21]. Bulk structures were used as the simplest test systems for the advanced simulators.

As examples of MC calculations performed with the most accurate available physical models Fig. 3.9 shows the electron distribution as a function of energy at room temperature for an applied electric field of 30 kV cm^{-1} along the $\langle 100 \rangle$ direction as obtained from different simulators. It can be noted that the results may differ significantly at energies above 1 eV due to the different physical models. Figure 3.10 shows data for the ionization coefficient and the quantum yield as obtained from the most advanced models of II.

A comparison among the results coming from the variety of the MC codes

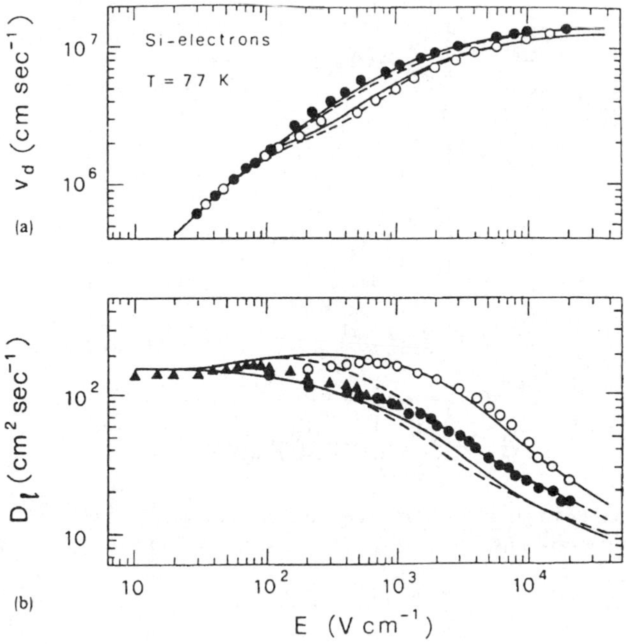

Figure 3.8 *(a) Drift velocity and (b) longitudinal diffusion coefficient of electrons in Si as functions of the electric field at 77 K. Points refer to experiments (open circles: E ∥ ⟨100⟩; solid circles: E ∥ ⟨111⟩; dashed and continuous curves show theoretical results obtained with different models [128].*

developed for different applications has recently been attempted [91] for a simple test situation (homogeneous intrinsic silicon at room temperature at different electric fields). Differences, sometimes very large, in the physical models produced a large spread in the obtained results (Fig. 3.11). This fact can be misleading about the reliability of the MC procedure. In fact it must be kept in mind that each code cannot be used to examine experiments well outside the limits of the adopted physical model. In contrast, this fact shows how the MC method is a formidable tool to discriminate among different physical models in the interpretation of experimental data.

The MC method has also been successfully applied to the study of electrical noise and fluctuations. Noise is usually considered as a limiting factor in the performances of electronic devices. On the other hand, the microscopic analysis and interpretation of nonequilibrium noise spectra and fluctuations on transport quantities usually provides new details of the transport picture. Noise and fluctuations can be directly extracted from the MC analysis of the carrier dynamics inside the sample. From the time of the first MC applications to noise problems [92] a lot of work has appeared in the literature both for bulk systems and more

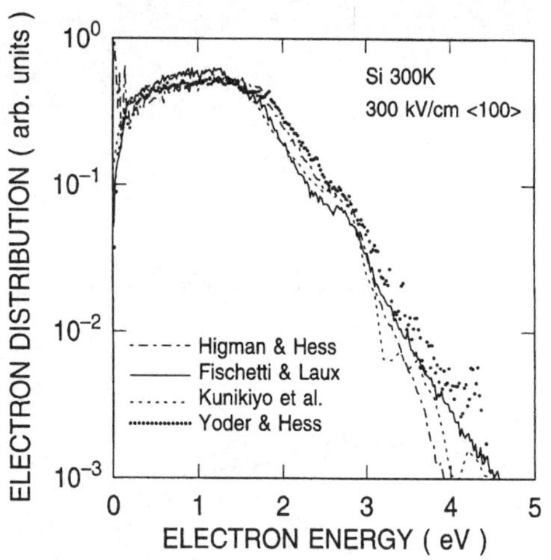

Figure 3.9 *Comparison of electron energy distributions obtained from homogeneous-Si simulations at room temperature for* $E \parallel \langle 100 \rangle$ *and* $E = 300\ kV\ cm^{-1}$ *using the scattering rates of Fig. 3.1 [21].*

complex structures, such as Si n^+nn^+ junctions and devices [93]. Usually the analysis is limited to high frequencies where $1/f$ noise can be neglected. The major sources of nonequilibrium electrical noise are thus recognized to be velocity fluctuations, generation–recombination and shot noise. Transport parameters such as carrier diffusion coefficient and thermal conductivity can be calculated from the MC estimate of appropriate correlation functions even for situations where the drift effect is very small (and the direct estimate of average quantities is affected by a large statistical uncertainty) [94]. A major achievement of MC simulations is the analysis of the different physical sources of fluctuations and their correlations to the total noise spectrum, often neglected in numerical calculations [95, 93]. Figure 3.12 shows an example of a calculation for the case of velocity fluctuations. In Fig. 3.12(a) the velocity autocorrelation functions due to fluctuations of the carrier wave vector (kk), energy ($\varepsilon - \varepsilon$) and valley (vv) are shown. In Fig. 3.12(b) the cross-correlation terms associated with the same fluctuations are shown.

3.4.2 Low-dimensional systems

The MC method has been widely used to study the electronic properties of semiconductor heterostructures under the effect of both an external electric field and laser excitation [69, 96, 67, 68, 97, 54, 43, 98].

As it regards structures based on Si, an interesting effect of mobility enhancement in Si/SiGe heterostructures has been found and confirmed by MC simulations, due to the presence of strain-induced subband separation [97, 99, 54, 100–

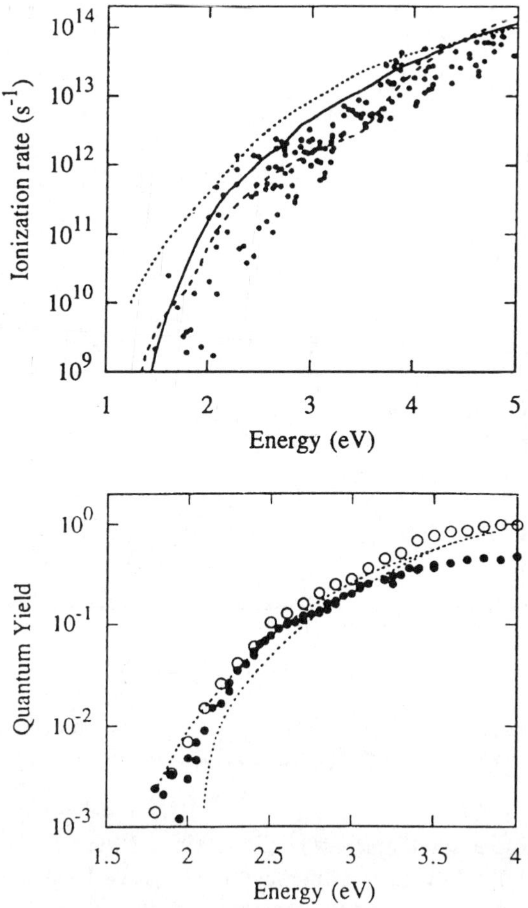

Figure 3.10 *(a) Ionization rate as a function of electron energy. Dots, dotted and continuous curves show theoretical results obtained with different models; the dashed curve represents experimental data [129]. (b) Quantum yield as a function of electron energy. Open circles represent theoretical results, whereas experimental results are shown by solid circles and dotted curves [129].*

103]. Furthermore, an interesting attempt at accounting for the variation of the confining potential along the interface through a discretization procedure has been presented for the case of electrons in Si inversion layers [53].

However, most of the attention has been focused on GaAs/AlAs structures and, in particular, the MC method has been applied to the analysis of the role of the electron–phonon interaction in the interpretation of the experimental data of photo-excited electron relaxation in quantum wells [104–106]. The collected data show a slow decay which was originally ascribed to phonon-confinement

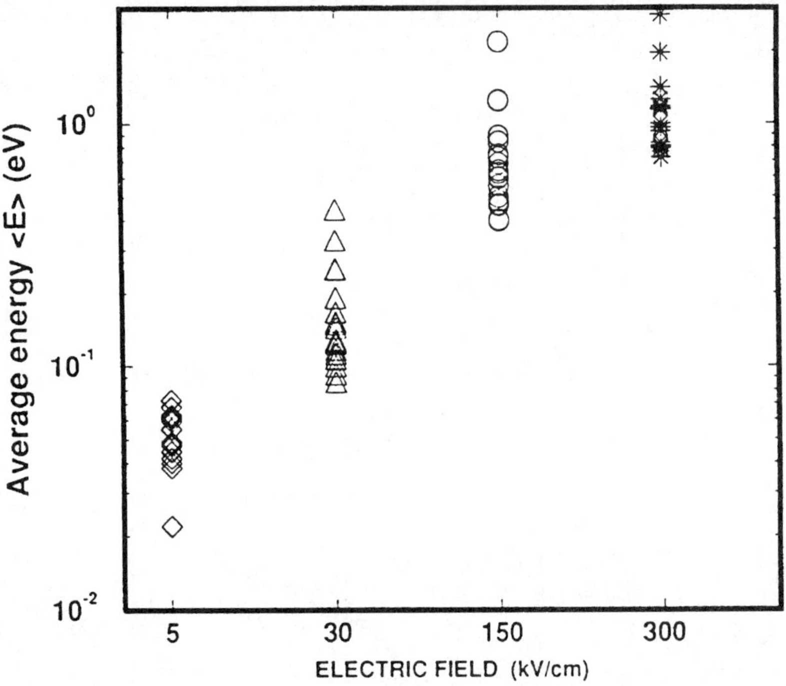

Figure 3.11 *Average energy as a function of the electric field for electrons in bulk Si at* $T = 300$ *K as obtained from different MC simulators [91].*

phenomena present in quantum wells. To verify this hypotheses the interactions of electrons with confined and interface modes have been investigated selectively and the increasing relative importance of the interface-mode interaction with decreasing well width has been pointed out [107, 70, 108]. However, MC simulations show that confined and interface-mode contributions, when summed, roughly reproduce the bulk phonon effect. Rather the reduced cooling has been shown to be a consequence of the reabsorption of nonequilibrium phonons emitted during the earlier stage of the relaxation process [69, 109, 108].

Relaxation phenomena of electrons in quantum wires show similar features: here the energy relaxation rate is further reduced with respect to the bulk and quantum-well cases [32, 44, 110, 43, 105, 106]. In 1D systems the reduction in the cooling rate can be also ascribed, besides the phonon effect, to the different features of short-range Coulomb interaction. In fact a strong reduction in the efficiency of the carrier–carrier interaction has been found [43], due to the fact that the EE intrasubband scattering mechanism is inhibited by momentum conservation. The thermalizing effect of this mechanism turns out to be significantly reduced, which, in turn, results in a reduction of efficiency of phonon emission.

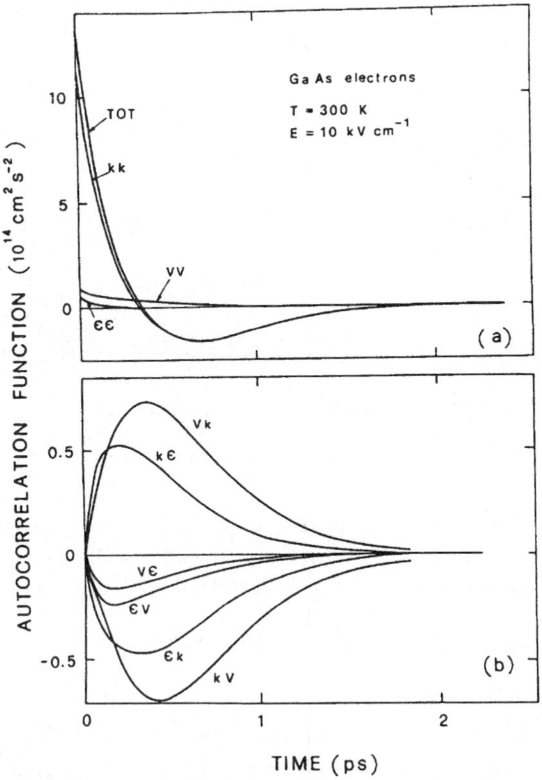

Figure 3.12 *(a) Autocorrelation function of velocity fluctuations and its diagonal terms for the case of electrons in GaAs at the indicated temperature and field; (b) off-diagonal terms. In (b) a different vertical scale has been used [95].*

As an example, Fig. 3.13 shows experimental and MC results for time-dependent carrier cooling in a GaAs V-groove quantum-wire structure. The inclusion of the hot-phonon effect in the MC simulator is fundamental to interpret the experiments. Figure 3.14 shows a comparison between the carrier cooling rates in quantum wires and bulk systems at low carrier density. The slower cooling in the wire has been justified invoking also the reduced efficiency of the intersubband-phonon scattering, related to the fact that the intersubband matrix elements are smaller because of the confinement of the electronic wave function in two directions.

In summary we can conclude that the different features of electron-phonon and EE interactions of low-dimensional systems with respect to the bulk case (at least for quantum wells and relatively large quantum wires) can be better displayed from optical rather than electrical measurements.

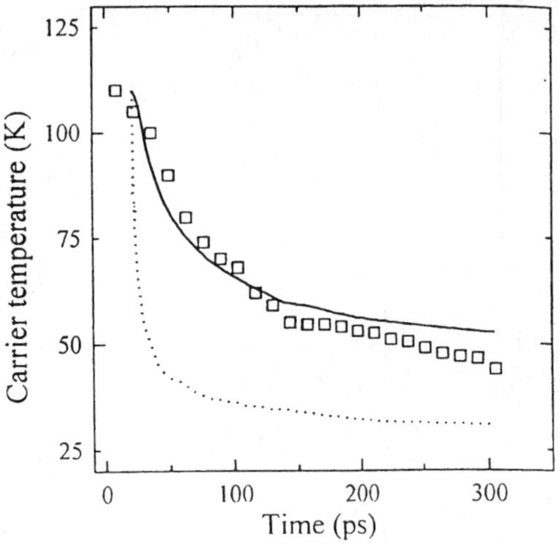

Figure 3.13 *Time-dependent carrier cooling in a GaAs V-groove quantum wire structure. The squares indicate the experimental data. The solid and dashed curves indicate the MC results with and without the hot-phonon effect, respectively [105].*

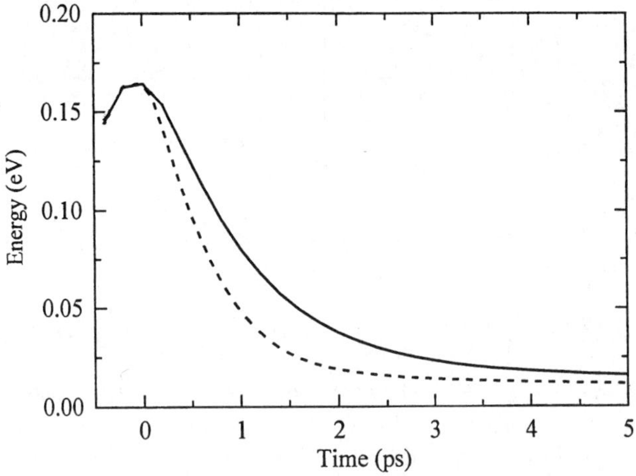

Figure 3.14 *Comparison between carrier cooling in a GaAs quantum wire (solid curve) for an electron density $n = 10^4$ cm^{-1} and in bulk GaAs for an electron density $n = 3.3 \times 10^{14}$ cm^{-3} (dashed curve) [43].*

3.4.3 Devices

As computers become faster and more powerful, MC simulation of devices approaches the status of being a possible tool for industrial applications. In our opinion it will not be, in any case, the standard CAD tool, but rather a so-

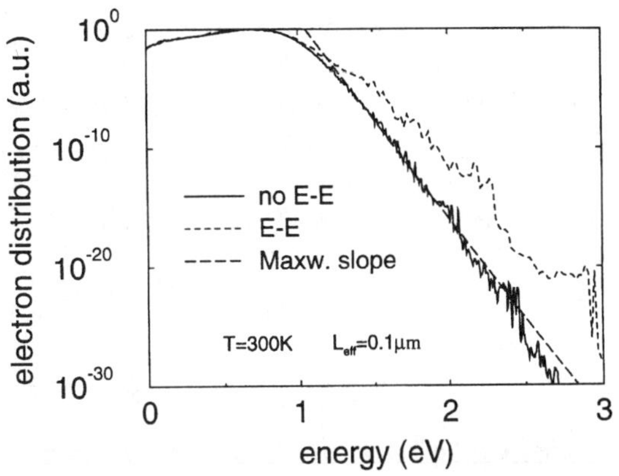

Figure 3.15 *Electron energy distribution at the drain junction of a* 0.1 μm *MOSFET biased at* $V_{GS} = 1$ *V,* $V_{DS} = 1.5$ *V. Note the relevance of the short-range Coulomb interaction in the enhancement of the high-energy tail of the distribution function [73].*

phisticated theoretical instrument for the investigation of particular situations of interest in any special device. For reasons of space, in what follows we will briefly report on a single example, i.e. that of the simulations of MOS transistors for low-power/low-voltage regimes. This is a topic of great importance from the technological point of view, particularly in connection with the diffusion of portable electronics. The reduction of bias voltages offers the possibility of reducing power dissipation and, at the same time, opens new fundamental questions on the correct operation of MOS transistors at such voltages. In particular, the device design (threshold voltage, noise margins, etc.) must be reviewed to match new requirements. In addition, some concerns arise on the possible realization of specific devices, such as nonvolatile memories which base their operation on the physical phenomenon of carrier injection through (and above) SiO_2 barriers.

In view of gate and substrate currents, which are typical effects of carrier heating, questions on the possible evidence of these phenomena at applied voltages lower than the thresholds of the causing mechanisms (SiO_2 injection (≈ 3.1 eV) and impact-ionization (≈ 1 eV)) had been raised. However, recent experimental works [111, 112] have confirmed the presence of substrate and gate currents at applied voltages as low as 0.6 V and 1.5 V, respectively.

The use of MC simulations has suggested possible explanations of this evidence. In particular, short-range carrier–carrier scattering has demonstrated a very effective mechanism for the enhancement of the high-energy electron population ([73]; Fig. 3.15), while further investigations [21] have clarified the role of II, also showing how the temperature dependence of the energy gap can explain the anomalous temperature behaviour of ionization rates in MOSFETs.

A novel effect must be added to this picture, namely oxide injection driven by secondary impact-ionization [113]. It has been demonstrated both by experiments and MC simulations that a primary-II generated hole, accelerated by the drain-to-bulk potential can produce a high-energy electron through a secondary-II process. The generated electron is then injected into a floating gate.

Very recent investigations on the physics of carrier injection in MOSFET oxides can be found in [114–116], where a very relevant example of MC simulation of a complex device is presented.

3.5 FROM SEMICLASSICAL TO QUANTUM TRANSPORT

3.5.1 Quantum effects into the semiclassical Monte Carlo simulation

As we mentioned in the introduction, semiclassical transport is based on a number of hypotheses which are no longer justified in an increasing number of structures and devices recently realized thanks to the tremendous improvement in material growth and processing technologies.

When very intense electric fields are applied (typically of the order of 100 kV cm^{-1} or larger) collisions become so active that the hypothesis of single point-like events fails. The finite collision duration cannot be neglected and the effect of the electric field during the scattering process (intracollisional field effect (ICFE)) must be accounted for. Furthermore, the collision rate can be so high that the time between two successive collisions becomes shorter than the time necessary to achieve the strict energy conservation required by the Fermi golden rule and the scattering energy balance must account for quantum collision broadening (CB) [7].

Furthermore, when in pure samples at very low temperatures the sample length is comparable with the coherent mean-free path, then the description based on semiclassical dynamics cannot be used. A full quantum propagation of the electron wave function across the sample must be considered and the scattering events must be treated accordingly.

The correct approach to charge transport when the quantum regime is achieved can no longer be based on the BTE.

There are several possible formulations of the quantum problem, but usually the solution of the associated quantum evolution equations and the interpretation of the results are very difficult. Furthermore, the number of simplifications in the physical model required to obtain numerical results forces us to consider simple ideal systems, often far from real devices.

As a consequence some attempts at including specific quantum effects into the semiclassical Boltzmann scheme have appeared in the literature. Obviously they are not rigorous approaches and their justifications are only based on their predictivity.

Some authors [117, 118] included the effect of CB into the semiclassical MC procedure by introducing the broadening of the electron state before [117] and/or after [118] phonon scattering. Both approaches are based on special modellings of the joint spectral density replacing the Dirac delta function in the Fermi golden rule. The model spectral density used in [118] also accounts for ICFE, but its strong oscillations have been smeared analytically in order to be able to include it into a MC simulation. The most relevant effects on the results are a significant increase in the population in the high-energy region of the distribution function [118], even though it is still not completely clear if these effects are physical or artefacts related to the particular approximations inherent to the theoretical approaches and a lowering and smear-out effect of the threshold energy for II [117].

Another attempt at including a realistic model for the carrier penetration of a quantum potential barrier has been made in connection with the physical problem of electron injection from Si to SiO_2 [119, 120]. The transfer probability from Si to the gate through SiO_2 has been evaluated as the transmission coefficient through the oxide barrier and included into a traditional MC simulator. The result obtained provides a smooth transition between tunnelling and thermionic emission conditions and exibits wiggles due to quantum reflections at the interface between SiO_2 and the gate. Good agreement is obtained between the simulative results and the experimental results in real devices [119, 120].

Another important intrinsically quantum phenomenon takes place in quantum-well structures when electrons of the 2D gas confined in the well are heated up by an applied electric field and gain enough energy to overcome the confining potential and escape the well. This effect is called 'real-space transfer' (RST) and its importance is also related to possible device applications [121]. As it regards the inclusion of this quantum effect into a semiclassical MC simulation, an important question arises about where an electron must be located in space soon after a RST. According to the standard perturbation theory, the final electron state is, in fact, extended everywhere inside the sample while the semiclassical MC simulation requires a particular position. A rigorous quantum theory of RST has been developed [122], even though it has not been included in a MC simulator, where only semiclassical models have been implemented so far [123].

3.5.2 Analytical formulation of the semiclassical Monte Carlo method

Until now we have examined how to obtain a numerical solution of the BTE through a direct simulation of the transport process from a microscopic point of view: the direct MC simulation is by far the most popular technique to solve the BTE. In this form MC can be hardly generalized to treat quantum transport.

However, there is an alternative way to formulate the MC approach to semiclassical problems particularly suitable for its quantum generalization to the solution of the Liouville–von Neumann equation. This alternative formulation also provides the theoretical grounds for a numerical method able to efficiently sample rare events (section 3.3.7).

In order to briefly review the basic principles of this approach we must recall the integral version of the transport equation, called the Chambers equation [124]. From BTE, through the use of the following phase-space transformation:

$$\begin{aligned} \boldsymbol{r}^*(\boldsymbol{r},\boldsymbol{k},t) &= \boldsymbol{r} - \int_{t_0}^{t} \dot{\boldsymbol{r}}(\tau)\,\mathrm{d}\tau \\ \boldsymbol{k}^*(\boldsymbol{r},\boldsymbol{k},t) &= \boldsymbol{k} - \int_{t_0}^{t} \dot{\boldsymbol{k}}(\tau)\,\mathrm{d}\tau \\ t^*(\boldsymbol{r},\boldsymbol{k},t) &= t \end{aligned}$$

we obtain the following integral equation [74]:

$$\begin{aligned} f(\boldsymbol{r},\boldsymbol{k},t) &= f(\boldsymbol{r}(t_0),\boldsymbol{k}(t_0),t_0)S(t,t_0) + \int_{t_0}^{t} \mathrm{d}t'\, S(t,t') \\ &\quad \times \frac{V}{(2\pi)^3}\int \mathrm{d}\boldsymbol{k}'\, f(\boldsymbol{r}(t'),\boldsymbol{k}',t')P(\boldsymbol{k}',\boldsymbol{k}(t')). \end{aligned} \tag{3.5.1}$$

Here $S(t,t') = \mathrm{e}^{-\int_{t'}^{t}\gamma(\boldsymbol{k}(t''))\,\mathrm{d}t''}$ and $\gamma(\boldsymbol{k}(t)) = \frac{V}{(2\pi)^3}\int P(\boldsymbol{k}(t),\boldsymbol{k}')\mathrm{d}\boldsymbol{k}'$; t_0 is the initial time, at which $f(\boldsymbol{r},\boldsymbol{k},t_0)$ is supposed to be known; $\boldsymbol{r}(t')$ and $\boldsymbol{k}(t')$ are the ballistic position and momentum of a particle at time t', that is at $\boldsymbol{r}$ and $\boldsymbol{k}$ at time t; $P(\boldsymbol{k},\boldsymbol{k}')$ is the transition probability from a state $\boldsymbol{k}$ to a state $\boldsymbol{k}'$.

This integral version of the transport equation has a very straightforward physical interpretation: the distribution function $f(\boldsymbol{r},\boldsymbol{k},t)$ is given by two contributions. The first one is given by the electrons that at $t = t_0$ are already in the 'right' trajectory and are not scattered away from it before t; the second contribution is given by the electrons that are put in the right trajectory in $\boldsymbol{r}'$ at time t' with the right momentum $\boldsymbol{k}'$ and are not scattered away before time t.

If equation (3.5.1) is iteratively substituted into its right-hand side, we obtain a series expansion (Neumann series) for the unknown f in powers of the scattering operators. We introduce the integral in-scattering operator P_i defined in such a way that, for any given function $g(\boldsymbol{r}^*,\boldsymbol{k}^*,t)$:

$$P_i(t)\,g = \frac{V}{(2\pi)^3}\int \mathrm{d}\boldsymbol{k}'\, P(\boldsymbol{k}',\boldsymbol{k}(t))g(\boldsymbol{r}(t),\boldsymbol{k}',t)$$

and express the expansion as:

$$\begin{aligned} f(t) &= S(t,t_0)f(t_0) + \int_{t_0}^{t}\mathrm{d}t_1\, S(t,t_1)P_i(t_1)S(t_1,t_0)f(t_0) \\ &\quad + \int_{t_0}^{t}\mathrm{d}t_1\int_{t_0}^{t_1}\mathrm{d}t_2\, S(t,t_1)P_i(t_1)S(t_1,t_2)P_i(t_2)S(t_2,t_0)f(t_0) + \cdots \\ &= f^{(0)}(t) + \Delta f^{(1)}(t) + \Delta f^{(2)}(t) + \cdots . \end{aligned}$$

Each term of the above expansion can be regarded as a particular electron trajectory in phase space from the initial time t_0 to the current time t, as discussed extensively in [74]. It can be shown that it is possible to give a similar iterative expansion directly starting from the BTE. In the Chambers approach, however, the role of the out-scattering processes is replaced by the presence of the damping factors S, which by themselves describe the effect of all possible sequences of out processes. This infinite summation, which can be performed in semiclassical transport, is the classical analogue of the infinite sum of Feynman diagrams which leads to the definition of the electron self-energy in quantum mechanics. The imaginary part of the self-energy is associated to the carrier lifetime, that reduces to the quantity γ in the semiclassical limit.

The MC method can be applied to evaluate the infinite sum of multiple integrals obtained from the iterative expansion of the Chambers equation by means of random selections of the various terms in the expansion [74]. These terms correspond to possible electron paths and can be chosen with arbitrary probabilities even quite different from the 'natural' ones. Such probabilities must then be properly taken into account in the statistical estimators. In practice either the initial state $(\boldsymbol{r}_0, \boldsymbol{k}_0)$ or the final state $(\boldsymbol{r}, \boldsymbol{k})$ can be selected; then the final (initial) state of the electron simulation is determined by the choices of the intermediate states associated to all scattering events along the generated path. In particular, by fixing the final state we obtain a method (backward MC) with the advantage that all the computational effort can be devoted to the evaluation of the distribution function at a given position in phase space. In this way it is possible to sample with great precision the distribution function in regions of the phase space where it has very small values, i.e. where the standard MC simulation yields very poor estimates, since electrons in their paths visit these regions very rarely.

3.5.3 The Monte Carlo approach to quantum transport

The analytical formulation sketched in section 3.5.2 is suitable for an extension to an equivalent MC treatment of quantum transport. We have already mentioned that in some of the present-day nanostructures the dimensions and the timescales are such that the assumptions on which semiclassical transport is based fail. One aspect of quantum transport is the coherent propagation of the electron waves across the system and the discretization of the electron energy. These features are responsible for many interesting phenomena on which quantum systems are today based (e.g. resonant-tunnelling diodes, T transistors, Coulomb-blockade oscillations). As long as the analysed quantum phenomenon is based on a fully coherent propagation, i.e. not influenced by phonon scattering, the theoretical difficulties are not so relevant and a self-consistent solution of Schrödinger and Poisson equations yields a proper solution.

Impurity and roughness scatterings make transport no more ballistic, but do not destroy the wave function coherence. Thus, in principle, they can be included in the above picture with the addition of the corresponding potentials into the Schrödinger equation [125]. The problem becomes much more difficult and is still waiting for a complete solution when phonon scattering must be accounted for, owing to the finite temperature of the system or to the need of dissipation of the electron energy. The inclusion of the phonon interaction makes the transport equation not separable for electron and phonon coordinates and a many-body problem has to be solved even for single-electron systems.

Several approaches have been attempted and can be found in the literature on the above problem, all of them are implicitly or explicitly dealing with Green's functions. We know, in fact, that the most elementary approach to quantum statistics is based on the concept of the density-matrix operator, a particular case of the $G^<$ Green's function and the Wigner function itself is a special Fourier transform of $G^<$. In what follows we shall adopt this last approach, since it is most useful for a description of quantum transport with the proper boundary conditions in electronic nanodevices in terms analogous to those of classical physics and an attempt has been made to extend the MC algorithm to such a description which has proved to be so useful in the semiclassical case [126].

In order to include phonons in the Wigner-function approach a generalized Wigner function including a single electron (or, equivalently, many independent electrons) and phonons has been defined as:

$$f_w(\boldsymbol{r},\boldsymbol{p};n_q,n'_q)=\frac{1}{h^3}\int \mathrm{d}\boldsymbol{r}'\,\mathrm{e}^{-\mathrm{i}\frac{\boldsymbol{p}\boldsymbol{r}'}{\hbar}}\rho\left(\boldsymbol{r}+\frac{\boldsymbol{r}'}{2},n_q;\boldsymbol{r}-\frac{\boldsymbol{r}'}{2},n'_q\right) \tag{3.5.2}$$

where ρ is the density matrix of the electron–phonon system.

Starting from the Liouville–von Neumann equation for the density matrix, an integral equation has been obtained for the Wigner function in the interaction picture:

$$\begin{aligned}\tilde{f}_w(\boldsymbol{r},\boldsymbol{p};n_q,n_{q'};t)&=\tilde{f}_w(\boldsymbol{r},\boldsymbol{p};n_q,n_{q'};0)+\int_0^t \mathrm{d}t'\sum_{nn'} f_{nn'}(\boldsymbol{r},\boldsymbol{p})\\&\quad\times\sum_{m,m_q}\hbar^3\int\mathrm{d}\boldsymbol{r}'\int\mathrm{d}\boldsymbol{p}'\,\{\tilde{\mathcal{H}}'(n,n_q;m,m_q;t')f^*_{m,n'}(\boldsymbol{r}',\boldsymbol{p}')\\&\quad\times\tilde{f}_w(\boldsymbol{r}',\boldsymbol{p}';m_q,n_{q'};t')-f^*_{n,m}(\boldsymbol{r}',\boldsymbol{p}')\tilde{f}_w(\boldsymbol{r}',\boldsymbol{p}';n_q,m'_q;t')\\&\quad\times\tilde{\mathcal{H}}'(m,m_q;n',n_{q'};t')\}\end{aligned} \tag{3.5.3}$$

where $\tilde{f}_w$ and $\tilde{\mathcal{H}}'(t)=\tilde{H}_{ep}/\mathrm{i}\hbar$ are the Wigner function and the normalized interaction

Hamiltonian in the interaction picture, respectively. The coefficients f_{lm} are defined, for any given basis $\{|\phi_l\rangle\}$ for the electron states, as:

$$f_{lm}(\boldsymbol{r},\boldsymbol{p})=\frac{1}{\hbar^3}\int\mathrm{d}\boldsymbol{r}'\,\mathrm{e}^{-\mathrm{i}\frac{\boldsymbol{p}\boldsymbol{r}'}{\hbar}}\phi_l^*\left(\boldsymbol{r}+\frac{\boldsymbol{r}'}{2}\right)\phi_m\left(\boldsymbol{r}-\frac{\boldsymbol{r}'}{2}\right). \tag{3.5.4}$$

A MC sampling of the resulting Neumann series can be performed which is the direct quantum generalization of the semiclassical algorithm described in section 3.5.2. The trace over the phonon coordinates is automatically performed by substituting the phonon occupation number with its average value [127],

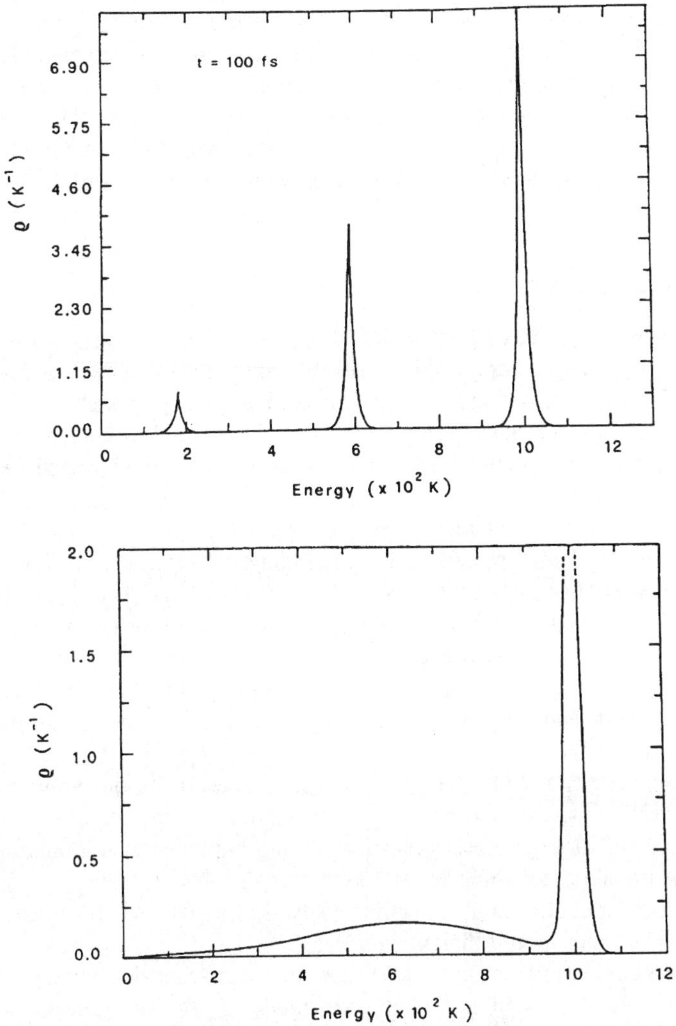

Figure 3.16 *(a) Classical and (b) quantum electron distribution as a function of energy for a simplified GaAs model at $t = 100\,fs$ after excitation. The highest peak at 1000 K is the initial distribution at $t = 0$ [127].*

under the assumption that the phonon population is not perturbed by the transport process. The above technique in now being applied to simple ideal systems for the validation of the numerical procedure. The investigations are usually performed at very short times after the initial condition.

As an example of application of the MC method to quantum transport (in the density matrix formalism), Fig. 3.16 shows a comparison between semiclassical and quantum electron distributions at 100 fs after an initial photo-excitation obtained from MC simulations. The interaction Hamiltonian includes polar–optical phonon coupling in both cases. The phonon replicas corresponding to electrons having emitted one or two optical phonons, present in the semiclassical result, are absent in the quantum result at the same time: at this short time electrons are spread over a very wide range of energies because of the uncertainty relation.

3.6 CONCLUSIONS

In this chapter we have briefly described how the Monte Carlo method has recently progressed. Sophisticated, reliable and predictive MC codes for the study of transport problems in semiconductors are today available. The parallel improvement of computer performance has allowed us, in general, to contain the computational burden usually associated to the method within reasonable limits.

The recent inclusion of accurate physical models allows us to reproduce quantitatively a very large amount of data, including transport data for bulk systems and devices at electron energies above 1 eV. Some essential details of the physical model are well understood and achieved a large consensus among different simulators (e.g. the use of the full band, the treatment of impact ionization phenomena). Some other relevant details are still open matters for scientific debate (e.g. short- and long-range Coulomb interactions and the role of dynamical screening).

MC simulators already play a fundamental role in the development of advanced semiconductor devices and, in the future it is expected that they will be used more and more often as sophisticated tools aiming at discriminating among different possible technological solutions through the realization of 'simulated experiments' and/or through a detailed analysis of a given technological solution provided by other more efficient simulators.

MC simulation of transport and relaxation processes in low-dimensional systems required a revision of the physical model in the MC procedure and the inclusion of typical features such as real-space transfer, surface and interface phonon modes, roughness scattering.

The limits of the semiclassical MC procedure seem to be defined by the limits intrinsic to the semiclassical theoretical scheme based on the validity of the BTE. However, full quantum transport can still profitably use the MC method as it was the case for semiclassical transport 30 years ago.

ACKNOWLEDGEMENTS

The authors thank all their colleagues who provided useful information on their work on the subjects treated in this paper. In particular, A. Abramo, L. Rota and M. Rudan are gratefully acknowledged for useful discussions.

REFERENCES

[1] Binder, K. (1984) *Applications of the Monte Carlo Method in Statical Physics*, Springer, New York.

[2] Kurosawa, T. (1966) Monte Carlo calculation of hot-electron problems *J. Phys. Soc. Jpn.* **21**, 424–6.

[3] Ferry, D. K., Barker, J. R. and Jacoboni, C. (1980) *Physics of Non-linear Transport in Semiconductors*, Plenum, New York.

[4] Hockney, R. W., Warriner, R. A. and Reiser, M. (1974) Two-dimensional particle models in semiconductor-device analysis *Electron. Lett.* **10**, 484–6.

[5] Baccarani, G., Jacoboni, C. and Mazzone, A. M. (1977) Current transport in narrow-base transistors *Solid State Electron.* **20**, 5–10.

[6] Jacoboni, C. and Lugli, P. (1989) *The Monte Carlo Method for Semiconductor Device Simulation*, Springer, New York.

[7] Grubin, H. L., Ferry D. K. and Jacoboni, C. (1980) *The Physics of Submicron Semiconductor Devices*, Plenum, New York.

[8] Brunetti, R., Jacoboni, C., Venturi, F., Sangiorgi, E. and Riccò, B. (1989) A many-band silicon model for hot-electron transport at high energies *Solid State Electron.* **32**, 1663–7.

[9] Abramo, A., Venturi, F., Sangiorgi, A., Higman, J. M. and Riccò, B. (1993) A numerical method to compute isotropic band models from anisotropic semiconductor band structures *IEEE Trans. Comput.-Aided Design Integrated Circuits* **12**, 1327–36.

[10] Jungemann, C., Keith, S., Meinerzhagen, B. and Engl, W. L. (1995) On the influence of band structure and scattering rates on hot electron modeling *SISDEP Tech. Dig.* Springer, Erlangen, pp. 222–5

[11] Shichijo, H., Tang, J. Y., Bude, J. and Yoder, D. (1991) Full band Monte Carlo program for electrons in silicon *Monte Carlo Device Simulation: Full Band and Beyond* (ed. K. Hess), Kluwer, Boston, pp. 285–307

[12] Ando, T., Fowler, A. B. and Stern, F. (1982) Electronic properties of two-dimensional systems *Rev. Mod. Phys.* **54**, 437–672.

[13] Kelly, M. J. (1995) *Low-dimensional Semiconductors*, Clarendon, Oxford.

[14] Burt, M. G. (1992) The justification for applying the effective-mass approximation to microstructures *J. Phys. Condens. Matter* **4**, 6651–90.

[15] Jacoboni, C. and Reggiani, L. (1983) The Monte Carlo method for the solution of charge transport in semiconductors with applications to covalent materials *Rev. Mod. Phys.* **55**, 645–705.

[16] Ziman, J. M. (1961) A theory of the electrical properties of liquid metals I: the monovalent metals. *Phil. Mag.* **6**, 1013–34.

[17] Herbert, D. C. (1973) Electron-phonon interaction and inter-valley scattering in semiconductors *Journal de Physique* **6**, 2788–810.

[18] Zollner, S., Gopolan, S. and Cardona, M. (1990) Microscopic theory of intervalley scattering in GaAs: $\vec{k}$-dependence of deformation potentials and scattering rates *J. Appl. Phys.* **68**, 1682–93.

[19] Hess, K. (1991) *Monte Carlo Device Simulation: Full Band and Beyond*, Kluwer, Boston.

[20] Yoder, P. D. (1994) *First-principle Monte Carlo Simulation of Charge Transport in Semiconductors* PhD thesis (Urbana-Champaign).

[21] Fischetti, M. V. and Laux, S. E. (1995) Monte Carlo study of sub-band gap impact ionization in small silicon field-effect transistors *IEDM Tech. Dig.*, pp. 305–8.

[22] Menéndez, J. (1989) Phonons in GaAs-$Al_xGa_{1-x}As$ superlattices *J. of Luminescence* **44**, 285.

[23] Molinari, E. (1995) Phonons and electron-phonon interaction in low-dimensional structures. *Confined Electrons and Photons: New Physics and Applications* (eds E. Burstein and C. Weisbuch), Plenum, New York, p. 161.

[24] Fuchs, R. and Kliewer, K. L. (1965) Optical modes of vibration in an ionic crystal slab *Phys. Rev.* **140**, A2076.

[25] Huang, K. and Zhu, B. (1988) Dielectric continuum model and Fröhlich interaction in superlattices *Phys. Rev.* B **38**, 13 377.

[26] Gerecke, H. and Bechstedt, F. (1991) Dielectric continuum model and Fröhlich interaction in superlattices *Phys. Rev.* B **43**, 7053.

[27] Mori, N. and Ando, T. (1989) Electron–optical-phonon interaction in single and double heterostructures *Phys. Rev.* B **40** 6175.

[28] Rücker, H., Molinari, E. and Lugli, P. (1992) Microscopic calculation of the electron-phonon interaction in quantum wells *Phys. Rev.* B **45**, 6747.

[29] Rossi, F., Rota, L., Bungaro, C., Lugli, P. and Molinari, E. (1993) Phonons in thin GaAs quantum wires *Phys. Rev.* B **47**, 1695.

[30] Molinari, E., Bungaro, C., Rossi, F., Rota, L. and Lugli, P. (1993) Phonons in thin GaAs/AlAs nanostructures: from two-dimensional to one-dimensional systems (eds J. P. Leburton, J. Pascual and C. M. Sotomayor Torres) *Phonons in Semiconductors Nanostructures*, Kluwer, Boston, p. 39

[31] Rota, L., Rossi, F., Gulia, M., Lugli, P. and Molinari, E. (1992) Monte Carlo simulation of a 'true' quantum wire *Advanced Semiconductor Epitaxial Growth Processes and Lateral and Vertical Fabrication* (ed. R. J. Malik), SPIE, Bellingham, p. 161.

[32] Lugli, P., Rota, L. and Rossi, F. (1992) Thermalization of photoexcited carriers in bulk and quantum wire semiconductors *Phys. Stat. Sol.* B **173**, 229.

[33] Mayer, J. R. and Bartoli, F. J. (1981) Phase-shift calculation of ionized impurity scattering in semiconductors *Phys. Rev.* B **23**, 5413–27.

[34] Chattopadhyay, D. and Queisser, H. J. (1981) Electron scattering by ionized impurities in semiconductors *Rev. Mod. Phys.* **53**, 745–68

[35] Norton, P. and Levinstein, H. (1972) Determination of compensation densities by Hall and mobility analysis in copper-doped germanium *Phys. Rev.* B **6**, 470–7.

[36] Pines, D. and Bohm, D. (1952) A collective description of electron interaction: II. collective vs. individual particle aspects of the interactions *Phys. Rev.* **85**, 338–53.

[37] Abramo, A., Brunetti, R., Jacoboni, C., Venturi, F. and Sangiorgi, E. (1994) A multiband Monte Carlo approach to Coulomb interaction for device analysis *J. Appl. Phys.* **76**, 5786–94.

[38] Bacchelli, L. and Jacoboni, C. (1972) Electron-electron interactions in Monte Carlo transport calculations *Solid State Comm.* **10**, 71–4.

[39] Matulionis, A., Pozela, J. and Reklaitis, A. (1975) Monte Carlo treatment of electron-electron collisions *Solid State Comm.* **16**, 1133–7.

[40] Lugli, P. and Ferry, D. K. (1985) Electron-electron interaction and high-field transport in silicon *Appl. Phys. Lett.* **46**, 594–6.

[41] Brunetti, R., Jacoboni, C., Matulionis, A. and Dienys, V. (1985) Effect of interparticle collisions on energy relaxation of carriers in semiconductors

Physica B **134**, 369–73.

[42] Siggia, E. D. and Kwok, P. C. (1970) Properties of electrons in semiconductor inversion layers with many occupied electric subbands. I. screening and impurity scattering *Phys. Rev.* B **2**, 1024.

[43] Rota, L., Rossi, F., Lugli, P. and Molinari, E. (1995) Ultrafast relaxation of photoexcited carriers in semiconductor quantum wires: a Monte Carlo approach *Phys. Rev.* B **52**, 5183.

[44] Rota, L., Rossi, F., Goodnick, S. M., Lugli, P., Molinari, E. and Porod, W. (1993) Reduced carrier cooling and thermalization in semiconductor quantum wires *Phys. Rev.* B **47**, 1632.

[45] Mosko, M. and Cambel, V. (1994) Thermalization of a one-dimensional electron gas by many-body Coulomb scattering: Molecular-dynamics model for quantum wires *Phys. Rev.* B **50**, 8864.

[46] Bosi, S. and Jacoboni, C. (1976) Monte Carlo high-field transport in degenerate GaAs *J. Phys. C: Solid State Phys.* **9**, 315–9.

[47] Goodnick, S. M., Ferry, D. K., Wilmsen, C. W., Lilental, Z., Fathy, D. and Krivanek, O. L. (1985) Surface roughness at the $Si(100) - SiO_2$ interface *Phys. Rev.* B **32**, 8171–86

[48] Sangiorgi, E. and Pinto, M. R. (1992) A semi-empirical model of surface scattering for Monte Carlo simulation of silicon n-MOSFETS *IEEE Trans. Electron Devices* **39**, 356–61.

[49] Laux, S. E., Fischetti, M. V. and Frank, D. J. (1990) The DAMOCLES program *IBM J. Res. Dev.* **34**, 466.

[50] Pinto, M. R., Sangiorgi, E. and Bude, J. (1993) Silicon MOS transconductance scaling into the overshoot regime *IEEE Electron Device Lett.* **14**, 375–8.

[51] Sabnis, A. G. and Clemens, J. T. (1979) Characterization of the electron mobility in the inverted ⟨100⟩ Si surface *IEDM Tech. Dig.*, pp. 18–21.

[52] Ishizaka, M., Iizuka, T., Ohi, S., Fukuma, M. and Mikoshiba, H. (1990) Advanced electron mobility model of MOS inversion layer considering 2D-degenerated electron gas physics *IEDM Tech. Dig.*, pp. 763–6.

[53] Fischetti, M. V. and Laux, S. (1993) Monte Carlo study of electron transport in silicon inversion layers *Phys. Rev.* B **48**, 2244–74.

[54] Abramo, A., Bude, J., Venturi, F. and Pinto, M. R. (1994) Mobility simulation in Si/SiGe heterostructure FETs *IEDM Tech. Dig.*, p. 731.

[55] Reggiani, L., Lugli, P. and Mitin, V. (1987) Monte Carlo algorithm for generation-recombination noise in semiconductors *Appl. Phys. Lett.* **51**, 925–7.

[56] Thoma, T., Peifer, H. J., Engl, W. L., Quade, W., Brunetti, R. and Jacoboni, C. (1991) A generalized impact-ionization model for high-energy electron transport in Si with Monte Carlo simulation *J. Appl. Phys.* **69**, 2300–11.

[57] Kunikiyo, T., Takenaka, M., Kamakura, Y., Yamaji, M., Mizuno, H., Morifuji, M., Taniguchi, K. and Hamaguchi, C. (1994) A Monte Carlo simulation of anisotropic electron transport in silicon including full band structure and anisotropic impact-ionization model *J. Appl. Phys.* **75**, 297–312.

[58] Cartier, E., Fischetti, M. V., Eklund, E. A. and McFeely, F. R. (1993) Impact ionization in silicon *Appl. Phys. Lett.* **62**, 3339–41.

[59] Sano, N. and Yoshii, A. (1995) Impact-ionization model consistent with the band structure of semiconductors *J. Appl. Phys.* **77**, 2020–5.

[60] Sangiorgi, E., Riccò, B. and Venturi, F. (1988) MOS^2: an efficient Monte Carlo Simulator for MOS devices *IEEE Trans. Comput.-Aided Design Integrated Circuits* **7**, 259–71.

[61] Landau, L. D. and Lifshitz, E. M. (1958) *Quantum Mechanics*, Pergamon, Oxford.

[62] Price, P. J. (1970) The theory of hot electrons *IBM J. Res. Dev.* **14**, 12–14.

[63] Lebwohl, P. A. and Price, P. J. (1971) Direct microscopic simulation of Gunn-domain phenomena *Appl. Phys. Lett.* **19**, 530–2.

[64] Kocevar, P. (1985) Hot phonons dynamics *Phys. Rev.* B **134**, 155.

[65] Lugli, P. (1988) Hot phonon dynamics *Solid State Electron.* **31**, 667.

[66] Rieger, M., Kocevar, P., Bordone, P., Lugli, P. and Reggiani, L. (1988) Transient hot-phonon effects on the velocity overshoot of GaAs: a Monte Carlo analysis *Solid State Electron.* **31**, 687.

[67] Lugli, P., Bordone, P., Reggiani, L., Rieger, M., Kocevar, P. and Goodnick., S. M. (1989) Monte Carlo studies of nonequilibrium phonon effects in polar semiconductors and quantum wells. I. Laser photoexcitation *Phys. Rev.* B **39**, 7852.

[68] Rieger, M., Kocevar, P., Lugli, P., Bordone, P., Reggiani, L. and Goodnick, S. M. (1989) Monte Carlo sudies of nonequilibrium phonon effects in polar semiconductors and quantum wells. II. Non-ohmic transport in *n*-type gallium arsenide *Phys. Rev.* B **39**, 7866.

[69] Lugli, P. and Goodnick, S. M. (1987) Nonequilibrium longitudinal-optical phonon effects in GaAs–AlGaAs quantum wells *Phys. Rev. Lett.* **59**, 716.

[70] Bordone, P. and Lugli, P. (1994) Effect of half-space and interface phonons on the transport properties of $Al_xGa_{1-x}As$/GaAs single heterostructures *Phys. Rev.* B **49**, 8178.

[71] Phillips, A. and Price, P. J. (1977) Monte Carlo calculations of hot electron energy tails. *Appl. Phys. Lett.* **30**, 528–30.

[72] Pacelli, A., Duncan, A. W. and Ravaioli, U. (1996) A multiplication scheme with variable weights for ensemble Monte Carlo simulation of hot-electron tails *Hot Carriers in Semiconductors*, Plenum, New York, pp. 409–12

[73] Abramo, A. and Fiegna, C. (1996) Electron energy distributions in silicon structures at low applied voltages and high electric fields *J. Appl. Phys.* **80**, 889–93.

[74] Rossi, F., Poli, P. and Jacoboni, C. (1992) Weighted Monte Carlo approach to electron transport in semiconductors *Semicond. Sci. Technol.* **7**, 1017–35.

[75] Venturi, F., Sangiorgi, E., Luryi, S., Poli, P., Rota, L. and Jacoboni, C. (1991) Energy oscillations in electron transport across a triangular barrier *IEEE Trans. Electron Devices* **38**, 611–18.

[76] Abramo, A., Venturi, F., Sangiorgi, E., Higman, J. and Riccò, B. (1992) A numerical method to compute isotropic band models from anisotropic semiconductor band structures *NUPAD Tech. Dig.*, pp. 85–90.

[77] Bude, J. and Smith, R. K. (1994) Phase-space simplex Monte Carlo for semiconductor transport *Semicond. Sci. Technol.* **9**, 840–3.

[78] Venturi, F., Smith, R. K., Sangiorgi, E., Pinto, M. R. and Riccò, B. (1988) A new coupling scheme for a self-consistent Poisson and Monte Carlo device simulator *SISDEP Tech. Dig.*, Tecnoprint, Bologna, pp. 383–6

[79] Woolard, D. L., Tian, W., Littlejohn, M. A. and Kim, K. W. (1994) The implementation of physical boundary conditions in the Monte Carlo simulation of electron devices *IEEE Trans. Comput.-Aided Design Integrated Circuits* **13**, 1241–6.

[80] Gonzales, T. and Pardo, D. (1996) Physical models of ohmic contact for Monte Carlo device simulation *Solid State Electron.* **39**, 555–62.

[81] Fischetti, M. V. and Laux, S. E. (1988) Monte Carlo analysis of electron transport in small semiconductor devices including band-structure and space-charge effects *Phys. Rev.* B **38**, 9721–45.

[82] Venturi, F., Smith, R. K., Sangiorgi, E., Pinto, M. R. and Riccò, B. (1989) A general purpose device simulator coupling Poisson and Monte Carlo transport with applications to deep submicron MOSFETs *IEEE Trans. Comput.-Aided Design Integrated Circuits* **8**, 360–9.

[83] Rambo, P. W. and Denavit, J. (1993) Time stability of Monte Carlo device simulation *IEEE Trans. Comput.-Aided Design Integrated Circuits* **12**, 1734–41.

[84] Ghetti, A., Wang, X., Venturi, F. and Leon, F. A. (1995) Stability issues in self-consistent Monte Carlo-Poisson simulations *SISDEP Tech. Dig.*, Springer, Erlangen, pp. 388–91.

[85] Kometer, K., Zandler, G. and Vogl, P. (1992) Lattice-gas cellular-automaton method for semiclassical transport in semiconductors *Phys. Rev.* B **46**, 1382–94.

[86] Ventura, D., Gnudi, A. and Baccarani, G. (1992) Multidimensional spherical harmonics expansion of Boltzmann equation for transport in semiconductors *Appl. Math. Lett.* **5**, 85–90.

[87] Vecchi, M. C., Ventura, D., Gnudi, A. and Baccarani, G. (1994) Incorporating full band-structure effects in the spherical-harmonics expansion of the Boltzmann transport equation *Proc. of the NUPAD V Conference, Honolulu* (eds H. S. Bennet and M. E. Law), pp. 55–8.

[88] Iizuka, T. and Fukuma, M. (1990) Carrier transport simulator for silicon based on carrier distribution function evolution *Solid State Electron.* **33**, 27–34.

[89] Fiegna, C., Venturi, F., Sangiorgi, E. and Riccò, B. (1990) Efficient non-local modeling of the electron energy distribution in sub-micron MOSFETS *IEDM Tech. Dig.*, pp. 451–4

[90] Tanaka, S. and Lundstrom, M. S. (1993) A flux-based approach to HBT device modeling *IEDM Tech. Dig.*, pp. 505–8.

[91] Abramo, A., Baudry, L., Brunetti, R., Castagné, R., Charef, M., Dessenne, F., Dolfus, P., Dutton, R., Engl, W. L., Fauquembergue, R., Fiegna, C., Fischetti, M. V., Galdin, S., Goldsman, N., Hackel, M., Hamaguchi, C., Hess, K., Hennacy, K., Hesto, P., Higman, J. M., Iizuka, T., Jungemann, C., Kamakura, Y., Kosina, H., Kunikiyo, T., Laux, S., Lin, H., Maziar, C., Mizuno, H., Peifer, H. J., Ramaswamy, S., Sano, N., Scrobohaci, P. G., Selberherr, S., Takenaka, M., Tang, T.-W., Taniguchi, K., Thobel, J. L., Thoma, R., Tomizawa, K., Tomizawa, M., Vogelsang, T., Wang, S.-L., Wang, X., Yao, C.-S., Yoder, P. D. and Yoshii, A. (1994) A comparison of numerical solutions of the Boltzmann transport equation for high-energy electron transport silicon *IEEE Trans. Electron Devices* **41**, 1646–54.

[92] Jacoboni, C., Gagliani, G., Reggiani, L. and Turci, O. (1978) Noise and diffusion of hot holes in Si *Solid State Electron.* **21**, 315–18.

[93] Reggiani, L., Golinelli, P., Varani, L., Gonzáles, T., Pardo, D., Starikov, E., Shiktorov, P. and Gruzinskis, V. (1996) Monte Carlo analysis of electronic noise in semiconductor materials and devices *Microelectronics Journal*, submitted.

[94] Golinelli, P., Brunetti, R., Varani, L., Reggiani, L. and Rudan, M. (1995) Monte Carlo calculation of hot-carrier thermal conductivity in semiconductors *Hot Carriers*

in Semiconductors (eds K. Hess, J. P. Leburton and U. Ravaioli), Plenum, New York, pp. 405–407.

[95] Brunetti, R. and Jacoboni, C. (1984) Analysis of the stationary and transient autocorrelation function in semiconductors *Phys. Rev.* B **29**, 5739–43.

[96] Goodnick, S. M. and Lugli, P. (1988) Effect of electron–electron scattering on nonequilibrium transport in quantum-well systems *Phys. Rev.* B **37**, 2578.

[97] Yamada, T., Miyata, H., Zhou, J.-R. and Ferry, D. K. (1994) Monte Carlo study of the low-temperature mobility of electrons in strained Si layers grown on a $Si_{1-x}Ge_x$ substrate *Phys. Rev.* B **49**, 1875–81.

[98] Thobel, J. L., Sleiman, A., Boural, P. and Dessenne, F. (1996) Monte Carlo study of electron transport in III-V heterostructures with doped quantum wells *J. Appl. Phys.* **80**, 928–35.

[99] Yamada, T., Zhou, J-R., Miyata, H. and Ferry, D. K. (1994) Velocity overshoot in a modulation-doped $Si/Si_{1-x}Ge_x$ structure *Semicond. Sci. Technol.* **9**, 775–7.

[100] Yamada, T. and Ferry, D. K. (1995) Monte Carlo simulation of hole transport in strained $Si_{1-x}Ge_x$ *Solid State Electron.* **38**, 881–90.

[101] Abramo, A., Bude, J., Venturi, F. and Pinto, M. R. (1996) Mobility simulation of a novel Si/SiGeFET structure *IEEE Electron Device Lett.* **17**, 59–61.

[102] Ismail, K., Arafa, M., Saenger, K. L., Chu, J. O. and Meyerson, B. S. (1995) Extremely high electron mobility in Si/SiGe modulation-doped heterostructures *Appl. Phys. Lett.* **66**, 1077–9.

[103] Ismail, K., Rishton, S., Chu, J. O., Chan, K. and Meyerson, B. S. (1993) High performance Si/SiGe n-type modulation-doped transistors *IEEE Electron Device Lett.* **14**, 348–50.

[104] Shah, J., Pinczuk, A., Gossard, A. C. and Wiegmann, W. (1985) Energy-loss rates for hot electrons and holes in GaAs quantum wells *Phys. Rev. Lett.* **54**, 2045–8.

[105] Maciel, A. C., Kiener, C., Rota, L., Ryan, J. F., Marti, U., Martin, D., Morier-Gemoud, F. K. and Reinhart, F. K. (1995) Hot carrier relaxation in GaAs V-groove quantum wires *Appl. Phys. Lett.* **66**, 3039–41.

[106] Freyland, J. M., Turner, K., Kiener, C., Rota, L., Ryan, J. F., Marti, U., Martin, D., Morier-Gemoud, F. and Reinhart, F. K. (1996) Reduced carrier cooling in GaAs V-groove quantum wires due to non-equilibrium phonon population *Hot carriers in semiconductors* (eds K Hess, J. P. Leburton and U. Ravaioli), Plenum, New York, pp. 323–5.

[107] Lugli, P., Bordone, P., Molinari, E., Rücker, H., de Paula, A. M., Maciel, A. C., Ryan, J. F. and Shayegan, M. (1992) Interaction of electrons with interface phonons in GaAs/AlAs and GaAs/AlGaAs heterostructures *Semicond. Sci. Technol.* **7**, 116–19.

[108] Turner, K., Rota, L., Taylor, R. A. and Ryan, J. F. (1996) Ultrafast optical absorption measurements of electron-phonon scattering in GaAs quantum well *Hot carriers in semiconductors* (eds K Hess, J. P. Leburton and U. Ravaioli), Plenum, New York, pp. 23–6.

[109] Lugli, P., Bordone, P., Gualdi, S., Poli, P. and Goodnick, S. M. (1989) Hot phonons in quantum wells systems *Solid State Electron.* **32**, 1881.

[110] Rota, L., Ryan, J. F., Rossi, F., Lugli, P. and Molinari, E. (1994) Hot phonons in quantum wires: a Monte Carlo investigation *Europhys. Lett.* **28**, 277.

[111] Mizuno, T., Toriumi, A., Iwase, M., Takahashi, M., Niiyama, H., Fukumoto, M. and Yoshimi, M. (1992) Hot carrier effects in 0.1 μm gate length CMOS devices

IEDM Tech. Dig., pp. 695–8.

[112] Esseni, D., Selmi, L., Sangiorgi, E., Bez, R. and Riccò, B. (1994) Bias and temperature dependence of gate and substrate currents in n-MOSFETs at low drain voltage *IEDM Tech. Dig.* pp. 307–10.

[113] Bude, J. D., Frommer, A., Pinto, M. R. and Weber, G. R. (1995) EEPROM/Flash sub 3.0V drain-source bias hot carrier writing *IEDM Tech. Dig.*, pp. 989–91.

[114] Fischer, B., Selmi, L., Ghetti, A. and Sangiorgi, E. (1996) Electron injection into the gate oxide of MOS structures at liquid nitrogen temperature: measurement and simulation *Journal de Physique IV* **6**, 19–24.

[115] Ellis-Monaghan, J. J., Hulfachor, R. B., Kim, K. W. and Littlejohn, M. A. (1996) Ensemble Monte Carlo study of interface-state generation in low-voltage scaled silicon MOS devices *IEEE Trans. Electron Devices* **43**, 1123–32.

[116] Lee, C. H., Ravaioli, U., Hess, K., Mead, C. A. and Hasler, P. (1995) Simulation of a long term memory device with full bandstructure Monte Carlo approach *IEEE Electron Device Lett.* **16**, 360–2.

[117] Tang, J. and Hess, K. (1983) Impact ionization of hot electrons in silicon (steady state) *J. Appl. Phys.* **54**, 5139–45.

[118] Reggiani, L., Lugli, P. and Jauho, A. P. (1988) Monte Carlo algorithms for collisional broadening and intracollisional field effect in semiconductor high-field transport *J. Appl. Phys.* **64**, 3072–8.

[119] Ghetti, A., Selmi, L., Sangiorgi, E., Abramo, A. and Venturi, F. (1994) A combined transport-injection model for hot-electron and hot-hole injection in the gate oxide of MOS structures *IEDM Tech. Dig.*, pp. 363–6.

[120] Fischetti, M. V., Laux, S. E. and Crabbè, E. (1995) Understanding hot-electron transport in silicon devices: is there a shortcut? *J. Appl. Phys.* **78**, 1058–87.

[121] Hess, K. (1988) Real space transfer: generalized approach to transport in confined geometries *Solid State Electron.* **31** 319–24.

[122] Brunetti, R., Jacoboni, C. and Price, P. J. (1994) Quantum-mechanical evolution of real-space transfer *Phys. Rev.* B **50**, 11 872–8.

[123] Kizilyalli, I., Hess, K., Higman, T., Emanuel, M. and Coleman, J. (1988) Ensemble Monte Carlo simulation of real space transfer (NERFET/CHINT) devices *Solid State Electron.* **31**, 355–8.

[124] Chambers, R. (1952) The kinetic formulation of the transport problem *Proc. Phys. Soc.* A **65**, 458–9.

[125] Abramo, A., Casarini, P. and Jacoboni, C. (1995) Transmission properties of resonant cavities and rough quantum wells *Hot Carriers in Semiconductors* (eds K. Hess, J. P. Leburton and U. Ravaioli), Plenum, New York, pp. 509–12.

[126] Brunetti, R., Jacoboni, C. and Nedjalkov, M. (1996) Quantum transport with electron-phonon interaction in the Wigner-function formalism *Hot Carriers in Semiconductors* (eds K. Hess, J. P. Leburton and U. Ravaioli), Plenum, New York, pp. 417–20.

[127] Brunetti, R., Jacoboni, C. and Rossi, F. (1989) Quantum theory of transient transport in semiconductors: A Monte Carlo approach *Phys. Rev.* B **39**, 10 781–90.

[128] Brunetti, R., Jacoboni, C., Nava, F., Reggiani, L., Bosman, G. and Zijlstra, R. J. J. (1981) Diffusion coefficient of electrons in silicon. *J. Appl. Phys.* **52**, 6713.

[129] Sano, N. and Yoshii, A. (1994) Impact ionization rate near thresholds in Si. *J. Appl. Phys.* **75**, 5102.

CHAPTER 4

Cellular automaton approach for semiconductor transport

P. Vogl*, G. Zandler*, A. Rein* and M. Saraniti†

*Physics Department and Walter Schottky Institut, TU München, D-85748 Garching, Germany

†Department of Electrical Engineering, Arizona State University, Tempe, AZ 85287, USA

4.1 INTRODUCTION

Recently, a new simulation strategy has been adopted for the realistic modelling of transport processes in a variety of problems. The physical systems are replaced by fictitious microworld models obeying discrete cellular automata (CA) rules such that the macroscopic dynamics is recovered as the ensemble average over the microworld states [1–5]. In hydrodynamics, for example, Frisch and coworkers [6, 7] were able to show that lattice gas automata can provide an effective numerical technique for solving the Navier–Stokes and many other types of partial differential equations with complex boundary conditions.

In this chapter, we will focus on semiclassical semiconductor transport and show that the concept of CA provides a novel framework for accurate and numerically extremely efficient schemes to solve the full Boltzmann equation for modern nanometer devices [3–5, 8, 10, 21]. From the point of view of physics, this method is equivalent to the standard ensemble Monte Carlo (MC) method that was reviewed in Chapter 3 of this book. Since the cellular automaton (CA) method invokes concepts and ideas that are applicable to many dynamical systems, however, we hope that the reader will find this chapter of some interest nevertheless.

To demonstrate the practical use, high level of accuracy, and reliability of the CA method for realistic device modelling, we present two timely examples, namely a study of ultrashort channel vertical metal-oxide-semiconductor field effect transistors (MOSFET) and the prediction of contact resistances in high electron mobility transistors (HEMT).

The concept of CA goes back to von Neumann [11]. A CA is a discrete dynamical system that evolves in discrete time steps. It is defined at the nodes of a lattice, each site of which is characterized by a finite number of Boolean states. In the context of transport theory, these Boolean states are classified as being occupied or empty, and the former ones are considered as representing fictitious 'particles'. In each time step, a set of transition rules updates the state occupancies synchronously on all lattice sites. The important point is that these rules are local in the sense that they only act on the occupancies of a given site and its nearest (and possibly next-nearest) neighbours. It is precisely for this reason that CA constitute one of the very few algorithms for dynamical systems that can optimally utilize massively parallel hardware [1, 2]. In addition, the locality of the dynamical rules allows an efficient and flexible handling of complex geometries and lattice topologies. A noticeable example is a CA simulation of flow through porous media with fractal structure [12].

Perhaps the most crucial factor that sets CA apart from standard finite differencing methods for partial differential equations is the reduction of all physical variables to a finite set of discrete values. Microscopic transport equations contain scattering kernels that depend on many physical parameters. Often, however, the solution is insensitive to some of these parameters and it is unnecessary to calculate all of them with equally high precision. In low-speed diffusive transport regimes, for example, the detailed angular dependence of a particle–particle collision rate is irrelevant. By representing each parameter by only as many discrete values as necessary and treating them on an equal footing, lattice gas automata can achieve substantial reductions in both storage and computer time as will be elaborated later in this chapter.

In the context of semiclassical Boltzmann transport far from thermal equilibrium such as high-field carrier dynamics in modern submicron semiconductor devices, we will show that CA simulations offer some striking advantages in situations with pronounced spatial inhomogeneities, complex geometries and large density variations.

4.2 EXAMPLES OF CELLULAR AUTOMATA IN FLUID DYNAMICS

4.2.1 The TM-gas model for the diffusion equation

To illustrate the concept of CA, we first discuss an exceedingly simple example, the so-called HPP-gas or TM-gas [13–16]. Consider a two-dimensional square lattice with unit lattice constant. Particles of unit mass and unit speed are moving along the lattice links and are located at the nodes at integer times. Not more than one particle is to be found at a given time and node, moving in a given direction (exclusion principle). When two and exactly two particles arrive at a node from opposite directions (head-on collisions), they immediately leave the node in the two other, previously unoccupied, directions (Fig. 4.1(a)). In every other case, the particles just keep moving along their direction without interacting with one another (Fig. 4.1(b)). These deterministic collision laws obviously conserve mass (particle number) and momentum.

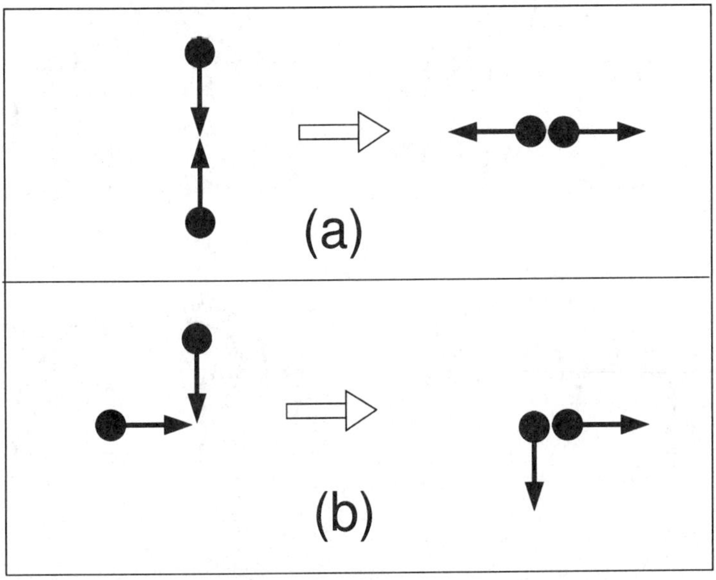

Figure 4.1 *Dynamical rules for the HPP gas. (a) Only a head-on collision of two particles causes scattering, that is, the particles change direction by* 90°. *The particles then continue to move at a constant speed in the new direction. (b) In all other configurations, particles do not scatter but are simply transported at a constant speed.*

In order to check the *macroscopic* dynamics that this single scattering rule leads to, we put this lattice gas in a 'bottle' (Fig. 4.2) and add one rule, namely specular reflection at any boundary in order to make the particles bounce off the walls. We start with a dense cloud of particles in the middle (Fig. 4.2(a)). The leading particles travel in a vacuum, all at the same speed, until they hit a wall (Fig. 4.2(b)). As they bounce off the walls, the head-on collision rule leads to a disruption of the paths of the particles that come after them (Figs. 4.2(c) and 4.2(d); see left bottle). This disruption rapidly extends in scope (Fig. 4.2(e)). Eventually the container will be filled with rarefied gas in a state of thermodynamic equilibrium (Fig. 4.2(f)). If we dump more and more gas into the container, the mean-free path of a particle is only a few cells. Thermal agitation is then driven by the totality of the remaining gas particles rather than by the boundary. In fact, the authors of [7, 15] were able to show the macroscopic ensemble average of this deterministic and invertible CA to be equivalent to the diffusion equation. Thus, surprisingly simple dynamical rules, defined on a microscopic scale, can mimic a nontrivial dynamics on a macroscale.

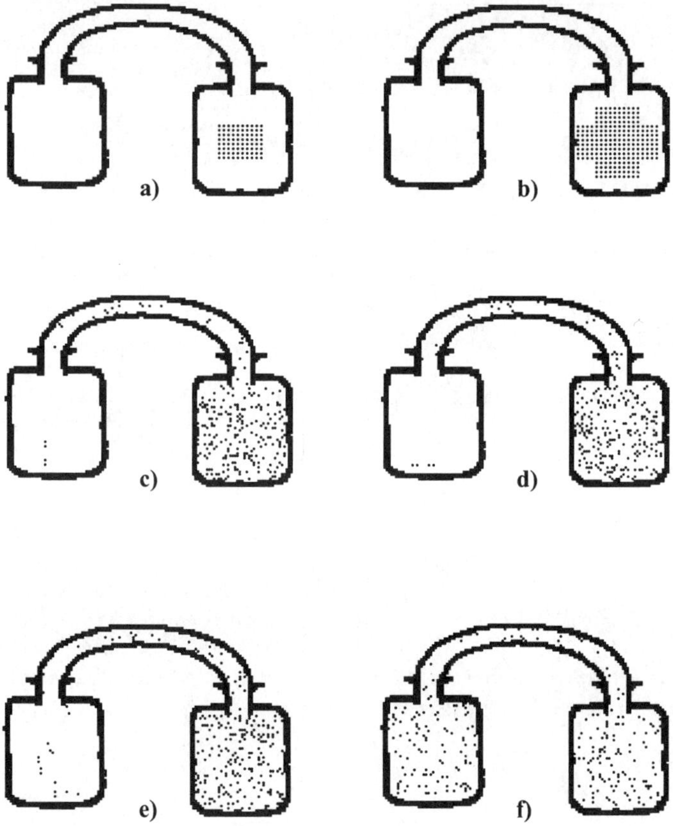

Figure 4.2 *Time evolution of the HPP gas on a two-dimensional lattice with bottle-shaped boundaries. The head-on scattering rule allows pairs of particles to propagate into previously unoccupied areas. This effect can be seen most distinctly in parts (c) and (d). Eventually, the HPP gas reaches complete thermalization, despite the deterministic and time reversible rules.*

4.2.2 Streaming motion

Before we turn to mapping the realistic Boltzmann dynamics onto a CA, we consider another simple example that elucidates the modelling of free streaming motion with CA. Consider a gas that has been set into uniform motion along a one-dimensional trajectory with constant velocity v. We wish to derive a CA algorithm that obeys the following three conditions:

- the continuous density distribution of the fluid moves with velocity v;
- the CA rules are identical in all cells;
- the CA rules couple only nearest-neighbour cells.

The first step in setting up a CA consists of subdividing the relevant spatial region of length L into cells of size $a \ll L$. The mesh size is to be set in such a way that all relevant spatial variations in the distribution are captured. In addition, the time steps Δt may be chosen to obey the condition $a/\Delta t \gg v$. The speed $c = a/\Delta t$ is conventionally called the 'speed of light' of the CA and should be larger than any physically relevant velocity. Finally, the continuous density is represented by an ensemble of CA particles.

Note that the condition $c \gg v$ implies the CA mesh size a to be much coarser than the motion of a CA particle during a single time step, $v\Delta t$. It is the smoothness of the distribution function rather than the time step that determines the mesh size. This is a crucial point that drastically reduces the memory requirements and computer time compared with standard finite differencing methods.

Let us now consider the following CA rule. In one time step, each CA particle hops to the next cell with probability $p = v/c$. Otherwise, with probability $q = 1 - p$, it stays in its cell. Obviously, this rule fulfils the three requirements stated above and guarantees that the *average* velocity of each particle is indeed v. However, this rule has a serious flaw. If all particles start in the same cell at time equal zero, they will not stay together but form a diffuse distribution. The mean value of a particle's position after N time steps is guaranteed to be $N\Delta t v$ but there is a mean variance that increases linearly in time and is given by $Npqa^2$. Thus, probabilistic rules lead to an undesirable artificial diffusion.

Fortunately, one can set up an alternative CA that obeys strictly deterministic rules and leads to a correct long-time behaviour. In our example, the real fluid particle needs $m = a/(v\Delta t)$ time steps to traverse one cell of length a. We map this continuous trajectory onto a discrete one by initially assigning m 'colours' to the CA particles in a random way and imposing the following rules that are applied to each CA particle per time step.

1. Increase the colour index by 1 (reset the colour index $m + 1$ to 1).

2. Move the particle to the next cell if the particle possesses colour index m, otherwise do not move it.

These rules apparently guarantee that the macroscopic dynamics of the ensemble is reproduced in the long-time limit. The continuous particle motion within the cell is replaced by a simple counter, the 'colour'. This gives an error in the position of the particle that never exceeds one cell dimension.

Thus, we have constructed a CA on a lattice that is chosen in a way that resolves the spatial variations in the density distribution but not the individual particles' trajectories. The CA rules couple only nearest neighbours, and the detailed intracell dynamics is taken care of in a global way by employing a scalar counter.

4.3 FULL BOLTZMANN TRANSPORT EQUATION AS CELLULAR AUTOMATON

4.3.1 Overview

The Boltzmann equation (BE) for the one-particle distribution function $f(\boldsymbol{x}, \boldsymbol{k}, t)$ in $(6+1)$-dimensional phase space, i.e. in position, momentum and time, can be rigorously mapped onto a CA with spatially local dynamical rules [3, 17, 18, 8]. Clearly, these CA rules are not as simple as those discussed in the previous examples and require several thousand states in each real space cell as will be discussed in this section. However, two key advantages of the CA method still apply: first, the strict discretization of time, space and momentum allows one to store all physical scattering processes and rates in hierarchical reference tables that can be evaluated once for all for a given device. Secondly, the spatial locality of the CA rules can be maintained. We note that the use of scattering tables is not unique to the CA method, of course, and is widely employed in other methods as well, for example in full-band MC schemes [19].

In principle, it is straightforward to convert the BE or any other integrodifferential equation into a CA, simply by introducing finite differences in space, momentum, and time. For a typical sub-μm semiconductor device such as a Si MOSFET, this would require an order of 10^6 $\boldsymbol{R}$-cells in real space. Attached to each of these cells is a three-dimensional momentum space lattice that must accommodate at least another 10^6 $\boldsymbol{K}$-cells [20–22]. Since, in general, the drift and scattering terms in the BE couple many different momenta and positions, such a discrete variant of the BE contains nonzero matrix elements between many of these cells. Thus, a straightforward discretization leads to a totally unmanageable matrix problem of dimension 10^{12}.

Let us discuss the phase space of high-field carrier transport in a semiconductor device in more detail. In the conventional self-consistent ensemble MC approach as well as in the present CA method, the time step is typically chosen to be smaller than the inverse of the plasma frequency, $\Delta t \lesssim 1/\omega_p$ [23]. During this time, the momentum of a particle changes by $\Delta k \sim eE\Delta t/\hbar$, where E is the electric field. The distribution function, however, changes on a much coarser scale in k-space, $\Delta k \sim eE\tau_{\text{free}}/\hbar$, where τ_{free} is the mean time of free flight. Similarly, the change in position of a particle during one time step is much smaller than the Debye length that determines the scale of the distribution function in position space, at least for the vast majority of carriers in a device. Consequently, it is wasteful to calculate and keep track of the detailed position or momentum of each particle within a phase space cell of size that suffices to capture changes in the distribution function. A key point of the CA method is its ability to fully exploit this redundancy.

As a further important point, below we will discuss the fact that the drift and diffusion terms in the BE can be mapped onto a CA that couples only nearest-neighbouring cells. In this way, one obtains a method for solving the BE that is mathematically equivalent to the standard ensemble MC method.

Numerically, however, the locality of the CA rules makes it optimally adapted to problems that are difficult to handle in ordinary approaches, such as complicated geometrical boundaries, irregular contact profiles, large density gradients or very large spatial variations in the number of scattering events (e.g. in regions of low or high doping).

4.3.2 Cellular automata rules

The BE for the single particle distribution function $f(\boldsymbol{r}, \boldsymbol{k}, t)$ can be written as follows

$$\partial f/\partial t = \partial f/\partial t|_{\text{kin}} + \partial f/\partial t|_{\text{coll}}. \tag{4.3.1}$$

The collision term on the right-hand side contains all of the quantum-mechanical scattering processes of interest,

$$\partial f/\partial t|_{\text{coll}} = \frac{V}{(2\pi)^3} \int \mathrm{d}\boldsymbol{k}' \{\omega(\boldsymbol{r}, \boldsymbol{k}', \boldsymbol{k}) f(\boldsymbol{r}, \boldsymbol{k}', t) - \omega(\boldsymbol{r}, \boldsymbol{k}, \boldsymbol{k}') f(\boldsymbol{r}, \boldsymbol{k}, t)\},$$

where, for simplicity, nondegenerate statistics has been assumed. The kinetic term on the right-hand side of equation (4.3.1) represents the drift and diffusion terms,

$$\partial f/\partial t|_{\text{kin}} = -\dot{\boldsymbol{k}} \nabla_k f(\boldsymbol{r}, \boldsymbol{k}, t) - \dot{\boldsymbol{r}} \nabla_x f(\boldsymbol{r}, \boldsymbol{k}, t), \tag{4.3.2}$$

where the Bloch momentum $\boldsymbol{k}$ and real space position $\boldsymbol{r}$ obey the semiclassical equations of motion. Within a given bulk material, they read

$$\dot{\boldsymbol{r}} = \frac{1}{\hbar} \nabla_k \varepsilon(\boldsymbol{k}), \tag{4.3.3}$$

$$\dot{\boldsymbol{k}} = -\frac{e}{\hbar} \boldsymbol{E}(\boldsymbol{r}). \tag{4.3.4}$$

Here, e is the elementary charge, $\varepsilon(\boldsymbol{k})$ is the energy dispersion and $\boldsymbol{E}(\boldsymbol{r})$ is the electric field. The first step of the conversion of the BE into a CA consists of discretizing real space, momentum, and time. We introduce discrete lattices in real and momentum space, with cells of diameter Δr and Δk that are labelled by indices $\boldsymbol{R}$ and $\boldsymbol{K}$, respectively, and replace the spatial and momentum space derivatives by corresponding finite differences. Analogously, we choose small discrete time steps Δt. By integrating the distribution function over the cell volume of the discrete phase space cell $(\boldsymbol{R}, \boldsymbol{K})$ of volume $V(\boldsymbol{R})V(\boldsymbol{K})$,

$$\begin{aligned} N(\boldsymbol{R}, \boldsymbol{K}, t_i) &= \int_{V(\boldsymbol{R})} \mathrm{d}\boldsymbol{r} \int_{V(\boldsymbol{K})} \frac{\mathrm{d}\boldsymbol{k}}{(2\pi)^3} f(\boldsymbol{r}, \boldsymbol{k}, t_i) \\ &\approx \frac{1}{(2\pi)^3} V(\boldsymbol{R}) V(\boldsymbol{K}) f(\boldsymbol{R}, \boldsymbol{K}, t_i), \end{aligned} \tag{4.3.5}$$

one can transform the BE into the following discrete master equation for the dynamics of the cell occupancies $N(\boldsymbol{R}, \boldsymbol{K}, t_i)$, [3, 8, 24]

$$\begin{aligned} N(\boldsymbol{R}, \boldsymbol{K}, t_i + \Delta t) &= N(\boldsymbol{R}, \boldsymbol{K}, t_i) + \sum_{\boldsymbol{R}' \neq \boldsymbol{R}} [P_T(\boldsymbol{R}', \boldsymbol{R}, \boldsymbol{K}) N(\boldsymbol{R}', \boldsymbol{K}, t_i) \\ &\quad - P_T(\boldsymbol{R}, \boldsymbol{R}', \boldsymbol{K}) N(\boldsymbol{R}, \boldsymbol{K}, t_i)] \\ &\quad + \sum_{\boldsymbol{K}' \neq \boldsymbol{K}} [P_E(\boldsymbol{R}, \boldsymbol{K}', \boldsymbol{K}) N(\boldsymbol{R}, \boldsymbol{K}', t_i) \\ &\quad - P_E(\boldsymbol{R}, \boldsymbol{K}, \boldsymbol{K}') N(\boldsymbol{R}, \boldsymbol{K}, t_i)] \\ &\quad + \sum_{\boldsymbol{K}' \neq \boldsymbol{K}} [P_{\mathrm{coll}}(\boldsymbol{R}, \boldsymbol{K}', \boldsymbol{K}) N(\boldsymbol{R}, \boldsymbol{K}', t_i) \\ &\quad - P_{\mathrm{coll}}(\boldsymbol{R}, \boldsymbol{K}, \boldsymbol{K}') N(\boldsymbol{R}, \boldsymbol{K}, t_i)]. \end{aligned} \tag{4.3.6}$$

There are two different types of scattering probabilities. First, P_{coll} denotes the quantum-mechanical scattering probability, integrated over the cell volumes, and is given by

$$P_{\mathrm{coll}}(\boldsymbol{R}, \boldsymbol{K}', \boldsymbol{K}) = \frac{V_g}{V(\boldsymbol{R}) V(\boldsymbol{K}')} \int_{V(\boldsymbol{R})} \mathrm{d}\boldsymbol{r} \int_{V(\boldsymbol{K})} \mathrm{d}\boldsymbol{k} \int_{V(\boldsymbol{K}')} \frac{\mathrm{d}\boldsymbol{k}'}{(2\pi)^3} \omega(\boldsymbol{r}, \boldsymbol{k}', \boldsymbol{k}) \Delta t, \tag{4.3.7}$$

where V_g denotes the crystal volume. This expression follows from equation (4.3.2). The crucial point in the CA method consists of the conversion of the kinetic drift and diffusion terms, equation (4.3.2), into effective collision probabilities P_E and P_T, respectively, that allow one to treat the kinetic terms in exactly the same manner as the quantum-mechanical scattering. These probabilities are ensemble averages of Boolean hopping terms p_E and $p_T \in \{0, 1\}$ for individual particles that populate the phase space cells. The dynamical rule for the motion of each CA particle in $\boldsymbol{K}$-space is given by a product of two discrete Kronecker δ-functions,

$$p_E(\boldsymbol{K}, \boldsymbol{K}', t_i) = \gamma_E(\boldsymbol{R}, \boldsymbol{K}, t_i) \delta \left(\boldsymbol{K}' - \boldsymbol{K} - \Delta k \frac{\boldsymbol{E}(\boldsymbol{R}, t_i)}{E_i} \right), \tag{4.3.8}$$

where

$$\gamma_E(\boldsymbol{R}, \boldsymbol{K}, t_i) = \delta \left(\sum_{j=i-\alpha}^{i} \frac{e}{\hbar} \left| \boldsymbol{E}(\boldsymbol{R}, t_j) \right| \Delta t - \Delta k \right). \tag{4.3.9}$$

The second δ-function in equation (4.3.8) takes into account the direction of the electric field and maps it onto one of the discrete directions of the lattice. It is unity if the momenta of the cells $\boldsymbol{K}, \boldsymbol{K}'$ differ by one lattice constant Δk along the direction of the electric field and is zero otherwise. This expression assumes that the direction of the electric field $\boldsymbol{E}$ can be chosen to be parallel to one of the nearest-neighbour directions. The term γ_E, on the other hand, integrates the equation of motion in $\boldsymbol{k}$-space within a $\boldsymbol{K}$-cell. It is zero until the summed change in the wave vector has reached a lattice constant Δk. The time t_α marks the time where the particle has entered cell $\boldsymbol{K}$. At time $t = 0$, each CA particle is assigned a random wave vector within its $\boldsymbol{K}$-cell so that t_α is different for each particle. This has the effect to suppress large fluctuations in the cell occupancies that would occur if all particles in a cell hop at the same time. One can show that the ensemble average of this deterministic scattering rate indeed reproduces the field-induced drift term in the BE [3, 21, 24]. We note that the deterministic hopping rule equation (4.3.8) avoids the problem of artificial diffusion that has plagued our original version of the CA [3], for reasons that are analogous to the streaming motion example that we discussed in the previous section.

Similarly, the diffusion term in the BE, i.e. the second term on the right-hand side of equation (4.3.2), can be mapped onto the following dynamical rule for the motion of each CA particle in the real space lattice with lattice constant Δr,

$$p_T(\boldsymbol{R}, \boldsymbol{R}', \boldsymbol{K}) = \gamma_T(\boldsymbol{R}, \boldsymbol{K}, t_i)\delta\left(\boldsymbol{R}' - \boldsymbol{R} - \Delta r \frac{\boldsymbol{K}}{|\boldsymbol{K}|}\right), \tag{4.3.10}$$

where

$$\gamma_T(\boldsymbol{R}, \boldsymbol{K}, t_i) = \delta\left(\sum_{j=i-\beta}^{i} |\boldsymbol{v}(\boldsymbol{K})|\Delta t - \Delta r\right). \tag{4.3.11}$$

The index β is defined analogously to the previously introduced index α. For an individual particle, this deterministic rule leads to an error in position that never exceeds half a cell diameter. Consequently, the ensemble average reproduces the Boltzmann dynamics in the long-time limit exactly.

4.3.3 Computational details and practical implementation

In this section, we discuss the practical implementation of the CA algorithm for semiconductor device simulations [9, 21, 22, 25, 26]. The momentum space is mapped onto a three-dimensional hexagonal close-packed (HCP) lattice. In this high-symmetry structure, each cell is surrounded by 12 nearest neighbours. The kinetic terms in the CA only couple nearest neighbours. If one chooses a lattice constant of $\Delta k = 1.5 \times 10^6$ cm^{-1} and a time step of $\Delta t = 1$ fs, this nearest-neighbour coupling is adequate for electric fields up to 1 MV cm^{-1}. The total number of $\boldsymbol{K}$-cells is determined by the maximum kinetic energy that needs to be included in the simulation; this is typically 3–4 eV. For two-dimensional device simulations, the real space is mapped onto a two-dimensional hexagonal

Bravais lattice. Each cell has six nearest neighbours. The cell diameter Δr is chosen to be a fraction of the Debye length λ_{Debye} and typically amounts to $\Delta r \lesssim 1.0$ nm which also guarantees nearest-neighbour interactions in real space up to velocities of the order of 10^8 cm s^{-1}. On the other hand, the quantum-mechanical scattering processes ($P_{\mathrm{coll}}(\boldsymbol{K}, \boldsymbol{K}')$) couple not only nearest-neighbour cells. Therefore, we have grouped together $\boldsymbol{K}$-cells into sets $\boldsymbol{C}$ and defined hierarchical scattering tables. This is a very efficient procedure that has been used in other schemes before, such as in full-band MC calculations [19]. The memory requirements for these tables are moderate and lie in the range of 10–100 MB [22, 24].

In a spatially inhomogeneous situation, the electric field that enters the BE must be calculated self-consistently with the Poisson equation. Since the statistical fluctuations in the particle density within one cell can be appreciable, it is both physically justified and numerically advantageous to lump together several real space cells into entities of dimension λ_{Debye} and calculate the electric field via the Poisson equation on this macroscale only, assuming a constant value of the field within each macrocell. This greatly facilitates the storage of field- or density-dependent scattering tables for quantum-mechanical scattering processes as well.

The full CA algorithm in the present particle representation [21, 25] can be summarized as follows.

- Choose a suitable phase space and time discretization Δr, Δk, Δt.
- Choose a sufficiently large number of CA particles, and determine geometry-dependent parameters such as nearest-neighbour tables.
- Choose macrocells for the field and other observables such as density and current.
- Generate hierarchical tables for the quantum-mechanical scattering rates.
- Initialize the (thermal) particle distribution on the $\boldsymbol{K}$-cells. In real space, occupy the $\boldsymbol{R}$-cells according to the impurity doping. Assign random intracell counters to each particle in $\boldsymbol{R}$- and $\boldsymbol{K}$-space.
- Carry out quantum-mechanical scattering between clusters of cells. The scattering probabilites $P(\boldsymbol{K}, \boldsymbol{K}')$ are to be averaged over the initial cells and integrated over the final cells. Pick out the final state $\boldsymbol{K}' \in \boldsymbol{C}'$ randomly.
- Add $eE(t_i)\hbar\Delta t/\Delta k$ to the $\boldsymbol{K}$-cell counter and determine final $\boldsymbol{K}$-cell according to nearest-neighbour tables in $\boldsymbol{K}$-space.
- Add $v\Delta t/\Delta r$ to $\boldsymbol{R}$-cell counter and determine final $\boldsymbol{R}$-site according to nearest-neighbour tables in $\boldsymbol{R}$-space.

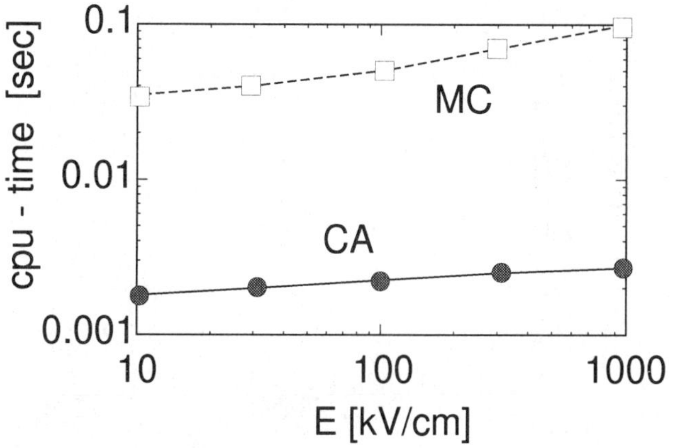

Figure 4.3 *Comparison of computer time, in seconds, required for 1 MC and 1 CA time step for the field-dependent carrier dynamics in bulk silicon that is represented by an ensemble of* 10^4 *particles, respectively. The electric field is in units of kV cm*$^{-1}$. *The computations were carried out on 1 vector CPU of a Cray YMP8/8-128 1004/445*

4.3.4 Speed of cellular automaton versus Monte Carlo

The numerical efficiency of our present CA implementation has been tested on several computer architectures and compared with MC codes of similar complexity.

As a first example, we study the dynamics of 10^4 pseudoparticles in bulk *n*-silicon at room temperature with both methods. In this case, the Poisson equation does not enter. The resulting computer time per time step on one vector processor of a Cray YMP8/8-128 are shown in Fig. 4.3 as a function of field strength. Starting with a speed-up of 25 at low electric field strengths, the CA method performs up to 40 times faster than the MC method at high fields. This increase in speed-up with increasing electric field can be attributed to the increasing number of scattering events, whose realization is less time consuming in the CA method due to the possibility of precalculated scattering tables in momentum space. The speed-up at low fields, on the other hand, shows the numerical efficiency of the deterministic transition rules for the free flight. The high numerical efficiency of the CA method on single processor machines is further supported by its ideal parallel performance. On the parallel architectures of a Cray YMP [17] as well as of a PowerPC 'Parsytec X'plorer' [27], one finds a nearly linear scaling of computational time with the number of processors used (Fig. 4.4). As discussed in [10], the reduced data exchange between processors required for the CA compared with MC allows one to maintain the CA/MC speed-up for the particle dynamics in device simulations, utilizing a geometrical decomposition of device geometries.

The second example involves a realistic device. We compare self-consistent

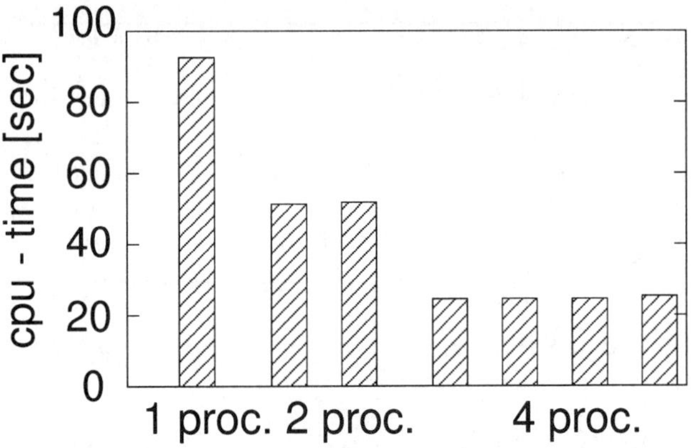

Figure 4.4 *Computer time, in seconds, required for 1000 CA time steps for the carrier dynamics in bulk silicon at a field strength of 10 kV cm*$^{-1}$ *with* 10^4 *particles on a parallel processor PowerPC Parsytec X'plorer computer. The time per processor is shown when one, two and four nodes are used, respectively.*

MC and CA calculations for a 200 nm gate planar MOSFET. Both calculations are based on the same quantum-mechanical scattering mechanisms and rates and the Poisson equation is solved iteratively with the same efficient multigrid algorithm [28]. In this example, the real space lattice constant is given by $\Delta r = 1$ nm. In addition, 400×200 $\boldsymbol{R}$-cells are used to span the whole relevant two-dimensional area of the device. The Poisson equation is solved after each CA time step of 1 fs, but on a coarser grid of 200×100 cells, with four CA cells forming one macrocell. The lattice constant in $\boldsymbol{K}$-space was $\Delta k = 1.5 \times 10^6$ cm^{-1} and 10^5 $\boldsymbol{K}$-cells are used to account for all scattering mechanisms. On an HP 9000/735 workstation, the computer time per iteration for an ensemble of 500 000 particles amounts to 1.2 and 6.9 s for the CA and MC simulations, respectively. This time includes the 0.5 s that are spent solving the Poisson equation. Thus, a complete CA calculation of one point of the I–V curve takes less than 1 h on a moderate workstation.

We would like to point out that one can easily think of the CA method as a *variant* of the MC method since both methods solve the same set of equations. Even more, most of the ideas and concepts of the CA that we sketched above can be used to augment or replace parts of the standard MC algorithms [29] by more efficient ones as demonstrated in [4, 5]. Thus, by optimally adapting the ensemble MC method to modern computers and focusing on the physically most relevant properties of carrier dynamics in devices, one can achieve fast yet microscopic and reliable device simulations at costs that are comparable with commercial hydrodynamic solvers.

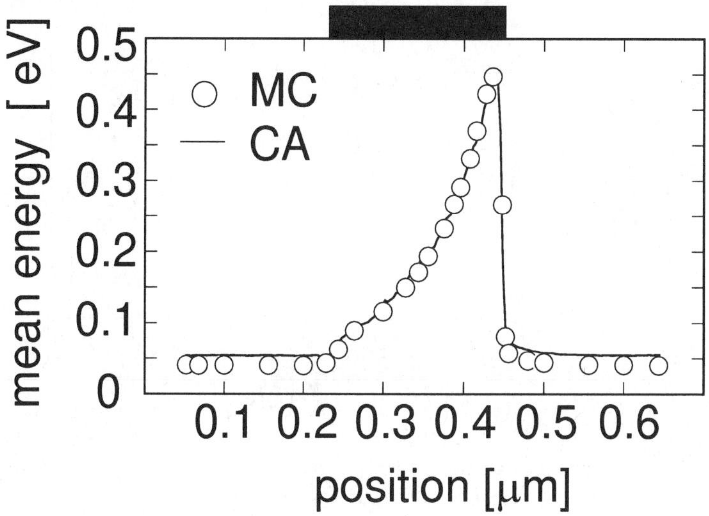

Figure 4.5 *Mean kinetic energy of charge carriers, in eV, as a function of position in μm along the channel in a 200 nm gate length planar MOSFET, calculated by the ensemble MC and CA methods.*

4.4 VALIDATION AND COMPARISON WITH MONTE CARLO RESULTS

In this section we wish to demonstrate the reliability and accuracy of the CA method and compare the results obtained by either the CA method or standard ensemble MC calculations.

4.4.1 Planar Si MOSFET

In this example, we study a Si MOSFET with a gate length of 200 nm, an oxide thickness of 20 nm and a weakly doped p-buffer ($p = 2 \times 10^{17}$ cm^{-3}). Starting from some initial distribution for the carriers, we have carried out CA simulations with 10^5 particles for 10 000 time steps (10 ps) in order to guarantee a well converged and stationary carrier distribution. In Fig. 4.5, we show the mean kinetic energy of the electrons in the channel region for a gate and drain voltage of 2 V [20].

The agreement with the MC results is excellent. In the low-field regions, the CA calculations slightly overestimate the temperature of the carriers. This is a consequence of the crude k-space resolution at low kinetic energies in the CA that prevents a more precise thermalization of the carriers.

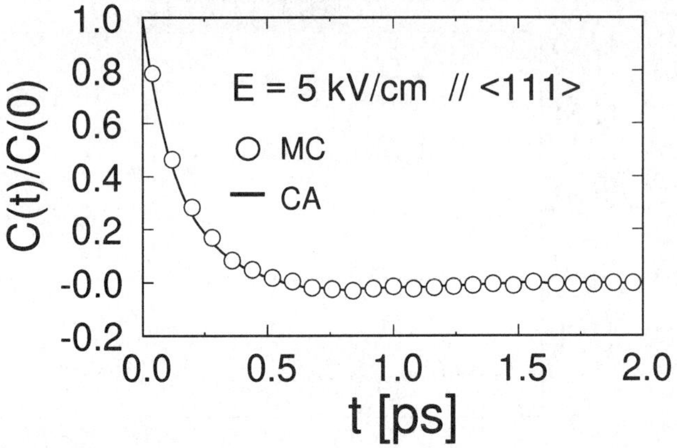

Figure 4.6 *Velocity autocorrelation function for electrons in* $\langle 111 \rangle$ *Si at a field strength of 5 kV cm*$^{-1}$*, calculated by the ensemble MC and CA methods.*

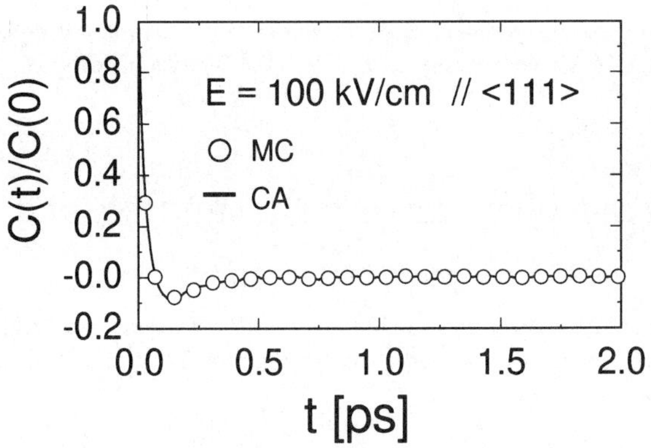

Figure 4.7 *Velocity autocorelation function for electrons in* $\langle 111 \rangle$ *Si at a field strength of 100 kV cm*$^{-1}$*, calculated by the ensemble MC and CA methods.*

4.4.2 Autocorrelation functions

Since statistical methods such as the ensemble MC or CA method yield fluctuating physical observables, fairly large particle ensembles and/or integrations over time are needed to obtain well-converged ensemble averages. The calculation of products of fluctuating quantities such as velocity or density correlation functions is an extra difficult step and allows a particularly critical assessment of a particle simulation technique.

The autocorrelation function of a stochastic variable $A(t) = \delta A(t) + \langle A \rangle$ is

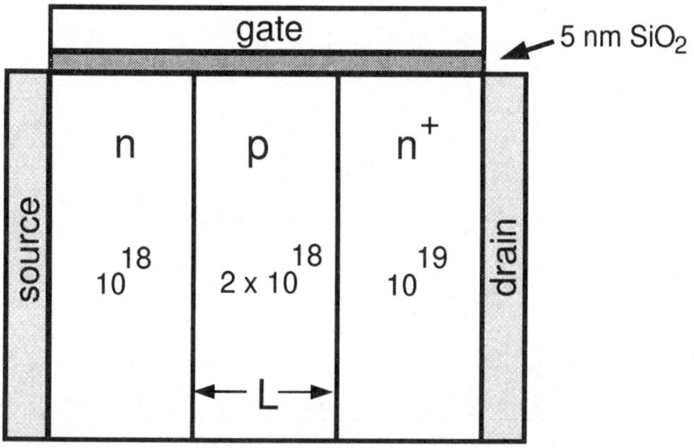

Figure 4.8 *Geometry of vertically grown NMOS transistor. The height of the p-layer L can be controlled on a nm scale by the growth process. The doping concentrations shown are in units of* cm^{-3}. *The thickness of the oxide layer between the doped regions and the gate contact is 5 nm.*

defined by

$$C_A(t) = \langle \delta A(t) \delta A(0) \rangle \tag{4.4.1}$$

where the average is taken over time or particle ensemble [30]. We have calculated the correlation function of the longitudinal drift velocity $C_v(t)$ in bulk Si at room temperature for various electric fields along the $\langle 111 \rangle$ direction [26]. In Figs. 4.6 and 4.7, the normalized velocity autocorrelation function is depicted as a function of time, as obtained by the CA and MC methods, respectively. For small fields, $C_v(t)$ depends exponentially on time, as is to be expected near thermal equilibrium. In the high-field regime, however, the momentum relaxation time of the hot carriers becomes significantly shorter than the energy relaxation time. The latter gives the asymptotic decay of the correlation function while the former leads to a slight dip below zero at short times due to enhanced backscattering. As can be seen from Fig. 4.7, the CA method fully captures this hot carrier effect and agrees excellently with standard MC simulations [26].

4.5 COMPARISON WITH EXPERIMENT

4.5.1 Ultrashort vertical MOSFETs

We have employed the CA method to study a large variety of devices and have studied their characteristics [18, 20, 21, 26, 27]. Here we present results for vertically grown MOSFETs with ultrashort channels that demonstrate the method's capability to predict high-field transport in modern submicron devices in a numerically efficient and accurate way.

The geometry of a vertical MOSFET device is shown in Fig. 4.8 [31]. It

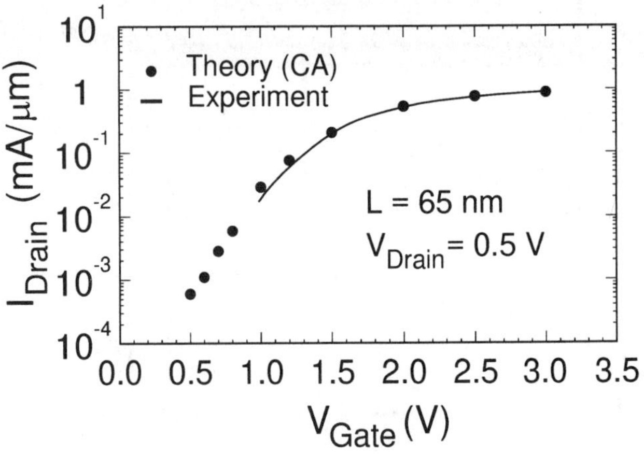

Figure 4.9 *Experimental and calculated sub-threshold behaviour of the drain current, in mA* μm^{-1}*, as a function of the gate voltage, in V, for the vertical n-MOSFET with gate length* $L = 65$ *nm. The theory is based on the present CA technique. It includes all standard quantum mechanical scattering processes and contains no parameters that are adjusted to fit the data. The interface roughness scattering at the oxide interface is set to zero to show its negligible influence on the results.*

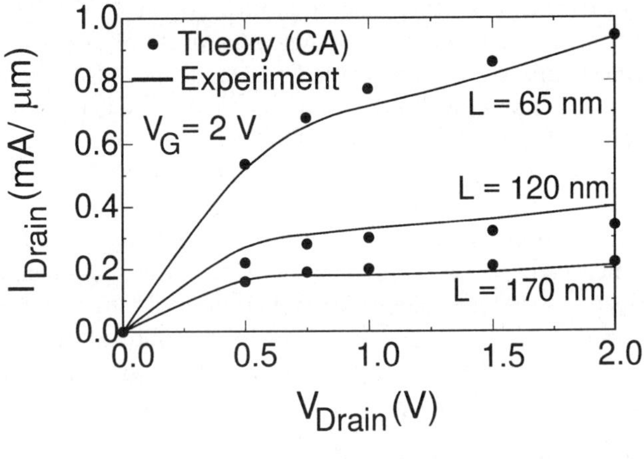

Figure 4.10 *Experimental and calculated drain current, in mA* μm^{-1}*, as a function of the drain voltage, in V, for the vertical n-MOSFET as a function of the gate length. The theory is based on the present CA technique.*

consists of a p-layer, sandwiched between two n-doped layers that form the source and drain dopings. The active channel length is determined by the width of the p-layer which is controlled by the epitaxial CVD process that is used to

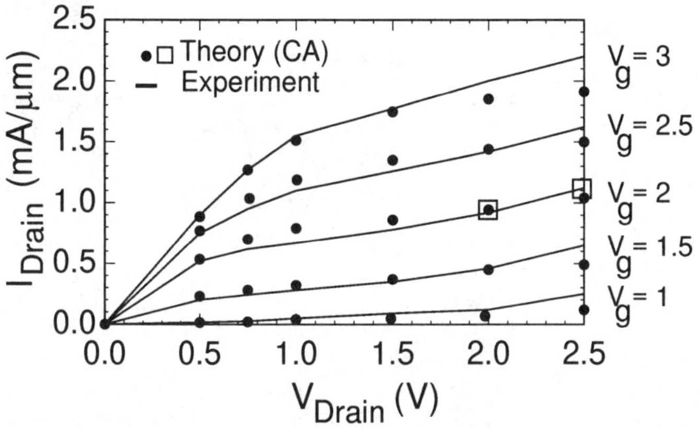

Figure 4.11 *Experimental and calculated drain current, in mA μm^{-1}, as a function of the drain voltage, in V, for the vertical NMOSFET with gate length L = 65 nm. The drain characteristics are shown for several gate voltages V_g that is given in V. The theory is based on the present CA technique. The calculated results shown as dots contain no impact ionization, whereas this effect is fully taken into account in the results shown as open squares.*

grow these structures. The gate contact is separated by 5 nm of SiO_2, which is thermally grown after a selective etch process. The sub-threshold behaviour of the drain current depends sensitively on these geometrical settings (i.e. *p*-doping profile and oxide thickness). The excellent agreement between theoretical [27] and experimental [31] data, that is exemplified in Fig. 4.9, indicates that the actual and nominal geometry indeed agree well with one another. Figure 4.10 shows the experimental and calculated drain current characteristics for three different gate lengths down to 65 nm. There is very good agreement with experimental data, despite the fact that we purposely ignored surface roughness scattering at the Si/SiO_2 interface. This indicates a high interface quality that is comparable with standard planar MOSFET technology. Pronounced short-channel effects cause a strong increase of the drain output conductance with decreasing channel length. Besides electrostatic effects, velocity overshoot enhances the drain current as well as the output conductance by reducing the gate length. For a channel length of 65 nm, this results in an average channel velocity of more than twice the saturation velocity of silicon. Figure 4.11 shows the complete I–V characteristics of the 65 nm structure. Again, impressive overall agreement with experimental data is achieved. At the highest applied drain voltages, impact ionization causes a small systematic increase of the drain current. To single out this effect, we have performed simulations with and without impact ionization. The drain currents that are calculated without this scattering mechanism (dots in Fig. 4.11) are systematically lower than the experimental

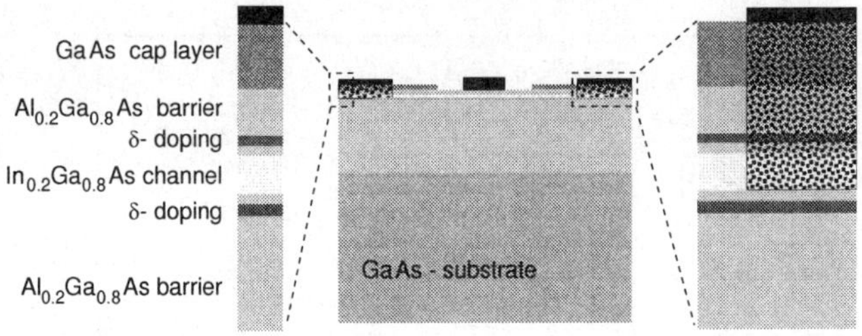

Figure 4.12 *Schematic drawing of the layer structure and doping profile of a double delta-doped* $In_{0.2}Ga_{0.8}As/Al_{0.2}Ga_{0.8}As$ *pseudomorphic HEMT with a gate length of* 0.24 μm *and a channel width of 12 nm. Nominally, the contact metallization stops at the n-doped GaAs cap layer. However, we assume the thermal treatment to induce a low-effective n-doping concentration of the order of* 10^{18} cm^{-3} *beneath the source and drain contacts nevertheless that is indicated schematically on the right-hand side by a dotted shading.*

data at the highest drain voltage. Taking into account impact ionization (shown for one gate voltage) restores agreement with experiment.

4.5.2 Contact models in III–V high-electron mobility transistors

Contact resistances play an important role in modern devices. Yet, most device simulations assume source and drain contacts to be ideal ohmic contacts and ignore the detailed dynamics of carriers close to the contact regions [32]. In planar high-electron mobility transistors (HEMT), there is a long-standing controversy about contact resistances [33–35]. In pseudomorphic InGaAs HEMTs with AlGaAs barriers, the carriers must cross energetically very high barriers to reach the channel beneath the source contact and to leave it again near the drain contact (Fig. 4.12). Often, it has been assumed that the carriers somehow tunnel through the barriers into the channel but experimental data [36] indicate a very weak rather than exponential dependence of the drain current on the barrier thickness.

Technologically, the contacts on planar HEMTs are often fabricated by rapid thermal alloying of various metal layers [37]. Our main conjecture that will be substantiated below is as follows. These thermal processes lead to a moderate level of effective doping of less than 10^{19} cm^{-3} beneath the source and drain contacts. This assumption, combined with detailed CA simulations, is able to explain the experimental data, at least for the concrete HEMTs that we have analysed. Applying the CA method to this problem demonstrates the ability of this microscopic approach to help clarify a highly complex transport problem.

We have studied pseudomorphic $Al_{0.2}Ga_{0.8}As/In_{0.2}Ga_{0.8}As$:GaAs HEMT's with 0.24 μm gate lengths that have been fabricated and characterized at Siemens AG [38]. The layer sequence is depicted in Fig. 4.12. We have calculated the I–V characteristics of this device with the CA method, by assuming a homo-

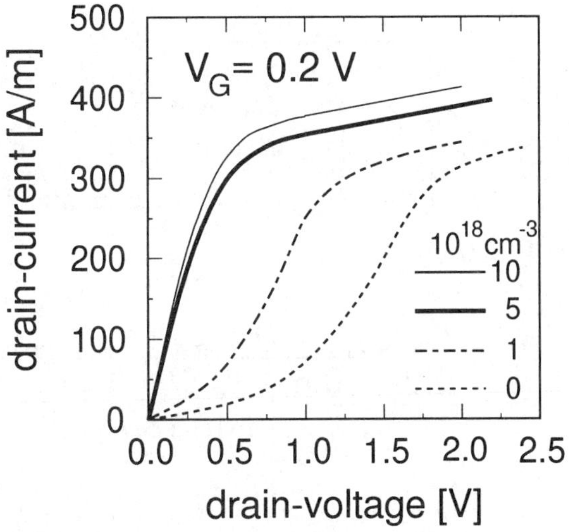

Figure 4.13 *Calculated influence of residual doping concentration beneath the contacts on drain characteristics in the HEMT of Fig. 4.12. The drain current is in A* m^{-1}, *the drain voltage in V, and the gate voltage is set to 0.2 V.*

geneous effective n-doping concentration beneath the contacts of 0, 1, 5, and 10×10^{18} cm^{-3}, respectively. This doping region is indicated in Fig. 4.12. Importantly, we have fully taken into account the degeneracy of the carrier distribution in the channel by incorporating the Pauli principle into the scattering rules.

Figure 4.13 shows the drain current characteristics obtained for the various contact doping levels. In the absence of such doping, the calculations give an activated current for low bias that directly reflects the barrier heights. This leads to a diode rather than a transistor characteristics in the I–V curve, as can be seen in Fig. 4.13. Since the cap layer is highly doped, its Fermi energy lies 80 meV within the conduction band. This lowers the GaAs:AlGaAs barrier ($\sim$180 meV) to a net amount of 100 meV that the carriers have to cross from the source contact into the channel. However, the barrier from the InGaAs channel across the AlGaAs barrier is 340 meV and therefore much higher.

Nevertheless, the calculations already yield acceptable transistor characteristics for very moderate doping concentrations below the contact regions, as indicated in Fig. 4.13. The reason lies in the Pauli principle and can be understood by studying the charge density and average kinetic energy in the InGaAs channel (Figs. 4.14 and 4.15). With increasing doping concentration, the carrier density in the channel beneath the contact regions increases from 2.4×10^{18} to more than 10^{19} cm^{-3}. Consequently, the Fermi level rises and increases the kinetic energy of the carriers by almost 100 meV, as shown in Fig. 4.15, which

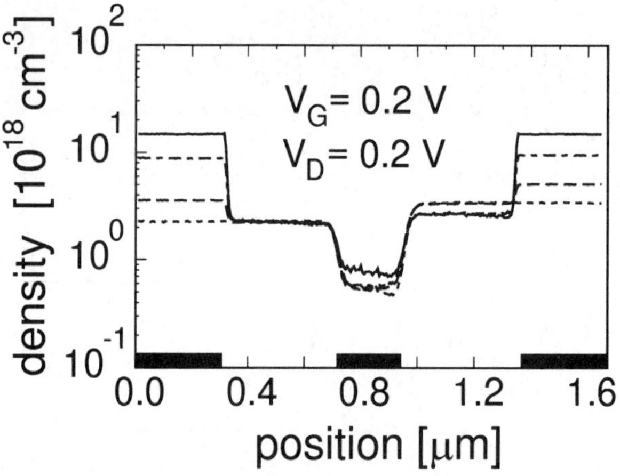

Figure 4.14 *Calculated electron density, in units of* 10^{18} cm^{-3}*, along the* $In_{0.2}Ga_{0.8}As$ *channel near the interface to the upper supply layer in the HEMT of Fig. 4.12 as a function of the residual doping concentration beneath the contacts. The position along the channel is given in μm, gate and drain voltage are 0.2 V.*

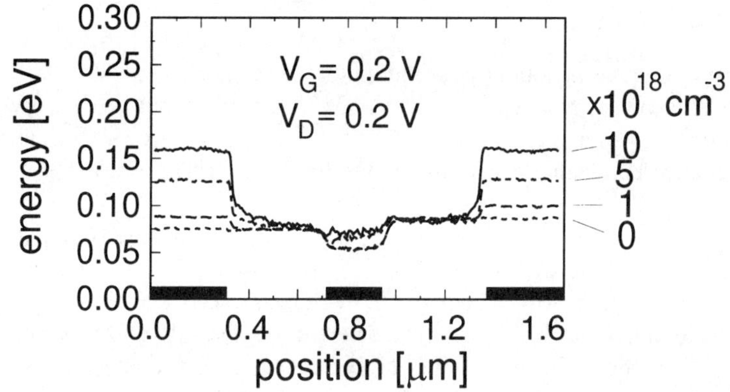

Figure 4.15 *Channel electron energy, in eV, along the* $In_{0.2}Ga_{0.8}As$ *channel near the interface to the upper supply layer in the HEMT of Fig. 4.12 as a function of the residual doping concentration beneath the contacts. The position along the channel is given in μm, both gate and drain voltage are equal to 0.2 V.*

suffices to produce a typical transistor I–V characteristics for reasonably high drain voltages.

In Fig. 4.16 we compare the experimental transfer characteristics with the calculated one. If we assume a doping concentration of 5×10^{18} cm^{-3} beneath the contacts, we obtain excellent agreement with the experimental data.

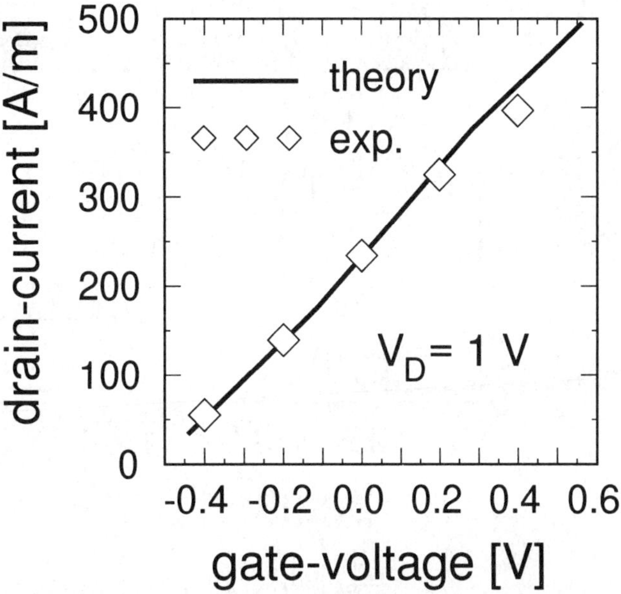

Figure 4.16 *Experimental and calculated transfer characteristics of the HEMT of Fig. 4.12. The drain current is in A* m^{-1}, *the gate voltage is in V and the drain voltage is 1 V. The theory assumes a residual doping concentration beneath the contacts of* 5×10^{18} cm^{-3}.

The average velocity of the carriers in the channel is shown in Fig. 4.17 for drain and gate voltages of $V_D = 1$ V, $V_G = 0.2$ V. There is a very pronounced velocity overshoot that exceeds the saturation velocity by a factor of 4. This demonstrates that the carrier transport in this submicron device can only be understood by a full solution of the BE that incorporates quantum-mechanical effects such as the exclusion principle and takes into account the interplay between hot carriers and nonlocal transport.

4.6 SUMMARY

In this chapter, we have presented an overview of a novel concept for solving dynamical equations in physics, which is the concept of CA. The basic idea is to map a complicated real-world dynamics onto a fictitious microworld dynamics that is governed by many but individually very simple Boolean rules that are optimally adapted to modern computer hardware. We have shown that this method is ideally suited for high-field transport simulations in modern nanometer-scale devices. Since the method solves the full BE, it is equally accurate and suitable to predict nonlocal hot-carrier effects as standard implementations of the ensemble MC method. However, the computational effort involved is grossly

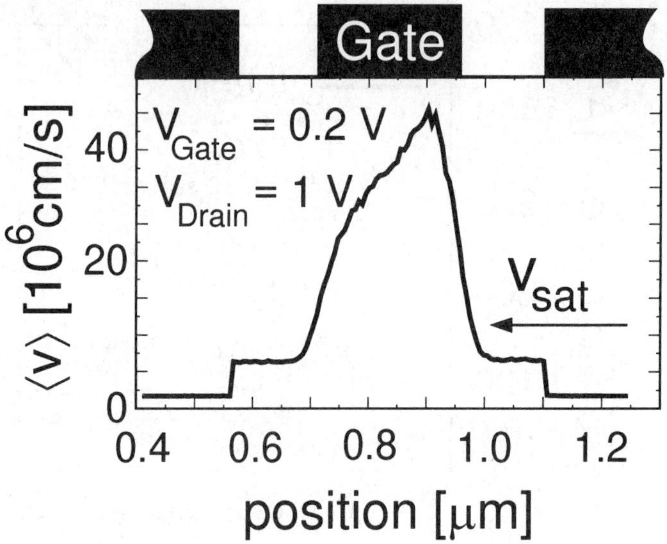

Figure 4.17 *Calculated average electron velocity, in units of* 10^6 *cm* s^{-1}*, along the channel of the HEMT of Fig. 4.12, for a source-drain voltage of 1 V, and a gate voltage of 0.2 V. The average is performed vertically to the channel plane and includes the total device in order to capture all carriers. The saturation velocity is indicated by an arrow.*

reduced and comparable with standard device simulation tools that invoke much simplified, empirical transport models. This remarkable computational efficiency of the CA method is based on four concepts that have been discussed in the previous sections.

- Optimal size discretization of phase space.
- Reduction of particle motion to spatially local, deterministic dynamical rules.
- Usage of precalculated, hierarchical scattering tables.
- Sole usage of integer arithmetics allows us to restrict the range of the dynamical variables.

ACKNOWLEDGEMENTS

Financial support from the Deutsche Forschungsgemeinschaft (SFB 348) and Siemens AG (Sonderforschungseinheit Mikrostrukturierte Bauelemente) is gratefully acknowledged.

REFERENCES

[1] For recent reviews, see Doolen, G. D. (ed.) (1990) Lattice gas methods for PDE's: Theory, application, and hardware *Physica* D **45**, p. 1 ff.

[2] See articles in Manneville, P., Boccara, N., Vichniac, G. Y. *et al.* (eds) (1989) *Cellular Automata and Modeling of Complex Physical Systems*, Springer, Berlin.
[3] Kometer, K., Zandler, G. and Vogl, P. (1992) *Phys. Rev.* B **46**, 1238.
[4] Fukuda, K. and Nishi, K. (1996) *Simulation of Semiconductor Devices and Processes* (eds H. Ryssel and P. Pichler), Springer Verlag, Wien, Vol. 6, p. 90.
[5] Liebig, D. (1996) *Simulation of Semiconductor Devices and Processes* (eds H. Ryssel and P. Pichler) Springer Verlag, Wien, Vol. 6, p. 74.
[6] Frisch, U., Hasslacher, B. and Pomeau, Y. (1986) *Phys. Rev. Lett.* **56**, 1505.
[7] Frisch, U., d'Humières, D., Hasslacher, B. *et al.* (1987) *Complex Systems* **1**, 649.
[8] Zandler, G., Di Carlo, A. and Kometer, K. *et al.* (1993) *IEEE Electron Dev. Lett.* **14**, 77.
[9] Rein, A., Zandler, G., Saraniti, M. *et al.* (1994) *Proceedings of the 3rd International Workshop on Computational Electronics* (ed. S. M. Goodnick), Oregon State University Press, Corvallis, p. 7.
[10] Zandler, G., Rein, A., Saraniti, M. *et al.* (1993) *Proceedings of the 23rd European Solid State Device Research Conference* (eds J. Borel, P. Gentil, J. P. Noblanc *et al.*), Edition Frontières, 91192 Gif-sur-Yvette Cedex-France, p. 21.
[11] von Neumann, J. (1966) *Theory of Self-reproducing Automata*, University of Illinois Press, Chicago.
[12] Chen, S., Diemer, K., Doolen, G. D. *et al.* (1991) *Physica* D **47**, 72.
[13] Hardy, J. and Pomeau,Y. (1972) *J. Math. Phys.* **13**, 1042.
[14] Hardy, J., Pomeau, Y. and de Pazzis, O (1973) *J. Math. Phys.* **14**, 1746.
[15] Hardy, J,. de Pazzis, O. and Pomeau, Y. (1976) *Phys. Rev.* A **13**, 1949.
[16] Toffoli, T. and Margolus, N. (1989) *Cellular Automata Machines*, 4th edn, MIT Press, Cambridge.
[17] Kometer, K., Zandler, G. and Vogl, P. (1992) *Semicond. Sci. Technol.* **7**, B559.
[18] Kometer, K., Zandler, G. and Vogl, P. *et al.* (1992) *Proceedings of the 1st International Workshop on Computational Electronics* (eds R. W. Dutton, K. Hess and U. Ravaioli), University of Illinois Press, Urbana, p. 25.
[19] Bude, J. and Smith, R. K. (1994) *Semicond. Sci. Technol.* **9**, 840.
[20] Rein, A., Zandler, G. and Lugli, P. (1993) *Proceedings of the 2nd International Workshop on Computational Electronics* (ed. C. Snowden), University of Leeds Press, Leeds, p. 121.
[21] Rein, A., Zandler, G., Saraniti, M. *et al.* (1994) *Proceedings of the 1994 IEEE International Electron Devices Meeting* (Electron Device Society of IEEE), p. 351.
[22] Rein, A. (1995) *PhD Thesis* TU München, Zelluläre Automaten in der Transporttheorie: Konzepte und Anwendungen, unpublished.
[23] Jacoboni, C. and Lugli, P. (1989) *The Monte Carlo Method for Semiconductor Device Simulation*, Springer-Verlag, Wien.
[24] Rein, A., Zandler, G. and Vogl, P. *Phys. Rev.* to be published.
[25] Rein, A., Zandler, G., Saraniti, M. *et al.* (1994) *Proceedings of the 24th European Solid State Device Research Conference* (eds C. Hill and P. Ashburn) Edition Frontières, 91192 Gif-sur-Yvette Cedex-France, p. 775.
[26] Rein, A., Zandler, G., Saraniti, M. *et al.* (1996) *Hot Carriers in Semiconductors IX*, (eds K. Hess, J. P. Leburton and U. Ravaioli), Plenum, New York, p. 497.
[27] Zandler, G., Saraniti, M., Rein, A. *et al.* (1996) *Proceedings of 1996 Int. Conf. on Simulation of Semiconductor Processes and Devices*, Business Center for Academic

Societies Japan, Tokyo, 1996, ISBN 0-7803-2745-4, p. 39.

[28] Saraniti, M., Rein, A., Zandler, G. *et al.* (1996) *IEEE Trans. on Computer-aided Design of Integrated Circuits and Systems* **15**, 141.

[29] See articles in Hess, K. (ed.) (1991) *Monte Carlo Device Simulation: Full Band and Beyond*, Kluwer Academic Publishers Group, Boston and Dordrecht.

[30] Varani, L. and Reggiani, L. (1994), *La Rivista del Nuovo Cimento* **17**, 1.

[31] Risch, L., Krautschneider, W., Hofmann F. *et al.* (1996) *IEEE Trans. of Electron Dev.* **43**, 1495–8 and references therein.

[32] Babiker, S., Asenov, A., Cameron, N. *et al.* (1996) *IEEE Trans. Electron Dev.* **43**, 2032.

[33] Hawksworth, S. J., Chamberlain, J. M., Cheng, T. S. *et al* (1992) *Semicond. Sci. Technol.* **7**, 1085.

[34] Look, D. C. (1988) *J. Electrochem. Soc. Solid State Science and Technology* **135**, 2054.

[35] Dingfen, W., Dening, W. and Heim, K. (1986) *Solid State Electron.* **29**, 489.

[36] Grave, T. and Kellner, W., 1996, private communication.

[37] Plauth, J., Kempter, R., Grigull, S. *et al.* (1994) Gallium Arsenide and Related Compound 1993 (eds H. S. Rupprecht and G. Weimann) *Inst. of Physics Conference Series* IOP Publishing Ltd., Bristol and Philadelphia **136**, 53.

[38] Brech, H., Simlinger, T., Grave, T. *et al.* (1996) *26th European Solid State Device Research Conference* (eds G. Baccarani and M. Rudan), Edition Frontières, 91192 Gif-sur-Yvette Cedex-France, p. 873.

CHAPTER 5

Quantum transport theory

A. P. Jauho

Mikroelektronik Centret, Technical University of Denmark, bldg. 345 east, DK-2800 Lyngby, Denmark

5.1 INTRODUCTION

Quantum transport theory applied to mesoscopic structures is one of the most active fields within theoretical solid state physics. As the sample size diminishes, screening may become less effective and consequently the interactions between the system's constituents increase in importance and it is necessary to provide a good description of the various interaction mechanisms. A point in case are one-dimensional systems (quantum wires): there electron–electron interactions may lead to a behaviour, which is qualitatively different from the standard Fermi liquid picture (Luttinger liquids). Further, the system is often far from equilibrium and it is thus necessary to develop a quantum many-body theory valid for nonequilibrium situations. Many different theoretical approaches have been developed for this purpose and a number of them are reviewed in other chapters of this book. In this chapter, two approaches will be outlined: the Kubo formula and the nonequilibrium Green's function technique. We shall illustrate these two methods in terms of two physical examples: in the case of the Kubo formula we introduce and analyse in detail the phenomenon of Coulomb drag, while the nonequilibrium Green's function technique will be applied to the case where a small, possibly strongly interacting mesoscopic region is coupled to noninteracting leads. The study of these two examples is motivated by the steady flow of new experimental data which challenges our ability to understand and interpret these measurements.

5.2 COULOMB DRAG

5.2.1 Basic phenomenology

Consider two systems containing mobile charge carriers so close to each other that the charges in the two respective subsystems experience the Coulomb forces originating from the other subsystem and yet far enough away from each other

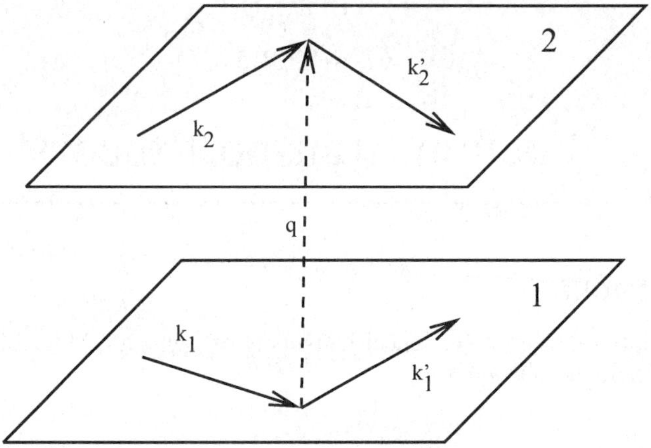

Figure 5.1 *Coulomb drag geometry. A carrier in the drive layer ('1') scatters from state* $\boldsymbol{k}_1$ *to a state* $\boldsymbol{k}'_1 \equiv \boldsymbol{k}_1 - \boldsymbol{q}$ *transferring momentum* $\boldsymbol{q}$ *to the probe layer ('2'), where a carrier scatters from state* $\boldsymbol{k}_2$ *to* $\boldsymbol{k}'_2 \equiv \boldsymbol{k}_2 + \boldsymbol{q}$. *In general, the scattering probability may also depend on the transferred energy,* $\Delta E = \epsilon(\boldsymbol{k}'_1) - \epsilon(\boldsymbol{k}_1)$.

that direct charge transfer between the two subsystems is not possible. Examples of such systems are, for example Coulomb coupled double-quantum-well systems [1, 2], arrangements where a three-dimensional system is close to a two-dimensional system [3], or two nearby quantum wires (to our knowledge, to date there are no measurements for this extremely interesting geometry). A scattering event between a carrier in one system and a carrier in the other system leads to momentum transfer between the two subsystems (Fig. 5.1).

Thus, if a current is driven through one of the systems (henceforth the driven system is denoted by '1'), then an induced current is dragged in the other subsystem (system '2'). Alternatively, if no current is allowed to flow in layer 2, a voltage is induced. Due to momentum conservation the two-particle number currents flow in the same direction. Since the mechanism for the Coulomb drag is due to carrier–carrier scattering, the drag current is proportional to the square of the effective interaction between the subsystems. The available phase space for electron–electron scattering tends to zero at low temperatures and consequently one would expect Coulomb drag to decrease with decreasing temperature. This expectation follows from the Fermi golden rule expression for the electron–electron scattering rate, which can qualitatively be written as

$$\frac{1}{\tau(\boldsymbol{k}_1)} \propto \sum_{\boldsymbol{k}'_1\boldsymbol{k}_2\boldsymbol{k}'_2} |W|^2 f_1^0 f_2^0 (1 - f_{1'}^0)(1 - f_{2'}^0)\delta(\epsilon_1 + \epsilon_2 - \epsilon_{1'} - \epsilon_{2'})\delta_{\boldsymbol{k}_1+\boldsymbol{k}_2,\boldsymbol{k}'_1+\boldsymbol{k}'_2}. \tag{5.2.1}$$

Here $f_i^0 \equiv f^0(\epsilon(\boldsymbol{k}_i))$ is the equilibrium Fermi–Dirac distribution in subsystem

i and $W(\boldsymbol{q}, \Delta E)$ is the (screened) Coulomb matrix element corresponding to momentum transfer $\boldsymbol{q}$ and energy transfer $\Delta E = \epsilon(\boldsymbol{k}_1) - \epsilon(\boldsymbol{k}_1')$. At low temperatures and in three spatial dimensions, the two Pauli factors $(1 - f_i^0)$ lead to a T^2-dependence. The situation is slightly different in two spatial dimensions, where one finds a $T^2 \log T$-behaviour†. Naturally, equation (5.2.1) can only give a qualitative estimate of the actual drag rate, to be computed later. Nevertheless, the T^2-behaviour is approximately seen in experiments [1]. Note, however, that there are small, but important deviations from the simple T^2-law; these deviations have been the topic of much recent interest [1, 5] and we shall comment on them below.

The possibility for Coulomb drag was realized long ago [6, 7] and the recent experimental advances [1–3] have brought about a flurry of theoretical works. A number of different theoretical approaches have been proposed. These include: 1. calculations based on the Boltzmann equation [1, 8]; 2. the memory function approach of [9]; 3. the momentum balance equation method [5]; 4. diagrammatic linear response (Kubo formula) approaches [10, 11]. In the present review we shall discuss two of these approaches: the semiclassical Boltzmann equation approach [1, 8] and the Kubo formula [10]. The Boltzmann approach has the advantage of being quite transparent, while the Kubo approach is needed if one wishes to address effects of quantum mechanical origin, such as weak localization or quantizing magnetic fields [12]. What makes these theories particularly interesting is their predictive power: they have suggested some surprising properties of the drag rate as a function of temperature or magnetic field; later experiments have confirmed certain aspects of these predictions but left open some others thus providing an incentive for further study.

5.2.2 Boltzmann equation description

The momentum transfer from the drive system to the drag system can be described by a drag rate $1/\tau_D$, which determines the experimentally measurable drag resistance (or transresistance) via $\rho_{21} = m/(n_1 e^2 \tau_D)$. We calculate $1/\tau_D$ by setting up two coupled Boltzmann equations for systems 1 and 2, respectively and use them to derive a balance equation between the induced electric field and the drag due to the drive current [8]. In the presence of an electric field the linearized Boltzmann equations read

$$\begin{aligned} -(e/\hbar)\boldsymbol{E}_1 \cdot \frac{\partial f_1^0}{\partial \boldsymbol{k}} &= \left[\frac{\partial f_1}{\partial t}\right]_{\text{coll}} \\ -(e/\hbar)\boldsymbol{E}_2 \cdot \frac{\partial f_2^0}{\partial \boldsymbol{k}} &= \left[\frac{\partial f_2}{\partial t}\right]_{\text{coll}} . \end{aligned} \tag{5.2.2}$$

† This interesting result, which can be viewed as a precursor of the breakdown of normal Fermi liquid theory in reduced dimensions, has been discovered several times [4].

We assume that the current flow in the drive system is limited by impurity scattering with a characteristic relaxation time τ_1. One then finds $f_1 = f_1^0 + f_1^0(1 - f_1^0)\psi_1$ with $\psi_1 = -\tau_1 e v_{1x} E_1$, where the driving electric field points in the x-direction. On the other hand, system 2 is coupled to a voltmeter and hence no current flows there implying $\psi_2 = 0$. This means that the linearized Coulomb collision term of equation (5.2.2) becomes

$$\left[\frac{\partial f_2}{\partial t}\right]_{\text{coll}} = -\sum_{\sigma_1,\sigma_{2'},\sigma_{1'}} \int \frac{d\boldsymbol{k}_1}{(2\pi)^2} \int \frac{d\boldsymbol{k}_{2'}}{(2\pi)^2} |W|^2(\psi_1 - \psi_{1'}) \times f_1^0 f_2^0 (1 - f_{1'}^0)(1 - f_{2'})\delta(\epsilon_1 + \epsilon_2 - \epsilon_{1'} - \epsilon_{2'}). \tag{5.2.3}$$

Here σ's indicate spin summations. By combining Equations (5.2.2) and (5.2.3) one can express the induced field E_2 in terms of the drive field E_1 [8]: $E_2/E_1 \equiv \tau_1/\tau_D$. One thus obtains the momentum transfer rate $1/\tau_D$:

$$\frac{1}{\tau_D} = \frac{\hbar^2}{2\pi^2 n_2 m k T} \int_0^\infty q\, dq \int_0^\infty d\omega \frac{q^2 |W(q,\omega)|^2 \operatorname{Im}\chi_0^{(1)}(q,\omega) \operatorname{Im}\chi_0^{(2)}(q,\omega)}{\sinh^2(\hbar\omega/2kT)}, \tag{5.2.4}$$

where the density response function is defined by

$$\chi_0^{(i)}(q,\omega) = -\int \frac{d\boldsymbol{k}_i}{(2\pi)^2} \frac{f^0(\epsilon_i) - f^0(\epsilon_{i'})}{\epsilon_i - \epsilon_{i'} + \hbar\omega + i\delta}, \tag{5.2.5}$$

where we used the notation $\epsilon_i = \epsilon(\boldsymbol{k}_i)$ and $\epsilon_{i'} = \epsilon(\boldsymbol{k}_i + \boldsymbol{q})$. We now wish to interpret the different terms in this result in physical terms: 1. q^2 is the momentum transfer; 2. W^2 is the effective interaction; 3. $\operatorname{Im}\chi$ combined with the sinh factor is the phase space (this term can be linked, via the fluctuation dissipation theorem, to the structure factor). Several authors have performed numerical calculations based on equation (5.2.4) and they are summarized in Fig. 5.2. The following features are noteworthy.

- The overall drag rate, when scaled by T^2, is not a constant, as the simple arguments presented in the introduction would suggest.
- When compared with experiments of [1], one sees that the calculated low-temperature $1/\tau_D$ ($T < 6$ K, or, referring to Fig. 5.2, $T < T_F/10$) is about one-half of the experimentally observed rate. The probable cause for this underestimation are phonon-mediated interactions, which are not included in the calculations of Fig. 5.2 [1, 5, 67]. Note, however, that the form of equation (5.2.4) is quite general: the particular form of the effective interaction has not yet been specified.
- More importantly, the drag rate is a very sensitive function of the screening model; thus one must use a fairly sophisticated polarization function (neither zero-temperature nor static-response functions are sufficient) and at low densities even RPA may fail [15].

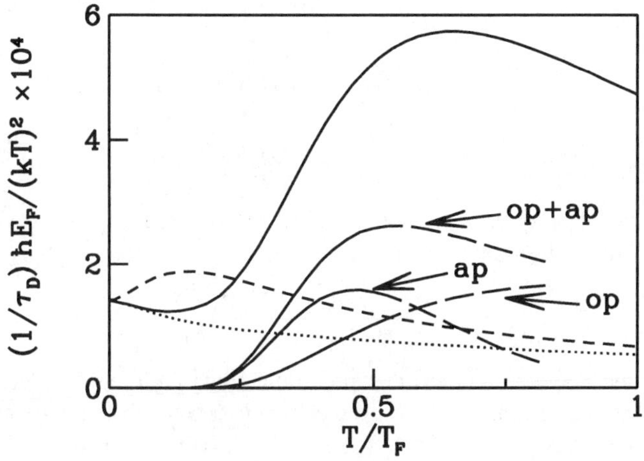

Figure 5.2 *Temperature dependence of the drag rate scaled by* T^2. *The solid heavy curve corresponds to calculations using the finite-T form of* $\chi(q, \omega)$, *the dotted curve to using the* $T = 0$ *form of* χ *and the short-dashed curve is based on the static screening approximation. Also shown are the plasmon-pole approximation estimates for the acoustic plasmon (ap) and optic plasmon (op) contributions to* τ_D^{-1} *and the sum of the two (op+ap). For* $T \geq 0.6T_F$, *this approximation becomes less reliable due to large Landau damping of the modes and hence we have plotted the results with dashed lines (adapted from [14]).*

- Finally, Fig. 5.2 contains a striking prediction [13, 14]: the drag rate is strongly enhanced at about $T = 0.5T_F$. The enhancement is due to plasmons; thus Coulomb drag offers a means of directly probing the collective excitations of the double-layer system. Recent experiments appear to agree quantitatively with these predictions [16].

We have described the semiclassical Boltzmann approach in some detail because it provides a pedagogical starting point for more elaborate theories. The theoretical description of many interesting effects, such as weak localization or quantizing magnetic field (which suggests the experimental study of Coulomb drag in the (F)QHE regime) necessarily requires a quantum transport theory and in the next section we outline how this can be achieved in the Kubo formula approach.

5.3 KUBO FORMULA FOR TRANSCONDUCTIVITY

In the previous section we discussed the Coulomb drag in terms of the momentum transfer rate $1/\tau_D$ (or the drag resistance ρ_D). The Kubo approach leads to a closely related object: the transconductivity σ_{21}. The inter-relationships between these objects are as follows [10]

$$\rho_{21} = \frac{E_2}{J_1}, \qquad \text{with } J_2 = 0; \tag{5.3.1}$$

$$\sigma_{21} = \frac{J_2}{E_1}, \qquad \text{with } E_2 = 0, \tag{5.3.2}$$

where E_i and J_i are, respectively, the electric field and current density in subsystem i. These two quantities are related via

$$\rho_{21} = \frac{-\sigma_{21}}{\sigma_{11}\sigma_{22} - \sigma_{12}\sigma_{21}} \approx \frac{-\sigma_{21}}{\sigma_{11}\sigma_{22}}. \tag{5.3.3}$$

In (5.3.3) the diagonal σ's are the individual subsystem conductivities and we note that the transconductivity is always much smaller than intralayer conductivity, because it is caused by a screened interaction between spatially separated systems (e.g. the data of [1] gives $\sigma_{21}/\sigma_{11} \simeq 10^{-6}$).

The Kubo formula [17] gives the linear response of a quantum-mechanical system to an external perturbation. In the case of a (generalized) electrical conductivity one often uses a vector potential to describe the external electric field, $H_{\text{ext}} = \int \mathrm{d}\boldsymbol{x} \boldsymbol{j}(\boldsymbol{x}) \cdot \boldsymbol{A}(\boldsymbol{x})$. Then the conductivity tensor can be expressed in terms of the retarded current–current correlation function,

$$\sigma_{ij}^{\alpha\beta}(\boldsymbol{x}-\boldsymbol{x}';\Omega) = \frac{\mathrm{i}e^2}{\Omega}\Pi_{ij}^{\alpha\beta,r}(\boldsymbol{x}-\boldsymbol{x}';\Omega) + \frac{\mathrm{i}e^2}{m\Omega}\delta(\boldsymbol{x}-\boldsymbol{x}')\delta_{ij}\delta_{\alpha\beta}\rho_i(\boldsymbol{x}), \tag{5.3.4}$$

where (hereafter we set $\hbar = 1$)

$$\Pi_{ij}^{\alpha\beta,r}(\boldsymbol{x}-\boldsymbol{x};t-t') = -\mathrm{i}\Theta(t-t')\langle[j_i^\alpha(\boldsymbol{x},t), j_j^\beta(\boldsymbol{x}',t')]\rangle. \tag{5.3.5}$$

Here $\{ij\}$ indicate the subsystem, $\{\alpha\beta\}$ in the superscripts label the Cartesian coordinates, $\rho_i(\boldsymbol{x})$ is the particle density in subsystem i and $\boldsymbol{j}(\boldsymbol{x},t)$ is the particle current operator. We have assumed that the subsystems are translationally invariant.

We emphasize the flexibility of the Kubo formulation: equation (5.3.4) is a standard textbook expression, but it applies equally well to the drag geometry! The only generalization is the subsystem indices $\{ij\}$. Our task consists of calculating the transconductivity $\sigma_{21}^{\alpha\beta}$; this is achieved by a systematic expansion in the interaction between the subsystems.

We employ the imaginary-time formalism to evaluate the retarded current–current correlation function, starting with the (imaginary-)time-ordered correlation function

$$\Pi_{21}^{\alpha\beta}(\boldsymbol{x}-\boldsymbol{x}';\tau-\tau') = -\langle T_\tau\{j_1^\alpha(\boldsymbol{x},\tau)j_2^\beta(\boldsymbol{x}',\tau')\}\rangle. \tag{5.3.6}$$

The retarded function then follows as

$$\Pi_{21}^{\alpha\beta,r}(\boldsymbol{x}-\boldsymbol{x}';\Omega) = \lim_{\mathrm{i}\Omega_n \to \Omega + \mathrm{i}\delta} \Pi_{21}^{\alpha\beta}(\boldsymbol{x}-\boldsymbol{x}';\mathrm{i}\Omega_n), \tag{5.3.7}$$

where

$$\Pi_{21}^{\alpha\beta}(\boldsymbol{x}-\boldsymbol{x}'; \mathrm{i}\Omega_n) = \int_0^\beta \mathrm{d}\tau\, \mathrm{e}^{\mathrm{i}\Omega_n\tau}\Pi_{21}^{\alpha\beta}(\boldsymbol{x}-\boldsymbol{x}'; \tau) \tag{5.3.8}$$

$$\Pi_{21}^{\alpha\beta}(\boldsymbol{x}-\boldsymbol{x}'; \tau) = \frac{1}{\beta}\sum_n \mathrm{e}^{-\mathrm{i}\Omega_n\tau}\Pi_{21}^{\alpha\beta}(\boldsymbol{x}-\boldsymbol{x}'; \mathrm{i}\Omega_n). \tag{5.3.9}$$

Here $\beta = 1/k_\mathrm{B}T$ and $\mathrm{i}\Omega_n = n(2\pi/\beta)\mathrm{i}$ is the imaginary Matsubara frequency.

The interaction Hamiltonian between the subsystems is given by

$$H_{12} = \int \mathrm{d}\boldsymbol{r}_1 \int \mathrm{d}\boldsymbol{r}_2\, \rho_1(\boldsymbol{r}_1)U_{12}(\boldsymbol{r}_1-\boldsymbol{r}_2)\rho_2(\boldsymbol{r}_2). \tag{5.3.10}$$

Here U_{12} is the bare Coulomb interaction between the systems. We note that other interaction processes, which couple the charge carriers in the two subsystems, can be treated similarly. An important example is the phonon mediated interaction, which plays an important role in the low-temperature behaviour of the momentum transfer rate [1, 5, 67].

The τ-dependence of the current operators in (5.3.6) is determined by the full Hamiltonian $H = H_1+H_2+H_{12}$, where H_i are the subsystem Hamiltonians. In order to develop a perturbation expansion, we must isolate the H_{12}-dependence. Following the standard many-body prescription [17], we transform it into the interaction representation and obtain

$$\Pi_{21}^{\alpha\beta}(\boldsymbol{x}-\boldsymbol{x}', \tau-\tau') = -\frac{\langle T_\tau\{S(\beta)\hat{j}_1^\alpha(\boldsymbol{x},\tau)\hat{j}_2^\beta(\boldsymbol{x}',\tau')\}\rangle}{\langle S(\beta)\rangle}, \tag{5.3.11}$$

where the carets indicate that the time dependence is now governed by the individual subsystem Hamiltonians and the operator $S(\beta)$ is

$$S(\beta) = T_\tau\left\{\exp\left[-\int_0^\beta \mathrm{d}\tau_1 \hat{H}_{12}(\tau_1)\right]\right\}. \tag{5.3.12}$$

As usual, only connected diagrams need to be included. It is now straightforward to expand $S(\beta)$ in powers of $\hat{H}_{12}$,

$$\begin{aligned} S(\beta) \approx\ & 1 - T_\tau\left\{\int_0^\beta \mathrm{d}\tau_1 \hat{H}_{12}(\tau_1)\right\} \\ & + \frac{1}{2}T_\tau\left\{\int_0^\beta \mathrm{d}\tau_1 \int_0^\beta \mathrm{d}\tau_2 \hat{H}_{12}(\tau_1)\hat{H}_{12}(\tau_2)\right\} + \cdots . \end{aligned} \tag{5.3.13}$$

The zeroth-order term leads to a vanishing contribution to the transconductivity because the two current operators are decoupled and hence commute. In the following sections we discuss the higher-order terms.

5.3.1 *Linear expansion*

The linear-order term in H_{12} leads to the correlation function

$$\Pi_{21}^{\alpha\beta}(\boldsymbol{x}-\boldsymbol{x}',\tau-\tau')^{(1)} = \int_0^\beta \mathrm{d}\tau_1 \int \mathrm{d}\boldsymbol{r}_1\,\mathrm{d}\boldsymbol{r}_2 \langle T_\tau\{\hat{j}_1^\alpha(\boldsymbol{x},\tau)\hat{\rho}_1(\boldsymbol{r}_1,\tau_1)\}\rangle \times U_{12}(\boldsymbol{r}_1-\boldsymbol{r}_2)\langle T_\tau\{\hat{\rho}_2(\boldsymbol{r}_2,\tau_1)\hat{j}_2^\beta(\boldsymbol{x}',\tau')\}\rangle. \tag{5.3.14}$$

Use of the continuity equation, $\mathrm{i}\Omega\rho + \nabla\cdot\boldsymbol{j} = 0$, allows us to eliminate the number of density operators and to express the density–current correlators in terms of the subsystem conductivities. After some simplification we find

$$\sigma_{21}^{\alpha\beta}(\boldsymbol{q},\Omega)^{(1)} = \frac{1}{\mathrm{i}e^2\Omega}\sum_{\gamma,\delta}\sigma_{22}^{\alpha\gamma}(\boldsymbol{q},\Omega)q^\gamma U_{12}(\boldsymbol{q})q^\delta\sigma_{11}^{\delta\beta}(\boldsymbol{q},\Omega). \tag{5.3.15}$$

This expression is exact and it can be used to calculate the first-order transconductivity for any system, once the subsystem conductivities σ_{ii} are known. From (5.3.15) we also infer that the first-order transconductivity vanishes in the DC-limit.

5.3.2 *Quadratic expansion*

To evaluate $\Pi_{21}^{\alpha\beta}$ to second order in H_{12}, we substitute the third term of (5.3.13) on the right-hand side of (5.3.11) and find that the current–current correlation function is given by

$$\Pi_{21}^{\alpha\beta}(\boldsymbol{x}-\boldsymbol{x}';\tau-\tau')^{(2)} = -\frac{1}{2}\int_0^\beta \mathrm{d}\tau_1\int_0^\beta \mathrm{d}\tau_2\int \mathrm{d}\boldsymbol{r}_1\int \mathrm{d}\boldsymbol{r}_2\int \mathrm{d}\boldsymbol{r}_1'\int \mathrm{d}\boldsymbol{r}_2' \times U_{12}(\boldsymbol{r}_1-\boldsymbol{r}_2)U_{12}(\boldsymbol{r}_1'-\boldsymbol{r}_2') \times \Delta_1^\alpha(\boldsymbol{x}\tau,\boldsymbol{r}_1\tau_1,\boldsymbol{r}_1'\tau_2)\Delta_2^\beta(\boldsymbol{x}'\tau',\boldsymbol{r}_2\tau_1,\boldsymbol{r}_2'\tau_2), \tag{5.3.16}$$

where we have defined the function

$$\Delta_i^\alpha(\boldsymbol{x}\tau,\boldsymbol{x}'\tau',\boldsymbol{x}''\tau'') = -\langle T_\tau\{\hat{j}_i^\alpha(\boldsymbol{x}\tau)\hat{\rho}_i(\boldsymbol{x}'\tau')\hat{\rho}_i(\boldsymbol{x}''\tau'')\}\rangle. \tag{5.3.17}$$

Just as in section 5.3.1, we factorized the time-ordered expectation value involving two current and four density operators; this step is justified because the two subsystems are decoupled after the formal expansion in H_{12}. The remainder of the calculation proceeds in two steps [10].

1. Due to the assumed translational invariance, Δ depends on only two coordinate differences and we transform into a Fourier-space.
2. By studying the analytic properties of Δ, one can locate its branch-cuts and perform the required frequency summations as contour integrals and finally

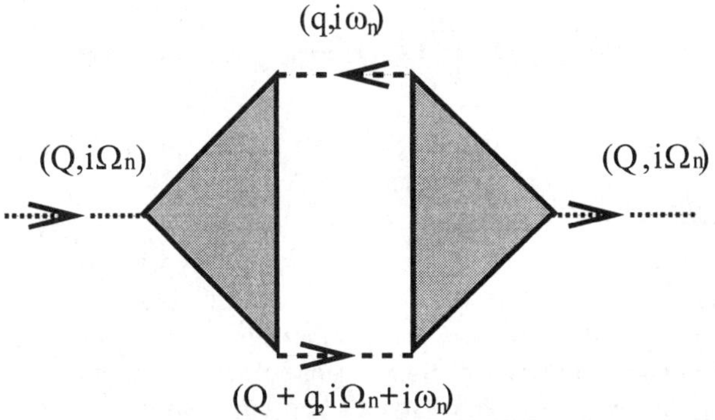

Figure 5.3 *Diagram corresponding to the current–current correlation function to second order in the interlayer Coulomb interaction. The shaded triangles correspond to the Δ given in equation (5.3.17), the dashed lines correspond to the interaction, the dotted lines correspond to the external current operators and the arrowheads correspond to the direction of momentum and energy transfer.*

analytically continue to the real axis. We omit the details and give the final answer in the DC-limit:

$$\sigma_{21}^{\alpha\beta(2)} = -\frac{e^2}{2\hbar}\frac{1}{\nu}\sum_{\boldsymbol{q}}\int_{-\infty}^{\infty}\frac{\mathrm{d}\omega}{2\pi}\,|U_{12}(q)|^2\,[\partial_\omega n_{\mathrm{B}}(\omega)]$$
$$\times\Delta_1^{\alpha}(\boldsymbol{q},\boldsymbol{q};\omega^+,\omega^-)\Delta_2^{\beta}(-\boldsymbol{q},-\boldsymbol{q};-\omega^-,-\omega^+). \quad (5.3.18)$$

Here we employ the notation $\omega^{\pm} = \omega \pm \mathrm{i}\delta$, where δ is a positive infinitesimal. Later we shall discuss the evaluation of this expression in various levels of approximation. Figure 5.3 gives the diagrammatic representation of the current–current correlation function evaluated to second order in the interaction H_{12}. This kind of diagram can generally be classified as a fluctuation diagram and similar structures have been studied in several other physical situations, such as superconductivity [18], or in the microscopic theory of van der Waals interactions [19, 20].

5.3.3 Higher-order terms

The S-matrix expansion (5.3.13) can be used to generate higher-order terms. To proceed systematically one must apply the techniques of the many-body formalism. As usual, the most important processes should be identified and the corresponding diagrams should be summed to infinite order. This procedure may then lead to an integral equation for the effective interaction, for example in the ladder approximation one obtains the Bethe–Salpeter equation. We do not pursue this line of argument further in this chapter, but note that a particularly useful resummation can be obtained, if one includes the 'bubble'

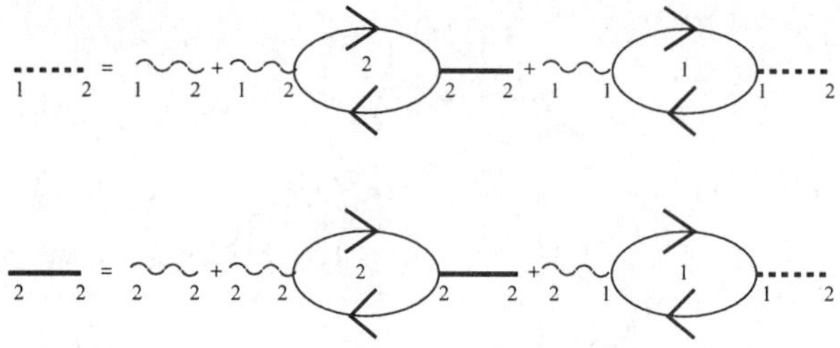

Figure 5.4 *Diagrams which lead to the screened interlayer interaction within the random phase approximation. The 'bubbles' are the bare polarizabilities of the subsystems, the light wavy lines are the bare interactions, the heavy lines are the screened interactions and numbers indicate the subsystem.*

diagrams (Fig. 5.4), which leads to an effective screened interaction, $U_{12}(q) \to W_{12}(q, \omega) = U_{12}(q)/\epsilon_{12}(q, \omega)$, where the dielectric function is given by

$$\begin{aligned} \epsilon_{12}(q,\omega) &= [1 - U_1(q)\chi^{(1)}(q,\omega)][1 - U_2(q)\chi^{(2)}(q,\omega)] \\ &\quad - U_{12}(q)^2\chi^{(1)}(q,\omega)\chi^{(2)}(q,\omega), \end{aligned} \tag{5.3.19}$$

where the $\chi^{(i)}$'s are the polarization functions given in equation (5.2.5) and the U_i's are the intrasystem Coulomb interactions. We observe that an energy-dependent $W(q, \omega)$ can be used in the above expressions, (5.3.15) and (5.3.18), for transconductivity with no additional difficulty. This form of effective interaction was actually tacitly assumed in the Boltzmann equation derivation of the drag rate, but strictly speaking it can be justified only in a many-body language, as used in this section. Most existing works [2, 5, 8, 14, 21] on drag problems have used (5.3.19) (or simplified versions of it).

5.4 IMPURITY SCATTERING

In section 5.3 we showed that the transconductivity can be expressed in terms of the general three-body correlation function Δ. We will next consider a specific example in order to calculate this three-body function, namely noninteracting electrons scattering against random impurities. The Hamiltonian representing impurity scattering is quadratic, and hence Wick's theorem is applicable, which means that the expectation value can be factorized into pairwise contractions, i.e. expressed in terms of Green's functions. Impurity averaging, which is now implicit in the expectation value, re-introduces correlations between the particles, which implies that one must introduce vertex functions. However, we do not allow impurity correlations between the two subsystems, i.e. we assume that $\langle \Delta_1 \Delta_2 \rangle_{\mathrm{imp}} = \langle \Delta_1 \rangle_{\mathrm{imp}} \langle \Delta_2 \rangle_{\mathrm{imp}}$. The particular choice for the impurity self-energy used in the calculation of the impurity averaged Green's function fixes the choice of the vertex function; in what follows we use the self-consistent Born approximation for the self-energy and the corresponding vertex function

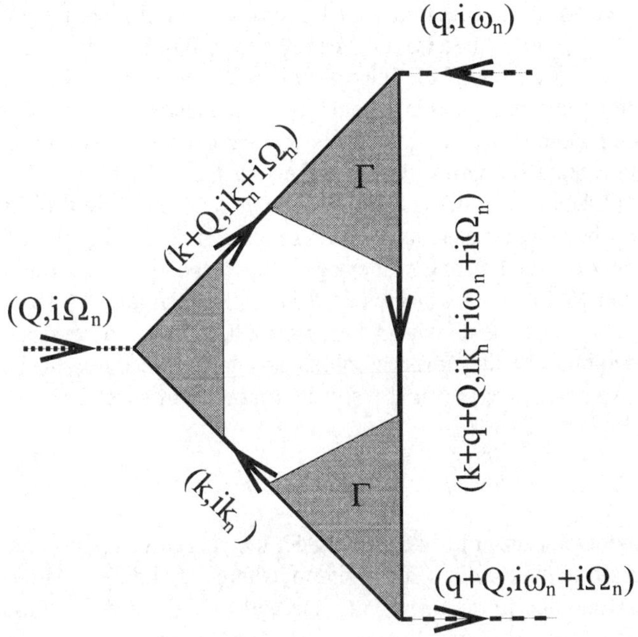

Figure 5.5 *The function* Δ *for the case in which vertex corrections are included at each of the individual charge and current vertices. Here* $k^\alpha\gamma$ *is the current vertex,* Γ *is the charge vertex, the dashed line is in-coming momentum and frequency, the dotted lines are the interaction* $U_{12}(q,\omega)$ *and the full lines with arrows are the Green's function. Normal momentum and energy conservation rules apply at the vertices.*

consists either of the ladder diagrams (section 5.4.2) or of the maximally crossed diagrams (section 5.4.3). The form for impurity-Δ giving the dominant contribution is shown in Fig. 5.5. We consider only uniform systems and set the external wave vector to zero, $\boldsymbol{Q}=\boldsymbol{0}$. We also denote fermionic complex frequencies by $\mathrm{i}k_m$ in contrast to bosonic frequencies $\mathrm{i}\omega_n$ and the external frequency Ω. Thus we have

$$\Delta^\alpha(\boldsymbol{q},\boldsymbol{q};\mathrm{i}\Omega_n+\mathrm{i}\omega_n,\mathrm{i}\omega_n) = -\frac{2}{m\nu}\sum_{\boldsymbol{k}}\frac{1}{\beta}\sum_{\mathrm{i}k_m}k^\alpha\Big[\mathcal{K}(\boldsymbol{k},\boldsymbol{q},\mathrm{i}k_m,\mathrm{i}\Omega_n,\mathrm{i}\omega_n) + \mathcal{K}(\boldsymbol{k},-\boldsymbol{q},\mathrm{i}k_m,\mathrm{i}\Omega_n,-(\mathrm{i}\omega_n+\mathrm{i}\Omega_n))\Big], \quad (5.4.1)$$

where

$$\mathcal{K}(\boldsymbol{k},\boldsymbol{q},\mathrm{i}k_m,\mathrm{i}\Omega_n,\mathrm{i}\omega_n) = \mathcal{G}(\boldsymbol{k},\mathrm{i}k_m)\gamma(\boldsymbol{k},\boldsymbol{k};\mathrm{i}k_m,\mathrm{i}k_m+\mathrm{i}\Omega_n)\mathcal{G}(\boldsymbol{k},\mathrm{i}k_m+\mathrm{i}\Omega_n) \times\Gamma(\boldsymbol{k},\boldsymbol{k}+\boldsymbol{q};\mathrm{i}k_m+\mathrm{i}\Omega_n,\mathrm{i}k_m+\mathrm{i}\omega_n+\mathrm{i}\Omega_n)\mathcal{G}(\boldsymbol{k}+\boldsymbol{q},\mathrm{i}k_m+\mathrm{i}\Omega_n+\mathrm{i}\omega_n) \times\Gamma(\boldsymbol{k}+\boldsymbol{q},\boldsymbol{k};\mathrm{i}k_m+\mathrm{i}\omega_n+\mathrm{i}\Omega_n,\mathrm{i}k_m). \quad (5.4.2)$$

The factor of 2 in equation (5.4.1) comes from the spin sum. In equation (5.4.2), $k^\alpha \gamma/m$ is the current (vector) vertex function and Γ is the charge (scalar) vertex function. In labelling the variables in the vertex functions, we use the convention that the incoming momentum (frequency) is the first variable and the outgoing momentum (frequency) is the second variable. The second term in the square brackets corresponds to reversing the order of the two U_{12} lines: Equations (5.4.1) and (5.4.2) need to be analytically continued to the real axis, after which they can be used as a starting point for evaluating the transconductance in the weak and strong scattering limits, respectively. It should be noted that equation (5.4.2) does not include all possible diagrams. An example of a diagram not included is shown in Fig. 5.6(c). The required frequency summations are tedious but they proceed along standard procedures [10] and in the following we shall give the final result in a number of special cases.

5.4.1 The Boltzmann limit ($\omega\tau > 1$ and/or $Dq^2\tau > 1$)

In the weak scattering limit, we can neglect the charge vertex corrections, since Γ differs from unity only in a small region, where $\omega\tau$ and $Dq^2\tau$ are small. Here τ is the lifetime due to impurity scattering and D is the diffusion constant. The corresponding current vertex function γ^{B} is well known [17]:

$$\gamma^{\mathrm{B}}_{-+}(\boldsymbol{k}, \boldsymbol{k}; \epsilon_{\boldsymbol{k}}, \epsilon_{\boldsymbol{k}}) = \frac{\tau_{\mathrm{tr}}(\boldsymbol{k})}{\tau(\boldsymbol{k})}. \tag{5.4.3}$$

Using these vertex functions it is possible to show that the Boltzmann limit of the triangle function, Δ_{B}, is given by (we suppress the subsystem index i)

$$\begin{aligned}\Delta^\alpha_{\mathrm{B}}(\boldsymbol{q}, \boldsymbol{q}; \omega^+, \omega^-) \quad &\approx \quad \frac{2}{m\nu}\sum_{\boldsymbol{k}}\int_{-\infty}^{\infty}\frac{d\epsilon}{2\pi}\left[(\boldsymbol{k}+\boldsymbol{q})^\alpha \tau_{\mathrm{tr}}(\boldsymbol{k}+\boldsymbol{q}) - k^\alpha \tau_{\mathrm{tr}}(\boldsymbol{k})\right] \\ &\times A(\boldsymbol{k}+\boldsymbol{q}, \epsilon+\omega)A(\boldsymbol{k}, \epsilon)[f^0(\epsilon+\omega) - f^0(\epsilon)].\end{aligned} \tag{5.4.4}$$

Here $A(\boldsymbol{k}, \omega) \equiv \mathrm{i}[G^r(\boldsymbol{k}, \omega) - G^a(\boldsymbol{k}, \omega)]$ is the spectral function. The Boltzmann limit is recovered by using free Green's functions in A, which implies that the spectral functions reduce to δ-functions. We find

$$\Delta^\alpha_{\mathrm{B}}(\boldsymbol{q}, \boldsymbol{q}; \omega^+, \omega^-) = \frac{2\bar{\tau}_{\mathrm{tr}}}{m} F^\alpha(\boldsymbol{q}, \omega), \tag{5.4.5}$$

where the transport polarization F^α is given by

$$F^\alpha(\boldsymbol{q}, \omega) = \frac{2}{\nu\bar{\tau}_{\mathrm{tr}}}\mathrm{Im}\sum_{\boldsymbol{k}}\frac{f^0(\epsilon_{\boldsymbol{k}+\boldsymbol{q}}) - f^0(\epsilon_{\boldsymbol{k}})}{\epsilon_{\boldsymbol{k}+\boldsymbol{q}} - \epsilon_{\boldsymbol{k}} - \omega - \mathrm{i}\delta}\left[(\boldsymbol{k}+\boldsymbol{q})^\alpha \tau_{\mathrm{tr}}(\boldsymbol{k}+\boldsymbol{q}) - k^\alpha \tau_{\mathrm{tr}}(\boldsymbol{k})\right]. \tag{5.4.6}$$

Here $\bar{\tau}_{\rm tr}$ determines the in-plane conductivity, $\sigma_{ii} = e^2 n_i \bar{\tau}_{{\rm tr},i}/m_i$. When (5.4.5) is inserted into the expression for transconductivity, equation (5.3.18), we obtain

$$\begin{aligned}\rho_{21} &= -\frac{\sigma_{21}}{\sigma_{11}\sigma_{22}} \qquad (5.4.7)\\ &= -\frac{1}{2n_1 n_2}\frac{1}{\nu}\sum_{\boldsymbol{q}}\int_{-\infty}^{\infty}\frac{{\rm d}\omega}{2\pi}\frac{\beta}{\sinh^2[\beta\omega/2]}\left|W_{12}(\boldsymbol{q},\omega)\right|^2 F_1(\boldsymbol{q},\omega)F_2(\boldsymbol{q},\omega).\end{aligned}$$

Several comments are now in order. Without an applied magnetic field, the transconductivity and consequently also the transresistivity are diagonal in the Cartesian coordinates and we have suppressed the $\{\alpha\beta\}$ indices in (5.4.7). For constant τ's transport polarization is related to the (bare) RPA polarization function, $F^\alpha(\boldsymbol{q},\omega) = q^\alpha {\rm Im}\,\chi_0(\boldsymbol{q},\omega)$. In this limit equation (5.4.7) reproduces the Boltzmann result for transresistivity, equation (5.2.4). It is also possible to obtain equation (5.4.7) directly from the Boltzmann equation by allowing a k-dependent intralayer scattering time [13]. We also emphasize that in general the drag rate, or the transresistivity, cannot be expressed in terms of the polarization function; rather, one must use the more general object F^α defined above. Equation (5.4.7) provides the framework for analysing the temperature dependence of Coulomb drag at elevated temperatures, one should however notice that the k-dependent transport relaxation time goes in as an input and to determine this object may require substantial work due to the many possible intralayer scattering mechanisms.

5.4.2 *Diffusive limit ($\omega\tau < 1$, $Dq^2\tau < 1$)*

In this section we consider the transconductivity for weak scattering in the diffusive limit ($Dq^2\tau < 1$ and $\omega\tau < 1$), including vertex corrections. Specifically, we consider momentum-independent relaxation times, in which case $\tau_{\rm tr} = \tau$ and include vertex corrections due to ladder diagrams. Then, we have $\gamma \equiv 1$ and the charge vertex is given by [22]

$$\Gamma^{\pm}(q,\omega) = [\tau(Dq^2 \pm {\rm i}\omega)]^{-1} \qquad (5.4.8)$$

where $\pm$ refer to the signs of the imaginary parts of the two frequency arguments of equation (5.4.2).

The triangle function in the ladder approximation, $\Delta^\alpha_{\rm L}$ becomes

$$\begin{aligned}\Delta^\alpha_{\rm L}(\boldsymbol{q},\boldsymbol{q},\omega^+,\omega^-) &= \frac{2\tau}{m\nu}\sum_{\boldsymbol{kk}}^{\alpha}\int_{-\infty}^{\infty}\frac{{\rm d}\epsilon}{2\pi}[f^0(\epsilon+\omega)-f^0(\epsilon)]\\ &\times\Big\{2{\rm Im}\big[\Gamma^-(q,\omega)G^r(\boldsymbol{k}+\boldsymbol{q},\epsilon+\omega)\big]A(\boldsymbol{k},\epsilon)\\ &-2{\rm Im}\big[\Gamma^+(q,\omega)G^r(\boldsymbol{k}-\boldsymbol{q},\epsilon)\big]A(\boldsymbol{k},\epsilon+\omega)\Big\}.\end{aligned} \qquad (5.4.9)$$

We observe that the constant-τ Boltzmann result is readily recovered from (5.4.9) by replacing Γ's by unity. A generalization to energy-dependent scattering rates is straightforward, but we do not reproduce the cumbersome results here.

The next task is to establish a connection between the dressed polarization function $\chi(\boldsymbol{q}, \omega)$ and the triangle function Δ_{L}. It is easy to show that

$$\begin{aligned} \mathrm{Im}\chi(\boldsymbol{q}, \omega) &= -\frac{2}{\nu}\sum_{\boldsymbol{k}} \mathrm{Im}\left\{\int \frac{\mathrm{d}\epsilon}{2\pi \mathrm{i}}[f^0(\epsilon+\omega) - f^0(\epsilon)] \right. \\ &\qquad \left. \times \Gamma^-(q, \omega) G^r(\boldsymbol{k}+\boldsymbol{q}, \epsilon+\omega) G^a(\boldsymbol{k}, \epsilon)\right\}. \end{aligned} \tag{5.4.10}$$

A comparison of (5.4.9) with (5.4.10) reveals some similarity, but clearly a few more steps are required. We complete the connection by making the following observations. First, we express the spectral functions in (5.4.9) in terms of the retarded and advanced functions. The resulting integrals can be grouped in two classes: 1. integrals involving products of the type $G^r G^a$; 2. integrals involving products of the form $G^r G^r$ or $G^a G^a$. An analysis shows that type-2. integrals can be neglected in comparison with type-1. integrals, if the momentum variables and frequency variables differ by less than $D\tau^{-1/2}$ and τ^{-1} (the diffusive limit). Thus, keeping only the $G^r G^a$-terms in (5.4.9) allows us to express the quantity in curly brackets as

$$\begin{aligned} \{\cdots\} \to \quad & - \left[\Gamma^-(q, \omega) G^r(\boldsymbol{k}+\boldsymbol{q}, \epsilon+\omega) G^a(\boldsymbol{k}, \epsilon) + \text{c.c.}\right] \\ & + \left[\Gamma^-(q, \omega) G^r(\boldsymbol{k}, \epsilon+\omega) G^a(\boldsymbol{k}-\boldsymbol{q}, \epsilon) + \text{c.c.}\right]. \end{aligned}$$

In the above expression the second term can be made to coincide with the first one by shifting the summation variable $\boldsymbol{k} \to \boldsymbol{k}+\boldsymbol{q}$; however when doing this the prefactor $\boldsymbol{k}^\alpha$ in (5.4.9) generates an extra $\boldsymbol{q}^\alpha$. This is exactly what is needed to give the required result,

$$\Delta_{\mathrm{L}}^{\alpha}(\boldsymbol{q}, \boldsymbol{q}, \omega^+, \omega^-) = \frac{2\tau \boldsymbol{q}^\alpha}{m} \mathrm{Im}\chi(\boldsymbol{q}, \omega). \tag{5.4.11}$$

The above analysis shows the equivalence of the triangle function and the polarization function in the small q and ω limit, confirming the result obtained by Zheng and MacDonald [9] with a different method. As observed by these authors, in the high-mobility samples studied so far the replacement $\chi_0 \to \chi$ does not appear to be important; however, in dirtier samples the consequences of the vertex corrections (i.e. full χ) may well become detectable and, as we shall see, they are very important for finite magnetic fields.

5.4.3 *Weak localization correction to Coulomb drag*

In the previous sections we included the leading-order impurity scattering diagrams which gave us the Boltzmann equation result for the case of weak

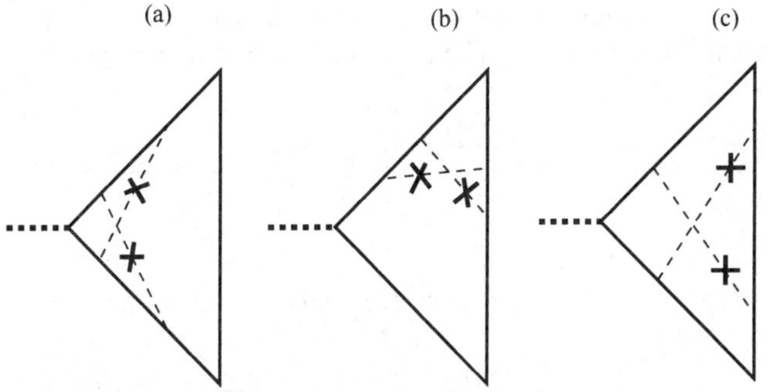

Figure 5.6 *The different types of crossed diagrams for the triangle diagram relevant for the weak localization correction to the transconductivity. Diagrams of the type in (a) give the leading-order contribution. Dressing the charge vertexes as in (b) gives a smaller contribution for moderately clean samples for the same reason that allows us to neglect the vertex corrections of the charge vertexes, discussed in section 5.4.2. The diagram in (c), which cannot be included using vertex functions alone, has an even smaller phase space and can hence also be neglected.*

scattering and showed how the bare polarization function in a certain parameter range must be replaced by the dressed polarization function. Here we develop the analysis further and calculate the quantum correction associated with weak localization (the basic physics of weak localization is reviewed, for example by Lee and Ramakrishnan [23]). The corrections will be of the order of $1/(k_F\ell) \ll 1$, where $\ell = v_F\tau$ is the elastic mean-free path.

In Fig. 5.6 we display the different types of crossed diagrams that exist for the function Δ. The maximally crossed one is the one shown in Fig. 5.6(c). This diagram is, however, smaller than the one shown in Fig. 5.6(a), because of the restricted phase space. The two Green's functions attached to the current vertex in the diagram in Fig. 5.6(a) have the same arguments because in the limit ($\boldsymbol{Q} = \mathbf{0}, \Omega = 0$) the current vertex leaves the momenta and energies of the entering and leaving Green's functions unchanged. Therefore there is the possibility of two overlapping spectral functions giving an overall factor of τ. This does not happen for the diagram in Fig. 5.6(c). Neither does the diagram in Fig. 5.6(b) lead to overlapping spectral functions except in a very small region of q, ω space, where q and ω are the incoming quantities at the charge vertices. Since we are integrating over q, ω the (logarithmic) singularity caused by the maximally crossed diagrams becomes regularized. In other words, the contribution from this diagram is small for the same reasons that the dressing of the charge vertexes, discussed in section 5.4.2 and can be neglected for experimentally relevant parameters [9]. We therefore conclude that diagrams of the Fig. 5.6(a)-type dominate the quantum correction to the drag rate.

The leading quantum correction is given as the sum of the maximally crossed diagrams, the Cooperon [24]. The resulting vertex function describing the weak localization correction, γ^{WL}, obeys

$$\gamma^{\mathrm{WL}}(\boldsymbol{k},\boldsymbol{k};\mathrm{i}k_m,\mathrm{i}k_m+\mathrm{i}\Omega_n) = \frac{1}{\nu}\sum_{\boldsymbol{k}'}\frac{\boldsymbol{k}\cdot\boldsymbol{k}'}{(k')^2}\mathcal{G}(\boldsymbol{k}',\mathrm{i}k_m)\mathcal{G}(\boldsymbol{k}',\mathrm{i}k_m+\mathrm{i}\Omega_n) \times\mathcal{C}(\boldsymbol{k},\boldsymbol{k}';\mathrm{i}k_m,\mathrm{i}k_m+\mathrm{i}\Omega_n), \tag{5.4.12}$$

where $\mathcal{C}$ stands for the Cooperon whose analytic continuation has the form

$$C_{-+} = \frac{1}{2\pi\rho\tau^2}\frac{1}{-\mathrm{i}\Omega+DQ^2}. \tag{5.4.13}$$

After integration over $\boldsymbol{Q}$ we find the weak localization vertex function,

$$\begin{aligned}\gamma^{\mathrm{WL}}(\boldsymbol{k},\boldsymbol{k};\epsilon-\Omega,\epsilon) &= \gamma^{\mathrm{WL}}(\boldsymbol{k},\boldsymbol{k};\epsilon,\epsilon+\Omega)\\ &\approx A(\boldsymbol{k},\epsilon)\frac{1}{\pi k_F\ell\tau}\ln(\Omega\tau)\\ &\equiv A(k,\epsilon)\eta^{\mathrm{WL}}(\Omega)/2\tau.\end{aligned} \tag{5.4.14}$$

Here η^{WL} is the ratio between the quantum correction and the classical conductivity: $\eta^{\mathrm{WL}}(\Omega)=\delta\sigma(\omega)/\sigma_0$ [24]. The weak localization correction $\delta\Delta_{\mathrm{WL}}$ to the triangle function is finally

$$\begin{aligned}\delta\Delta^{\alpha}_{\mathrm{WL}}(\boldsymbol{q},\boldsymbol{q};\omega^+,\omega^-,\Omega) &= \frac{2\tau\eta^{\mathrm{WL}}(\Omega)}{m}q^{\alpha}\mathrm{Im}\,\chi_0(\boldsymbol{q},\omega)\\ &= \eta^{\mathrm{WL}}(\Omega)\Delta^{\alpha}_{\mathrm{B}}(\boldsymbol{q},\boldsymbol{q};\omega^+,\omega^-),\end{aligned} \tag{5.4.15}$$

which immediately leads to the conclusion that to leading order the weak localization correction to transconductance is

$$\delta\sigma_{21}^{\mathrm{WL}}(\Omega) = [\eta_1^{\mathrm{WL}}(\Omega)+\eta_2^{\mathrm{WL}}(\Omega)]\sigma_{21}^0. \tag{5.4.16}$$

Consequently, the transresistance is unaffected by the weak localization correction because

$$\rho_{21}\approx -\frac{\sigma_{21}^0(1+\eta_1+\eta_2)}{\sigma_{11}^0(1+\eta_1)\sigma_{22}^0(1+\eta_2)}\approx\rho_{21}^0. \tag{5.4.17}$$

Weak localization is strongly affected by external magnetic fields. The formalism presented above can be extended to include magnetic fields; in particular, the topology of all diagrams remains unaltered. Below we shall elaborate in detail the theory in the limit of strong magnetic fields.

5.5 COULOMB DRAG IN A MAGNETIC FIELD

5.5.1 Introductory remarks

The combination of electron–electron interaction and strong magnetic fields in two-dimensional electron gases has provided an exciting venue of research for both experimentalists and theorists over the past few decades [25]. One well-known example of this is the fractional quantum Hall effect, where the physics is determined by the subtle interplay between interactions and the large density of states caused by all the electrons being confined to the lowest Landau level. Thus, phenomena involving interelectron interactions in a B-field often produce surprising and interesting results.

Since Coulomb drag displays rich physics, it is quite natural to ask whether an applied magnetic field would introduce some further interesting aspects. It turns out, however, that experiments are not easy and it is only very recently that the first preliminary results have begun to emerge [66, 28]. A similar situation exists on the theoretical side: it is clear that Boltzmann theory is not sufficient and that the quantum theory becomes rather involved due to the singular density of states in a quantizing magnetic field. Nevertheless, the linear response theory described in the previous sections provides a way of approaching these problems and we will next outline the required generalizations, and also present some numerical results [12, 26].

A pure system in a quantizing magnetic field has a singular density of states and it is necessary to include impurities to avoid unphysical results. In our diagrammatic language this implies that both screening (the 'bubble' diagram) and the triangle diagram must be evaluated with vertex functions (Fig. 5.7). This leads to quite involved calculations [12, 26]; however, many simplifications occur in the weak-scattering limit $\omega_c\tau \gg 1$ (here $\omega_c = eB/m$ is the cyclotron frequency and τ is the impurity scattering time). In particular it is possible to link Δ with the $\chi(\boldsymbol{q}, \omega)$ shown in Fig. 5.7(a):

$$\Delta(\boldsymbol{q}, \boldsymbol{q}; \omega^{\pm}, \omega^{\mp}) = \pm 2\hbar^2 e^{-1} \boldsymbol{q} \times \boldsymbol{B} \frac{\mathrm{Im}[\chi(\boldsymbol{q}, \omega)]}{B^2} + \mathcal{O}[(\omega_c\tau)^{-1}]. \tag{5.5.1}$$

Equations (5.3.18) and (5.5.1) form the basis of our numerical calculations: the response function χ is found by solving the appropriate vertex equation in the ladder approximation (to be discussed in the following section) and Δ, as given by (5.5.1), can be used in (5.3.18) to get the transconductivity and hence the transresistivity.

5.5.2 $\chi(q, \omega)$ in high magnetic fields

The vertex functions for the $B = 0$ case, discussed in the previous sections, are rather well documented in the literature and hence we did not give the detailed derivations. However, the finite-B case requires some special considerations, which we sketch next. The density response function of Fig. 5.7(a) reads

$$\begin{aligned} \chi(\boldsymbol{q}, \omega) \quad = \quad & -\frac{1}{2\pi \ell_{\mathrm{B}}^2} \sum_{n,m,\mathrm{i}\omega_p} f_{n,m}(\boldsymbol{q}) \Gamma(-\boldsymbol{q}, m, n, \mathrm{i}\omega_p, \mathrm{i}\omega_p - \mathrm{i}\omega) \\ & \times \mathcal{G}(n, \mathrm{i}\omega_p - \mathrm{i}\omega) \mathcal{G}(m, \mathrm{i}\omega_p), \end{aligned} \tag{5.5.2}$$

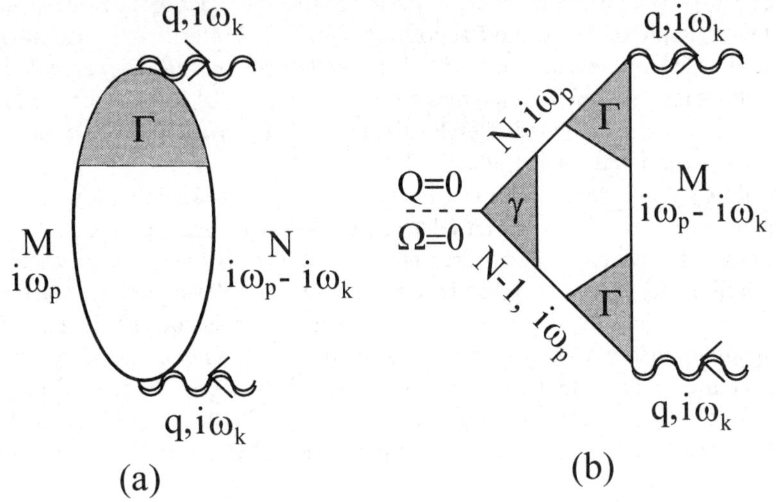

Figure 5.7 *(a) The diagram corresponding to the density response function* χ. *(b) The triangle diagram contribution to the 3-body response function* Δ. *The* Γ *and* γ *are the charge and current vertices, respectively and the labels M and N denote the LL's.*

where n and m label the Landau levels, $\ell_B = \sqrt{\hbar/eB}$ is the magnetic length and the factors $f_{n,m}(\boldsymbol{q})$ are current operator matrix elements (the rather cumbersome expressions involving Laguerre polynominals are given in [26]). The vertex function Γ satisfies

$$\begin{aligned}\Gamma(\boldsymbol{q}, n, m, \mathrm{i}\epsilon_1, \mathrm{i}\epsilon_2) \quad &= \quad f_{n,m}(\boldsymbol{q}) + \rho_{\mathrm{imp}} \sum_{a,b} \int \frac{\mathrm{d}\boldsymbol{p}}{(a\pi)^2} U^2(p) f_{b,m}(\boldsymbol{p}) f_{n,a}(-\boldsymbol{p}) \\ &\times \mathcal{G}(a, \mathrm{i}\epsilon_1)\mathcal{G}(b, \mathrm{i}\epsilon_2)\Gamma(\boldsymbol{q}, a, b, \mathrm{i}\epsilon_1, \mathrm{i}\epsilon_2), \qquad (5.5.3)\end{aligned}$$

and the Green's functions are calculated in the self-consistent Born approximation [27]:

$$G(n, \epsilon) = \frac{2}{\Gamma_0}\left[(\epsilon - \epsilon_n)/\Gamma_0 \pm \sqrt{(\epsilon - \epsilon_n)^2/\Gamma_0^2 - 1}\right]^{-1}, \qquad (5.5.4)$$

where $\epsilon_n = \hbar\omega_c(n + \frac{1}{2})$ and $\Gamma_0^2 = (2/\pi)\omega_c m \rho_{\mathrm{imp}} U^2/\hbar$ is the level width. The sign of the square root is determined whether one needs the retarded or the advanced function in the analytic continuation of equation (5.5.3).

In the high-magnetic field limit one can ignore the coupling between the Landau levels and the integral equation for the vertex function reduces to an algebraic equation which is trivially solved with the result

$$\Gamma(\boldsymbol{q}, n, m, \mathrm{i}\epsilon + \mathrm{i}\omega, \mathrm{i}\epsilon) = \frac{f_{n,m}(\boldsymbol{q})}{1 - I(\boldsymbol{q}, n, m)\mathcal{G}(n, \mathrm{i}\epsilon + \mathrm{i}\omega)\mathcal{G}(m, \mathrm{i}\epsilon)}, \qquad (5.5.5)$$

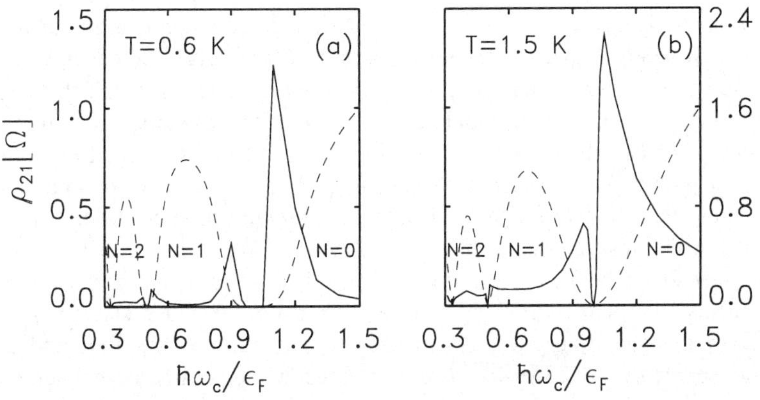

Figure 5.8 *Transresistivity* ρ_{21} *(solid curves) and the thermally averaged DOS* $g = \partial n/\partial\mu$ *(dashed, in arbitrary units) for (a)* $T = 0.6$ *K and (b)* $T = 1.5$ *K as a function of magnetic field in GaAs for density* $n = 1.5 \times 10^{11}\,\mathrm{cm}^{-2}$ *(*$E_F/k_B \approx 60\,\mathrm{K}$*), well separation* $d = 350$ Å *and zero well widths. N is the LL index and* $\hbar\omega_c = \epsilon_F$ *corresponds to* $B = 3.1$ T. *While the* $g(B)$ *peaks in the middle of the Landau level, the interlayer coupling is weakest there (due to large screening), pushing the peaks in* ρ_{21} *towards the edges of the Landau level bands (from [12]).*

where the form factor $I(\boldsymbol{q}, n, m)$ has, for short-range scatterers, the explicit form:

$$I(\boldsymbol{q}, n, m) = \frac{U^2 \rho_{\mathrm{imp}}}{a\pi \ell_{\mathrm{B}}^2 \hbar^2} (-1)^{n+m} \mathrm{e}^{-(\ell_{\mathrm{B}} q)^2} L_n^{m-n}[(\ell_{\mathrm{B}} q)^2/2] L_m^{n-m}[(\ell_{\mathrm{B}} q)^2/2]. \tag{5.5.6}$$

Thus, with this solution for the vertex function we have an explicit expression for the triangle function (5.5.1), and we can numerically evaluate the transconductivity (5.3.18), and hence the transresistivity.

5.5.3 Numerical results

Qualitatively we would expect the low-temperature transresistivity for fixed T to be directly proportional to the product of the thermally averaged density of states of both layers, $g_1(B) \times g_2(B)$, around the chemical potential μ. Since the density of states is strongly enhanced over its zero-field value, one might expect that: 1. $|\rho_{21}^{xx}(\omega_c \tau \gg 1)| \gg |\rho_{21}^{xx}(B = 0)|$; 2. $\rho_{21}^{xx}(B)$ should more or less simply reflect the shape of $g_1(B) g_2(B)$.

Figure 5.8 shows the results of a calculation for $\rho_{21}^{xx}(B)$ for two identical layers at fixed densities. For comparison, we also show $g(B)$. As expected, ρ_{21}^{xx} is very large: approximately 50–100 times larger than at $B = 0$. This is in agreement with recent experimental results [66, 28]. Also, ρ_{21}^{xx} is largest when

μ is in the bands of extended states and suppressed when it is in between the extended bands. However, the shape of ρ_{21}^{xx} is markedly different from $g^2(B)$.

We can understand this surprising behaviour as follows. Recall that ρ_{21} also depends on the interlayer coupling, i.e. also on the screening properties of the system. For two-dimensional electron gases the screened interaction varies inversely with g [27]. Therefore, increasing $g(B)$ weakens the interlayer coupling, implying that the terms $g_1 g_2$ and $|W|^2$ tend to work in opposition. This results in the following scenario when B is changed. When μ lies in the region of localized states below a Landau level band, ρ_{21} is very small because very few electrons have sufficient energy to be excited into extended states where they contribute to the drag. As B is increased so that μ moves into the Landau level band, the density of extended states increases, while the interlayer interaction is strong due to weak screening, resulting in a sharp rise in ρ_{21}. However, as the magnetic field is further increased so that μ moves closer towards the centre of the Landau level and the density of states further increases, the screening becomes so effective that it more than compensates for the increase in density of states, leading to a reduction in ρ_{21}. This competition of density of states and screening produces the unique shape of $\rho_{21}(B)$. It should be remarked that while the recent data [66, 28] is in qualitative agreement with our calculations and, in fact, also shows a double peak structure, it cannot necessarily be viewed as being a quantitative confirmation of our predictions. The reason is that our simplified calculation ignores spin-splitting, which is a possible reason for the experimental double peaks. The experiment of [28], however, seems to be best interpreted in terms the competition between screening and phase-space, as suggested by our theory. Clearly at this point more systematic studies are needed. It is an open question whether an inclusion of the spin-splitting in the theory would change the results. Likewise, coupling between the Landau levels should be accounted for. New, unexplained data is rapidly becoming available, both for low and extremely high magnetic fields [66], underlining the vitality of this research area.

5.6 SUMMARY OF COULOMB DRAG

We have used the phenomenon of Coulomb drag as a benchmark case to illustrate the concepts involved in linear response quantum transport theory. This has allowed us to both get hands-on experience with the theoretical tools (Matsubara formalism, self-consistent equations for the Green's functions, vertex equations etc.), and has at the same time brought us to contact with an experimentally lively research area. There are many open theoretical considerations, the most important of which are, in our opinion, a detailed evaluation of the phonon–mediated interaction* and an extension of the present theory to one-dimensional

* See, however, the very recent [67], which appears to resolve the remaining problems related to phonon-mediated interactions.

coupled systems, where interactions within the subsystems may give rise to qualitative changes to the Fermi liquid picture underlying the ideas presented in this chapter.

5.7 NONEQUILIBRIUM GREEN'S FUNCTION TECHNIQUES

5.7.1 Introductionary remarks

We stressed the importance of considering interactions in section 5.1 and here we re-emphasize it: a rigorous discussion of transport in an interacting mesoscopic system requires a formalism which is capable of explicitly including the interactions. Obvious candidates for such a theoretical tool are various techniques based on Green's functions. Since many problems of interest involve systems far from equilibrium, we cannot use linear-response methods, such as those based on the Kubo formula illustrated in section 5.3, but must use an approach capable of addressing the full nonequilibrium situation. The nonequilibrum Green's function techniques, as developed about 30 years ago by Kadanoff and Baym [29] and Keldysh [30], have recently gained increasing attention in the analysis of transport phenomena in mesoscopic semiconductors systems [31, 33]. In particular, the steady state situation has been addressed by a large number of papers [34–39]. Among the central results obtained in these papers is that under certain conditions (to be discussed below) a Landauer-type conductance formula [40] can be derived. This is quite appealing in view of the wide-spread success of conductance formulae in the analysis of transport in mesoscopic systems. More recently, the attention of many researchers has been drawn towards time-dependent situations, which is witnessed by the rapidly growing literature, e.g. [41–59].

We next briefly outline the physical background of the Baym–Kadanoff–Keldysh nonequilibrium Green's functions. Readers interested in further details can find them, for example in [24, 29, 30, 32, 33]. The basic difference between construction of equilibrium and nonequilibrium perturbation schemes is that in nonequilibrium one cannot assume that the system returns to its ground state (or its thermodynamic equilibrium state at finite temperatures) as $t \to +\infty$. Irreversible effects break the symmetry between $t = -\infty$ and $t = +\infty$ and this symmetry is heavily exploited in the derivation of the equilibrium perturbation expansion. In nonequilibrium situations one can circumvent this problem by allowing the system to evolve from $-\infty$ to the moment of interest (for definiteness, let us call this instant t_0) and then continues the time evolvement from $t = t_0$ back to $t = -\infty$ [60, 61]. When dealing with quantities which depend on two time variables, such as Green's functions, the time evolution must be continued to the later time.

The advantage of this procedure is that all expectation values are defined with respect to a well-defined state, i.e. the state in which the system was prepared

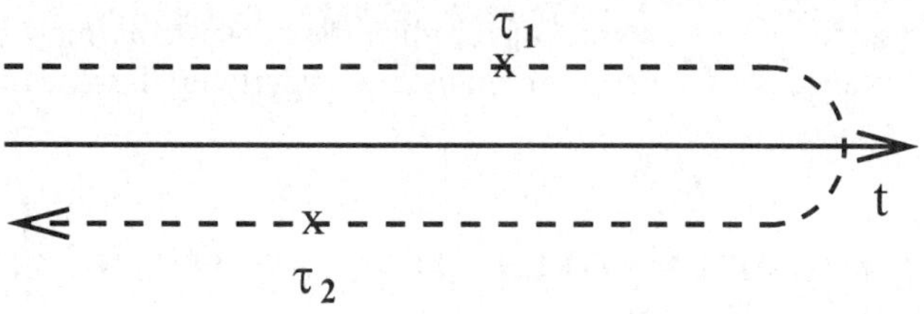

Figure 5.9 *The complex-time contour on which the nonequilibrium Green's function theory is constructed. In the contour sense, the time τ_1 is earlier than τ_2 even though its real-time projection appears larger.*

in the remote past. The price is that one must treat the two time branches on an equal footing (Fig. 5.9).

A typical object of interest would be a two-time Green's function; the two times can be located on either of the two branches of the complex time path (e.g. τ and τ' in Fig. 5.9). One is thus led to consider 2×2 Green's function matrices and the various terms in the perturbation theory can be evaluated by matrix multiplication. Since the internal time-integrations run over the complex time path, a method of book-keeping for the time labels is required, and there are various ways of doing this. In the present work we employ a version of the Keldysh technique.

In the context of tunnelling problems the time-independent Keldysh formalism works as follows. In the remote past the contacts (i.e. the left and right lead) and the central region are decoupled, and each region is in thermal equilibrium. The equilibrium distribution functions for the three regions are characterized by their respective chemical potentials; these do not have to coincide nor are the differences between the chemical potentials necessarily small. The couplings between the different regions are then established and treated as perturbations via the standard techniques of perturbation theory, albeit on the two branch time contour. It is important to notice that the coupling do not have to be small, for example with respect level spacings or $k_B T$ and typically must be treated to all orders.

A central question that one must address is: under which conditions are the nonequilibrium techniques, applied successfully to the steady-state problem, transferrable to time-dependent situations, such as the experiments mentioned above?

The time-dependent problem has to be formulated carefully, particularly with respect to the leads. It is essential to a Landauer type of approach that the electrons in the leads be noninteracting. In practice, however, the electrons in the leads near the mesoscopic region contribute to the self-consistent potential. We approach this problem by dividing the transport physics in two steps [53]. 1. The self-consistent determination of charge pile-up and depletion in the contacts, the resulting barrier heights and single-particle energies in the interacting region. 2. Transport in a system defined by these self-consistent parameters. Step 1 requires a capacitance calculation for each specific geometry [53] and we do not address it in this chapter. The interested reader is referred to Chapter 7 of this book, where Büttiker and Christen present several examples of how the self-consistency can be achieved. For the present purposes we assume the results of 1. as time-dependent input parameters and give a full treatment of the transport through the mesoscopic region 2. In practice, the interactions in the leads are absorbed into a time-dependent potential and from then on the electrons in the leads are treated as noninteracting. This means that when relating our results to actual experiments some care must be exercised. Specifically, we calculate only the current flowing into the mesoscopic region, while the total time-dependent current measured in the contacts includes contributions from charge flowing in and out of accumulation and depletion regions in the leads. In the time-averaged (DC) current, however, these capacitive contributions vanish and the corresponding time-averaged theoretical formulae, such as equation (5.9.16) below, are directly relevant to experiment. It should be noted, though, that these capacitive currents may influence the effective time-dependent parameters in step 1. above. Therefore, we have the following construction in mind when we address time-dependent problems. Before the couplings between the various regions are turned on, the single-particle energies acquire rigid time-dependent shifts, which, in the case of the noninteracting contacts, translate into extra phase factors for the propagators (but not in changes in occupations). The perturbation theory with respect to the couplings has the same diagrammatic structure as in the stationary case. The calculations, of course, become more complicated because of the broken time translational invariance.

5.8 MODEL HAMILTONIAN

We split the total Hamiltonian in three pieces: $H = H_c + H_T + H_{\rm cen}$, where H_c describes the contacts, H_T is the tunnelling coupling between contacts and the interacting region and $H_{\rm cen}$ models the interacting central region, respectively. Below we discuss each of these terms. Figure 5.10 shows a representative experimental configuration.

5.8.1 Contacts, H_c

Guided by the typical experimental geometry in which the leads rapidly broaden into metallic contacts, we view electrons in the leads as noninteracting except for an overall self-consistent potential. Physically, applying a time-dependent bias between the source and drain contacts corresponds to accumulating or depleting charge to form a dipole around the central region. The resulting electrostatic-potential difference means that the single-particle energies become time dependent: $\epsilon^0_{k\alpha} \to \epsilon_{k\alpha}(t) = \epsilon^0_{k\alpha} + \Delta_\alpha(t)$ (here α labels the channel in the left (L) or right (R) lead). The occupation of each state $k\alpha$, however, remains unchanged.

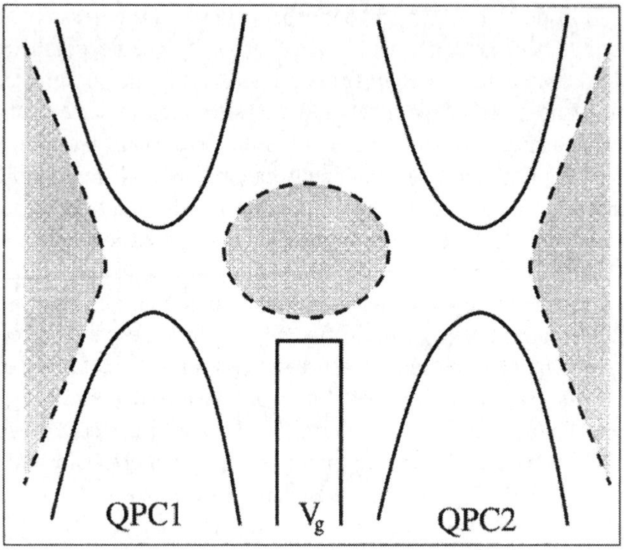

Figure 5.10 *Coulomb island, which consists of two tunable quantum point contacts QPC1 and QPC2 and a side gate which allows one to vary the chemical potential and hence the charge density in the central region. The two-dimensional electron gas underlying the gate structures is depleted outside the hatched regions. Time-dependent voltages attached to the QPCs, or to the side gate can be modelled by the Hamiltonian discussed in the text.*

The occupation, for each contact, is determined by an equilibrium distribution function established in the distant past, before the time dependence or tunnelling matrix elements are turned on. Thus, the contact Hamiltonian is

$$H_c = \sum_{k,\alpha \in L,R} \epsilon_{k\alpha}(t) c_{k\alpha}^{\dagger} c_{k\alpha} \tag{5.8.1}$$

and the exact time-dependent Green's functions in the leads for the uncoupled system are

$$\begin{aligned} g_{k\alpha}^{<}(t,t') &\equiv \mathrm{i}\langle c_{k\alpha}^{\dagger}(t') c_{k\alpha}(t)\rangle \\ &= \mathrm{i} f(\epsilon_{k\alpha}^{0}) \exp\left[-\mathrm{i}\int_{t'}^{t} \mathrm{d}t_1\, \epsilon_{k\alpha}(t_1)\right] \\ g_{k\alpha}^{r,a}(t,t') &\equiv \mp \mathrm{i}\theta(\pm t \mp t')\langle\{c_{k\alpha}(t), c_{k\alpha}^{\dagger}(t')\}\rangle \\ &= \mp \mathrm{i}\theta(\pm t \mp t') \exp\left[-\mathrm{i}\int_{t'}^{t} \mathrm{d}t_1\, \epsilon_{k\alpha}(t_1)\right]. \end{aligned} \tag{5.8.2}$$

5.8.2 *Coupling between leads and central region, H_T*

The coupling between the leads and the central (interacting) region can be modified with time-dependent gate voltages, as is the case in single-electron pumps. The precise functional form of the time dependence is determined by the detailed geometry and self-consistent response of charge in the contacts to external driving. We assume that these parameters are known and simply write

$$H_T = \sum_{\substack{k,\alpha\in L,R \\ n}} [V_{k\alpha,n}(t)\boldsymbol{c}^{\dagger}_{k\alpha}\boldsymbol{d}_n + \text{h.c.}]. \tag{5.8.3}$$

Here $\{\boldsymbol{d}^{\dagger}_n\}$ and $\{\boldsymbol{d}_n\}$ form a complete orthonormal set of single-electron creation and annihilation operators in the interacting region.

5.8.3 *The central region Hamiltonian H_{cen}*

The form chosen for H_{cen} in the central interacting region depends on geometry and the physical behaviour being investigated. Our results relating the current to local properties, such as densities of states and Green's functions, are valid generally. To make the results more concrete, we will discuss two particular examples in detail. In the first, the central region is taken to consist of noninteracting, but time-dependent levels,

$$H_{\text{cen}} = \sum_m \epsilon_m(t)\boldsymbol{d}^{\dagger}_m\boldsymbol{d}_m. \tag{5.8.4}$$

Here $\boldsymbol{d}^{\dagger}_m(\boldsymbol{d}_m)$ creates (destroys) an electron in state m. The choice (5.8.4) represents a simple model for time-dependent resonant tunnelling. Below we shall present general results for an arbitrary number of levels and analyse the case of a single level in detail. The latter is interesting both as an exactly solvable example and for predictions of coherence effects in time-dependent experiments.

The second example that we shall discuss is resonant tunnelling with electron–phonon interaction,

$$H^{\text{el-ph}}_{\text{cen}} = \epsilon_0\boldsymbol{d}^{\dagger}\boldsymbol{d} + \boldsymbol{d}^{\dagger}\boldsymbol{d}\sum_{\boldsymbol{q}} M_{\boldsymbol{q}}[\boldsymbol{a}^{\dagger}_{\boldsymbol{q}} + \boldsymbol{a}_{-\boldsymbol{q}}]. \tag{5.8.5}$$

In the above, the first term represents a single site, while the second term represents interaction of an electron on the site with phonons: $\boldsymbol{a}^{\dagger}_{\boldsymbol{q}}(\boldsymbol{a}_{\boldsymbol{q}})$ creates (destroys) a phonon in mode $\boldsymbol{q}$, and $M_{\boldsymbol{q}}$ is the interaction matrix element. The full Hamiltonian of the system must also include the free-phonon contribution $H_{\text{ph}} = \sum_{\boldsymbol{q}} \hbar\omega_{\boldsymbol{q}}\boldsymbol{a}^{\dagger}_{\boldsymbol{q}}\boldsymbol{a}_{\boldsymbol{q}}$. This example, while not exactly solvable, is helpful to show how interactions influence the current. Furthermore, we can do a direct comparison with previous time-independent results [57] using (5.8.5) to demonstrate the power of the present formalism. We also outline a new technique of how to calculate the central region nonequilibrium Green's function for the central region [62].

5.9 CALCULATION OF THE TUNNELLING CURRENT

The current from the left contact through the left barrier to the central region can be calculated from the time evolution of the occupation number operator of the left contact:

$$J_{\mathrm{L}}(t) = -e\langle \dot{N}_{\mathrm{L}} \rangle = -\frac{\mathrm{i}e}{\hbar}\langle [H, N_{\mathrm{L}}] \rangle, \tag{5.9.1}$$

where $N_{\mathrm{L}} = \sum_{k,\alpha\in L} c_{k\alpha}^{\dagger} c_{k\alpha}$ and $H = H_c + H_T + H_{\mathrm{cen}}$. Since H_c and H_{cen} commute with N_{L}, one readily finds

$$J_{\mathrm{L}} = \frac{\mathrm{i}e}{\hbar} \sum_{\substack{k,\alpha\in L \\ n}} \left[V_{k\alpha,n} \langle c_{k\alpha}^{\dagger} d_n \rangle - V_{k\alpha,n}^{*} \langle d_n^{\dagger} c_{k\alpha} \rangle \right]. \tag{5.9.2}$$

Now define two Green's functions:

$$G_{n,k\alpha}^{<}(t,t') \equiv \mathrm{i}\langle c_{k\alpha}^{\dagger}(t') d_n(t) \rangle \tag{5.9.3}$$

$$G_{k\alpha,n}^{<}(t,t') \equiv \mathrm{i}\langle d_n^{\dagger}(t') c_{k\alpha}(t) \rangle. \tag{5.9.4}$$

Using $G_{k\alpha,n}^{<}(t,t) = -[G_{n,k\alpha}^{<}(t,t)]^{*}$ and inserting the time labels, the current can be expressed as

$$J_{\mathrm{L}}(t) = \frac{2e}{\hbar} \mathrm{Re} \left\{ \sum_{\substack{k,\alpha\in L \\ n}} V_{k\alpha,n}(t) G_{n,k\alpha}^{<}(t,t) \right\}. \tag{5.9.5}$$

One next needs an expression for $G_{n,k\alpha}^{<}(t,t')$. For the present case, with non-interacting leads, a general relation for the contour-ordered Green's function $G_{n,k\alpha}$ can be derived rather easily, either with the equation-of-motion technique, or by a direct expansion of the S-matrix [54]. The result is

$$G_{n,k\alpha}(\tau,\tau') = \sum_m \int \mathrm{d}\tau_1 \, G_{nm}(\tau,\tau_1) V_{k\alpha,m}^{*}(\tau_1) g_{k\alpha}(\tau_1,\tau'). \tag{5.9.6}$$

Here $G_{nm}(\tau,\tau_1)$ is the contour-ordered Green's function for the central region and the τ-variables are now defined on the contour of Fig. 5.9. Note that the time dependence of the tunnelling matrix elements and single-particle energies has broken the time-translational invariance. The standard analytic continuation rules [32, 33] can now be applied and we find

$$\begin{aligned} G_{n,k\alpha}^{<}(t,t') &= \sum_m \int \mathrm{d}t_1 \, V_{k\alpha,m}^{*}(t_1) [G_{nm}^{r}(t,t_1) g_{k\alpha}^{<}(t_1,t') \\ &\quad + G_{nm}^{<}(t,t_1) g_{k\alpha}^{a}(t_1,t')], \end{aligned} \tag{5.9.7}$$

where the Green's functions $g^{<,a}$ for the leads are defined in (5.8.2) above. Combining (5.8.2), (5.9.5) and (5.9.7), yields

$$J_{\rm L}(t) = -\frac{2e}{\hbar}{\rm Im}\Big\{\sum_{\substack{k,\alpha\in L\\ n,m}} V_{k\alpha,n}(t)\int_{-\infty}^{t} {\rm d}t_1\, {\rm e}^{{\rm i}\int_{t_1}^{t}{\rm d}t_2\,\epsilon_{k\alpha}(t_2)} \times V^{*}_{k\alpha,m}(t_1)[G^{r}_{nm}(t,t_1)f_L(\epsilon_{k\alpha})+G^{<}_{nm}(t,t_1)]\Big\}. \quad (5.9.8)$$

The discrete sum over k in $\sum_{k\alpha}$ can be expressed in terms of densities of states in the leads: $\int {\rm d}\epsilon\,\rho_\alpha(\epsilon)$. Then it is useful to define

$$[\Gamma^L(\epsilon,t_1,t)]_{mn} = 2\pi\sum_{\alpha\in L}\rho_\alpha(\epsilon)V_{\alpha,n}(\epsilon,t)V^{*}_{\alpha,m}(\epsilon,t_1) \times \exp[{\rm i}\int_{t_1}^{t}{\rm d}t_2\,\Delta_\alpha(\epsilon,t_2)], \quad (5.9.9)$$

where $V_{k\alpha,n}=V_{\alpha,n}(\epsilon_k)$. In terms of this generalized line-width function (5.9.9), the general expression for the current is

$$J_{\rm L}(t) = -\frac{2e}{\hbar}\int_{-\infty}^{t}{\rm d}t_1\int\frac{{\rm d}\epsilon}{2\pi}{\rm Im}{\rm Tr}\Big\{{\rm e}^{-{\rm i}\epsilon(t_1-t)}\boldsymbol{\Gamma}^L(\epsilon,t_1,t) \times[\boldsymbol{G}^{<}(t,t_1)+f_{\rm L}(\epsilon)\boldsymbol{G}^{r}(t,t_1)]\Big\}. \quad (5.9.10)$$

Here the bold-face notation indicates that the level-width function $\boldsymbol{\Gamma}$ and the central-region Green's functions $\boldsymbol{G}^{<,r}$ are matrices in the central-region indices m,n. An analogous formula applies for $J_R(t)$, the current flowing into the central region through the right barrier.

Equation (5.9.10) provides a very general starting point for specific applications and in subsequent sections we discuss its evaluation in several different levels of approximation. Here we note the following general properties. The current is expressed in terms of local quantities: Green's functions of the central region. The first term in equation (5.9.10), which is proportional to the lesser function $G^<$, suggests an interpretation as the out-tunnelling rate (recalling ${\rm Im}\,G^<(t,t)=N(t)$). Likewise, the second term, which is proportional to the occupation in the leads and to the density of states in the central region, can be associated to the in-tunnelling rate. However, one should bear in mind that all Green's functions in equation (5.9.10) are to be calculated in the presence of tunnelling. Thus, $G^<$ will depend on the occupation in the leads. Furthermore, in the presence of interactions G^r will depend on the central region occupation. Consequently, the current can be a nonlinear function of the occupation factors. Recently, this issue has also been discussed by other authors [63].

5.9.1 Time-independent case

In the time-independent case equation (5.9.10) reduces to

$$J = \frac{ie}{2\hbar}\int\frac{d\epsilon}{2\pi}\mathrm{Tr}\{[\mathbf{\Gamma}^L(\epsilon)-\mathbf{\Gamma}^R(\epsilon)]\mathbf{G}^<(\epsilon) + [f_L(\epsilon)\mathbf{\Gamma}^L(\epsilon) - f_R(\epsilon)\mathbf{\Gamma}^R(\epsilon)][\mathbf{G}^r(\epsilon)-\mathbf{G}^a(\epsilon)]\}. \quad (5.9.11)$$

This result was reported in [34] and applied to the out-of-equilibrium Anderson impurity problem. If the left and right line-width functions are proportional to each other, i.e. $\mathbf{\Gamma}^L(\epsilon) = \lambda\mathbf{\Gamma}^R(\epsilon)$, further simplification can be achieved:

$$J = \frac{ie}{\hbar}\int\frac{d\epsilon}{2\pi}[f_L(\epsilon)-f_R(\epsilon)] \times \mathrm{Tr}\left\{\frac{\mathbf{\Gamma}^L(\epsilon)\mathbf{\Gamma}^R(\epsilon)}{\mathbf{\Gamma}^L(\epsilon)+\mathbf{\Gamma}^R(\epsilon)}\left[\mathbf{G}^r(\epsilon)-\mathbf{G}^a(\epsilon)\right]\right\}. \quad (5.9.12)$$

The ratio is well defined because the $\mathbf{\Gamma}$-matrices are proportional. The difference between the retarded and advanced Green's functions is essentially the density of states. Despite the apparent similarity of (5.9.12) to the Landauer formula, it is important to bear in mind that in general there is no immediate connection between the quantity $\mathrm{Tr}\{(\mathbf{\Gamma}^L(\epsilon)\mathbf{\Gamma}^R(\epsilon)/[\mathbf{\Gamma}^L(\epsilon)+\mathbf{\Gamma}^R(\epsilon)])[\mathbf{G}^r(\epsilon)-\mathbf{G}^a(\epsilon)]\}$ and the transmission coefficient. In particular, when inelastic scattering is present, we do not believe that such a connection exists. To further illustrate this point, suppose now that the Green's function for the interacting central region can be solved: $G^{r,a}(\epsilon) = [\epsilon - \epsilon_0 - \lambda(\epsilon) \pm i\gamma(\epsilon)/2]^{-1}$, where λ and $\gamma/2$ are the real and imaginary parts of the self-energy (including interactions and tunnelling). Then the interacting result for proportionate coupling (5.9.12) becomes

$$J = \frac{e}{\hbar}\int\frac{d\epsilon}{2\pi}[f_L(\epsilon)-f_R(\epsilon)]\frac{\Gamma^L(\epsilon)\Gamma^R(\epsilon)}{\Gamma^L(\epsilon)+\Gamma^R(\epsilon)} \times\frac{\gamma(\epsilon)}{[\epsilon-\epsilon_0-\lambda(\epsilon)]^2+[\gamma(\epsilon)/2]^2}. \quad (5.9.13)$$

The factors multiplying the difference of the lead distribution functions reduce to the transmission coefficient only in the noninteracting limit, $\lambda(\epsilon)\to\Lambda(\epsilon)$ and $\gamma(\epsilon)\to\Gamma(\epsilon) = \Gamma^R(\epsilon)+\Gamma^L(\epsilon)$. In a phenomenological model, where the total level width is expressed as a sum of elastic and inelastic widths, $\gamma = \gamma_e + \gamma_i$, one recovers the results of Jonson and Grincwajg and Weil and Vinter [64].

5.9.2 Average current

Measurements are often performed at low frequencies and it is therefore important to consider time averages of the various measurable quantities. In analogy with section 5.9.1, the time-dependent case allows further simplification, if assumptions are made on the line-width functions. In particular, we assume a generalized proportionality condition:

$$\Gamma^L(\epsilon, t_1, t) = \lambda \Gamma^R(\epsilon, t_1, t). \tag{5.9.14}$$

One should note that in general this condition can only be satisfied if $\Delta_\alpha^L(t) = \Delta_\alpha^R(t) = \Delta(t)$. However, in the wide-band limit (WBL), to be considered in detail below, the time-variations of the energies in the leads do not have to be equal.

The time average of a time-dependent object $F(t)$ is defined by

$$\langle F(t) \rangle = \lim_{T\to\infty} \frac{1}{T} \int_{-T/2}^{T/2} dt\, F(t). \tag{5.9.15}$$

If $F(t)$ is a periodic function of time, it is sufficient to average over the period. It can then be shown [54] that the time-averaged current can be computed from

$$\begin{aligned} \langle J_L(t) \rangle &= -\frac{2e}{\hbar} \int \frac{d\epsilon}{2\pi} [f_L(\epsilon) - f_R(\epsilon)] \\ &\quad \times \mathrm{ImTr} \left\{ \frac{\Gamma^L(\epsilon)\Gamma^R(\epsilon)}{\Gamma^L(\epsilon) + \Gamma^R(\epsilon)} \langle u(t) A(\epsilon, t) \rangle \right\}, \end{aligned} \tag{5.9.16}$$

where

$$A(\epsilon, t) = \int dt_1\, u(t_1) G^r(t, t_1) \exp[i\epsilon(t - t_1) + i \int_{t_1}^{t} dt_2\, \Delta(t_2)]. \tag{5.9.17}$$

Here we assumed that the energy and time dependences of the tunnelling coupling can be factorized: $V_{k\alpha,n}(t) = u(t) V_{\alpha,n}(\epsilon_k)$. Due to equation (5.9.14) we do not have to distinguish between L/R in the definition of $A(\epsilon, t)$; below we shall encounter situations where this distinction is necessary.

Expression (5.9.16) is of the Landauer type: it expresses the current as an integral over a weighted density of states times the difference of the two contact occupation factors. It is valid for arbitrary interactions in the central region, but it was derived with the somewhat restrictive assumption of proportional couplings to the leads.

5.10 NONINTERACTING RESONANT-LEVEL MODEL

5.10.1 General formulation

The real strength and importance, of the nonequilibrium Green's function formalism lies in its ability to treat interacting systems. However, it is useful to study a noninteracting case, first in order to get some general familiarity with the techniques. There are two additional bonuses: a number of interesting results can be obtained, and analysed in detail, for the noninteracting case and, secondly, the complicated many-body physics can under certain circumstances be described with judiciously chosen noninteracting models and thus the results given below may have a wider range of applications than one might think at first sight. In the noninteracting case the Hamiltonian is $H = H_c + H_T + H_{\mathrm{cen}}$, where $H_{\mathrm{cen}} = \sum_n \epsilon_n \boldsymbol{d}_n^\dagger \boldsymbol{d}_n$. Following standard analysis, one can derive the Dyson equation for the retarded Green's function,

$$\boldsymbol{G}^r(t,t') = \boldsymbol{g}^r(t,t') + \int \mathrm{d}t_1 \int \mathrm{d}t_2\, \boldsymbol{g}^r(t,t_1)\boldsymbol{\Sigma}^r(t_1,t_2)\boldsymbol{G}^r(t_2,t'), \qquad (5.10.1)$$

where

$$\Sigma^r_{nn'}(t_1,t_2) = \sum_{k\alpha\in L,R} V^*_{k\alpha,n}(t_1) g^r_{k\alpha}(t_1,t_2) V_{k\alpha,n'}(t_2), \qquad (5.10.2)$$

and $g^r_{k\alpha}$ is given by equation (5.8.2). The Keldysh equation for $G^<$ is

$$\begin{aligned} \boldsymbol{G}^<(t,t') &= \int \mathrm{d}t_1 \int \mathrm{d}t_2\, \boldsymbol{G}^r(t,t_1)\boldsymbol{\Sigma}^<(t_1,t_2)\boldsymbol{G}^a(t_2,t') \\ &= \mathrm{i}\int \mathrm{d}t_1 \int \mathrm{d}t_2\, \boldsymbol{G}^r(t,t_1)\Bigg[\sum_{\mathrm{L},R}\int \frac{\mathrm{d}\epsilon}{2\pi} \mathrm{e}^{-\mathrm{i}\epsilon(t_1-t_2)} \\ &\qquad \times f_{\mathrm{L}/R}(\epsilon)\boldsymbol{\Gamma}^{L/R}(\epsilon,t_1,t_2)\Bigg]\boldsymbol{G}^a(t_2,t'). \end{aligned} \qquad (5.10.3)$$

Provided that the Dyson equation for the retarded Green's function can be solved, equation (5.10.3) together with the current expression equation (5.9.10) provides the complete solution to the noninteracting resonant-level model. Below we examine special cases where analytic progress can be made.

5.10.2 Time-independent case

In the time-independent case the time-translational invariance is restored, and it is advantageous to go over to energy variables:

$$\begin{aligned} \boldsymbol{G}^r(\epsilon) &= [(\boldsymbol{g}^r)^{-1} - \boldsymbol{\Sigma}^r(\epsilon)]^{-1} \\ \boldsymbol{G}^<(\epsilon) &= \boldsymbol{G}^r(\epsilon)\boldsymbol{\Sigma}^<(\epsilon)\boldsymbol{G}^a(\epsilon). \end{aligned} \qquad (5.10.4)$$

In general, the Dyson equation for the retarded Green's function requires matrix inversion. In the case of a single level, the scalar equations can be readily solved. The retarded (advanced) self-energy is

$$\Sigma^{r,a}(\epsilon) = \sum_{k\alpha\in L,R} \frac{|V_{k\alpha}|^2}{\epsilon - \epsilon_{k\alpha} \pm \mathrm{i}\eta} = \Lambda(\epsilon) \mp \frac{\mathrm{i}}{2}\Gamma(\epsilon), \qquad (5.10.5)$$

where the real and imaginary parts contain 'left' and 'right' contributions: $\Lambda(\epsilon) = \Lambda^L(\epsilon) + \Lambda^R(\epsilon)$ and $\Gamma(\epsilon) = \Gamma^L(\epsilon) + \Gamma^R(\epsilon)$. The lesser self-energy is

$$\begin{aligned} \Sigma^<(\epsilon) &= \sum_{k\alpha\in L,R} |V_{k\alpha}|^2 g^<_{k\alpha}(\epsilon) \\ &= \mathrm{i}[\Gamma^L(\epsilon) f_{\mathrm{L}}(\epsilon) + \Gamma^R(\epsilon) f_R(\epsilon)]. \end{aligned} \qquad (5.10.6)$$

Using the identities $G^r G^a = (G^r - G^a)/(1/G^a - 1/G^r) = a(\epsilon)/\Gamma(\epsilon)$ (here $a(\epsilon) = \mathrm{i}[G^r(\epsilon) - G^a(\epsilon))$ is the spectral function], one can write $G^<$ in a 'pseudoequilibrium' form:

$$G^<(\epsilon) = \mathrm{i}a(\epsilon)\bar{f}(\epsilon), \qquad (5.10.7)$$

where

$$\begin{aligned} \bar{f}(\epsilon) &= \frac{\Gamma^L(\epsilon) f_{\mathrm{L}}(\epsilon) + \Gamma^R(\epsilon) f_R(\epsilon)}{\Gamma(\epsilon)} \\ a(\epsilon) &= \frac{\Gamma(\epsilon)}{[\epsilon - \epsilon_0 - \Lambda(\epsilon)]^2 + [\Gamma(\epsilon)/2]^2}. \end{aligned} \qquad (5.10.8)$$

With these expressions the evaluation of the current (5.9.11) is straightforward:

$$\begin{aligned} J &= -\frac{e}{2\hbar}\int \frac{d\epsilon}{2\pi} a(\epsilon)\left\{\left[\Gamma^L(\epsilon) - \Gamma^R(\epsilon)\right]\bar{f}(\epsilon)\right. \\ &\quad \left. - \left[f_{\mathrm{L}}(\epsilon)\Gamma^L(\epsilon) - f_R(\epsilon)\Gamma^R)\right]\right\} \\ &= \frac{e}{\hbar}\int \frac{d\epsilon}{2\pi}\frac{\Gamma^L(\epsilon)\Gamma^R(\epsilon)}{[\epsilon - \epsilon_0 - \Lambda(\epsilon)]^2 + [\Gamma(\epsilon)/2]^2}\left[f_{\mathrm{L}}(\epsilon) - f_R(\epsilon)\right]. \end{aligned} \qquad (5.10.9)$$

The factor multiplying the difference of the Fermi functions is the elastic transmission coefficient $T(\epsilon)$ and thus this model yields a conductance of the Landauer type, $g = (e^2/h)T(\epsilon_{\mathrm{F}})$.

5.10.3 Wide-band limit

(a) Basic formulae

For simplicity, we continue to consider only a single level in the central region. As in the previous section, we assume that one can factorize the momentum and time dependence of the tunnelling coupling, but allow for different time dependence for each barrier: $V_{k\alpha}(t) \equiv u_{\mathrm{L}/R}(t)V_{\alpha,n}(\epsilon_k)$. Referring to equation (5.10.5), the WBL consists of: 1. neglecting the level-shift $\Lambda(\epsilon)$; 2. assuming that the line-widths are energy-independent constants, $\sum_{\alpha\in L,R}\Gamma_\alpha = \Gamma^{L/R}$; 3. allowing a single time dependence, $\Delta_{\mathrm{L}/R}(t)$, for the energies in each lead.

Let us comment on the relevance of the WBL. This approximation captures the main physics in a range of applications and has the great advantage of yielding explicit analytic results. In particular, transport is often dominated by states close to the Fermi level and since $\Gamma(\epsilon)$ and $\Lambda(\epsilon)$ are generally slowly varying functions of energy, the WBL for this case is an excellent approximation. The WBL also allows asymmetric barriers ($\Gamma_{\mathrm{L}} \neq \Gamma_R$). Consequently, it is possible to describe resonant-tunnelling systems under high bias by using a suitable model for the bias dependence of the level widths and/or shifts. Finally, while the simplest WBL leads to an unphysical monotonic I–V curve for a resonant-tunnelling diode (because the model lacks band edges), it is relatively simple to generalize the WBL so that it does yield negative differential resistance (section 5.10.5).

The retarded self-energy in equation (5.10.1) thus becomes

$$\Sigma^r(t_1, t_2) = -\frac{\mathrm{i}}{2}[\Gamma^L(t_1) + \Gamma^R(t_1)]\delta(t_1 - t_2). \tag{5.10.10}$$

With this self-energy, the retarded (advanced) Green's function becomes [44, 57]

$$G^{r,a}(t, t') = g^{r,a}(t, t') \exp\left\{\mp \int_{t'}^{t} \mathrm{d}t_1 \tfrac{1}{2}[\Gamma^L(t_1) + \Gamma^R(t_1)]\right\} \tag{5.10.11}$$

with

$$g^{r,a}(t, t') = \mp \mathrm{i}\theta(\pm t \mp t') \exp\left[-\mathrm{i}\int_{t'}^{t} \mathrm{d}t_1 \epsilon_0(t_1)\right]. \tag{5.10.12}$$

This solution can now be used to evaluate the lesser function equation (5.10.3) and further in equation (5.9.10) to obtain the time-dependent current. We write the current as (using $\mathrm{Im}\{G^<(t, t)\} = N(t)$, where $N(t)$ is the occupation of the resonant level)

$$\begin{aligned} J_{\mathrm{L}}(t) &= -\frac{e}{\hbar}\Bigg[\Gamma^L(t)N(t) + \int \frac{\mathrm{d}\epsilon}{\pi} f_{\mathrm{L}}(\epsilon) \\ &\quad \times \int_{-\infty}^{t} \mathrm{d}t_1\, \Gamma^L(t_1, t)\mathrm{Im}\{\mathrm{e}^{-\mathrm{i}\epsilon(t_1 - t)} G^r(t, t_1)\}\Bigg]. \end{aligned} \tag{5.10.13}$$

For a compact notation we introduce

$$A_{\mathrm{L}/R}(\epsilon, t) = \int \mathrm{d}t_1 u_{\mathrm{L}/R}(t_1) G^r(t, t_1) \exp[\mathrm{i}\epsilon(t - t_1) - \mathrm{i}\int_{t}^{t_1} \mathrm{d}t_2 \Delta_{\mathrm{L}/R}(t_2)]. \tag{5.10.14}$$

Obviously, in the time-independent case $A(\epsilon)$ is just the Fourier transform of the retarded Green's function $G^r(\epsilon)$. In terms of $A(\epsilon, t)$ the occupation $N(t)$ (using equation (5.10.3) for $G^<$) is given by

$$N(t) = \sum_{\mathrm{L},R} \Gamma^{L/R} \int \frac{\mathrm{d}\epsilon}{2\pi} f_{\mathrm{L}/R}(\epsilon) |A_{\mathrm{L}/R}(\epsilon, t)|^2. \tag{5.10.15}$$

We write the current as a sum of currents flowing out from the central region into the left (right) contact and currents flowing into the central region from the left (right) contact, $J_{\mathrm{L}/R}(t) = J^{\mathrm{out}}_{\mathrm{L}/R}(t) + J^{\mathrm{in}}_{\mathrm{L}/R}(t)$:

$$J^{\mathrm{out}}_{\mathrm{L}/R}(t) = -\frac{e}{\hbar}\Gamma^{L/R}(t)N(t) \tag{5.10.16}$$

$$J^{\mathrm{in}}_{\mathrm{L}/R}(t) = -\frac{e}{\hbar}\Gamma^{L/R} u_{\mathrm{L}/R}(t) \int \frac{\mathrm{d}\epsilon}{\pi} f_{\mathrm{L}/R}(\epsilon)\mathrm{Im}\{A_{\mathrm{L}/R}(\epsilon, t)\}. \tag{5.10.17}$$

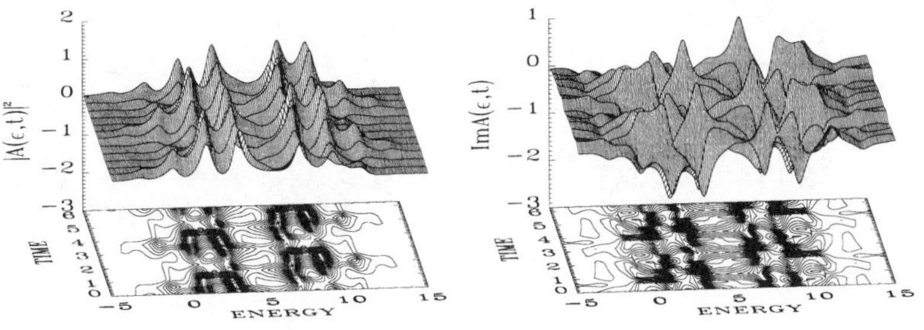

Figure 5.11 *(a)*$|A(\epsilon,t)|^2$ *as a function of time for harmonic modulation for a symmetric structure,* $\Gamma_L = \Gamma_R = \Gamma/2$. *The unit for the time-axis is* $\hbar/\Gamma$ *and all energies are measured in units of* Γ, *with the values* $\mu_L = 10$, $\mu_R = 0$, $\epsilon_0 = 5$, $\Delta = 5$, $\Delta_L = 10$ *and* $\Delta_R = 0$. *The modulation frequency is* $\omega = 2\Gamma/\hbar$. *(b) The time-dependence of* $\mathrm{Im}A(\epsilon,t)$ *for the case shown in (a).*

It is readily verified that these expressions coincide with equation (5.10.9) if all time-dependent quantities are replaced by constants.

The time-averaged current in the WBL is given by

$$\langle J\rangle = -\frac{2e}{\hbar}\frac{\Gamma^L\Gamma^R}{\Gamma^L+\Gamma^R}\int\frac{d\epsilon}{2\pi}\mathrm{Im}\{f_L(\epsilon)\langle u(t)A_L(\epsilon,t)\rangle - f_R(\epsilon)\langle u(t)A_R(\epsilon,t)\rangle\}. \tag{5.10.18}$$

Unlike the general case of equation (5.9.16), there is no restriction in the WBL that the time dependence be the same in the two leads. Equation (5.10.18) can therefore be used for the case of a time-dependent bias, where $\Delta_L(t)$ and $\Delta_R(t)$ will be different. In the next section we consider an example for the time variation, which is relevant for experimental situations.

5.10.4 Response to harmonic modulation

Harmonic time modulation is probably the most commonly encountered example of time dependence. Here we treat the case when the contact and site energy levels vary as

$$\Delta_{L/R,0}(t) = \Delta_{L/R,0}\cos(\omega t) \tag{5.10.19}$$

It is easy to generalize the treatment to situations where the modulation frequencies and/or phases are different in different parts of the device. Assuming that the barrier heights do not depend on time ($u_{L/R} = 1$) and substituting (5.10.19) in the expression (5.10.14) for $A(\epsilon, t)$, one finds

$$\begin{aligned} A_{L/R}(\epsilon, t) &= \exp\left[-\mathrm{i}\frac{(\Delta_0 - \Delta_{L/R})}{\omega}\sin(\omega t)\right] \\ &\times \sum_{k=-\infty}^{\infty} J_k\left(\frac{\Delta_0 - \Delta_{L/R}}{\omega}\right)\frac{\mathrm{e}^{\mathrm{i}k\omega t}}{\epsilon - \epsilon_0 - k\omega + \mathrm{i}\Gamma/2}, \end{aligned} \tag{5.10.20}$$

where $J_{-k}(x) = (-1)^k J_k(x)$. Figures 5.11($a$) and ($b$) show $|A(\epsilon, t)|^2$ and $\mathrm{Im}A(\epsilon, t)$ as a function of time, respectively. We recall from equations (5.10.15) and (5.10.16) that the current at a given time is determined by integrating $|A(\epsilon, t)|^2$ and $\mathrm{Im}A(\epsilon, t)$ over energy and thus an examination of Fig. 5.11 helps us to understand to complicated time dependence discussed below. (We show only A_L; similar results hold for A_R.) The three-dimensional plot (top part of Fig. 5.11) is projected down on a plane to yield a contour plot in order to help to visualize the time dependence. As expected, the time variation is periodic with period $T = 2\pi/\omega$. The time dependence is strikingly complex. The most easily recognized features are the maxima in the plot for $|A|^2$; these are related to photon side-bands occurring at $\epsilon = \epsilon_0 \pm k\omega$ (see also equation (5.10.21) below) [65].

The current is computed using the methods described in [54] and is shown in Fig. 5.12. We also display the drive voltage as a broken line. Bearing in mind the complex time dependence of $|A|^2$ and $\mathrm{Im}A$, which determine the out- and in-currents, respectively, it is not surprising that the current displays a nonadiabatic time dependence. The basic physical mechanism underlying the secondary maxima and minima in the current is the line-up of a photon-assisted resonant-tunnelling peak with the contact chemical potentials. The rapid time variations are due to J^{in} (or, equivalently, due to $\mathrm{Im}A$): the out-current J^{out} is determined by the occupation $N(t)$ and hence varies only on a timescale $\Gamma/\hbar$, which is the timescale for charge density changes.

We next consider the time average of the current. For the case of harmonic time dependence, we find

$$\langle \mathrm{Im}A_{L/R}(\epsilon, t)\rangle = -\frac{\Gamma}{2}\sum_{k=-\infty}^{\infty} J_k^2\left(\frac{\Delta_0 - \Delta_{L/R}}{\omega}\right)\frac{1}{(\epsilon - \epsilon_0 - k\omega)^2 + (\Gamma/2)^2}. \tag{5.10.21}$$

Figure 5.13 shows the resulting time-averaged current J_{DC}. A consequence of the complex harmonic structure of the time-dependent current is that for temperatures $k_B T < \hbar\omega$ the average current oscillates as function of period $2\pi/\omega$. The oscillation can be understood by examining the general expression for average current equation (5.9.16) together with (5.10.21): whenever a photon-assisted peak in the effective density of states, occurring at $\epsilon = \epsilon_0 \pm k\omega$ in the time-averaged density of states $\langle \mathrm{Im}A_{L/R}\rangle$, moves in or out of the allowed energy

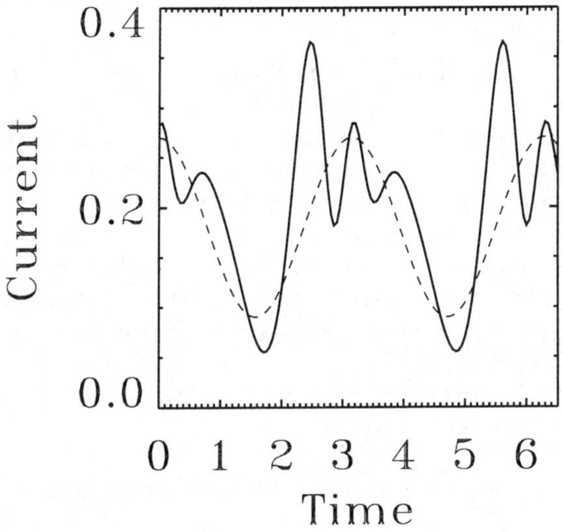

Figure 5.12 *The time-dependent current $J(t)$ for harmonic modulation corresponding to the parameters of Fig. 5.11. The DC bias is defined via $\mu_L = 10$ and $\mu_R = 0$, respectively. The dashed curve shows (not drawn to scale) the time dependence of the drive signal. The temperature is $k_B T = 0.1\Gamma$.*

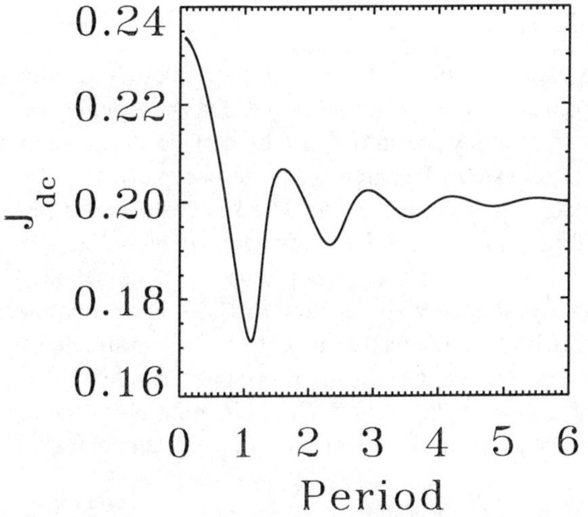

Figure 5.13 *Time-averaged current J_{DC} as a function of the AC oscillation period $2\pi/\omega$. The DC amplitudes are the same as those in Fig. 5.12.*

range, determined by the difference of the contact occupation factors, a maximum (or minimum) in the average current results.

5.10.5 Linear response

For circuit modelling purposes it would often be desirable to replace the mesoscopic device with a conventional circuit element, with an associated complex impedance $Z(\omega)$, or admittance $Y(\omega)$. Our results for the nonlinear time-dependent current form a very practical starting point for such a calculation. For the noninteracting case, the current is determined by $A(\epsilon, t)$ (see equation (5.10.15) and (5.10.16)) and all one has to do is to linearize A (equation (5.10.14)) with respect to the amplitude of the drive signal, i.e. $\Delta - \Delta_{L/R}$. It is important to notice that we do not linearize with respect to the chemical potential difference: the results given below apply to an arbitrary static bias voltage.

Performing the linearization, one finds [54] (here we give, for simplicity, just the result for zero external bias)

$$J^{(1)}(t) = -\frac{e}{\hbar}\frac{\Gamma}{4}\frac{\Delta_L - \Delta_R}{2\pi\omega}[\cos(\omega t)F(\omega) + \sin(\omega t)G(\omega)], \tag{5.10.22}$$

where

$$G_{L/R}(\omega) = \log \frac{|\mu_{L/R} - \epsilon_0 + i\Gamma/2|^2}{|(\mu_{L/R} - \epsilon_0 + i\Gamma/2)^2 - \omega^2|} \tag{5.10.23}$$

and

$$F_{L/R}(\omega) = \tan^{-1}\left[\frac{\mu_{L/R} - \epsilon_0 - \omega}{\Gamma/2}\right] - \tan^{-1}\left[\frac{\mu_{L/R} - \epsilon_0 + \omega}{\Gamma/2}\right]. \tag{5.10.24}$$

This result exactly coincides with the recent calculation of Fu and Dudley [52], which employed the AC Landauer–Büttiker linear-response theory.

We now wish to apply the formal results derived in this section to an experimentally relevant system. The archetypal mesoscopic with potential for applications is the resonant-tunnelling diode. The key feature of a resonant-tunnelling diode is its ability to show negative differential resistance (NDR). The WBL model studied in this section does not have this feature: its IV-characteristic, which is readily evaluated with equation (5.10.9), is a monotonically increasing function. A much more interesting model can be constructed by considering a model where the contacts have a finite occupied bandwidth; this can achieved by introducing a low-energy cut-off $D_{L/R}$ (in addition to the upper cut-off provided by the electrochemical potential). The zero-temperature IV-characteristic is now

$$\begin{aligned} J_{DC}(V) &= \frac{e}{h}\frac{2\Gamma_L\Gamma_R}{\Gamma}\left\{\tan^{-1}\left[\frac{\mu_L - \epsilon_0(V)}{\Gamma/2}\right] - \tan^{-1}\left[\frac{\mu_L - D_L - \epsilon_0(V)}{\Gamma/2}\right]\right. \\ &\quad \left. - \tan^{-1}\left[\frac{\mu_R(V) - \epsilon_0(V)}{\Gamma/2}\right] + \tan^{-1}\left[\frac{\mu_R(V) - D_R - \epsilon_0(V)}{\Gamma/2}\right]\right\}. \end{aligned} \tag{5.10.25}$$

Here we assume that the right chemical potential is field dependent: $\mu_R(V) =$

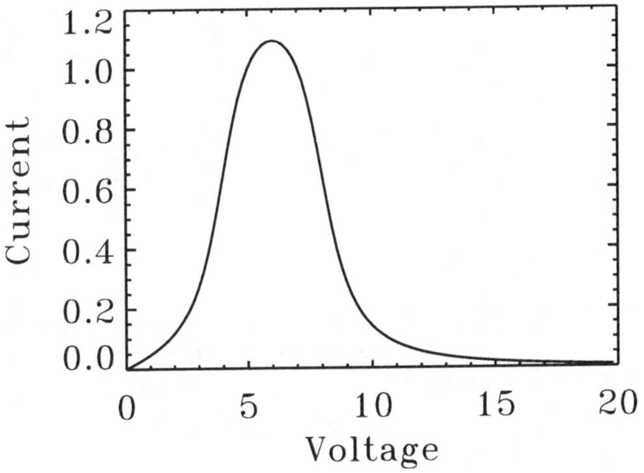

Figure 5.14 *IV-characteristic for a model resonant-tunnelling device (quantum dot). The system is defined by parameters* $\epsilon_0(V = 0) = 2$, $\mu_L = \mu_R(V = 0) = 0$ *and* $D_L = D_R = 2$ *and the current is given in units of* $e\Gamma/h$.

$\mu_R - eV$ and that the field dependence of the central region level is given by $\epsilon_0(V) = \epsilon_0 - V/2$. (A self-consistent calculation of the field dependence of the level $\epsilon_0(V)$ is performed in Chapter 7.) The resulting current–voltage characteristic is depicted in Fig. 5.14. We note that the strong increase in current, which is observed in experimental systems at very high voltages, is not present in our model: this is because we have ignored the bias dependence of the barrier heights as well as any higher-lying resonances. In Fig. 5.15, we show the resulting linear-response admittance $Y(\omega)$ for a symmetric structure ($\Gamma_L = \Gamma_R$). Several points are worth noticing. For DC bias $eV = 5$ (energies are given in units of Γ) the calculated admittance resembles qualitatively the results reported by Fu and Dudley for zero external bias, except that the change in sign for the imaginary part of $Y(\omega)$ is not seen. For zero external bias (not shown in the figure) our finite bandwidth model leads to an admittance, whose imaginary part changes sign and thus the behaviour found by Fu and Dudley cannot be ascribed to an artefact of their infinite bandwidth model. More interestingly, for DC bias in the NDR regime, the real part is negative for small frequencies. This simply reflects the fact that the device is operating under NDR bias conditions. At higher frequencies the real part becomes positive, thus indicating that further modelling along the lines sketched here may lead to important implications on the high-frequency response of resonant-tunnelling structures.

To conclude this section, we wish to emphasize that the linear-response analysis presented above is only a special case of the general results of section 5.9, which seem to have the potential for many applications.

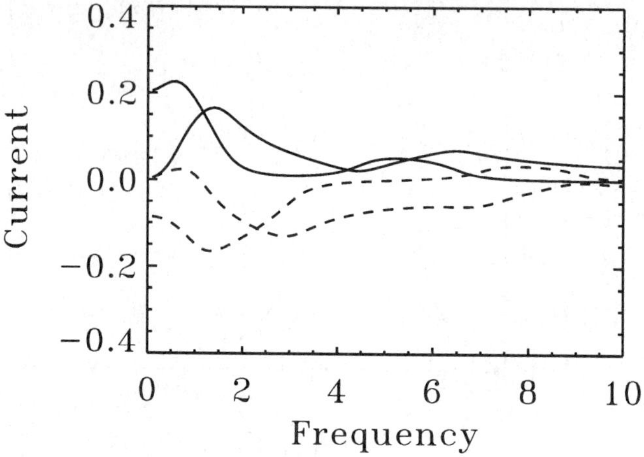

Figure 5.15 *In and out-of-phase components of the linear response current (in units of $e\Gamma/h$ and normalized with the amplitude of the drive signal Δ_L to yield admittance) for two bias points, $eV = 5$ (solid curve) and $eV = 10$ (dashed curve). Other parameters are as in Fig. 5.14. The out-of-phase components (or, equivalently, imaginary parts) always tend to zero for vanishing frequency, while the in-phase component can have either a positive or negative zero-frequency limit depending on the DC bias.*

5.11 RESONANT TUNNELLING WITH ELECTRON–PHONON INTERACTIONS

5.11.1 Single-particle theory

As a final application, we study the effect of phonons on resonant tunnelling. We shall first establish a connection to previous results [57] and then outline a new approach [62] which seems to have the potential of a wide range of applications. For simplicity, we consider a single resonant level with energy-independent level widths Γ_L and Γ_R (i.e. the WBL). The expression for the current equation (5.9.12) now becomes

$$J = \frac{e}{\hbar} \frac{\Gamma^L \Gamma^R}{\Gamma^L + \Gamma^R} \int \frac{d\epsilon}{2\pi} [f_L(\epsilon) - f_R(\epsilon)] \int_{-\infty}^{\infty} dt e^{i\epsilon t} a(t), \tag{5.11.1}$$

where $a(t) = i[G^r(t) - G^a(t)]$ is the interacting spectral density. In general, an exact evaluation of $a(t)$ is not possible: the combined effect of tunnelling to the contact Fermi seas and electron–phonon interaction is to induce complicated many-body interactions between the particles, even though no such term explicitly appears in the Hamiltonian. However, if one considers a *single* particle interacting with the phonons, or, in other words, ignores the Fermi sea, $G^r(t)$

(and hence $a(t)$) can be calculated exactly*

$$G^r(t) = -\mathrm{i}\theta(t)\exp[-\mathrm{i}t(\epsilon_0 - \Delta) - \Phi(t) - \Gamma t/2], \tag{5.11.2}$$

where $\Delta = \sum_q \frac{M_q^2}{\omega_q}$ and

$$\Phi(t) = \sum_q \frac{M_q^2}{\omega_q^2}[N_q(1 - \mathrm{e}^{\mathrm{i}\omega_q t}) + (N_q + 1)(1 - \mathrm{e}^{-\mathrm{i}\omega_q t})], \tag{5.11.3}$$

and the electron–phonon interaction is given by equation (5.8.5). When substituted into the expression for current, one recovers the result of [57], which was originally derived by analysing the much more complex two-particle Green's function $G(\tau, s, t) = \theta(s)\theta(t)\langle \mathrm{d}(\tau - s)\mathrm{d}^\dagger(\tau)\mathrm{d}(t)\mathrm{d}^\dagger(0)\rangle$. The advantage of the method presented here is that one only needs the single-particle Green's function to use the interacting current formula (5.9.12). Other systematic approaches to the single-particle Green's function can therefore be directly applied to the current (e.g. perturbation theory in the tunnelling Hamiltonian) and the next section illustrates how the linked-cluster expansion [62] works within the present formulation.

5.11.2 Nonequilibrium linked-cluster expansion

(a) Equilibrium linked-cluster expansion

Here we briefly review some basic features of the linked cluster expansion (LCE) for one-particle Matsubara Green functions [17] and briefly comment on the application of LCE on many-particle systems. Consider a quantum many-particle system described by the Hamiltonian $H = H_0 + \lambda H_1$, where H_0 is diagonalizable and H_1 contains the interactions. λ is a formal device used to enumerate the order of the perturbation expansion; at the end of the calculation one sets $\lambda = 1$. In equilibrium this system can be described by the Matsubara Green functions, which can be perturbatively expanded in terms of the interaction part of the Hamiltonian H as follows [17]

$$\begin{aligned} G(1,2) &= -\frac{\langle T_\tau[S(\beta)c(1)c^\dagger(2)]\rangle_0}{\langle S(\beta)\rangle_0} = \sum_{n=0}^{\infty} \lambda^n W_n(1,2) \\ &= G_0(1,2)\exp\left[\sum_{n=1}^{\infty} \lambda^n F_n(1,2)\right]. \end{aligned} \tag{5.11.4}$$

Here the subscript '0' indicates that the expectation value must be evaluated with respect to H_0, i.e. the noninteracting state. The terms generated by the S-matrix expansion can be represented, after an application of Wick's theorem, with Feynman diagrams. The disconnected diagrams originating from

* See, for example [17, pp. 285–324]

$\langle T_\tau[S(\beta)c(1)c^\dagger(2)]\rangle_0$ are cancelled, just as in the linear-response formulation for the Coulomb drag problem in section 5.3, by the vacuum polarization diagrams contained in the term $\langle S(\beta)\rangle_0$. The remaining connected diagrams can be arranged according to the perturbation order n and the corresponding clusters W_n can be summed as indicated by the second equality in (5.11.4). The resummed series in the last equality in (5.11.4), expressed in terms of the as of yet undetermined objects F_n, is called the LCE for the Green's function. The expansion coefficients F_n are evaluated by equating the coefficients of the interaction parameter λ of the two expansions in (5.11.4):

$$F_1(1,2) = \frac{W_1(1,2)}{G_0(1,2)}, \qquad F_2(1,2) = \frac{W_2(1,2)}{G_0(1,2)} - \frac{1}{2}F_1(1,2)^2, \ldots. \tag{5.11.5}$$

The higher-order expressions rapidly get quite complicated. Importantly, however, it is often sufficient to consider in expansion (5.11.4) only the first- and second-order terms, F_1 and F_2, equation (5.11.5). The advantage of the method is that it provides a convenient way of performing an infinite resummation: in contrast to the Dyson equation which often involves the solution of a complicated integral equation in several variables, here the evaluation of a few diagrams is sufficient. The disadvantage is that one does not have an *a priori* knowledge about the convergence radius of the expansion. Thus, an uncritical application may occasionally lead to unphysical results, in particular for strongly interacting systems.

(b) Nonequilibrium LCE

Here we present a generalization of LCE to nonequilibrium [62]. The analytical continuation and reduction to real times can be directly performed for the term $\sum_n \lambda^n W_n$ in (5.11.4) for the Matsubara Green function. The final LCE resummation of the series can be performed in the same way as in equilibrium, so that the nonequilibrium linked cluster expansion (NLCE) for the correlation function $G^<(1,2) \equiv \mathrm{i}\langle c^\dagger(1)c(2)\rangle$ in real times is

$$G^<(1,2) = \sum_{n=0}^{\infty} \lambda^n W_n^<(1,2) = G_0^<(1,2)\exp\left[\sum_{n=1}^{\infty} \lambda^n F_n^<(1,2)\right]. \tag{5.11.6}$$

A similar expansion holds for the correlation function $G^>(1,2)$. The expansion coefficients $F_n^<$ are given by a nonequilibrium generalization of (5.11.5):

$$F_1^<(1,2) = \frac{W_1^<(1,2)}{G_0^<(1,2)}, \quad F_2^<(1,2) = \frac{W_2^<(1,2)}{G_0^<(1,2)} - \frac{1}{2}F_1^<(1,2)^2, \ldots. \tag{5.11.7}$$

Finally, the correlation functions for the cluster diagrams $W_n^{<,>}(1,2)$ can be found by application of the Langreth rules [32, 33].

Another possibility of how to combine the NGF and LCE methods is to perform the analytical continuation in the LCE expansion for the spectral function A. This version of the NLCE method does not give all the information con-

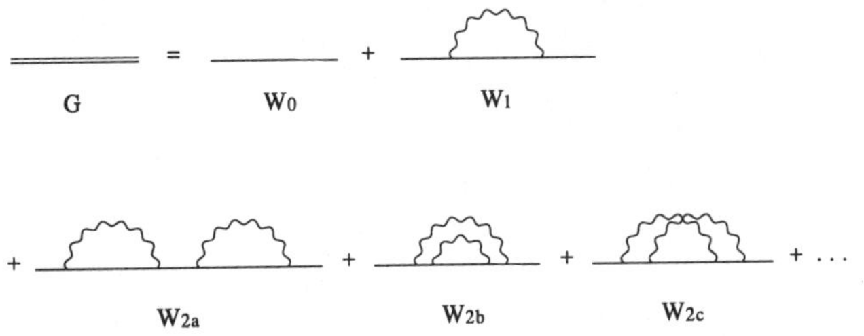

Figure 5.16 *Diagrammatic expansion of the Dyson equation for the Green's function G. All diagrams with noncrossed phonon lines contribute to the expansion of the Green's function with the Migdal self-energy. The remaining diagrams give rise to vertex corrections to this expansion. Diagrams contributing to the zeroth-, first- and second-order clusters are denoted here by* W_0, W_1 *and* $W_{2a,b,c}$ *[62].*

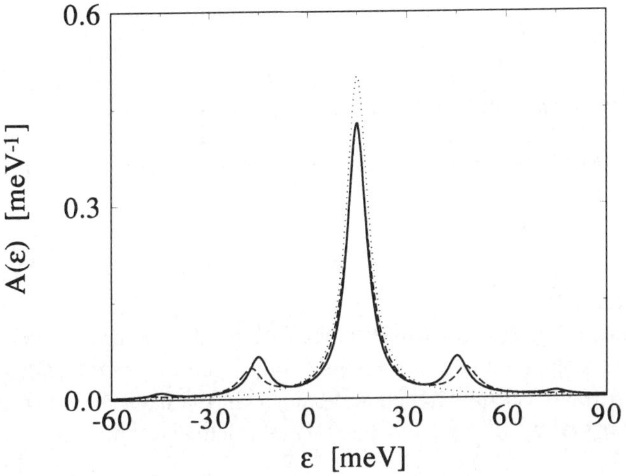

Figure 5.17 *The nonequilibrium spectral function for parameters* $\epsilon_0 = 15$ *meV,* $\Gamma = 8$ *meV,* $\hbar\omega_0 = 30$ *meV,* $g = 0.25$ *and* $T = 70$ *K. The bias is* $\mu_L = -\mu_R = 60$ *meV* ($\mu_0 = 0$)*. Legend: NCLE=solid curve; Migdal approximation, equation (5.11.8)= dashed curve, noninteracting mode=dots. Note that the spectral function is symmetric both for NLCE and NGF in this relatively high DC bias [62].*

tained in the correlation functions $G^<$ and $G^>$, but in some situations, such as the WBL limit of the electron–phonon problem, cf. equation (5.11.1), this information is not necessary. We do not reproduce the expansion here, but its derivation is entirely equivalent to the one leading to equation (5.11.7) and it has the advantange of being, in certain situations, numerically more stable. In the examples presented below the results for the spectral function A calculated either from (5.11.6) or the linked cluster expansion for A itself are numerically practically indistinguishable.

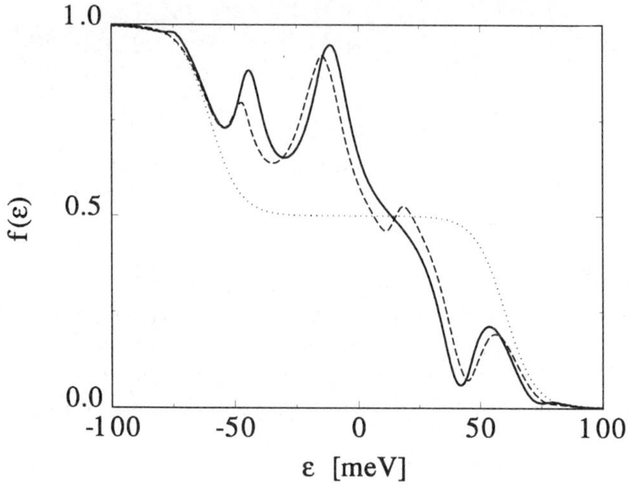

Figure 5.18 *The distribution function $\bar{f}(\omega)$ calculated for the parameters of Fig. 5.17. The solution for the noninteracting model, presented by the light dotted curve, is formed by two Fermi–Dirac distributions shifted with the amount $|\mu_L - \mu_R|$. The NGF (dashed curve) and NLCE (solid curve), show additional oscillatory structures due to phonon satellites [63].*

(c) Numerical example

We conclude by showing some numerical results obtained with the NLCE method and compare them to standard results obtained with the Migdal approximation for the electron–phonon self-energy. In the latter approach the spectral function appearing in equation (5.11.1) is calculated from the self-consistent solution of

$$\begin{aligned} G^r(\omega) &= \left[\omega - \epsilon_0 + \mathrm{i}\Gamma/2 - \Sigma^r_{\text{el-ph}}(\omega)\right]^{-1} \\ \Sigma^r_{\text{el-ph}}(\omega) &= \mathrm{i}\int \frac{\mathrm{d}\omega'}{2\pi} G^r(\omega - \omega')D^r(\omega'), \end{aligned} \tag{5.11.8}$$

where the phonon-propagator D^r is evaluated for free optical phonons. Once G^r is determined, we get the current by evaluating equation (5.11.1).

As far as the NLCE method is concerned, one must evaluate the diagrams shown in Fig. 5.16. We do not reproduce the rather cumbersome formulae here (the details are given in [62]), and instead display a numerical example in Figs. 5.17 and 5.18. The nonequilibrium Green's function can again be represented in the 'pseudo-equilibrium' form of equation (5.10.8), $G^<(\omega) = \mathrm{i}\bar{f}(\omega)a(\omega)$ and it is instructive to compare the results in the various approximations. The noninteracting result, shown as dots, is seen to coincide with equation (5.10.8), which shows that for constant and equal Γ's the distribution function is just the average of the contact distribution functions. There are clear differences between the results obtained for the standard Migdal approximation and the NLCE method. A comparison of these two approximations in cases where exact solutions are available indicates that NLCE is more accurate [62]. However, due to the preliminary character of these investigations the final assessment cannot yet be done, but the indications are certainly promising. This would be very encouraging, because the NLCE approach can be, with relative ease, applied to situations where other schemes have severe difficulties.

ACKNOWLEDGEMENTS

The work described in this chapter was done in form of various collaborations with H. Smith, Ben Yu-Kuang Hu, Karsten Flensberg, Jari Kinaret, Martin Bønsager, Ned S. Wingreen, Yigal Meir and Petr Král. The author expresses his sincere thanks to these colleagues for sharing their insights and for providing a pleasant working atmosphere. The hospitality of ISI, Torino, during the final preparation of this manuscript is gratefully acknowledged.

REFERENCES

[1] Gramila, T. J., Eisenstein, J. P., MacDonald, A. H. *et al.* (1991) *Phys. Rev. Lett.* **66**, 1261.
Gramila, T. J., Eisenstein, J. P., MacDonald, A. H. *et al.* (1993) *Phys. Rev.* B **47**, 12 957.
Gramila, T. J., Eisenstein, J. P., MacDonald, A. H. *et al.* (1994) *Physica* B **197**, 442.
[2] Sivan, U., Solomon, P. M. and Shtrikman, H. (1992) *Phys. Rev. Lett.* **68**, 1196.
[3] Solomon, P. M., Price, P. J., Frank, D. J. *et al.* (1989) *Phys. Rev. Lett.* **63**, 2508.
[4] Hodges, C., Smith, H. and Wilkins J. W. (1971) *Phys. Rev.* B **4**, 302.
Chaplik, A. V. (1971) *Zh. Eksp. Teor. Fiz.* **60**, 1845 [(1971) *Sov. Phys.–JETP* **33**, 997)].
Giuliani, G. F. and Quinn, J. J. (1982) *Phys. Rev.* B **26**, 4421.
[5] Tso, H. C., Vasilopoulos, P. and Peeters, F. M. (1992) *Phys. Rev. Lett.* **68**, 2516.
Tso, H. C., Vasilopoulos, P. and Peeters, F. M. (1993) *Phys. Rev. Lett.* **70**, 2146.
Tso, H. C. and Vasilopoulos, P. (1992) *Phys. Rev.* B **45**, 1333.
[6] Pogrebinskii, M. B. (1977) *Fiz. Tekh. Poluporovdn.* **11**, 637 [(1977) *Sov. Phys. Semicond.* **11**, 372].
[7] Price, P. J. (1983) *Physica* B **117**, 750.
[8] Jauho, A. P. and Smith, H. (1993) *Phys. Rev.* B **47**, 4420.
[9] Zheng, L. and MacDonald, A. H. (1993) *Phys. Rev.* B **48**, 8203.
[10] Flensberg, K., Hu, B. Y.-K., Jauho, A. P. *et al.* (1995) Phys. Rev. B **52**, 14 761.
[11] Kamenev, A. and Oreg, Y. (1995) *Phys. Rev.* B **52**, 7516.
[12] Bønsager, M. C., Flensberg, K., Hu, B. Y.-K. *et al.* (1996) *Phys. Rev. Lett.* **77**, 1366.
[13] Flensberg, K. and Hu, B. Y.-K. (1995) *Phys. Rev.* B **52**, 14 796.
[14] Flensberg, K. and Hu, B. Y.-K. (1994) *Phys. Rev. Lett.* **73**, 3572.
[15] Świerkowski, L., Szymański, J. and Gortel, Z. W. (1995) *Phys. Rev. Lett.* **74**, 3245.

[16] Hill, N. P. R, Nicholls, J. T., Linfield, E. H. *et al.* (1997) *Phys. Rev. Lett.* **78**, 2204.
[17] Mahan, G. D. (1990) *Many-Particle Physics*, Plenum Press, New York.
[18] Aslamazov, L. G. and Larkin, A. I. (1968) *Fiz. Tverd. Tela* **10**, 1104 [(1968) *Sov. Phys.–Solid State* **10**, 875].
[19] Rapcewicz, K. and Ashcroft, N. W. (1991) *Phys. Rev.* B **44**, 4032.
[20] Goldstein, R. E., Parola, G. and Smith, A. P. (1989) *J. Chem. Phys.* **91**, 1843.
[21] Laikhtman, B. and Solomon, P. M. (1990) *Phys. Rev.* B **41**, 9921.
Solomon, P. M. and Laikhtman, B. (1991) *Superlatt. Microstruct.* **10**, 89.
[22] Al'tshuler, B. L. and Aronov, A. G. (1979) *Zh. Eksp. Teor. Fiz.* **77**, 2028 [(1979) *Sov. Phys.–JETP* **50**, 968].
[23] Lee, P. A. and Ramakrishnan, T. V. (1985) *Rev. Mod. Phys.* **57**, 287.
[24] Rammer, J. and Smith, H. (1986) *Rev. Mod. Phys.* **58**, 323.
[25] See, for example, Prange, R.E. and Girvin, S.M. (eds) (1990) *The Quantum Hall Effect*, 2nd edn, Springer-Verlag, New York.
[26] Bønsager, M. C, Flensberg, K. and Hu, B. Y.-K. *et al.* (1997) *Phys. Rev.* B **56**, 10314.
Bønsager, M. C., *M.Sc. Thesis*, Technical University of Denmark (unpublished).
[27] Ando, T., Fowler, A. B. and Stern, F. (1982) *Rev. Mod. Phys.* **54**, 437.
[28] Rubel, H., Fischer, A., Dietsche, W. *et al.* (1997) *Phys. Rev. Lett.* **78**, 1763.
[29] Kadanoff, L. P. and Baym, G. (1962) *Quantum Statistical Mechanics*, Benjamin, New York.
[30] Keldysh, L. V. (1964) *Zh. Eksp. Teor. Fiz.* **47**, 1515 [(1965) *Sov.–Phys. JETP* **20**, 1018].
[31] An early reference to tunnelling problems vs. nonequilibrium Green's functions is the series of papers by Caroli, C., Combescot, R., Nozieres, P. *et al.* (1971) *J. Phys. C: Solid. St. Phys.* **4**, 916.
Caroli, C., Combescot, R., Lederer, D. *et al.* (1971) *J. Phys. C: Solid. St. Phys.* **4** 2598.
Combescot, R. (1971) *J. Phys. C: Solid. St. Phys.* **4**, 2611.
Caroli, C., Combescot, R., Nozieres, P *et al.* (1972) *J. Phys. C: Solid. St. Phys.* **5**, 21.
[32] Langreth, D. C. (1976) *Linear and Nonlinear Electron Transport in Solids*, vol. 17 of *Nato Advanced Study Institute, Series B: Physics* (eds J. T. Devreese and V. E. van Doren) Plenum, New York.
[33] Haug, H. and Jauho, A. P. (1996) *Quantum Kinetics in Transport and Optics of Semiconductors, Springer Series in Solid-State Sciences* **123**, Springer Heidelberg, New York.
[34] Meir, Y. and Wingreen, N. S. (1992) *Phys. Rev. Lett.* **68**, 2512.
[35] Davies, J. H., Hershfield, S., Hyldgaard, P. *et al.* (1993) *Phys. Rev.* B **47**, 4603.
[36] Anda, E. V. and Flores, F. (1991) *J. Phys.: Condens. Matter* **3**, 9087.
[37] Hershfield, S., Davies, J. H. and Wilkins, J. W. (1991), *Phys. Rev. Lett.* **67**, 3720.
[38] Meir, Y., Wingreen, N. S. and Lee, P. A. (1993) *Phys. Rev. Lett.* **70**, 2601.
[39] Lake, R. and Datta, S. (1992) *Phys. Rev.* B **45**, 6670.
[40] Landauer, R. (1957) *IBM J. Res. Dev.* **1**, 233.
Landauer, R. (1970) *Philos. Mag.* **21**, 863.
[41] Blandin, A., Nourtier, A. and Hone, D. W. (1976) *J. de Physique* **37**, 369.
[42] Jacoboni, C. and Price, P. (1990) *Solid State Commun.* **75**, 193.

[43] Chen, L. Y. and Ting, C. S. (1991) *Phys. Rev.* B **43**, 2097.
[44] Langreth, D. C. and Nordlander, P. (1991) *Phys. Rev.* B **43**, 2541.
[45] Wingreen, N. S., Jauho, A. P. and Meir, Y. (1993) *Phys. Rev.* B **47**, 8487.
[46] Bruder, C. and Schoeller, H. (1994) *Phys. Rev. Lett.* **72**, 1076.
[47] Runge, E. and Ehrenreich, H. (1992) *Annals of Physics* **219**, 55.
Runge, E. and Ehrenreich, H. (1992) *Phys. Rev.* B **45**, 9145.
[48] Pastawski, H. M. (1992) *Phys. Rev.* B **46**, 4053.
[49] Hung, K.-M. and Wu, G. Y. (1993) *Phys. Rev.* B **48**, 14 687.
[50] Ivanov, T., Marvakov, D. and Valtchinov, V. *et al.* (1993) *Phys. Rev.* B **48**, 4679.
[51] Liu, C. and Niu, G. (1993) *Phys. Rev.* B **47**, 13 031.
[52] Fu, Y. and Dudley, S. C. (1993) *Phys. Rev. Lett.* **70**, 65.
[53] Büttiker, M., Prêtre, A. and Thomas, H. (1993) *Phys. Rev. Lett.* **70**, 4114.
Büttiker, M., Thomas, H. and Prêtre, A. (1994) *Z. Phys.* B **94**, 133.
Büttiker, M. (1993) *J. Phys. Condens. Matter* **5**, 9361.
[54] Jauho, A. P., Wingreen, N. S. and Meir, Y. (1994) *Phys. Rev.* B **50**, 5528.
[55] Hettler, M. H. and Schoeller, H. (1995) *Phys. Rev. Lett.* **74**, 4907.
[56] Ivanov, T. (1995) *Phys. Rev.* B **52**, 2838.
[57] Wingreen, N. S., Jacobsen, K. W. and Wilkins, J. W. (1989) *Phys. Rev.* B **40**, 11 834.
See also Glazman, L. I. and Shekter, R. I. (1987) *Zh. Eksp. Teor. Fiz.* **94**, 292 [(1988) *Sov. Phys.–JETP* **67**, 163].
[58] Ng, T.-K. (1996) *Phys. Rev. Lett.* **76**, 487.
[59] Stafford, C. and Wingreen, N. S. (1996) *Phys. Rev. Lett.* **76**, 11 961.
[60] Schwinger, J. (1961) *J. Math. Phys.* **2**, 407.
Martin, P. C. and Schwinger, J. (1959) *Phys. Rev.* **115**, 1342.
[61] Bakshi, P. M. and Mahanthappa, K. T. (1963) *J. Math. Phys.* **4**, 1.
Bakshi, P. M. and Mahanthappa, K. T. (1963) *J. Math. Phys.* **4**, 12.
[62] Král, P. (1997) *Phys. Rev.* B **56**, 7293.
[63] Lake, R., Klimeck, G., Anantram, M. P. *et al.* (1993) *Phys. Rev.* B **48**, 15 132.
See also Landauer, R. (1992) *Phys. Scr.* T **42**, 110.
[64] Weil, T. and Vinter, B. (1987) *Appl. Phys. Lett.* **50**, 1281.
Jonson, M. and Grincwajg, A. (1987) *Appl. Phys. Lett.* **51**, 1729.
[65] Büttiker, M. and Landauer, R. (1982) *Phys. Rev. Lett.* **49**, 1739.
Büttiker, M. and Landauer, R. (1985) *Physica Scripta* **32**, 429.
[66] See, e.g. Hill, N. P. R., Nicholls, J. T. and Linfield, E. H. (1996) *J. Phys.: Cond. Matt.* **8**, L557.
Feng, X., Noh, H., Zelakiewicz, S. and Gramila, T. J. (1997), *Bull. Am. Phys. Soc* **42**, 487.
Eisenstein, J. P., Pfeiffer, L. N. and West, K. W. (1997) *Bull. Am. Phys. Soc* **42**, 486.
[67] Bønsager, M., Flensberg, K., Hu, B. Y.-K. *et al.* (1997), Preprint IUCM97-018, archived at cond-mat/9707111.

CHAPTER 6

Density matrix theory of coherent ultrafast dynamics

Tilmann Kuhn

Institut für Theoretische Physik II, Westfälische Wilhelms-Universität, Münster, Germany

6.1 INTRODUCTION

The carrier dynamics in semiconductor nanostructures on ultrashort timescales exhibits a variety of phenomena which cannot be understood on a semiclassical level based completely on the particle aspect of the elementary excitations. Instead, these phenomena are at least partially related to the wave aspect and therefore a quantum-mechanical theory has to be used. Recently there have been mainly two different approaches used for the theoretical analysis of the ultrafast carrier dynamics in systems far from the thermodynamic equilibrium. These are the nonequilibrium Green's function theory [1–4] as discussed in Chapter 5 of this book and the density matrix theory [5–8] which will be reviewed in this chapter. Both theories are quantum kinetic theories in the sense that the basic variables are some generalizations of the classical concept of a distribution function which may then be used to calculate expectation values of any observable quantity such as the electric current or polarization. This is in contrast to other approaches such as for example, the Kubo formalism [9] which are used in situations close to equilibrium where an expression directly for the desired observable is derived.

The derivation of a quantum kinetic theory as is shown in this chapter has two main objectives. The first one is to describe phenomena which are not included in a semiclassical kinetic theory. Among these are, in particular, all phenomena based on the fact that a carrier is not in an eigenstate of the ideal crystal but in some linear combination giving rise, for example to four-wave mixing [10–13], quantum beats [14–17] and THz emission [18–20]. Furthermore, there are phenomena related to the uncertainty principle between energy and time or momentum and position. The former limits the concept of instantaneous, energy-conserving scattering processes and requires us to take into account memory effects [21–25]. In the latter case the concept of a carrier distribution function depending on position and momentum has to be generalized, for example to a Wigner function [26–28] which is no longer positive definite and, therefore, cannot be interpreted in terms of a probability density. The second objective is to show how the semiclassical Boltzmann picture, discussed in Chapters 3 and 4 of this book, which has been very successful in explaining transport and relaxation properties on not too short length and timescales, emerges from this quantum-mechanical picture, and to analyse the conditions under which this description is valid. It will turn out that phenomena which can be explained qualitatively also on the semiclassical level, for example carrier photogeneration, exhibit deviations on ultrashort timescales as has been observed in band-to-acceptor luminescence spectra [29, 30].

In this chapter the density matrix approach to the carrier dynamics in semiconductors is presented in a notation which can be applied to bulk systems as well as to a variety of nanostructures. We will derive the contributions to the equations of motion due to three types of interactions: interaction with an external field, carrier–phonon and carrier–carrier interaction, which in many cases are the most important ones for the ultrafast dynamics. In each case we will analyse which phenomena go beyond the semiclassical limit and what the conditions are to recover the semiclassical Boltzmann equation. Then we will give a few examples of the carrier dynamics in bulk, quantum well and quantum wire structures showing features which cannot be interpreted purely in terms of a classical distribution function.

6.2 DENSITY MATRIX FORMALISM

The density matrix theory, like any quantum-mechanical theory, is independent of a particular representation. This equivalence, however, is removed if approximations are performed which are usually better fulfilled in one basis than in another. Here, we will formulate the theory in a notation which can be applied to semiconductor nanostructures of different dimensions including bulk, quantum well and quantum wire structures. Typically in such structures the three-dimensional real space can be divided into confined or structured directions and free directions, i.e. $\boldsymbol{r} = \boldsymbol{r}_\perp + \boldsymbol{r}_\parallel$. The single-particle states in the free directions are given by plane waves, characterized by a wave vector $\boldsymbol{k}$ with the same dimension as $\boldsymbol{r}_\parallel$, in the confined directions we have localized states characterized by an index i. Examples which will be discussed in this chapter are quantum wells where we have one confined dimension and a two-dimensional wave vector $\boldsymbol{k}$ and quantum wires where $\boldsymbol{r}_\perp$ is two dimensional and $\boldsymbol{k}$ is one dimensional. Of course, also bulk materials where $\boldsymbol{r}_\perp$ is three dimensional and quantum dots where $\boldsymbol{r}_\parallel$ is three dimensional are included in this classification. For superlattices in the presence of an applied bias, which are discussed in detail in Chapter 9, $\boldsymbol{k}$ is two dimensional. The structured direction is characterized either by a continuous index, the miniband momentum or a discrete index when taking a Wannier–Stark basis. Chapter 9 also shows the modifications of the density matrix approach if it is formulated in a time-dependent basis.

Because we are treating a many-body system where the number of electrons, holes and phonons is not conserved, it is convenient to use the language of second quantization. Here, in space representation, the electronic system is described by field creation and annihilation operators $\Psi^\dagger(\boldsymbol{r})$ and $\Psi(\boldsymbol{r})$. Using the mode representation for a two-band model in the electron–hole picture, the annihilation operator can be expanded according to

$$\Psi(\boldsymbol{r}) = \sum_{i,\boldsymbol{k}} \mathrm{e}^{\mathrm{i}\boldsymbol{k}\cdot\boldsymbol{r}_\parallel} \phi^e_{i,\boldsymbol{k}}(\boldsymbol{r}_\perp) u^e(\boldsymbol{r}) c_{i,\boldsymbol{k}} + \sum_{j,\boldsymbol{k}} \mathrm{e}^{\mathrm{i}\boldsymbol{k}\cdot\boldsymbol{r}_\parallel} \phi^h_{j,\boldsymbol{k}}(\boldsymbol{r}_\perp) u^h(\boldsymbol{r}) d^\dagger_{j,-\boldsymbol{k}} \tag{6.2.1}$$

where $c_{i,\boldsymbol{k}}(d_{j,\boldsymbol{k}})$ denotes the destruction of an electron (hole) in state $i, \boldsymbol{k}(j, \boldsymbol{k})$, $c^\dagger_{i,\boldsymbol{k}}, d^\dagger_{j,\boldsymbol{k}}$ are the corresponding creation operators, $u^{e/h}(\boldsymbol{r})$ are the lattice-periodic Bloch functions of conduction/valence band which, in effective mass and envelope function approximation, are taken at the band extrema and $\phi^{e/h}_{i,\boldsymbol{k}}(\boldsymbol{r}_\perp)$ denote the electron/hole envelope functions in the confined directions. Due to different masses in wells and barriers or due to band mixing, they may depend on the parallel momentum. However, to simplify the notation, in the following we will neglect this $\boldsymbol{k}$-dependence.

The phonon system, in general, is also determined by geometrical and material properties of the structure. In the following, however, we will limit ourselves to bulk LO-phonons interacting with the carriers via the polar Fröhlich coupling. Typically, this coupling mechanism is the most important one for the ultrafast carrier–phonon dynamics. Then, the phonons are described by creation (annihilation) operators $b^\dagger_{\boldsymbol{q}}(b_{\boldsymbol{q}})$ for a phonon with wave vector $\boldsymbol{q}$. A generalization to several phonon modes and types of interactions, as well as to more refined phonon models, as discussed in Chapter 11 of this book including, for example interface and confined optical phonons [31] or folded acoustic phonons [32], is straightforward.

For this system the Hamiltonian of the noninteracting carriers and phonons in the ideal structure is given by

$$H_0 = \sum_{i,\boldsymbol{k}} \epsilon^e_{i,\boldsymbol{k}} c^\dagger_{i,\boldsymbol{k}} c_{i,\boldsymbol{k}} + \sum_{j,\boldsymbol{k}} \epsilon^h_{j,\boldsymbol{k}} d^\dagger_{j,\boldsymbol{k}} d_{j,\boldsymbol{k}} + \sum_{\boldsymbol{q}} \hbar\omega_{LO} b^\dagger_{\boldsymbol{q}} b_{\boldsymbol{q}} \tag{6.2.2}$$

with $\epsilon^e_{i,\boldsymbol{k}}(\epsilon^h_{j,\boldsymbol{k}})$ being the energies of electron (holes) in subband $i(j)$ with wave vector $\boldsymbol{k}$ and $\hbar\omega_{LO}$ being the LO-photon energy.

Many experimentally observable quantities such as densities, phonon populations or electronic polarizations are single-particle quantities. They can be expressed in terms of single-particle density matrices. Therefore, the basic variables in the density matrix approach are the intraband density matrices of electrons and holes (distribution functions and intersubband polarizations),

$$f^e_{i_1 i_2,\boldsymbol{k}} = \langle c^\dagger_{i_1,\boldsymbol{k}} c_{i_2,\boldsymbol{k}} \rangle \quad \text{and} \quad f^h_{j_1 j_2,\boldsymbol{k}} = \langle d^\dagger_{j_1,\boldsymbol{k}} d_{j_2,\boldsymbol{k}} \rangle, \tag{6.2.3}$$

the interband density matrix (interband polarization)

$$p_{j_1 i_1,\boldsymbol{k}} = \langle d_{j_1,-\boldsymbol{k}} c_{i_1,\boldsymbol{k}} \rangle \qquad \text{and} \qquad p^*_{j_1 i_1,\boldsymbol{k}} = \langle c^\dagger_{i_1,\boldsymbol{k}} d^\dagger_{j_1,-\boldsymbol{k}} \rangle, \tag{6.2.4}$$

and the phonon density matrix (phonon distribution function)

$$n_{\boldsymbol{q}} = \langle b^\dagger_{\boldsymbol{q}} b_{\boldsymbol{q}} \rangle. \tag{6.2.5}$$

In this contribution we will restrict ourselves to the case of a system which is homogeneous in the free directions. To include inhomogeneities, density matrix elements off-diagonal in $\boldsymbol{k}$ have to be included. Alternatively, a Wigner representation may be used by performing a Fourier transformation with respect to the relative momentum. A detailed derivation of the equations of motion for a many-body system in the Wigner representation including a gradient expansion showing the origin as well as the limitations of the Boltzmann drift term can be found in [33–35].

Using the Heisenberg equations of motion for the mode operators of electrons, holes and phonons, the contributions to the equations of motion due to H_0 are easily obtained as the trivial unitary rotations

$$\frac{\mathrm{d}}{\mathrm{d}t} f^e_{i_1 i_2,\boldsymbol{k}} \bigg|^{(0)} = \frac{1}{\mathrm{i}\hbar}(\epsilon^e_{i_2,\boldsymbol{k}} - \epsilon^e_{i_1,\boldsymbol{k}}) f^e_{i_1 i_2,\boldsymbol{k}}, \tag{6.2.6a}$$

$$\frac{\mathrm{d}}{\mathrm{d}t} p_{j_1 i_1,\boldsymbol{k}} \bigg|^{(0)} = \frac{1}{\mathrm{i}\hbar}(\epsilon^h_{j_1,-\boldsymbol{k}} + \epsilon^e_{i_1,\boldsymbol{k}}) p_{j_1 i_1,\boldsymbol{k}}, \tag{6.2.6b}$$

$$\frac{\mathrm{d}}{\mathrm{d}t} n_{\boldsymbol{q}} = 0. \tag{6.2.6c}$$

Here, as in the following, we will not explicitly write the equations for f^h since it has the same structure as for f^e.

6.3 INTERACTION WITH AN EXTERNAL FIELD

Due to the gauge invariance of electrodynamics, the interaction of the carriers with a spatially homogeneous (static or time-dependent) external electric field $\boldsymbol{E}$ can be described both in terms of a scalar or vector potential. Any measurable quantity, of course, has to be independent of the gauge. In Chapter 9 both approaches are discussed and the resulting equations of motion are confronted, here we will consider only the case of the scalar potential gauge, where the Hamiltonian is given by

$$H^{cf} = \int \mathrm{d}^3 r\, \Psi^\dagger(\boldsymbol{r}) e\boldsymbol{E} \cdot \boldsymbol{r} \Psi(\boldsymbol{r}) \tag{6.3.1}$$

with e denoting the (positive) elementary charge. Neglecting the space dependence of the field is in many cases a good approximation since even for a light field in the visible range the wavelength is sufficiently long. In the case of two-pulse excitation the information on the direction of propagation has to be retained in order to separate signals in various directions [29, 33]. As the vector $\boldsymbol{r}$, the field vector can be separated into free and confined directions, $\boldsymbol{E} = \boldsymbol{E}_\parallel + \boldsymbol{E}_\perp$. Inserting the mode representation of the field operators, the interaction Hamiltonian is given by

$$\begin{aligned} H^{cf} =& \sum_{i,\boldsymbol{k}} c^\dagger_{i,\boldsymbol{k}} \mathrm{i}e\boldsymbol{E}_\parallel \cdot \frac{\partial}{\partial \boldsymbol{k}} c_{i,\boldsymbol{k}} - \sum_{j,\boldsymbol{k}} d^\dagger_{j,\boldsymbol{k}} \mathrm{i}e\boldsymbol{E}_\parallel \cdot \frac{\partial}{\partial \boldsymbol{k}} d_{j,\boldsymbol{k}} \\ & - \sum_{i_1,i_2,\boldsymbol{k}} \boldsymbol{E}_\perp \cdot \boldsymbol{M}^e_{i_1 i_2} c^\dagger_{i_1,\boldsymbol{k}} c_{i_2,\boldsymbol{k}} - \sum_{j_1,j_2,\boldsymbol{k}} \boldsymbol{E}_\perp \cdot \boldsymbol{M}^h_{j_1 j_2} d^\dagger_{j_1 \boldsymbol{k}} d_{j_2,\boldsymbol{k}} \\ & - \sum_{i,j,\boldsymbol{k}} [\boldsymbol{E} \cdot \boldsymbol{M}_{ij} c^\dagger_{i,\boldsymbol{k}} d^\dagger_{j,-\boldsymbol{k}} + \boldsymbol{E} \cdot \boldsymbol{M}^*_{ij} d_{j,-\boldsymbol{k}} c_{i,\boldsymbol{k}}] \end{aligned} \tag{6.3.2}$$

with the intraband dipole matrix elements

$$\boldsymbol{M}^e_{i_1 i_2} = - e \int \mathrm{d}^d r^e \, \phi^{e*}_{i_1}(\boldsymbol{r}_\perp) \boldsymbol{r}_\perp \phi^e_{i_2}(\boldsymbol{r}_\perp), \tag{6.3.3a}$$

$$\boldsymbol{M}^h_{j_1 j_2} = e \int \mathrm{d}^d r_\perp \, \phi^{h*}_{j_2}(\boldsymbol{r}_\perp) \boldsymbol{r}_\perp \phi^h_{j_1}(\boldsymbol{r}_\perp), \tag{6.3.3b}$$

where the superscript d denotes the dimension of the confined subspace and the interband dipole matrix element

$$\boldsymbol{M}_{ij} = -e \int_{v_c} \mathrm{d}^3 r \, u^{e*}(\boldsymbol{r}) \boldsymbol{r} u^h(\boldsymbol{r}) \int \mathrm{d}^d r_\perp \, \phi^{e*}_i(\boldsymbol{r}_\perp) \phi^h_j(\boldsymbol{r}_\perp), \tag{6.3.4}$$

where v_c denotes the elementary cell.

6.3.1 Equations of motion

With this Hamiltonian, the carrier–field contribution to the equations of motion is given by

$$\begin{aligned}\left.\frac{\mathrm{d}}{\mathrm{d}t} f^e_{i_1 i_2,\boldsymbol{k}}\right|^{(cf)} =& \frac{\mathrm{e}}{\hbar}\boldsymbol{E}_{\|}\cdot\frac{\partial}{\partial\boldsymbol{k}} f^e_{i_1 i_2,\boldsymbol{k}} \\ & -\frac{1}{\mathrm{i}\hbar}\sum_{i_3}\boldsymbol{E}_{\perp}\cdot(\boldsymbol{M}^e_{i_2 i_3} f^e_{i_1 i_3,\boldsymbol{k}} - \boldsymbol{M}^e_{i_3 i_1} f^e_{i_3 i_2,\boldsymbol{k}}) \\ & -\frac{1}{\mathrm{i}\hbar}\sum_{j_1}\boldsymbol{E}\cdot(\boldsymbol{M}_{i_2 j_1} p^*_{j_1 i_1,\boldsymbol{k}} - \boldsymbol{M}^*_{i_1 j_1} p_{j_1 i_2,\boldsymbol{k}}),\end{aligned} \tag{6.3.5a}$$

$$\begin{aligned}\left.\frac{\mathrm{d}}{\mathrm{d}t} p_{j_1 i_1,\boldsymbol{k}}\right|^{(cf)} =& \frac{\mathrm{e}}{\hbar}\boldsymbol{E}_{\|}\cdot\frac{\partial}{\partial\boldsymbol{k}} p_{j_1 i_1,\boldsymbol{k}} \\ & -\frac{1}{\mathrm{i}\hbar}\boldsymbol{E}_{\perp}\cdot\left(\sum_{j_2}\boldsymbol{M}^h_{j_1 j_2} p_{j_2 i_1,\boldsymbol{k}} - \sum_{i_2}\boldsymbol{M}^e_{i_1 i_2} p_{j_1 i_2,\boldsymbol{k}}\right) \\ & -\frac{1}{\mathrm{i}\hbar}\boldsymbol{E}\cdot\left(\boldsymbol{M}_{i_1 j_1} - \sum_{i_2}\boldsymbol{M}_{i_2 j_1} f^e_{i_2 i_1,\boldsymbol{k}} - \sum_{j_2}\boldsymbol{M}_{i_1 j_2} f^h_{j_2 j_1,-\boldsymbol{k}}\right).\end{aligned} \tag{6.3.5b}$$

Due to the single-particle character of the interaction Hamiltonian, a closed set of equations of motion for single-particle density matrices is obtained. The first term of the right-hand side of equations (6.3.5) describes the acceleration of the carriers in the free direction due to an electric field. It is the same for diagonal and off-diagonal parts and agrees with the Boltzmann acceleration term. It should be noted, however, that this agreement is removed if spatially varying electric fields are taken into account. The second term describes the renormalization of the states in the confined direction as well as transitions between the subbands. The last term leads to transitions between the valence and conduction band.

For the derivation of the equations of motion no specific time dependence of the electric field has been assumed. The importance of the different contributions, however, is in general strongly determined by the frequency of the field. For a static or low-frequency field usually the intraband terms are most important. They give rise to the conductivity in the free direction and to field-dependent changes of energies and localization of the wave functions in the confined direction leading, for example to the quantum-confined Stark effect. The interband terms give rise to Zener tunnelling which requires very high fields to become relevant. If the frequency of the field is of the order of the subband splitting, the second term becomes dominant leading to intersubband transitions while for a frequency of the order of the band gap the generation of electron–hole pairs due to the last term is most important.

6.3.2 Semiclassical limit

As mentioned in the introduction, one purpose of a quantum kinetic theory is to derive the semiclassical kinetic theory as a limiting case and to study the approximations which have to be performed in this derivation. The general procedure to obtain the semiclassical limit is the same for all types of interactions. It consists of an adiabatic elimination of variables involving quantum-mechanical correlations by means of a Markov approximation under the assumption that initially the system was uncorrelated. In this section we discuss this approach for the case of the dynamics induced by an external field.

In the semiclassical limit the system is completely determined by the distribution functions of electrons and holes. Thus, all off-diagonal elements, i.e. interband and intersubband contributions have to be eliminated. The acceleration terms do not couple different bands or subbands. They agree with the well known terms in the Boltzmann equation.

As discussed above, the dominant terms in equation (6.3.5) are determined by the frequency of the field. For a light field in the frequency range of the band gap the dominant process is the generation of electron–hole pairs. Keeping only the near-resonant parts of the equations of motion, i.e. performing a rotating wave approximation [36], we obtain

$$\frac{\mathrm{d}}{\mathrm{d}t} f^e_{ii,\boldsymbol{k}} = \sum_j g_{ji,\boldsymbol{k}}(t), \qquad \frac{\mathrm{d}}{\mathrm{d}t} f^h_{jj,-\boldsymbol{k}} = \sum_i g_{ji,\boldsymbol{k}}(t), \tag{6.3.6a}$$

$$\frac{\mathrm{d}}{\mathrm{d}t} p_{ji,\boldsymbol{k}} = \frac{1}{\mathrm{i}\hbar}(\epsilon^h_{j,-\boldsymbol{k}} + \epsilon^e_{i,\boldsymbol{k}}) p_{ji,\boldsymbol{k}} - \frac{1}{\mathrm{i}\hbar}\boldsymbol{M}_{ij} \cdot \boldsymbol{E}^{(+)}(t)[1 - f^e_{ii,\boldsymbol{k}} - f^h_{jj,-\boldsymbol{k}}] \tag{6.3.6b}$$

with the generation rate

$$g_{ji,\boldsymbol{k}}(t) = -\frac{1}{\mathrm{i}\hbar}[\boldsymbol{M}_{ij} \cdot \boldsymbol{E}^{(+)}(t) p^*_{ji,\boldsymbol{k}} - \boldsymbol{M}^*_{ij} \cdot \boldsymbol{E}^{(-)}(t) p_{ji,\boldsymbol{k}}]. \tag{6.3.7}$$

Here we have used the decomposition of the light field in the positive and negative frequency parts according to

$$\boldsymbol{E}(t) = \boldsymbol{E}^{(+)}(t) + \boldsymbol{E}^{(-)}(t) = \boldsymbol{E}_0(t)\mathrm{e}^{-\mathrm{i}\omega_L t} + \text{c.c.} \tag{6.3.8}$$

with ω_L being the central frequency and $\boldsymbol{E}_0(t)$ being the amplitude of the field. Then, the equation for the polarization can be formally integrated resulting in

$$p_{ji,\boldsymbol{k}}(t) = -\frac{1}{\mathrm{i}\hbar}\mathrm{e}^{-\mathrm{i}\omega_L t}\int_0^\infty \mathrm{d}\tau\, \mathrm{e}^{-\mathrm{i}(\omega_{ji,\boldsymbol{k}} - \omega_L)\tau} \boldsymbol{M}_{ij} \cdot \boldsymbol{E}_0(t-\tau)$$
$$\times [1 - f^e_{ii,\boldsymbol{k}}(t-\tau) - f^h_{jj,-\boldsymbol{k}}(t-\tau)] \tag{6.3.9}$$

with $\hbar\omega_{ji,\boldsymbol{k}} = \epsilon^h_{j,-\boldsymbol{k}} + \epsilon^e_{i,\boldsymbol{k}}$. The semiclassical limit is obtained by two assumptions. First, the field is adiabatically switched on according to $\boldsymbol{E}_0(t) = \tilde{\boldsymbol{E}}_0(t)\mathrm{e}^{\eta t/\hbar}$ and secondly, within the Markov approximation one assumes that the dominant time dependence is given by the exponential in equation (6.3.9) and that the distribution functions and field amplitude are sufficiently slowly varying functions of time so that their value at the time t can be taken out of the integral. Then, the polarization is an instantaneous function of the distribution functions and the field according to

$$p_{ji,\boldsymbol{k}}(t) = \boldsymbol{M}_{ij} \cdot \boldsymbol{E}_0(t) \mathrm{e}^{-\mathrm{i}\omega_L t}[1 - f^e_{ii,\boldsymbol{k}}(t) - f^h_{jj,-\boldsymbol{k}}(t)]\mathcal{D}(\hbar\omega_{ij,\boldsymbol{k}} - \hbar\omega_L) \tag{6.3.10}$$

where we have introduced the function

$$\mathcal{D}(x) = -\frac{\mathrm{i}}{\pi} \lim_{\eta\to 0} \frac{1}{x - \mathrm{i}\eta} = \delta(x) - \frac{\mathrm{i}}{\pi}\frac{\mathcal{P}}{x} \tag{6.3.11}$$

with $\mathcal{P}$ denoting the principal value. This leads to the semiclassical generation rate

$$g_{ji,\boldsymbol{k}}(t) = \frac{2\pi}{\hbar^2} |\boldsymbol{M}_{ij} \cdot \boldsymbol{E}_0(t)|^2 [1 - f^e_{ii,\boldsymbol{k}}(t) - f^h_{jj,-\boldsymbol{k}}(t)]\delta(\omega_{ij,\boldsymbol{k}} - \omega_L). \tag{6.3.12}$$

The general procedure to obtain a semiclassical rate which has been performed here for the generation process is the same for all interaction mechanisms: the interaction process introduces a new variable describing the correlation associated with this interaction. In the present case this is the electron–hole correlation due to the light field. This new variable is then adiabatically eliminated on the basis of a Markov approximation and the assumption of an initially uncorrelated system.

Neglecting completely the time dependence of the field amplitude has led us to a monochromatic generation rate. Any pulse with a finite duration, however, is characterized by a finite spectral width. Often this broadening is introduced by multiplying the generation rate by the spectral intensity of the pulse and integrating over the light frequency. More rigourously, this broadening is obtained if the Markov approximation is performed only for the distribution functions and not for the field amplitude. This leads to a semiclassical generation rate according to

$$\begin{aligned} g_{ji,\boldsymbol{k}}(t) =& \frac{1}{\hbar^2}[1 - f^e_{ii,\boldsymbol{k}}(t) - f^h_{jj,-\boldsymbol{k}}(t)] \\ &\times \int_{-\infty}^{\infty} \mathrm{d}t' \boldsymbol{M}^*_{ij} \cdot \boldsymbol{E}^*_0(t + \tfrac{1}{2}t')\boldsymbol{M}_{ij} \cdot \boldsymbol{E}_0(t - \tfrac{1}{2}t')\mathrm{e}^{-\mathrm{i}(\omega_{ji,\boldsymbol{k}} - \omega_L)t'}. \end{aligned} \tag{6.3.13}$$

For the important case of Gaussian pulse $\boldsymbol{E}_0(t) = \boldsymbol{E}_L \exp[-t^2/\tau_L^2]$ equation (6.3.13) leads to

$$\begin{aligned} g_{ji,\boldsymbol{k}}(t) =& \frac{\sqrt{2\pi}}{\hbar^2}[1 - f^e_{ii,\boldsymbol{k}}(t) - f^h_{jj,-\boldsymbol{k}}(t)] \\ &\times |\boldsymbol{M}_{ij} \cdot \boldsymbol{E}_L|^2 \tau_L \exp\left[-2\frac{t^2}{\tau_L^2}\right] \exp\left[-\frac{1}{2}\tau_L^2(\omega_{ij,\boldsymbol{k}} - \omega_L)^2\right], \end{aligned} \tag{6.3.14}$$

i.e. a generation according to the product of the temporal and the spectral shape of the pulse. In general, however, such a decomposition is not possible. Details of the derivation of equation (6.3.13) are given in appendix A.

6.4 CARRIER–PHONON INTERACTION

Optical phonons are characterized by a relative motion of different atoms in an elementary cell. In compound semiconductors such as GaAs, the bonds have a partly ionic component, therefore such a motion is associated with a lattice polarization which, in the case of a longitudinal phonon, leads to a dipole coupling with the electrons. This is the Fröhlich interaction which is typically the most important carrier–phonon interaction in a polar semiconductor.

By using the plane wave representation of the lattice polarization associated with an LO phonon [37],

$$\boldsymbol{P}_{lat}(\boldsymbol{r}) = \left[\frac{e^2\hbar\omega_{LO}}{8\pi\mathcal{V}}(\varepsilon_\infty^{-1} - \varepsilon_s^{-1})\right]^{1/2} \sum_{\boldsymbol{q}} \frac{\boldsymbol{q}}{q} e^{i\boldsymbol{q}\boldsymbol{r}}(b_{\boldsymbol{q}} - b_{-\boldsymbol{q}}^\dagger), \tag{6.4.1}$$

the Fröhlich Hamiltonian is given by

$$\begin{aligned} H^{cp} &= \sum_{i_1,i_2,\boldsymbol{k}} \sum_{\boldsymbol{q}} [g_{\boldsymbol{q}}^{i_1 i_2} c_{i_1,\boldsymbol{k}}^\dagger b_{\boldsymbol{q}} c_{i_2,\boldsymbol{k}-\boldsymbol{q}_\parallel} + g_{\boldsymbol{q}}^{*i_1 i_2} c_{i_2,\boldsymbol{k}-\boldsymbol{q}_\parallel}^\dagger b_{\boldsymbol{q}}^\dagger c_{i_1,\boldsymbol{k}}] \\ &\quad - \sum_{j_1,j_2,\boldsymbol{k}} \sum_{\boldsymbol{q}} [g_{\boldsymbol{q}}^{j_2 j_1} d_{j_1,\boldsymbol{k}}^\dagger b_{\boldsymbol{q}} d_{j_2,\boldsymbol{k}-\boldsymbol{q}_\parallel} + g_{\boldsymbol{q}}^{*j_2 j_1} d_{j_2,\boldsymbol{k}-\boldsymbol{q}_\parallel}^\dagger b_{\boldsymbol{q}}^\dagger d_{j_1,\boldsymbol{k}}] \end{aligned} \tag{6.4.2}$$

with the Fröhlich coupling matrix element

$$g_{\boldsymbol{q}}^{n_1 n_2} = g_{\boldsymbol{q}}^{3D} \mathcal{F}_{\boldsymbol{q}_\perp}^{n_1 n_2}, \tag{6.4.3a}$$

$$g_{\boldsymbol{q}}^{3D} = -i\left[\frac{2\pi e^{2\hbar\omega_{LO}}}{\mathcal{V}}(\varepsilon_\infty^{-1} - \varepsilon_s^{-1})\right]^{1/2} \frac{1}{q}, \tag{6.4.3b}$$

and the form factor

$$\mathcal{F}_{\boldsymbol{q}_\perp}^{n_1 n_2} = \int d^d r_\perp\, \phi_{n_1}^*(\boldsymbol{r}_\perp)\phi_{n_2}(\boldsymbol{r}_\perp) e^{i\boldsymbol{q}_\perp \boldsymbol{r}_\perp}. \tag{6.4.4}$$

Here, $\mathcal{V}$ denotes the normalization volume and ε_s and ε_∞ are the static and optical dielectric constant, respectively.

This Hamiltonian leads to the carrier–phonon contribution to the equation of motion, for example of f^e and n according to

$$\frac{\mathrm{d}}{\mathrm{d}t} f^e_{i_1 i_2,\boldsymbol{k}}\bigg|^{(cp)} = -\frac{1}{\mathrm{i}\hbar}\sum_{i_3,\boldsymbol{q}}[g_{\boldsymbol{q}}^{i_3 i_1}\langle c^\dagger_{i_3,\boldsymbol{k}+\boldsymbol{q}_\parallel} b_{\boldsymbol{q}} c_{i_2,\boldsymbol{k}}\rangle + g_{\boldsymbol{q}}^{*i_1 i_3}\langle c^\dagger_{i_3,\boldsymbol{k}-\boldsymbol{q}_\parallel} b^\dagger_{\boldsymbol{q}} c_{i_2,\boldsymbol{k}}\rangle$$
$$- g_{\boldsymbol{q}}^{i_2 i_3}\langle c^\dagger_{i_1,\boldsymbol{k}} b_{\boldsymbol{q}} c_{i_3,\boldsymbol{k}-\boldsymbol{q}_\parallel}\rangle - g_{\boldsymbol{q}}^{*i_3 i_2}\langle c^\dagger_{i_1,\boldsymbol{k}} b^\dagger_{\boldsymbol{q}} c_{i_3,\boldsymbol{k}+\boldsymbol{q}_\parallel}\rangle], \tag{6.4.5a}$$

$$\frac{\mathrm{d}}{\mathrm{d}t} n_{\boldsymbol{q}}\bigg|^{(cp)} = -\frac{1}{\mathrm{i}\hbar}\sum_{i_1,i_2,\boldsymbol{k}}[g_{\boldsymbol{q}}^{i_1 i_2}\langle c^\dagger_{i_1,\boldsymbol{k}} b_{\boldsymbol{q}} c_{i_2,\boldsymbol{k}-\boldsymbol{q}_\parallel}\rangle - g_{\boldsymbol{q}}^{*i_1 i_2}\langle c^\dagger_{i_2,\boldsymbol{k}-\boldsymbol{q}_\parallel} b^\dagger_{\boldsymbol{q}} c_{i_1,\boldsymbol{k}}\rangle]$$
$$+ \frac{1}{\mathrm{i}\hbar}\sum_{j_1,j_2,\boldsymbol{k}}[g_{\boldsymbol{q}}^{j_2 j_1}\langle d^\dagger_{j_1,\boldsymbol{k}} b_{\boldsymbol{q}} d_{j_2,\boldsymbol{k}-\boldsymbol{q}_\parallel}\rangle - g_{\boldsymbol{q}}^{*j_2,j_1}\langle d^\dagger_{j_2,\boldsymbol{k}-\boldsymbol{q}_\parallel} b^\dagger_{\boldsymbol{q}} d_{j_1,\boldsymbol{k}}\rangle]. \tag{6.4.5b}$$

Thus, in contrast to the case of interaction with an external field, the system of equations of motion for single-particle density matrices is not closed. New variables have to be introduced, the electron and hole phonon-assisted density matrices [21]

$$s^{e,i_1 i_2}_{\boldsymbol{k},\boldsymbol{q},\boldsymbol{k}'} = \langle c^\dagger_{i_1,\boldsymbol{k}} b_{\boldsymbol{q}} c_{i_2,\boldsymbol{k}'}\rangle \quad \text{and} \quad s^{h,j_1 j_2}_{\boldsymbol{k},\boldsymbol{q},\boldsymbol{k}'} = \langle d^\dagger_{j_1,\boldsymbol{k}} b_{\boldsymbol{q}} d_{j_2,\boldsymbol{k}'}\rangle, \tag{6.4.6}$$

as well as mixed electron–hole phonon-assisted density matrices which appear in the equation of motion for the interband polarization [24]. These quantities describe the correlations between carriers and phonons. Equations (6.4.5) is the starting point of an infinite hierarchy involving higher-order density matrices. To obtain a solution this hierarchy has to be truncated at some level. Truncation schemes based on different ideas have been discussed in the literature. For optically excited intrinsic semiconductors, an expansion in powers of the exciting light field has been proposed which in each order involves only a limited number of correlations [38–40]. Here we will use a correlation expansion based on the idea that correlations involving an increasing number of carriers or phonons are of decreasing importance. This method is applicable both to intrinsic and doped semiconductors. For carrier–carrier interactions it is the quantum-mechanical analogue of the BBGKY hierarchy in classical gas dynamics [41, 42].

6.4.1 First order: Coherent phonons

The lowest order in the hierarchy is obtained by neglecting all correlations between carriers and phonons. This corresponds to a factorization according to

$$s^{e,i_1 i_2}_{\boldsymbol{k}+\boldsymbol{q}_\parallel,\boldsymbol{q},\boldsymbol{k}} \approx \langle c^\dagger_{i_1,\boldsymbol{k}} c_{i_2,\boldsymbol{k}}\rangle\langle b_{\boldsymbol{q}_\perp}\rangle\delta_{\boldsymbol{q}_\parallel,0} = f^e_{i_1,i_2,\boldsymbol{k}} B_{\boldsymbol{q}_\perp}\delta_{\boldsymbol{q}_\parallel,0} \tag{6.4.7}$$

where we have introduced the coherent phonon amplitude $B_{\boldsymbol{q}} = \langle b_{\boldsymbol{q}}\rangle$ and the condition of spatial homogeneity in the free directions has been used leading to the Kronecker delta. The physical meaning of this quantity becomes evident from equation (6.4.1). A nonvanishing coherent phonon amplitude is equivalent to a nonvanishing Fourier component of the lattice polarization and thus to a displacement of the ions [43, 44]. This is in contrast to the usual number state of a phonon which is characterized only by a mean square displacement.

On this level of approximation the contribution of carrier–phonon interaction to the equations of motion of the single-particle density matrices can be expressed in terms of nondiagonal energy renormalizations according to

$$\frac{\mathrm{d}}{\mathrm{d}t} f^e_{i_1 i_2,\boldsymbol{k}}\bigg|^{(cp,1)} = \frac{1}{\mathrm{i}\hbar}\sum_{i_3 i_4}[\mathcal{E}^{e(cp,1)}_{i_2 i_4}\delta_{i_1 i_3} - \mathcal{E}^{e(cp,1)}_{i_3 i_1}\delta_{i_2 i_4}] f^e_{i_3 i_4,\boldsymbol{k}}, \tag{6.4.8a}$$

$$\frac{\mathrm{d}}{\mathrm{d}t} p_{j_1 i_1,\boldsymbol{k}}\bigg|^{(cp,1)} = \frac{1}{\mathrm{i}\hbar}\sum_{i_2 j_2}[\mathcal{E}^{h(cp,1)}_{j_1 j_2}\delta_{i_1 i_2} + \mathcal{E}^{e(cp,1)}_{i_1 i_2}\delta_{j_1 j_2}] p_{j_2 i_2,\boldsymbol{k}} \tag{6.4.8b}$$

with the $\boldsymbol{k}$-independent self-energy matrices

$$\mathcal{E}^{e(cp,1)}_{i_1 i_2} = \sum_{\boldsymbol{q}_\perp}[g^{i_1 i_2}_{\boldsymbol{q}_\perp} B_{\boldsymbol{q}_\perp} + g^{*i_2 i_1}_{\boldsymbol{q}_\perp} B^*_{\boldsymbol{q}_\perp}], \tag{6.4.9a}$$

$$\mathcal{E}^{h(cp,1)}_{j_1 j_2} = -\sum_{\boldsymbol{q}_\perp}[g^{j_2 j_1}_{\boldsymbol{q}_\perp} B_{\boldsymbol{q}_\perp} + g^{*j_1 j_2}_{\boldsymbol{q}_\perp} B^*_{\boldsymbol{q}_\perp}] \tag{6.4.9b}$$

and the phonon amplitude satisfies the equation of motion

$$\frac{\mathrm{d}}{\mathrm{d}t} B_{\boldsymbol{q}_\perp} = -\mathrm{i}\omega_{LO} B_{q_z} - \frac{1}{\mathrm{i}\hbar}\left[\sum_{j_2,j_3,\boldsymbol{k}} g^{*j_2 j_3}_{\boldsymbol{q}_\perp} f^h_{j_2 j_3,-\boldsymbol{k}} - \sum_{i_2,i_3,\boldsymbol{k}} g^{*i_2 i_3}_{\boldsymbol{q}_\perp} f^e_{i_3 i_2,\boldsymbol{k}}\right]. \tag{6.4.10}$$

It is interesting to analyse the long-wavelength limit of equation (6.4.10): The matrix element diverges as q^{-1}. Due to the orthogonality of the envelope functions, the off-diagonal form factors vanish in this limit, thus the divergence only occurs in the diagonal elements. Under the condition

$$\sum_{i_1,\boldsymbol{k}} f^e_{i_1 i_1,\boldsymbol{k}} = \sum_{j_1,\boldsymbol{k}} f^h_{j_1 j_1,\boldsymbol{k}}, \tag{6.4.11}$$

i.e. the condition of charge neutrality, this divergence cancels in equation (6.4.10) leading to a finite value of B_0 which is equivalent to a phononic dipole moment of the structure. By transforming equation (6.4.10) into a space representation it can be shown that coherent phonons are generated only in the case of a local charge imbalance between electrons and holes. An example of coherent phonon dynamics will be shown in section 7.3 where an oscillating electronic wave packet leads to such a charge imbalance.

6.4.2 Second order: Scattering and dephasing

The next step in the hierarchy is obtained by taking into account deviations of the phonon-assisted density matrices from the factorization according to

$$\delta s^{e,i_1 i_2}_{\boldsymbol{k}+\boldsymbol{q}_\parallel,\boldsymbol{q},\boldsymbol{k}} = s^{e,i_1 i_2}_{\boldsymbol{k}+\boldsymbol{q}_\parallel,\boldsymbol{q},\boldsymbol{k}} - f^e_{i_1 i_2,\boldsymbol{k}} B_{\boldsymbol{q}_\perp}\delta_{\boldsymbol{q}_\parallel,0}. \tag{6.4.12}$$

To determine the equation of motion of this phonon-assisted correlation we first derive the equation of motion for the phonon-assisted density matrix. It is given by

$$\begin{aligned}
\frac{\mathrm{d}}{\mathrm{d}t} s^{e,i_1 i_2}_{k+q_\parallel,q,k} = & -\frac{1}{\mathrm{i}\hbar}(\epsilon^e_{i_1,k+q_\parallel} - \epsilon^e_{i_2,k} - \hbar\omega_{LO}) s^{e,i_1 i_2}_{k+q_\parallel,q,k} \\
& -\frac{1}{\mathrm{i}\hbar}\sum_{i_3,q'}[g^{i_3 i_1}_{q'} \langle c^\dagger_{i_3,k+q_\parallel+q'_\parallel} c_{i_2,k} b_{q'} b_q\rangle \\
& + g^{*i_1 i_3}_{q'} \langle c^\dagger_{i_3,k+q_\parallel-q'_\parallel} c_{i_2,k} b^\dagger_{q'} b_q\rangle \\
& - g^{i_2 i_3}_{q'} \langle c^\dagger_{i_1,k+q_\parallel} c_{i_3,k-q'_\parallel} b_q b_{q'}\rangle - g^{*i_3 i_2}_{q'} \langle c^\dagger_{i_1,k+q_\parallel} c_{i_3,k+q'_\parallel} b_q b^\dagger_{q'}\rangle] \\
& + \frac{1}{\mathrm{i}\hbar}\sum_{\substack{i_3,i_4\\ k'}} g^{*i_3 i_4}_{q} \langle c^\dagger_{i_1,k+q_\parallel} c^\dagger_{i_4,k'-q_\parallel} c_{i_3,k'} c_{i_2,k}\rangle \\
& - \frac{1}{\mathrm{i}\hbar}\sum_{\substack{j_3,j_4\\ k'}} g^{*j_4 j_3}_{q} \langle c^\dagger_{i_1,k+q_\parallel} d^\dagger_{j_4,k'-q_\parallel} d_{j_3,k'} c_{i_2,k}\rangle, \qquad (6.4.13)
\end{aligned}$$

thus it involves expectation values of four operators. These are electron–phonon and electron–electron two-particle density matrices. In the spirit of the correlation expansion, these quantities have to be decomposed into all possible lower-order factorizations, the remaining part describing two-particle correlations. Such a decomposition is given by

$$\begin{aligned}
\langle c^\dagger_{i_3,k+q_\parallel-q'_\parallel} c_{i_2,k} b^\dagger_{q'} b_q\rangle = & f^e_{i_3 i_2,k} B^*_{q\perp} B_{q\perp} \delta_{q_\parallel,0} \delta_{q'_\parallel,0} \\
& + \delta s^{e*,i_2 i_3}_{k,q',k+q'_\parallel} B_{q\perp} \delta_{q_\parallel,0} + \delta s^{e,i_3 i_2}_{k+q_\parallel,q,k} B^*_{q'_\perp} \delta_{q'_\parallel,0} \\
& + f^e_{i_3 i_2,k} n_q \delta_{q,q'} + \delta\langle c^\dagger_{i_3,k+q_\parallel-q'_\parallel} c_{i_2,k} b^\dagger_{q'} b_q\rangle. \qquad (6.4.14)
\end{aligned}$$

The equation of motion for the phonon-assisted correlation is obtained by inserting these decompositions into equation (6.4.13) and subtracting the equations of motion for the second term on the right-hand side of equation (6.4.12). Then, the first and second terms on the right-hand side of equation (6.4.14) cancel, the third one leads to a renormalization of the free carrier energies by the coherent phonon contributions, the fourth one leads to the scattering part and the last one describes the influence of two-particle correlations. Again, the hierarchy can be truncated by neglecting these higher-order correlations resulting in the equation of motion

$$\frac{\mathrm{d}}{\mathrm{d}t}\delta s^{e,i_1 i_2}_{\boldsymbol{k}+\boldsymbol{q}_\parallel,\boldsymbol{q},\boldsymbol{k}} = -\frac{1}{\mathrm{i}\hbar}\sum_{i_3,i_4}(\mathcal{E}^e_{i_3 i_1,\boldsymbol{k}+\boldsymbol{q}_\parallel}\delta_{i_2 i_4} - \mathcal{E}^e_{i_2 i_4,\boldsymbol{k}}\delta_{i_1 i_3} - \hbar\omega_{LO}\delta_{i_1 i_3}\delta_{i_2 i_4})\delta s^{e,i_3 i_4}_{\boldsymbol{k}+\boldsymbol{q}_\parallel,\boldsymbol{q},\boldsymbol{k}}$$
$$+\frac{1}{\mathrm{i}\hbar}\sum_{i_3,i_4} g^{*i_4 i_3}_{\boldsymbol{q}}[(n_{\boldsymbol{q}}+1)f^e_{i_1 i_4,\boldsymbol{k}+\boldsymbol{q}_\parallel}(\delta_{i_3,i_2} - f^e_{i_3 i_2,\boldsymbol{k}})$$
$$- n_{\boldsymbol{q}} f^e_{i_3 i_2,\boldsymbol{k}}(\delta_{i_1,i_4} - f^e_{i_1 i_4,\boldsymbol{k}+\boldsymbol{q}_\parallel})]$$
$$-\frac{1}{\mathrm{i}\hbar}\sum_{j_1,j_2} g^{*j_1 j_2}_{\boldsymbol{q}} p^*_{j_1 i_1,\boldsymbol{k}+\boldsymbol{q}_\parallel} p_{j_2 i_2,\boldsymbol{k}}, \qquad (6.4.15)$$

where the renormalized energies are given by

$$\mathcal{E}^e_{i_1 i_2,\boldsymbol{k}} = \epsilon^e_{i_1 i_2,\boldsymbol{k}} + \mathcal{E}^{e(cp,1)}_{i_1 i_2}. \qquad (6.4.16)$$

This equation together with the equations for the single-particle density matrices constitutes the basis for the analysis of electron–phonon quantum kinetics which includes memory effects and fulfils the energy-time uncertainty principle. An application to a simple quantum wire model will be discussed in section 7.4.

The semiclassical limit is obtained by an adiabatic elimination of the phonon-assisted correlations. This can be performed by the same procedure as described for the carrier–light interaction if the first-order energy renormalizations are neglected. However, when performing the Markov approximation the fast oscillations of interband and intraband polarizations caused by H_0 have to be taken into account. With this approximation, the phonon-assisted correlation is given by

$$\delta s^{e,i_1 i_2}_{\boldsymbol{k}+\boldsymbol{q}_\parallel,\boldsymbol{q},\boldsymbol{k}} = -\mathrm{i}\pi\sum_{i_3,i_4}\mathcal{D}(-\epsilon^e_{i_4,\boldsymbol{k}+\boldsymbol{q}_\parallel} + \epsilon^e_{i_3,\boldsymbol{k}} + \hbar\omega_{LO})g^{*i_4 i_3}_{\boldsymbol{q}}$$
$$\times[(n_{\boldsymbol{q}}+1)f^e_{i_1 i_4,\boldsymbol{k}+\boldsymbol{q}_\parallel}(\delta_{i_3,i_2} - f^e_{i_3 i_2,\boldsymbol{k}}) - n_{\boldsymbol{q}} f^e_{i_3 i_2,\boldsymbol{k}}(\delta_{i_1 i_4} - f^e_{i_1,i_4,\boldsymbol{k}+\boldsymbol{q}_\parallel})]$$
$$+\mathrm{i}\pi\sum_{j_1,j_2}\mathcal{D}(\epsilon^h_{j_1,\boldsymbol{k}+\boldsymbol{q}_\parallel} - \epsilon^h_{j_2,\boldsymbol{k}} + \hbar\omega_{LO})g^{*j_1 j_2}_{\boldsymbol{q}} p^*_{j_1 i_1,\boldsymbol{k}+\boldsymbol{q}_\parallel} p_{j_2 i_2,\boldsymbol{k}}. \qquad (6.4.17)$$

If this phonon-assisted correlation is inserted into the equation of motion for the single-particle density matrix, it becomes evident that the principal value part of $\mathcal{D}$ is associated with renormalizations of the energies describing the polaron corrections to the band structure while the δ-function part is associated with irreversible scattering and dephasing processes. Typically, the dominant polaronic features are a rigid shift of the bands and a slight modification of the effective mass. In this case, these effects can be included in H_0 since in any experiment determining the band structure they are always present. Therefore, in the following the principal value contribution will be neglected. Then, the second-order carrier–phonon contribution in the Markov approximation can be written in the general form

$$\left.\frac{\mathrm{d}}{\mathrm{d}t} f^{e}_{i_1 i_2,\boldsymbol{k}}\right|^{(cp,2)} = \sum_{i_5} [-\Gamma^{e,out(cp,2)}_{i_2 i_5,\boldsymbol{k}} f^{e}_{i_1 i_5,\boldsymbol{k}} - \Gamma^{e,out(cp,2)*}_{i_1 i_5,\boldsymbol{k}} f^{e*}_{i_2 i_5,\boldsymbol{k}}$$
$$+ \Gamma^{e,in(cp,2)}_{i_2 i_5,\boldsymbol{k}} (\delta_{i_1,i_5} - f^{e}_{i_1 i_5,\boldsymbol{k}}) + \Gamma^{e,in(cp,2)*}_{i_1 i_5,\boldsymbol{k}} (\delta_{i_2,i_5} - f^{e*}_{i_2 i_5,\boldsymbol{k}})]$$
$$+ \frac{1}{\mathrm{i}\hbar} \sum_{j_5} [\mathcal{U}^{e(cp,2)}_{i_2 j_5,\boldsymbol{k}} p^{*}_{j_5 i_1,\boldsymbol{k}} - \mathcal{U}^{e(cp,2)*}_{i_1 j_5,\boldsymbol{k}} p_{j_5 i_2,\boldsymbol{k}}], \qquad (6.4.18a)$$

$$\left.\frac{\mathrm{d}}{\mathrm{d}t} p_{j_1 i_1,\boldsymbol{k}}\right|^{(cp,2)} = \sum_{i_5} [(-\Gamma^{e,out(cp,2)}_{i_1 i_5,\boldsymbol{k}} - \Gamma^{e,in(cp,2)}_{i_1 i_5,\boldsymbol{k}}) p_{j_1 i_5,\boldsymbol{k}}$$
$$+ \Gamma^{he,out(cp,2)}_{j_1 i_5,\boldsymbol{k}} f^{e}_{i_5 i_1,\boldsymbol{k}} + \Gamma^{he,in(cp,2)}_{j_1 i_5,\boldsymbol{k}} (\delta_{i_5,i_1} - f^{e}_{i_5 i_1,\boldsymbol{k}})]$$
$$+ \sum_{j_5} [(-\Gamma^{h,out(cp,2)}_{j_1 j_5,\boldsymbol{k}} - \Gamma^{h,in(cp,2)}_{j_1 j_5,\boldsymbol{k}}) p_{j_5 i_1,\boldsymbol{k}}$$
$$+ \Gamma^{eh,out(cp,2)}_{i_1 j_5,\boldsymbol{k}} f^{h}_{j_5 j_1,-\boldsymbol{k}} + \Gamma^{eh,in(cp,2)}_{i_1 j_5,\boldsymbol{k}} (\delta_{j_5,j_1} - f^{h}_{j_5 j_1,-\boldsymbol{k}})]. \qquad (6.4.18b)$$

The various terms appearing on the right-hand side of equations (6.4.18) are given in appendix B.

In the fully semiclassical limit, where all off-diagonal elements of the intraband density matrices and all interband polarizations are neglected, the well known Boltzmann-like scattering contributions due to phonon emission and absorption are recovered.

6.4.3 Third order: Collisional broadening

The correlation expansion can be continued by taking into account two-particle correlations. In this section we will briefly discuss the structure of these contributions. However, to simplify the notation we will limit ourselves to the case of a homogeneous single-band system by dropping all subband indices and neglecting coherent phonons. The generalization to the multiband, multisubband case is straightforward.

The equations of motion of two-particle correlations involve expectation values of five operators, the two-particle phonon-assisted density matrices. The hierarchy is truncated by a factorization into single-particle density matrices and phonon-assisted density matrices resulting in an equation of motion for the correlation in equation (6.4.14) according to

$$\begin{aligned}\frac{\mathrm{d}}{\mathrm{d}t}\delta\langle c^\dagger_{k+q-q'}c_k b^\dagger_{q'} b_q\rangle = &-\frac{1}{\mathrm{i}\hbar}(\epsilon^e_{k+q-q'}-\epsilon^e_k)\delta\langle c^\dagger_{k+q-q'}c_k b^\dagger_{q'} b_q\rangle\\ &-\frac{1}{\mathrm{i}\hbar}g_{q'}(1-f^e_{k+q-q'}+n_{q'})\delta s^e_{k+q,q,k}\\ &+\frac{1}{\mathrm{i}\hbar}g_{q'}(f^e_k+n_q)\delta s^e_{k+q-q',q,k-q'}\\ &-\frac{1}{\mathrm{i}\hbar}g^*_q(1-f^e_k+n_q)\delta s^{e*}_{k+q,q',k+q-q'}\\ &+\frac{1}{\mathrm{i}\hbar}g^*_q(f^e_{k+q-q'}+n_{q'})\delta s^{e*}_{k,q',k-q'}.\end{aligned} \tag{6.4.19}$$

For a further simplification of this contribution we keep in mind the fact that equation (6.4.13) involves a summation over $\boldsymbol{q}'$. Since the phonon-assisted correlation is a complex quantity, in a first approximation we can assume that all contributions where a summation over this quantity is performed are small due to random phases at different momenta. Thus, we neglect the last three terms in equation (6.4.19). Then, after performing a Markov approximation, the third-order contributions give rise to self-energy corrections in the equations of motion δs according to

$$\frac{\mathrm{d}}{\mathrm{d}t}\delta s^e_{k+q,q,k} = -\frac{1}{\mathrm{i}\hbar}(\mathcal{E}^{e(cp,2)}_{k+q}-\mathcal{E}^{e(cp,2)*}_{k})\delta s^e_{k+q,q,k} \tag{6.4.20}$$

with the complex second-order carrier–phonon self-energy

$$\begin{aligned}\mathcal{E}^{e(cp,2)}_k = &-\mathrm{i}\pi\sum_{q',\pm}|g_{q'}|^2\mathcal{D}(\epsilon^e_k-\epsilon^e_{k+q'}\pm\hbar\omega_{LO})\\ &\times[(n_{q'}+\tfrac{1}{2}\pm\tfrac{1}{2})f^e_{k+q'}+(n_{q'}+\tfrac{1}{2}\mp\tfrac{1}{2})(1-f^e_{k+q'})].\end{aligned} \tag{6.4.21}$$

The real part of the self-energy describes the fact that the scattering processes occur between renormalized polaronic states, the imaginary part describes a collisional broadening.

However, it turns out that this approximation leads to an overestimation of the role played by the two-particle correlations. In particular, the imaginary part leads to a violation of energy conservation as well as to an overestimation of the broadening. For a discussion of the problems associated with the self-energy approximation the reader is referred to [45].

6.5 CARRIER–CARRIER INTERACTION

Electrons and holes are charged particles and therefore they interact via the Coulomb potential. In a two-band model this Coulomb interaction can be decomposed into two different parts. One part conserves the number of carriers in each band. It comprises electron–electron, hole–hole and electron–hole interactions. The other part describes Coulomb-induced transitions between the valence and conduction band. It comprises impact ionization and Auger recombination. Impact ionization typically requires carriers at very high energies while Auger recombination requires high densities. In most experiments probing the ultrafast dynamics these conditions are not fulfilled, therefore we will neglect this part. A quantum kinetic theory of impact ionization can be found in [46, 47].

With these assumptions, the carrier–carrier Hamiltonian is given by

$$\begin{aligned} H^{cc} = & \frac{1}{2} \sum_{\substack{i_1,i_2,i_3,i_4 \\ \boldsymbol{k},\boldsymbol{k}',\boldsymbol{q}}} V_{\boldsymbol{q}}^{i_1 i_2 i_3 i_4} c^\dagger_{i_1,\boldsymbol{k}+\boldsymbol{q}} c^\dagger_{i_2,\boldsymbol{k}'-\boldsymbol{q}} c_{i_3,\boldsymbol{k}'} c_{i_4,\boldsymbol{k}} \\ & + \frac{1}{2} \sum_{\substack{j_1,j_2,j_3,j_4 \\ \boldsymbol{k},\boldsymbol{k}',\boldsymbol{q}}} V_{\boldsymbol{q}}^{j_4 j_3 j_2 j_1} d^\dagger_{j_1,\boldsymbol{k}+\boldsymbol{q}} d^\dagger_{j_2,\boldsymbol{k}'-\boldsymbol{q}} d_{j_3,\boldsymbol{k}'} d_{j_4,\boldsymbol{k}} \\ & - \sum_{\substack{i_1,i_2,j_1,j_2 \\ \boldsymbol{k},\boldsymbol{k}',\boldsymbol{q}}} V_{\boldsymbol{q}}^{i_1 j_2 j_1 i_2} c^\dagger_{i_1,\boldsymbol{k}+\boldsymbol{q}} d^\dagger_{j_1,\boldsymbol{k}'-\boldsymbol{q}} d_{j_2,\boldsymbol{k}'} c_{i_2,\boldsymbol{k}} \end{aligned} \tag{6.5.1}$$

with the Coulomb matrix element

$$V_{\boldsymbol{q}}^{n_1 n_2 n_3 n_4} = \sum_{q_\perp} V^{3D}_{\boldsymbol{q},q_\perp} \mathcal{F}^{n_1 n_4}_{q_\perp} F^{*n_3 n_2}_{q_\perp}, \tag{6.5.2a}$$

$$V^{3D}_{\boldsymbol{q},q_\perp} = \frac{4\pi e^2}{\mathcal{V}\epsilon_s} \frac{1}{q^2 + q_\perp^2}. \tag{6.5.2b}$$

Here, $\boldsymbol{q}$ denotes a wave vector with the dimension of $\boldsymbol{k}$, i.e. of $\boldsymbol{r}_\parallel$. This Hamiltonian leads to a carrier–carrier contribution in the equation of motion, for example for the electron density matrix according to

$$\begin{aligned} \frac{\mathrm{d}}{\mathrm{d}t} f^e_{i_1 i_2,\boldsymbol{k}}\bigg|^{(cc)} = & \frac{1}{\mathrm{i}\hbar} \sum_{\substack{i_3,i_4,i_5 \\ \boldsymbol{k}',\boldsymbol{q}}} [V_{\boldsymbol{q}}^{i_2 i_3 i_4 i_5} \langle c^\dagger_{i_1,\boldsymbol{k}} c^\dagger_{i_3,\boldsymbol{k}'} c_{i_4,\boldsymbol{k}'+\boldsymbol{q}} c_{i_5,\boldsymbol{k}-\boldsymbol{q}} \rangle \\ & - V_{\boldsymbol{q}}^{i_5 i_4 i_3 i_1} \langle c^\dagger_{i_5,\boldsymbol{k}-\boldsymbol{q}} c^\dagger_{i_4,\boldsymbol{k}'+\boldsymbol{q}} c_{i_3,\boldsymbol{k}'} c_{i_2,\boldsymbol{k}} \rangle] \\ & - \frac{1}{\mathrm{i}\hbar} \sum_{\substack{i_3,j_1,j_2 \\ \boldsymbol{k}',\boldsymbol{q}}} [V_{\boldsymbol{q}}^{i_2 j_2 j_1 i_3} \langle c^\dagger_{i_1,\boldsymbol{k}} d^\dagger_{j_1,\boldsymbol{k}'} d_{j_2,\boldsymbol{k}'+\boldsymbol{q}} c_{i_3,\boldsymbol{k}-\boldsymbol{q}} \rangle \\ & - V_{\boldsymbol{q}}^{i_3 j_1 j_2 i_1} \langle c^\dagger_{i_3,\boldsymbol{k}-\boldsymbol{q}} d^\dagger_{j_2,\boldsymbol{k}'+\boldsymbol{q}} d_{j_1,\boldsymbol{k}'} c_{i_2,\boldsymbol{k}} \rangle] \end{aligned} \tag{6.5.3}$$

involving two-particle density matrices. Again, this is the starting point of an infinite hierarchy of equations of motion for density matrices involving an increasing number of carriers. Here, it is directly the quantum-mechanical analogue of the classical BBGKY hierarchy, where the equation of motion for a N-particle distribution function involves $(N+1)$-particle distribution functions [48].

6.5.1 First order: Excitons and renormalization

The lowest-order contribution due to carrier–carrier interactions is obtained by factorizing the two-particle density matrices into single-particle density matrices, for example according to

$$\langle c^{\dagger}_{i_1,\boldsymbol{k}} c^{\dagger}_{i_3,\boldsymbol{k}'} c_{i_4,\boldsymbol{k}'+\boldsymbol{q}} c_{i_5,\boldsymbol{k}-\boldsymbol{q}} \rangle = f^{e}_{i_1 i_5,\boldsymbol{k}} f_{i_3 i_4,\boldsymbol{k}'} \delta_{\boldsymbol{q},0} - f^{e}_{i_1 i_4,\boldsymbol{k}} f^{e}_{i_3 i_5,\boldsymbol{k}'} \delta_{\boldsymbol{k}',\boldsymbol{k}-\boldsymbol{q}}. \quad (6.5.4)$$

This is the Hartree–Fock or mean-field level where all correlations between the carriers are neglected. The corresponding equations of motion are given by

$$\frac{\mathrm{d}}{\mathrm{d}t} f^{e}_{i_1 i_2,\boldsymbol{k}}\Big|^{(cc,1)} = \frac{1}{\mathrm{i}\hbar} \sum_{i_3 i_4} [\mathcal{E}^{e(cc,1)}_{i_2 i_4,\boldsymbol{k}} \delta_{i_1 i_3} - \mathcal{E}^{e(cc,1)}_{i_3 i_1,\boldsymbol{k}} \delta_{i_2 i_4}] f^{e}_{i_3 i_4,\boldsymbol{k}} + \frac{1}{\mathrm{i}\hbar} \sum_{j_1} [\mathcal{U}^{(cc,1)}_{i_2 j_1,\boldsymbol{k}} p^{*}_{j_1 i_1,\boldsymbol{k}} - \mathcal{U}^{*(cc,1)}_{i_1 j_1,\boldsymbol{k}} p_{j_1 i_2,\boldsymbol{k}}], \quad (6.5.5a)$$

$$\frac{\mathrm{d}}{\mathrm{d}t} p_{j_1 i_1,\boldsymbol{k}}\Big|^{(cc,1)} = \frac{1}{\mathrm{i}\hbar} \sum_{i_2 j_2} [\mathcal{E}^{h(cc,1)}_{j_1 j_2,-\boldsymbol{k}} \delta_{i_1 i_2} + \mathcal{E}^{e(cc,1)}_{i_1 i_2,\boldsymbol{k}} \delta_{j_1 j_2}] p_{j_2 i_2,\boldsymbol{k}} + \frac{1}{\mathrm{i}\hbar} \sum_{i_2 j_2} \mathcal{U}^{(cc,1)}_{i_2 j_2,\boldsymbol{k}} [\delta_{i_1 i_2} \delta_{j_1 j_2} - \delta_{j_1 j_2} f^{e}_{i_2 i_1,\boldsymbol{k}} - \delta_{i_1 i_2} f^{h}_{j_2 j_1,-\boldsymbol{k}}] \quad (6.5.5b)$$

with the self-energy matrices due to the Hartree and Fock contributions of electron–electron and hole–hole interaction

$$\mathcal{E}^{e(cc,1)}_{i_1 i_2,\boldsymbol{k}} = -\sum_{i_3,i_4,\boldsymbol{q}} V^{i_1 i_3 i_2 i_4}_{\boldsymbol{q}} f^{e}_{i_3 i_4,\boldsymbol{k}+\boldsymbol{q}} + \sum_{i_3,i_4,\boldsymbol{k}'} V^{i_1 i_3 i_4 i_2}_{0} f^{e}_{i_3 i_4,\boldsymbol{k}'} - \sum_{j_3,j_4,\boldsymbol{k}'} V^{i_1 j_4 j_3 i_2}_{0} f^{h}_{j_3 j_4,\boldsymbol{k}'}, \quad (6.5.6a)$$

$$\mathcal{E}^{h(cc,1)}_{j_1 j_2,\boldsymbol{k}} = -\sum_{j_3,j_4,\boldsymbol{q}} V^{j_2 j_4 j_1 j_3}_{\boldsymbol{q}} f^{h}_{j_3 j_4,\boldsymbol{k}+\boldsymbol{q}} - \sum_{i_3,i_4,\boldsymbol{k}'} V^{j_2 i_3 i_4 j_1}_{0} f^{e}_{i_3 i_4,\boldsymbol{k}'} + \sum_{j_3,j_4,\boldsymbol{k}'} V^{j_2 j_4 j_3 j_1}_{0} f^{h}_{j_3 j_4,\boldsymbol{k}'} \quad (6.5.6b)$$

and the internal field matrix due to the Fock contributions of electron–hole interaction

$$\mathcal{U}^{(cc,1)}_{i_1 j_1,\boldsymbol{k}} = -\sum_{i_2,j_2} \sum_{\boldsymbol{q}} V^{i_1 j_2 j_1 i_2}_{\boldsymbol{q}} p_{j_2 i_2,\boldsymbol{k}+\boldsymbol{q}}. \quad (6.5.7)$$

In a homogeneous system where the Coulomb matrix element does not depend on subband indices the Hartree terms i.e. the terms involving V_0, cancel due to charge neutrality. In a heterostructure, charge densities may build up in the directions $\boldsymbol{r}_\perp$ which will be screened by these Hartree contributions. As discussed above for the Fröhlich interaction, also in the present case the limit $\boldsymbol{q} \to 0$ in the Hartree terms has to be taken with some care. The Coulomb matrix elements diverge in this limit. The divergence, however, occurs only in the diagonal elements which again cancel due to charge neutrality. Thus, the difference between electron and hole Hartree contributions in equation (6.5.6) is finite.

6.5.2 Second order: Scattering and dephasing

As in the case of carrier–phonon interaction, the next step in the correlation hierarchy is obtained by including two-particle correlations like

$$\delta\langle c^\dagger_{i_1,\boldsymbol{k}} c^\dagger_{i_2,\boldsymbol{k}'} c_{i_3,\boldsymbol{k}'+\boldsymbol{q}} c_{i_4,\boldsymbol{k}-\boldsymbol{q}}\rangle = \langle c^\dagger_{i_1,\boldsymbol{k}} c^\dagger_{i_2,\boldsymbol{k}'} c_{i_3,\boldsymbol{k}'+\boldsymbol{q}} c_{i_4,\boldsymbol{k}-\boldsymbol{q}}\rangle - f^e_{i_1 i_4,\boldsymbol{k}} f^e_{i_2 i_3,\boldsymbol{k}'}\delta_{\boldsymbol{q},0} + f^e_{i_1 i_3,\boldsymbol{k}} f^e_{i_2 i_4,\boldsymbol{k}-\boldsymbol{q}}\delta_{\boldsymbol{k}',\boldsymbol{k}-\boldsymbol{q}} \tag{6.5.8}$$

describing deviations from the corresponding factorizations. The equations of motion for these quantities involve three-particle density matrices. The hierarchy can be truncated by factorizing the three-particle density matrices into products of three single-particle density matrices resulting in the equation of motion

$$\begin{aligned}
\frac{\mathrm{d}}{\mathrm{d}t}\delta\langle c^\dagger_{i_1,\boldsymbol{k}} c^\dagger_{i_2,\boldsymbol{k}'} c_{i_3,\boldsymbol{k}'+\boldsymbol{q}} c_{i_4,\boldsymbol{k}-\boldsymbol{q}}\rangle = \; & -\frac{1}{\mathrm{i}\hbar}(\epsilon^e_{i_1,\boldsymbol{k}} + \epsilon^e_{i_2,\boldsymbol{k}'} - \epsilon^e_{i_3,\boldsymbol{k}'+\boldsymbol{q}} - \epsilon^e_{i_4,\boldsymbol{k}-\boldsymbol{q}}) \\
& \times \delta\langle c^\dagger_{i_1,\boldsymbol{k}} c^\dagger_{i_2,\boldsymbol{k}'} c_{i_3,\boldsymbol{k}'+\boldsymbol{q}} c_{i_4,\boldsymbol{k}-\boldsymbol{q}}\rangle \\
& -\frac{1}{\mathrm{i}\hbar}\sum_{i_5,i_6,i_7,i_8}(V^{i_5 i_6 i_7 i_8}_{\boldsymbol{q}} - V^{i_6 i_5 i_7 i_8}_{\boldsymbol{k}-\boldsymbol{k}'-\boldsymbol{q}}) \\
& \times[(\delta_{i_1 i_8} - f^e_{i_1 i_8,\boldsymbol{k}})(\delta_{i_2 i_7} - f^e_{i_2 i_7,\boldsymbol{k}'}) \\
& \times f^e_{i_6 i_3,\boldsymbol{k}'+\boldsymbol{q}} f^e_{i_5 i_4,\boldsymbol{k}-\boldsymbol{q}} \\
& - f^e_{i_1 i_8,\boldsymbol{k}} f^e_{i_2 i_7,\boldsymbol{k}'}(\delta_{i_6 i_3} - f^e_{i_6 i_3,\boldsymbol{k}'+\boldsymbol{q}}) \\
& \times(\delta_{i_5 i_4} - f^e_{i_5 i_4,\boldsymbol{k}-\boldsymbol{q}})] \\
& +\frac{1}{\mathrm{i}\hbar}\sum_{i_5,i_8,j_1,j_2} V^{i_5 j_1 j_2 i_8}_{\boldsymbol{q}}[p^*_{j_2 i_1,\boldsymbol{k}} p_{j_1 i_4,\boldsymbol{k}-\boldsymbol{q}} \\
& \times(\delta_{i_2 i_8} f^e_{i_5 i_3,\boldsymbol{k}'+\boldsymbol{q}} - f^e_{i_2 i_8,\boldsymbol{k}'}\delta_{i_5 i_3}) \\
& + p^*_{j_2 i_2,\boldsymbol{k}'} p_{j_1 i_3,\boldsymbol{k}'+\boldsymbol{q}} \\
& \times(\delta_{i_1 i_8} f^e_{i_5 i_4,\boldsymbol{k}-\boldsymbol{q}} - f^e_{i_1 i_8,\boldsymbol{k}}\delta_{i_5 i_4})] \\
& -\frac{1}{\mathrm{i}\hbar}\sum_{i_5,i_8,j_1,j_2} V^{i_5 j_1 j_2 i_8}_{\boldsymbol{k}-\boldsymbol{k}'-\boldsymbol{q}}[p^*_{j_2 i_1,\boldsymbol{k}} p_{j_1 i_3,\boldsymbol{k}'+\boldsymbol{q}} \\
& \times(\delta_{i_2 i_8} f^e_{i_5 i_4,\boldsymbol{k}-\boldsymbol{q}} - f^e_{i_2 i_8,\boldsymbol{k}'}\delta_{i_5 i_4}) \\
& + p^*_{j_2 i_2,\boldsymbol{k}'} p_{j_1 i_4,\boldsymbol{k}-\boldsymbol{q}} \\
& \times(\delta_{i_1 i_8} f^e_{i_5 i_3,\boldsymbol{k}'+\boldsymbol{q}} - f^e_{i_1 i_8,\boldsymbol{k}}\delta_{i_5 i_3})].
\end{aligned} \tag{6.5.9}$$

This equation constitutes the starting point for an investigation of carrier–carrier quantum kinetics. An analysis of the intersubband quantum kinetics in quantum wires based on this approach can be found in [49, 50]. The semiclassical limit is obtained by an adiabatic elimination of the two-particle correlation exactly in the same way as discussed for the phonon-assisted correlations. Again, the result can be written in the form

$$\begin{aligned}\frac{\mathrm{d}}{\mathrm{d}t} f^{e}_{i_1 i_2,\boldsymbol{k}}\bigg|^{(cc,2)} &= \sum_{i_5} [-\Gamma^{e,out(cc,2)}_{i_2 i_5,\boldsymbol{k}} f^{e}_{i_1 i_5,\boldsymbol{k}} - \Gamma^{e,out(cc,2)*}_{i_1 i_5,\boldsymbol{k}} f^{e*}_{i_2 i_5,\boldsymbol{k}} \\ &\quad + \Gamma^{e,in(cc,2)}_{i_2 i_5,\boldsymbol{k}} (\delta_{i_1,i_5} - f^{e}_{i_1 i_5,\boldsymbol{k}}) + \Gamma^{e,in(cc,2)*}_{i_1 i_5,\boldsymbol{k}} (\delta_{i_2,i_5} - f^{e*}_{i_2 i_5,\boldsymbol{k}})] \\ &\quad + \frac{1}{\mathrm{i}\hbar} \sum_{j_5} [\mathcal{U}^{e(cc,2)}_{i_2 j_5,\boldsymbol{k}} p^{*}_{j_5 i_1,\boldsymbol{k}} - \mathcal{U}^{e(cc,2)*}_{i_1 j_5,\boldsymbol{k}} p_{j_5 i_2,\boldsymbol{k}}], \end{aligned} \tag{6.5.10a}$$

$$\begin{aligned}\frac{\mathrm{d}}{\mathrm{d}t} p_{j_1 i_1,\boldsymbol{k}}\bigg|^{(cc,2)} &= \sum_{i_5} [(-\Gamma^{e,out(cc,2)}_{i_1 i_5,\boldsymbol{k}} - \Gamma^{e,in(cc,2)}_{i_1 i_5,\boldsymbol{k}}) p_{j_1 i_5,\boldsymbol{k}} \\ &\quad + \Gamma^{he,out(cc,2)}_{j_1 i_5,\boldsymbol{k}} f^{e}_{i_5 i_1,\boldsymbol{k}} + \Gamma^{he,in(cc,2)}_{j_1 i_5,\boldsymbol{k}} (\delta_{i_5,i_1} - f^{e}_{i_5 i_1,\boldsymbol{k}})] \\ &\quad + \sum_{j_5} [(-\Gamma^{h,out(cc,2)}_{j_1 j_5,\boldsymbol{k}} - \Gamma^{h,in(cc,2)}_{j_1 j_5,\boldsymbol{k}}) p_{j_5 i_1,\boldsymbol{k}} \\ &\quad + \Gamma^{eh,out(cc,2)}_{i_1 j_5,\boldsymbol{k}} f^{h}_{j_5 j_1,-\boldsymbol{k}} + \Gamma^{eh,in(cc,2)}_{i_1 j_5,\boldsymbol{k}} (\delta_{j_5,j_1} - f^{h}_{j_5 j_1,-\boldsymbol{k}})], \end{aligned} \tag{6.5.10b}$$

i.e. the form of equation (6.4.18), however, with different functions Γ and $\mathcal{U}$. These functions are given in appendix C.

6.5.3 Second order: Screening

In the previous section the carrier–carrier scattering terms were derived in the second Born approximation. It is well known that in the semiclassical limit the total scattering rate for a bare Coulomb potential diverges due to the long-range nature of the potential. Usually this divergence is removed by taking a screened Coulomb potential as described by the Lindhard dielectric function. This dielectric function is obtained if the response of the electron gas to an external test charge is studied in the random phase approximation [36]. The aim of this section is to show how screening appears self-consistently in the density-matrix approach. We will restrict the discussion to a single-band single-subband case and follow the derivation in [51]. We will only give the basic arguments and the general procedure to obtain the screened potential. For details the reader is referred to [51].

The equation of motion for the distribution function in a one-band model is given by

$$\mathrm{i}\hbar \frac{\mathrm{d}}{\mathrm{d}t} f_{\boldsymbol{k}} = \sum_{\boldsymbol{k}',\boldsymbol{q}} V_{\boldsymbol{q}} [\delta \langle c^{\dagger}_{\boldsymbol{k}} c^{\dagger}_{\boldsymbol{k}'} c_{\boldsymbol{k}'+\boldsymbol{q}} c_{\boldsymbol{k}-\boldsymbol{q}} \rangle - \delta \langle c^{\dagger}_{\boldsymbol{k}-\boldsymbol{q}} c^{\dagger}_{\boldsymbol{k}'+\boldsymbol{q}} c_{\boldsymbol{k}'} c_{\boldsymbol{k}} \rangle]. \tag{6.5.11}$$

The equation of motion for the two-particle correlations is obtained in the same way as in the previous section leading to three-particle density matrices. While there these three-particle correlations have been factorized into single-particle density matrices, now all factorizations into lower-order correlations are included. Thus, additional contributions due to a factorization of a three-particle density matrix into a distribution function and a two-particle correlation appear. Neglecting only the three-particle correlations, the equation of motion is given by

$$i\hbar\frac{\mathrm{d}}{\mathrm{d}t}\delta\langle c_k^\dagger c_{k'}^\dagger c_{k'+q}c_{k-q}\rangle = (\mathcal{E}_{k-q}+\mathcal{E}_{k'+q}-\mathcal{E}_{k'}-\mathcal{E}_k)\delta\langle c_k^\dagger c_{k'}^\dagger c_{k'+q}c_{k-q}\rangle + \sum_{i=1}^{6} S_i \tag{6.5.12}$$

with the renormalized energies

$$\mathcal{E}_k = \epsilon_k - \sum_q V_q f_{k-q} \tag{6.5.13}$$

and

$$\begin{aligned} S_1 =&(V_q - V_{k-k'-q})[f_k f_{k'}(1-f_{k'+q})(1-f_{k-q}) \\ &- f_{k-q}f_{k'+q}(1-f_{k'})(1-f_k)], \end{aligned} \tag{6.5.14a}$$

$$\begin{aligned} S_2 =&V_q[(f_k - f_{k-q})\sum_{k''}\delta\langle c_{k''}^\dagger c_{k'}^\dagger c_{k'+q}c_{k''-q}\rangle \\ &+(f_{k'} - f_{k'+q})\sum_{k''}\delta\langle c_k^\dagger c_{k''}^\dagger c_{k''+q}c_{k-q}\rangle], \end{aligned} \tag{6.5.14b}$$

$$\begin{aligned} S_3 =&-V_{k-k'-q}[(f_{k'} - f_{k-q})\sum_{k''}\delta\langle c_k^\dagger c_{k''}^\dagger c_{k''+k-k'-q}c_{k'+q}\rangle \\ &+(f_k - f_{k'+q})\sum_{k''}\delta\langle c_{k''}^\dagger c_{k'}^\dagger c_{k-q}c_{k''-k+k'+q}\rangle], \end{aligned} \tag{6.5.14c}$$

$$\begin{aligned} S_4 =&(1 - f_{k-q} - f_{k'+q})\sum_{q'}V_{q'}\delta\langle c_k^\dagger c_{k'}^\dagger c_{k'+q+q'}c_{k-q-q'}\rangle \\ &-(1 - f_k - f_{k'})\sum_{q'}V_{q'}\delta\langle c_{k-q'}^\dagger c_{k'+q'}^\dagger c_{k'+q}c_{k-q}\rangle, \end{aligned} \tag{6.5.14d}$$

$$\begin{aligned} S_5 =&-(f_k - f_{k-q})\sum_{q'}V_{q'}\delta\langle c_{k-q'}^\dagger c_{k'}^\dagger c_{k'+q}c_{k-q-q'}\rangle \\ &-(f_{k'} - f_{k'+q})\sum_{q'}V_{q'}\delta\langle c_k^\dagger c_{k'+q'}^\dagger c_{k'+q+q'}c_{k-q}\rangle, \end{aligned} \tag{6.5.14e}$$

$$\begin{aligned} S_6 =&-(f_{k'} - f_{k-q})\sum_{q'}V_{q'}\delta\langle c_k^\dagger c_{k'+q'}^\dagger c_{k'+q}c_{k-q+q'}\rangle \\ &-(f_k - f_{k'+q})\sum_{q'}V_{q'}\delta\langle c_{k-q'}^\dagger c_{k'}^\dagger c_{k'+q-q'}c_{k-q}\rangle. \end{aligned} \tag{6.5.14f}$$

Apart from the term S_1 which leads to the Born approximation and has been discussed in the previous section, additional contributions appear: S_2 gives rise to the screening of the Coulomb potential; S_3 contains the exchange contributions to the screening; S_4 describes the repeated scattering of two particles from each other and leads to the exact T-matrix; S_5 and S_6 contain terms which in the language of Green's functions would be called vertex corrections to the screening terms S_2 and S_3.

Due to the divergence of $V_{\boldsymbol{q}}$ for small $\boldsymbol{q}$, S_1 and S_2 may be expected to dominate since they involve no summation over the Coulomb matrix element. Keeping only these contributions is equivalent to the random phase approximation. It can be shown that in the Markov limit they lead to scattering terms where the bare Coulomb matrix element is replaced by the dynamically screened potential as described by the Lindhard dielectric function. For details of the rather lengthy derivation we refer the reader to [51]. Equation (6.5.12) with the contributions S_1 and S_2 may serve as a starting point for a quantum kinetic investigation of carrier–carrier scattering including the build-up of the screening at very short times [52].

6.5.4 Third order: Collisional broadening

The next order in the hierarchy is obtained in the same way as for the case of the carrier–phonon interaction: the equations of motion for three-particle correlations have to be set up and the resulting four-particle density matrices have to be factorized into all kinds of lower-order terms. Among the many possible factorizations there is one class which has the structure of a self-energy correction to the second-order equation by introducing second-order energy corrections and a collisional broadening into the equation for the two-particle correlation. Here, however, due to the strong dominance of small-angle scattering, in particular at low densities, this approximation is expected to overestimate the broadening in a much more dramatic way than for carrier–phonon scattering. This strong overestimation of a collisional broadening due to carrier–carrier scattering has been studied in [53, 54] for the case of carrier generation and it has been shown how the inclusion of additional terms which have the structure of in-scattering terms leads to the correct physical behaviour.

6.6 MULTIPLE INTERACTIONS

In the previous sections three different interactions have been introduced. We have derived the respective contributions to the equations of motion of the basic variables on different levels of approximations. However, the interactions have been treated separately. Therefore we have obtained contributions which have to be added independently as in the case of the semiclassical Boltzmann equation where the scattering rates for different mechanisms are simply added. In a full quantum-mechanical treatment of the many-body system different interaction mechanisms in general interfere and we can no longer separate the contributions. These interference phenomena are easily obtained in the density matrix formalism. In this section we will briefly discuss a few of these effects.

If we have simultaneously a light field in the optical range and a strong static field in the free direction, the carrier generation is influenced by the static field giving rise to a tail in the absorption spectrum below the gap as well as to an oscillatory behaviour above the gap. This is the **Franz–Keldysh effect** [55]. If the static field is replaced by a low-frequency field, absorption sidebands occur [56]. If the static field is in the confined direction, it gives rise to a change of the resonance frequency and of the oscillator strength of the optical transitions, which is called the **quantum confined Stark effect** [57, 58]. Also a strong optical field near resonance gives rise to a change in the absorption, the optical or AC Stark effect [59, 21]. All these phenomena are included in equations (6.3.5). An alternative way to obtain such effects is shown in Chapter 9 for the case of a superlattice where the static field is already included in the basis.

If the external field is included in the derivation of the equations of motion for the second-order variables, i.e. in equation (6.4.13) for the phonon-assisted density matrix or the corresponding equation for the two-particle density matrix, scattering processes are influenced by the electric field. This is called the **intra-collisional field effect** and can be interpreted in two equivalent ways. Either the scattering processes occur between states which are renormalized by the field or the carriers are accelerated during the finite time required for a collision in a quantum-mechanical treatment [60, 61].

Including first-order Coulomb terms in the equations of motion for the phonon-assisted density matrices leads to phonon scattering between exciton states. Including second-order Coulomb terms leads to an additional collisional broadening due to carrier–carrier scattering of the carrier–phonon interaction. Vice versa, second-order carrier–phonon terms in the equations of motion for the two-particle density matrices lead to a collisional broadening of the carrier–carrier scattering processes. In addition, the coupling between carrier–carrier and carrier–phonon dynamics gives rise to a plasmon–phonon coupling and a combined screening due to both interaction mechanisms [62].

6.7 RESULTS

In the previous sections the general procedure to describe the carrier dynamics in semiconductor bulk and nanostructures based on the density matrix formalism and a correlation expansion has been presented. Now this theory will be applied to several structures where we will concentrate on various aspects of the dynamics. In particular, by comparing results obtained from the density matrix approach with those obtained from a semiclassical Boltzmann approach we will show the characteristic quantum features of the various contributions.

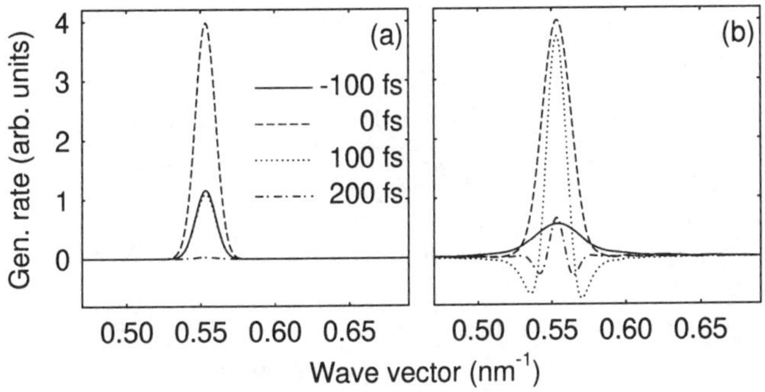

Figure 6.1 *Generation rate for (a) Boltzmann equation and (b) semiconductor Bloch equations. No scattering and dephasing processes are included.*

6.7.1 Carrier generation in bulk semiconductors

As a first application of the theory we will investigate the carrier generation in bulk GaAs in the case of a simple parabolic two-band model. We consider the case of a Gaussian laser pulse with $\tau_L = 150$ fs at a photon energy of 1.68 eV. In the semiclassical Boltzmann picture the generation is described by the rate according to equation (6.3.14). This rate is plotted in Fig. 6.1(a) at four different times. Except for the phase-space filling terms which are responsible for the slight difference between the curves at -100 and 100 fs, the spectral shape is independent of time. In contrast, Fig. 6.1(b) shows the generation rate according to equation (6.3.7) as obtained from the Bloch equations in the absence of carrier–phonon and carrier–carrier interactions. Now we notice a pronounced time dependence of the spectral shape. The generation starts very broad, by increasing time the shape becomes narrower and in the wings negative values of the rate appear. From a physical point of view this can be easily understood: as in the semiclassical case the spectral width is determined by energy-time uncertainty. Due to causality, however, only the time from the onset of the laser pulse up to the observation time determines the broadening leading to a very broad structure at early times. With increasing time, the energy uncertainty decreases and, since there is still a complete phase coherence between the carriers and laser field, those carriers generated off-resonance perform a stimulated recombination leading to the negative wings. As a result, the time-integrated generation rate agrees with the semiclassical case as long as phase-space filling is not important.

The situation changes if scattering processes are taken into account. The semiclassical rate is not affected by these processes, since all interaction processes are treated independently. In the Bloch case, on the other hand, the generation rate is still given by equation (6.3.7), the polarization, however, is strongly influenced via the second-order contributions of carrier–phonon and carrier–carrier interactions which lead to a dephasing and therefore to a loss of coherence. As a result, the stimulated recombination processes in the wings are inhibited and

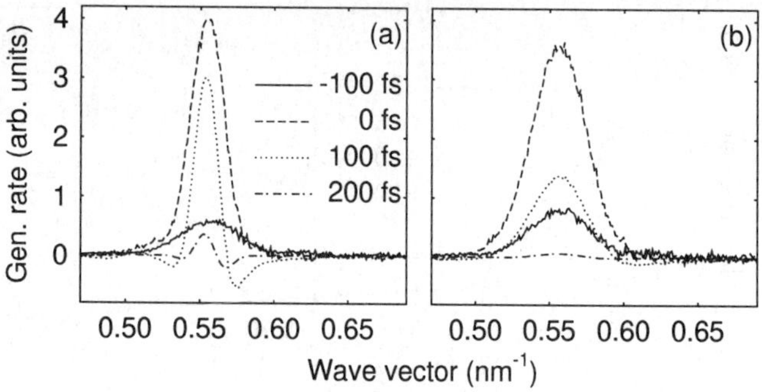

Figure 6.2 *Generation rate for a final density of (a)* 1×10^{14} cm^{-3} *and (b)* 2×10^{16} cm^{-3}. *Carrier–carrier and carrier–phonon scattering processes are included.*

the narrowing of the spectral shape is reduced. This is shown in Fig. 6.2 where the generation rate obtained from the semiconductor Bloch equations including first- and second-order interaction terms is plotted for two different densities. These results have been obtained by using a generalized Monte Carlo approach for the solution of the semiconductor Bloch equations [53, 54]. Here, the dominant process is carrier–carrier scattering. At the lower density the dephasing is not yet very efficient and we observe only a slight reduction of the negative parts. At the higher density, however, the loss of coherence is very strong and the generation rate remains broad during the pulse leading to a much broader carrier distribution as compared with the semiclassical case. It has been found that this density-dependent generation process has to be taken into account in order to explain quantitatively the spectral details observed in band-to-acceptor luminescence experiments [29, 30].

In the case of carrier generation high above the band gap the first-order (Hartree–Fock) terms are not very important for the qualitative features of the generation process. The situation is different close to the band gap as shown in Fig. 6.3 where the carrier density generated by a 1 ps pulse with fixed intensity is plotted as a function of the excess energy [63, 8]. Without Hartree–Fock terms (dashed curve) the square-root behaviour of the joint density of states is recovered. Including these terms (solid curve), a completely different result is obtained: we find a strong generation at an energy of 4.5 meV below the gap, i.e. at the 1s exciton energy, followed by a minimum between the 1s and the 2s exciton. Due to the spectral width of the pulse, the higher excitonic states are not resolved. Above the gap, the Coulomb enhancement leads to an increased generation with respect to the semiclassical result.

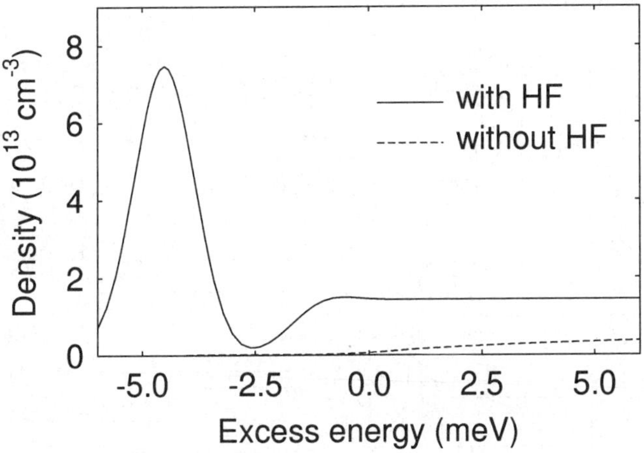

Figure 6.3 *Carrier density generated by a 1 ps laser pulse with fixed intensity as a function of the excess energy.*

6.7.2 Terahertz emission from quantum wells

Semiconductor nanostructures offer the fascinating opportunity to tailor the spectral properties as well as the localization behaviour of the wave functions by band gap engineering. As an example, in this section we will study the carrier–light interaction in an asymmetric double quantum-well structure where these features can be controlled not only by the growth of the specific structure but also by an external static electric field applied in the growth direction. The structure consists of two GaAs quantum wells with widths of 14.5 and 10 nm separated by an $Al_{0.2}Ga_{0.8}As$ barrier with a width of 2.5 nm. A schematic plot of the band edge of the conduction and valence band is shown for three different values of the external field in the inset of Fig. 6.4. Included are the envelope functions of the two lowest conduction band states and the highest valence band state. For the sake of clarity, here we will limit ourselves to these states since they are responsible for the essential features. A more detailed analysis including additional hole states can be found in [64].

Figure 6.4 shows the absorption spectrum of the structure in the band gap region for different values of the applied field calculated again for the case of excitation by a 1 ps pulse. At vanishing external field we observe qualitatively the same spectrum as in the bulk. Due to the confinement in the growth direction, the Coulomb correlation between electron and hole increases leading to an increased exciton binding energy and oscillator strength. This is an example of the well-known fact that Coulomb effects increase with decreasing dimensionality of the system. With increasing field we observe a second peak with increasing amplitude. This peak refers to the exciton of the hole with an electron of the second subband. Its height increases due to the increasing overlap

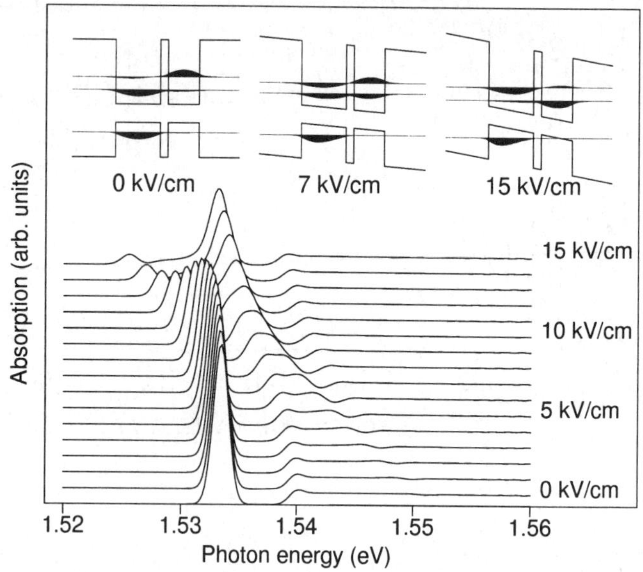

Figure 6.4 *Absorption spectrum for a 1 ps laser pulse at different values of the applied field. The inset schematically shows the band edges and envelope wave functions.*

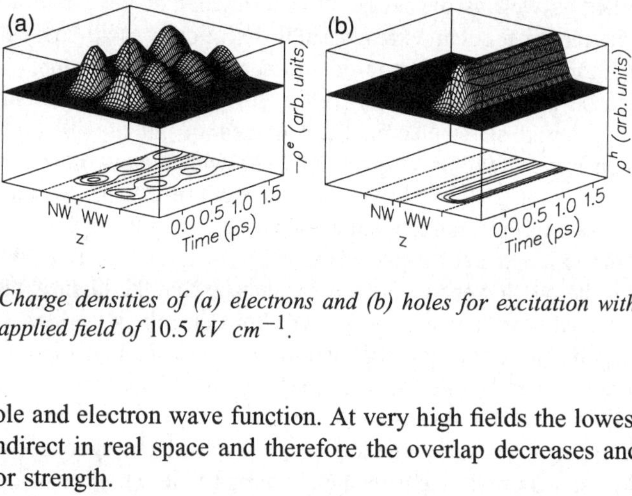

Figure 6.5 *Charge densities of (a) electrons and (b) holes for excitation with a 150 fs pulse at an applied field of* $10.5\ kV\ cm^{-1}$.

between hole and electron wave function. At very high fields the lowest exciton becomes indirect in real space and therefore the overlap decreases and it loses its oscillator strength.

An interesting situation occurs at an intermediate field strength where both excitons have approximately the same oscillator strength. In this case, the interaction with a sufficiently short laser pulse with a spectral width which covers both transitions leads to the creation of a superposition of both electronic states. This superposition can be observed as quantum beats in four-wave mixing experiments [65]. Due to the locality of the carrier–light coupling and the localization of the holes in the wide well, the laser pulse generates the electrons initially in the wide well. Then, the electrons oscillate between the two wells. Figure 6.5 shows the charge densities of electrons and holes as functions of space and time for the case of excitation with a 150 fs laser pulse at an applied field of 10.5 kV cm^{-1} in a three-dimensional representation as well as in a contour plot. The oscillations of the electronic wave packet are clearly visible.

Since the electrons move in space while the holes remain fixed, this system exhibits an oscillating dipole moment. According to classical electrodynamics, such an oscillating dipole emits electromagnetic radiation which, in the present case, is in the range of several THz. Such an emission of THz-radiation from the structure discussed here has been observed experimentally [18]. The emission of a THz-radiation due to spatially oscillating wave packets has also been observed in single quantum wells [20] and superlattices [19] in the presence of a static electric field. An extensive analysis of the microwave emission in the case of optically excited superlattices is given in Chapter 9 of this book.

6.7.3 Coherent phonons in quantum wells

The oscillating electronic wave packet in the asymmetric quantum-well structure discussed in the previous section has been found to be associated with an oscillating space charge. The space charge acts on the lattice ions via the polar Fröhlich interaction resulting in a lattice distortion. This phenomenon is described by coherent phonons. In this section we will analyse the coherent phonon dynamics and their consequence for the carrier dynamics. We use the same structure as in the previous section, however, we will restrict ourselves to the carrier–phonon interactions and neglect carrier–carrier interactions, i.e. in particular the Hartree–Fock terms leading to the excitonic effects, which are not essential for the qualitative features of the dynamics.

In Fig. 6.6(a) the space charge associated with the carrier dynamics, i.e. the sum of electron and hole charge density, is plotted as a function of space and time. Figure 6.6(b) shows the corresponding lattice polarization $\langle P_{lat} \rangle$ in the growth direction calculated from the coherent phonon amplitudes. As expected, the build up of the electronic dipole moment is associated with the build up of a lattice polarization due to the opposite electrostatic forces acting on the positive Ga and negative As ions.

The set of equations of motion including coherent phonons is nonlinear, therefore one should expect a density dependence in the coupling efficiency. This can be seen in Fig. 6.7 where the electronic dipole moment has been plotted as a function of time for two different excitation densities. The solid curves refer to a calculation including coherent phonons, the dashed curves refer to a calculation where they are absent. Without coherent phonons the oscillation frequency is given by the splitting of the two electron subbands and is therefore independent of the density. Including coherent phonons, we find that the oscillation frequency decreases with increasing density. This can be understood by

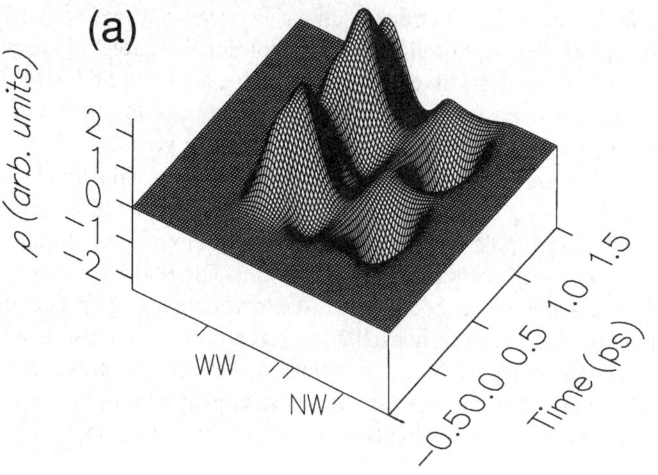

Figure 6.6 *(a) Electronic charge density.*

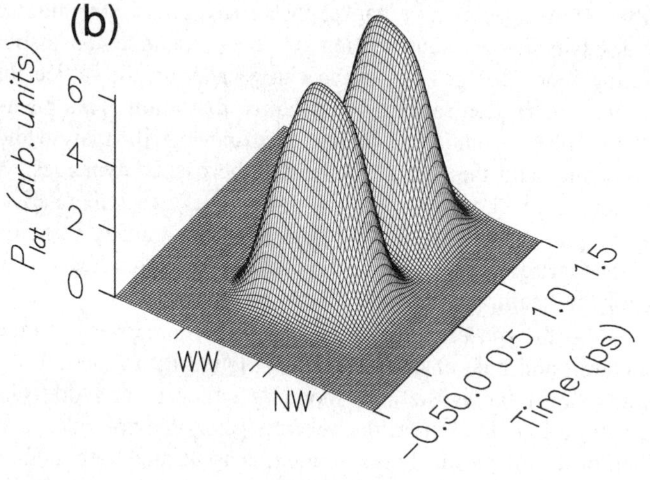

Figure 6.6 *(Continued.) (b) Lattice polarization as functions of time and the z-coordinate for excitation with a 150 fs pulse at an applied field of 7 $kV\ cm^{-1}$.*

the increased coupling: the electronic motion is increasingly slowed down by the back action of the ions which have to be accelerated.

In the case studied here the coherent phonons manifest themselves in a rather indirect way. Due to the strongly different frequencies of phonons and electrons, the phonon system exhibits a nearly adiabatic following behaviour. Oscillations of the phonons with their characteristic frequency have been observed after impulsive excitations in a variety of systems [66–69].

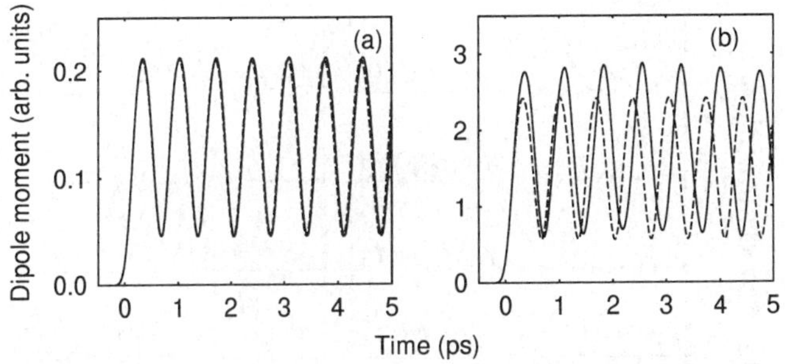

Figure 6.7 *Electronic dipole moments as functions of time for a density of (a)* 1.9×10^{11} *cm*$^{-2}$ *and (b)* 2.3×10^{12} *cm*$^{-2}$. *The solid curves are calculated including coherent phonons, while the dashed curves neglect carrier–phonon coupling.*

6.7.4 Electron–phonon quantum kinetics in quantum wires

In section 7.1 we saw how the generation of carriers is influenced by scattering processes. There, however, these processes have been treated on the semiclassical level, i.e. in the Markov limit as described by equation (6.3.18) for the carrier–phonon interaction. As for carrier generation, however, one might expect that on very short timescales the assumption of sufficiently slowly varying functions is no longer valid and that energy-time uncertainty plays an important role [21, 70–72, 24, 45]. Such quantum kinetic features are included if the phonon-assisted density matrices are treated as independent variables. In this section we will analyse the carrier–phonon quantum kinetics in a simple model of a cylindrical quantum wire with a cross section of 100 nm^2. The electronic states are given by Bessel functions and the Fröhlich matrix element is calculated according to equations (6.4.3). In order to concentrate on the quantum effects associated with the carrier–phonon interaction, we will restrict ourselves to a single-band single-subband model where we study the relaxation of an initial nonequilibrium distribution of electrons. Because there are no space charges, no coherent phonons are created and the dynamic variables are given by the electron and phonon distribution functions, f^e and n and the phonon-assisted density matrix s^e which satisfy the equations of motion (6.4.5), (6.4.15) without all quantities related to holes. Accordingly, the semiclassical limit which will serve us as a reference, is described by the Boltzmann equation for the electron distribution function (6.4.18a).

Figure 6.8 shows the electron energy distribution, i.e. the distribution function multiplied by the density of states, at different times for the case of an initial Gaussian distribution. The linear carrier density is 2.4×10^3 cm^{-1}. In the Boltzmann case (Fig. 6.8(a)) the subsequent emission of optical phonons leads to the appearance of replicas of the initial distribution. In the quantum kinetic case (Fig. 6.8(b)) we find pronounced differences, in particular at early times: initially, the relaxation leads to a very broad tail of the distribution on the low-

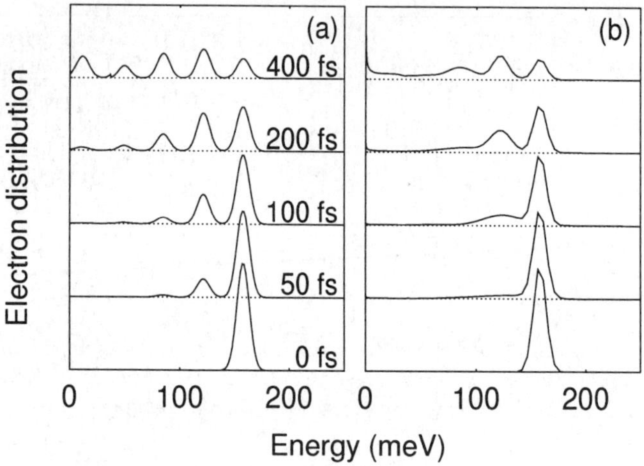

Figure 6.8 *Electron energy distribution at different times for (a) the Boltzmann and (b) the quantum kinetic model at a density of* 2.4×10^3 cm^{-1}.

energy side. With increasing time, the first replica one phonon energy below the initial distribution builds up and becomes narrower until it finally reaches the width of the initial distribution. At later times the same occurs for the second replica two phonon energies below the initial distribution.

The physical reason for this behaviour is the same as discussed above for the generation process: due to energy-time uncertainty, at early times the electron energy is not yet a well-defined quantity and transitions occur into a broad continuum of states, with increasing time the energy uncertainty is reduced and finally reaches a value below the energetic width of the initial distribution.

The qualitative behaviour of semiclassical and quantum kinetics shown here for a quantum wire is the same as for a bulk semiconductor. A detailed analysis for this latter case both for a one- and two-band model can be found in [24]. With increasing density, an additional qualitative difference between semiclassical and quantum kinetics occurs which is not present in the bulk. Figure 6.9 shows the electron energy distributions for the semiclassical and quantum kinetic case at the same times as in Fig. 6.8 but for a higher carrier density of 3.8×10^4 cm^{-1}. In the quantum kinetic case an additional peak above the initial distribution occurs which is not present in the Boltzmann case. Such a peak is well known in the bulk case even in the Boltzmann limit. It is a hot-phonon effect and is related to the absorption of phonons which have been previously emitted. In a one-dimensional system, however, energy and momentum conservation strongly restrict the phase space for such processes. A phonon which has been emitted by a carrier in the initial peak cannot be absorbed by a carrier in the same peak due to the curvature of the electronic band in k-space. The phonon distribution shown in Fig. 6.10 exhibits a series of peaks, each one associated

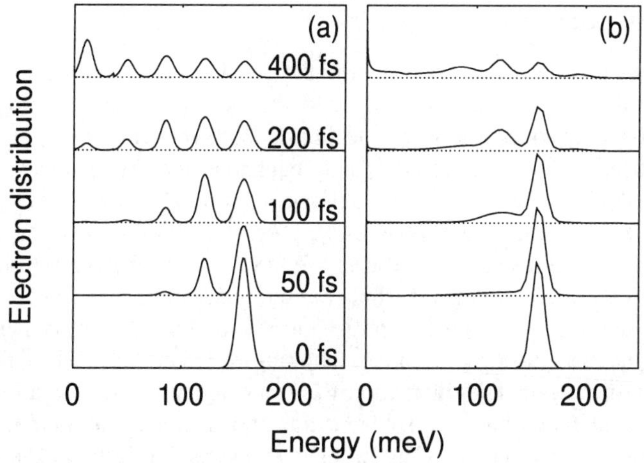

Figure 6.9 *Electron energy distribution at different times for (a) the Boltzmann and (b) the quantum kinetic model at a density of* 3.8×10^4 cm^{-1}.

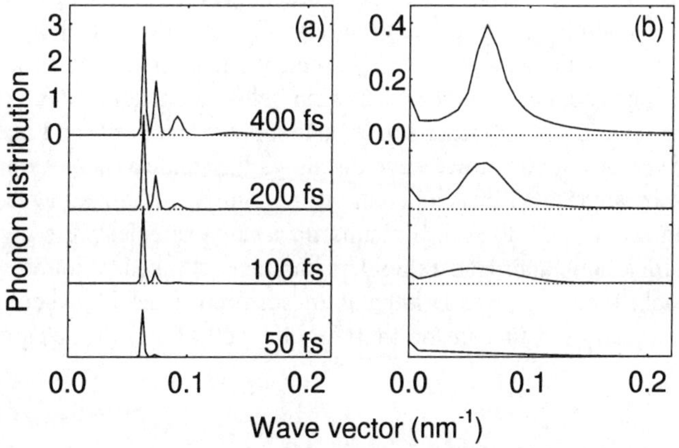

Figure 6.10 *Phonon distribution at different times for (a) the Boltzmann and (b) the quantum kinetic model at a density of* 3.8×10^4 cm^{-1}.

with one transition between two subsequent peaks of the electron distribution. An absorption process by a carrier in the initial peak would require a phonon at a lower wave vector than the first phonon peak. In the quantum kinetic case, on the other hand, energy-time uncertainty leads to a broadened phonon distribution and consequently to the possibility of absorption processes resulting in an electronic peak above the initial distribution.

6.8 CONCLUSIONS

The aim of this chapter was to review the density matrix approach to the carrier dynamics in semiconductors far from thermal equilibrium and to show that this approach provides a powerful technique both for a microscopic derivation and justification of semiclassical kinetic equations and for the analysis of phenomena which go beyond this semiclassical level. The basic idea to obtain a tractable formulation of the many-particle system was a correlation expansion of the many-particle density matrices. We have derived the contributions to the equations of motion for single-particle density matrices up to second order in this hierarchy and discussed some features entering in third-order terms. For various types of interactions the phenomena entering on a certain level of description have been discussed. The interaction with an external electric field leads to acceleration, interband and intersubband transitions, it also introduces interband and intersubband coherence. Carrier–phonon interaction leads to the excitation of coherent phonons in inhomogeneous systems (first order), scattering processes and polaron renormalizations (second order) and collisional broadening (third order). Carrier–carrier interaction leads to excitonic effects and plasma-induced band-gap renormalization (first order), scattering and screening (second order) and collisional broadening (third order). The simultaneous treatment of different interactions gives rise to a variety of cross effects such as the Franz–Keldysh effect (static and optical field), intracollisional field effect (external field and scattering processes), phonon scattering between excitonic states (first-order carrier–carrier and second-order carrier–phonon contributions) and many others. For each type of interaction we have discussed the semiclassical limit and, thus, the series of approximations inherent in the Boltzmann equation has become evident. In section 6.7 numerical results for a few examples were given, where the quantum kinetic approach exhibits phenomena not present in the semiclassical approach. Many other experimental and theoretical results where coherence plays a role can be found in the literature as well as in other chapters of this book.

ACKNOWLEDGEMENTS

The author thanks E. Binder, S. Haas, D. Preisser, F. Rossi and J. Schilp for stimulating collaboration. Financial support of the Deutsche Forschungsgemeinschaft in the framework of the Schwerpunktprogramm Quantenkohärenz in Halbleitern is gratefully acknowledged.

APPENDIX A. GENERATION RATE WITH SPECTRAL BROADENING

The approximate result for the semiclassical generation rate including the broadening due to the finite pulse duration can be obtained from the time-integrated generation

$$G_{ij,\boldsymbol{k}} = \int_{-\infty}^{\infty} \mathrm{d}t\, g_{ij,\boldsymbol{k}}(t). \tag{6.A.1}$$

Using definition (6.3.7) and inserting (6.3.9), after a transformation of variables the exact result is obtained

$$\begin{aligned} G_{ij,\boldsymbol{k}} =& \frac{1}{\hbar^2} \int_{-\infty}^{\infty} \mathrm{d}t \int_{-\infty}^{\infty} \mathrm{d}t' [1 - f^e_{ii,\boldsymbol{k}}(t - \tfrac{1}{2}t') - f^h_{jj,-\boldsymbol{k}}(t - \tfrac{1}{2}t')] \\ & \times \boldsymbol{M}^*_{ij} \cdot \boldsymbol{E}^*_0(t + \tfrac{1}{2}t')\boldsymbol{M}_{ij} \cdot \boldsymbol{E}_0(t - \tfrac{1}{2}t')\mathrm{e}^{-\mathrm{i}(\omega_{ji,\boldsymbol{k}} - \omega_L)t'}. \end{aligned} \tag{6.A.2}$$

If we identify the time of the generation by t and assume that the distribution functions are slowly varying functions of t', we can interpret the argument of the first integral as the semiclassical generation rate (6.3.13). It should be noted that this result cannot be obtained directly from equations (6.3.7) and (6.3.9) since the identification of the time t with the generation time is only an approximation. A consequence of this approximation is the fact that the generation rate (6.3.13) violates causality since it involves the electric field at later times or, in other words, the fact that the spectral shape of the pulse is known at any time.

APPENDIX B. SCATTERING TERMS FOR CARRIER–PHONON SCATTERING

The second-order carrier–phonon contributions to the equations of motion for intraband and interband density matrices as introduced in equation (6.4.18) are given by

$$\Gamma^{e,out(cp,2)}_{i_1 i_5,\boldsymbol{k}} = \frac{\pi}{\hbar} \sum_{\substack{i_4,i_6\\ \boldsymbol{q},\pm}} \delta(-\epsilon^e_{i_5,\boldsymbol{k}} + \epsilon^e_{i_6,\boldsymbol{k}-\boldsymbol{q}_\parallel} \mp \hbar\omega_{LO}) \times g^{*i_1 i_4}_{\boldsymbol{q}} g^{i_5 i_6}_{\boldsymbol{q}} (n_{\boldsymbol{q}} + \tfrac{1}{2} \mp \tfrac{1}{2})(\delta_{i_4,i_6} - f^e_{i_4 i_6,\boldsymbol{k}-\boldsymbol{q}_\parallel}) \tag{6.B.1a}$$

$$\Gamma^{e,in(cp,2)}_{i_1 i_5,\boldsymbol{k}} = \frac{\pi}{\hbar} \sum_{\substack{i_4,i_6\\ \boldsymbol{q},\pm}} \delta(-\epsilon^e_{i_5,\boldsymbol{k}} + \epsilon^e_{i_6,\boldsymbol{k}-\boldsymbol{q}_\parallel} \mp \hbar\omega_{LO}) \times g^{*i_1 i_4}_{\boldsymbol{q}} g^{i_5 i_6}_{\boldsymbol{q}} (n_{\boldsymbol{q}} + \tfrac{1}{2} \pm \tfrac{1}{2}) f^e_{i_4 i_6,\boldsymbol{k}-\boldsymbol{q}_\parallel} \tag{6.B.1b}$$

$$\Gamma^{h,in(cp,2)}_{j_1 j_5,-\boldsymbol{k}} = \frac{\pi}{\hbar} \sum_{\substack{j_4,j_6\\ \boldsymbol{q},\pm}} \delta(-\epsilon^h_{j_5,-\boldsymbol{k}} + \epsilon^h_{j_6,-\boldsymbol{k}+\boldsymbol{q}_\parallel} \mp \hbar\omega_{LO}) \times g^{j_1 j_4}_{\boldsymbol{q}} g^{*j_5 j_6}_{\boldsymbol{q}} (n_{\boldsymbol{q}} + \tfrac{1}{2} \pm \tfrac{1}{2}) f^h_{j_6 j_4,-\boldsymbol{k}+\boldsymbol{q}_\parallel} \tag{6.B.1c}$$

$$\Gamma^{h,out(cp,2)}_{j_1 j_5,-\boldsymbol{k}} = \frac{\pi}{\hbar} \sum_{\substack{j_4,j_6\\ \boldsymbol{q},\pm}} \delta(-\epsilon^h_{j_5,-\boldsymbol{k}} + \epsilon^h_{j_6,-\boldsymbol{k}+\boldsymbol{q}_\parallel} \mp \hbar\omega_{LO}) \times g^{j_1 j_4}_{\boldsymbol{q}} g^{*j_5 j_6}_{\boldsymbol{q}} (n_{\boldsymbol{q}} + \tfrac{1}{2} \mp \tfrac{1}{2})(\delta_{j_6,j_4} - f^h_{j_6 j_4,-\boldsymbol{k}+\boldsymbol{q}_\parallel}) \tag{6.B.1d}$$

$$\Gamma^{eh,out(cp,2)}_{i_1 j_5,\boldsymbol{k}} = -\frac{\pi}{\hbar} \sum_{\substack{i_4,j_6\\ \boldsymbol{q},\pm}} \delta(-\epsilon^h_{j_5,-\boldsymbol{k}} + \epsilon^h_{j_6,-\boldsymbol{k}+\boldsymbol{q}_\parallel} \mp \hbar\omega_{LO}) \times g^{i_1 i_4}_{\boldsymbol{q}} g^{j_5 j_6}_{\boldsymbol{q}} (n_{\boldsymbol{q}} + \tfrac{1}{2} \mp \tfrac{1}{2}) p_{j_6 i_4,\boldsymbol{k}-\boldsymbol{q}_\parallel} \tag{6.B.1e}$$

$$\Gamma^{eh,in(cp,2)}_{i_1 j_5,\boldsymbol{k}} = -\frac{\pi}{\hbar} \sum_{\substack{i_4,j_6\\ \boldsymbol{q},\pm}} \delta(-\epsilon^h_{j_5,-\boldsymbol{k}} + \epsilon^h_{j_6,-\boldsymbol{k}+\boldsymbol{q}_\parallel} \mp \hbar\omega_{LO}) \times g^{i_1 i_4}_{\boldsymbol{q}} g^{j_5 j_6}_{\boldsymbol{q}} (n_{\boldsymbol{q}} + \tfrac{1}{2} \pm \tfrac{1}{2}) p_{j_6 i_4,\boldsymbol{k}-\boldsymbol{q}_\parallel} \tag{6.B.1f}$$

$$\Gamma^{he,out(cp,2)}_{j_1 i_5,\boldsymbol{k}} = -\frac{\pi}{\hbar} \sum_{\substack{i_6,j_4\\ \boldsymbol{q},\pm}} \delta(-\epsilon^e_{i_5,\boldsymbol{k}} + \epsilon^e_{i_6,\boldsymbol{k}+\boldsymbol{q}_\parallel} \mp \hbar\omega_{LO}) \times g^{j_1 j_4}_{\boldsymbol{q}} g^{i_5 i_6}_{\boldsymbol{q}} (n_{\boldsymbol{q}} + \tfrac{1}{2} \mp \tfrac{1}{2}) p_{j_4 i_6,\boldsymbol{k}-\boldsymbol{q}_\parallel} \tag{6.B.1g}$$

$$\Gamma^{he,in(cp,2)}_{j_1 i_5,\boldsymbol{k}} = -\frac{\pi}{\hbar} \sum_{\substack{i_6,j_4\\ \boldsymbol{q},\pm}} \delta(-\epsilon^e_{i_5,\boldsymbol{k}} + \epsilon^e_{i_6,\boldsymbol{k}-\boldsymbol{q}_\parallel} \mp \hbar\omega_{LO}) \times g^{j_1 j_4}_{\boldsymbol{q}} g^{*i_6 i_5}_{\boldsymbol{q}} (n_{\boldsymbol{q}} + \tfrac{1}{2} \pm \tfrac{1}{2}) p_{j_4 i_6,\boldsymbol{k}-\boldsymbol{q}_\parallel} \tag{6.B.1h}$$

$$\mathcal{U}^{e(cp,2)}_{i_1 j_5,\boldsymbol{k}} = \mathrm{i}\pi \sum_{\substack{i_4,j_6\\ \boldsymbol{q},\pm}} \delta(-\epsilon^h_{j_5,-\boldsymbol{k}} + \epsilon^h_{j_6,-\boldsymbol{k}+\boldsymbol{q}_\parallel} \pm \hbar\omega_{LO}) \times g^{i_1 i_4}_{\boldsymbol{q}} g^{j_5 j_6}_{\boldsymbol{q}} p_{j_6 i_4,\boldsymbol{k}-\boldsymbol{q}_\parallel} \tag{6.B.1i}$$

$$\mathcal{U}^{h(cp,2)}_{j_1 i_5,\boldsymbol{k}} = \mathrm{i}\pi \sum_{\substack{j_3,i_6\\ \boldsymbol{q},\pm}} \delta(-\epsilon^e_{i_5,\boldsymbol{k}} + \epsilon^e_{i_6,\boldsymbol{k}-\boldsymbol{q}_\parallel} \pm \hbar\omega_{LO}) \times g^{j_1 j_3}_{\boldsymbol{q}} g^{i_5 i_6}_{\boldsymbol{q}} p_{j_3 i_6,\boldsymbol{k}-\boldsymbol{q}_\parallel}. \tag{6.B.1j}$$

APPENDIX C. SCATTERING TERMS FOR CARRIER–CARRIER SCATTERING

The second-order carrier–carrier contributions to the equations of motion for intraband and interband density matrices as introduced in equations (6.5.10) are given by

$$\begin{aligned}
\Gamma^{e,out(cc,2)}_{i_1 i_5,\boldsymbol{k}} =& \frac{\pi}{\hbar} \sum_{\substack{i_4,i_6,i_7,i_8\\ \boldsymbol{k}',\boldsymbol{q}}} \delta(-\epsilon^e_{i_5,\boldsymbol{k}} - \epsilon^e_{i_6,\boldsymbol{k}'} + \epsilon^e_{i_7,\boldsymbol{k}'+\boldsymbol{q}} + \epsilon^e_{i_8,\boldsymbol{k}-\boldsymbol{q}}) \\
&\times V^{i_5 i_6 i_7 i_8}_{\boldsymbol{q}} \left[\sum_{i_2,i_3} V^{i_1 i_2 i_3 i_4}_{\boldsymbol{q}} f^e_{i_2 i_6,\boldsymbol{k}'} (\delta_{i_7,i_3} - f^e_{i_7 i_3,\boldsymbol{k}'+\boldsymbol{q}}) \right. \\
&\left. + \sum_{j_2,j_3} V^{i_1 j_2 j_3 i_4}_{\boldsymbol{q}} p_{j_2 i_6,\boldsymbol{k}'} p^*_{j_3 i_7,\boldsymbol{k}'+\boldsymbol{q}} \right] (\delta_{i_8,i_4} - f^e_{i_8 i_4,\boldsymbol{k}-\boldsymbol{q}}) \\
&+ \frac{\pi}{\hbar} \sum_{\substack{i_4,j_6,j_7,i_8\\ \boldsymbol{k}',\boldsymbol{q}}} \delta(-\epsilon^e_{i_5,\boldsymbol{k}} - \epsilon^h_{j_7,-\boldsymbol{k}'-\boldsymbol{q}} + \epsilon^h_{j_6,-\boldsymbol{k}'} + \epsilon^e_{i_8,\boldsymbol{k}-\boldsymbol{q}}) \\
&\times V^{i_5 j_7 j_6 i_8}_{\boldsymbol{q}} \left[\sum_{j_2,j_3} V^{i_1 j_3 j_2 i_4}_{\boldsymbol{q}} f^h_{j_3 j_7,-\boldsymbol{k}'-\boldsymbol{q}} (\delta_{j_6,j_2} - f^h_{j_6 j_2,-\boldsymbol{k}'}) \right. \\
&\left. + \sum_{i_2,i_3} V^{i_1 i_3 i_2 i_4}_{\boldsymbol{q}} p_{j_7 i_3,-\boldsymbol{k}'+\boldsymbol{q}} p^*_{j_6 i_2,\boldsymbol{k}'} \right] (\delta_{i_8,i_4} - f^e_{i_8 i_4,\boldsymbol{k}-\boldsymbol{q}})
\end{aligned} \tag{6.C.1a}$$

$$\Gamma^{e,in(cc,2)}_{i_1 i_5,\boldsymbol{k}} = \frac{\pi}{\hbar} \sum_{\substack{i_4,i_6,i_7,i_8\\ \boldsymbol{k}',\boldsymbol{q}}} \delta(-\epsilon^e_{i_5,\boldsymbol{k}} - \epsilon^e_{i_6,\boldsymbol{k}'} + \epsilon^e_{i_7,\boldsymbol{k}'+\boldsymbol{q}} + \epsilon^e_{i_8,\boldsymbol{k}-\boldsymbol{q}})$$
$$\times V^{i_5 i_6 i_7 i_8}_{\boldsymbol{q}} \Bigg[\sum_{i_2,i_3} V^{i_1 i_2 i_3 i_4}_{\boldsymbol{q}} (\delta_{i_2,i_6} - f^e_{i_2 i_6,\boldsymbol{k}'}) f^e_{i_7 i_3,\boldsymbol{k}'+\boldsymbol{q}}$$
$$+ \sum_{j_2,j_3} V^{i_1 j_2 j_3 i_4}_{\boldsymbol{q}} p_{j_2 i_6,\boldsymbol{k}'} p^*_{j_3 i_7,\boldsymbol{k}'+\boldsymbol{q}} \Bigg] f^e_{i_8 i_4,\boldsymbol{k}-\boldsymbol{q}}$$
$$+ \frac{\pi}{\hbar} \sum_{\substack{i_4,j_6,j_7,i_8\\ \boldsymbol{k}',\boldsymbol{q}}} \delta(-\epsilon^e_{i_5,\boldsymbol{k}} - \epsilon^h_{j_7,-\boldsymbol{k}'-\boldsymbol{q}} + \epsilon^h_{j_6,-\boldsymbol{k}'} + \epsilon^e_{i_8,\boldsymbol{k}-\boldsymbol{q}})$$
$$\times V^{i_5 j_7 j_6 i_8}_{\boldsymbol{q}} \Bigg[\sum_{j_2,j_3} V^{i_1 j_3 j_2 i_4}_{\boldsymbol{q}} (\delta_{j_3,j_7} - f^h_{j_3 j_7,-\boldsymbol{k}'-\boldsymbol{q}}) f^h_{j_6 j_2,-\boldsymbol{k}'}$$
$$+ \sum_{i_2,i_3} V^{i_1 i_2 i_3 i_4}_{\boldsymbol{q}} p^*_{j_6 i_2,\boldsymbol{k}'} p_{j_7 i_3,\boldsymbol{k}'+\boldsymbol{q}} \Bigg] f^e_{i_8 i_4,\boldsymbol{k}-\boldsymbol{q}} \tag{6.C.1b}$$

$$\Gamma^{h,out(cc,2)}_{j_1 j_5,\boldsymbol{k}} = \frac{\pi}{\hbar} \sum_{\substack{j_4,j_6,j_7,j_8\\ \boldsymbol{k}',\boldsymbol{q}}} \delta(-\epsilon^h_{j_5,-\boldsymbol{k}} - \epsilon^h_{j_6,-\boldsymbol{k}'} + \epsilon^h_{j_7,-\boldsymbol{k}'-\boldsymbol{q}} + \epsilon^h_{j_8,-\boldsymbol{k}+\boldsymbol{q}})$$
$$\times V^{j_5 j_6 j_7 j_8}_{\boldsymbol{q}} \Bigg[\sum_{j_2,j_3} V^{j_1 j_2 j_3 j_4}_{\boldsymbol{q}} f^h_{j_2 j_6,-\boldsymbol{k}'} (\delta_{j_7,j_3} - f^h_{j_7 j_3,-\boldsymbol{k}'-\boldsymbol{q}})$$
$$+ \sum_{i_2,i_3} V^{j_1 i_2 i_3 j_4}_{\boldsymbol{q}} p_{j_6 i_2,\boldsymbol{k}'} p^*_{j_7 i_3,\boldsymbol{k}'+\boldsymbol{q}} \Bigg] (\delta_{j_8,j_4} - f^h_{j_8 j_4,-\boldsymbol{k}+\boldsymbol{q}})$$
$$+ \frac{\pi}{\hbar} \sum_{\substack{j_4,i_6,i_7,j_8\\ \boldsymbol{k}',\boldsymbol{q}}} \delta(-\epsilon^h_{j_5,-\boldsymbol{k}} - \epsilon^e_{i_7,\boldsymbol{k}'+\boldsymbol{q}} + \epsilon^e_{i_6,\boldsymbol{k}'} + \epsilon^h_{j_8,-\boldsymbol{k}+\boldsymbol{q}})$$
$$\times V^{j_5 i_7 i_6 j_8}_{\boldsymbol{q}} \Bigg[\sum_{i_2,i_3} V^{j_1 i_3 i_2 j_4}_{\boldsymbol{q}} f^e_{i_3 i_7,\boldsymbol{k}'+\boldsymbol{q}} (\delta_{i_6,i_2} - f^e_{i_6 i_2,\boldsymbol{k}'})$$
$$+ \sum_{j_2,j_3} V^{j_1 j_3 j_2 j_4}_{\boldsymbol{q}} p_{j_3 i_7,\boldsymbol{k}'+\boldsymbol{q}} p^*_{j_2 i_6,\boldsymbol{k}'} \Bigg] (\delta_{j_8,j_4} - f^h_{j_8 j_4,-\boldsymbol{k}+\boldsymbol{q}}) \tag{6.C.1c}$$

$$
\begin{aligned}
\Gamma^{h,in(cc,2)}_{j_1 j_5,\boldsymbol{k}} =& \frac{\pi}{\hbar} \sum_{\substack{j_4,j_6,j_7,j_8 \\ \boldsymbol{k}',\boldsymbol{q}}} \delta(-\epsilon^h_{j_5,-\boldsymbol{k}} - \epsilon^h_{j_6,-\boldsymbol{k}'} + \epsilon^h_{j_7,-\boldsymbol{k}'-\boldsymbol{q}} + \epsilon^h_{j_8,-\boldsymbol{k}+\boldsymbol{q}}) \\
& \times V_{\boldsymbol{q}}^{j_5 j_6 j_7 j_8} \Bigg[\sum_{j_2,j_3} V_{\boldsymbol{q}}^{j_1 j_2 j_3 j_4} (\delta_{j_2,j_6} - f^h_{j_2 j_6,-\boldsymbol{k}'}) f^h_{j_7 j_3,-\boldsymbol{k}'-\boldsymbol{q}} \\
& + \sum_{i_2,i_3} V_{\boldsymbol{q}}^{j_1 i_2 i_3 j_4} p_{j_6 i_2,\boldsymbol{k}'} p^*_{j_7 i_3,\boldsymbol{k}'+\boldsymbol{q}} \Bigg] f^h_{j_8 j_4,-\boldsymbol{k}+\boldsymbol{q}} \\
& + \frac{\pi}{\hbar} \sum_{\substack{j_4,i_6,i_7,j_8 \\ \boldsymbol{k}',\boldsymbol{q}}} \delta(-\epsilon^h_{j_5,-\boldsymbol{k}} - \epsilon^e_{i_7,\boldsymbol{k}'+\boldsymbol{q}} + \epsilon^e_{i_6,\boldsymbol{k}'} + \epsilon^h_{j_8,-\boldsymbol{k}+\boldsymbol{q}}) \\
& \times V_{\boldsymbol{q}}^{j_5 i_7 i_6 j_8} \Bigg[\sum_{i_2,i_3} V_{\boldsymbol{q}}^{j_1 i_3 i_2 j_4} (\delta_{i_3,i_7} - f^e_{i_3 i_7,\boldsymbol{k}'+\boldsymbol{q}}) f^e_{i_6 i_2,\boldsymbol{k}'} \\
& + \sum_{j_2,j_3} V_{\boldsymbol{q}}^{j_1 j_3 j_2 j_4} p_{j_3 i_7,\boldsymbol{k}'+\boldsymbol{q}} p^*_{j_2 i_6,\boldsymbol{k}'} \Bigg] f^h_{j_8 j_4,-\boldsymbol{k}+\boldsymbol{q}} \qquad (6.C.1d)
\end{aligned}
$$

$$
\begin{aligned}
\Gamma^{eh,out(cc,2)}_{i_1 j_5,\boldsymbol{k}} =& \frac{\pi}{\hbar} \sum_{\substack{i_4,j_6,j_7,j_8 \\ \boldsymbol{k}',\boldsymbol{q}}} \delta(-\epsilon^h_{j_5,-\boldsymbol{k}} - \epsilon^h_{j_6,-\boldsymbol{k}'} + \epsilon^h_{j_7,-\boldsymbol{k}'-\boldsymbol{q}} + \epsilon^h_{j_8,-\boldsymbol{k}+\boldsymbol{q}}) \\
& \times V_{\boldsymbol{q}}^{j_5 j_6 j_7 j_8} \Bigg[\sum_{j_2,j_3} V_{\boldsymbol{q}}^{i_1 j_2 j_3 i_4} f^h_{j_2 j_6,-\boldsymbol{k}'} (\delta_{j_7,j_3} - f^h_{j_7 j_3,-\boldsymbol{k}'-\boldsymbol{q}}) \\
& + \sum_{i_2,i_3} V_{\boldsymbol{q}}^{i_1 i_2 i_3 i_4} p_{j_6 i_2,\boldsymbol{k}'} p^*_{j_7 i_3,\boldsymbol{k}'+\boldsymbol{q}} \Bigg] p_{j_8 i_4,\boldsymbol{k}-\boldsymbol{q}} \\
& + \frac{\pi}{\hbar} \sum_{\substack{i_4,i_6,i_7,j_8 \\ \boldsymbol{k}',\boldsymbol{q}}} \delta(-\epsilon^h_{j_5,-\boldsymbol{k}} - \epsilon^e_{i_7,\boldsymbol{k}'+\boldsymbol{q}} + \epsilon^e_{i_6,\boldsymbol{k}'} + \epsilon^h_{j_8,-\boldsymbol{k}+\boldsymbol{q}}) \\
& \times V_{\boldsymbol{q}}^{j_5 i_7 i_6 j_8} \Bigg[\sum_{i_2,i_3} V_{\boldsymbol{q}}^{i_1 i_3 i_2 i_4} f^e_{i_3 i_7,\boldsymbol{k}'+\boldsymbol{q}} (\delta_{i_6,i_2} - f^e_{i_6 i_2,\boldsymbol{k}'}) \\
& + \sum_{j_2,j_3} V_{\boldsymbol{q}}^{i_1 j_3 j_2 i_4} p_{j_2 i_6,\boldsymbol{k}'} p^*_{j_3 i_7,\boldsymbol{k}'+\boldsymbol{q}} \Bigg] p_{j_8 i_4,\boldsymbol{k}-\boldsymbol{q}} \qquad (6.C.1e)
\end{aligned}
$$

$$
\begin{aligned}
\Gamma^{eh,in(cc,2)}_{i_1 j_5,\boldsymbol{k}} =& \frac{\pi}{\hbar} \sum_{\substack{i_4,j_6,j_7,j_8\\ \boldsymbol{k}',\boldsymbol{q}}} \delta(-\epsilon^h_{j_5,-\boldsymbol{k}} - \epsilon^h_{j_6,-\boldsymbol{k}'} + \epsilon^h_{j_7,-\boldsymbol{k}'-\boldsymbol{q}} + \epsilon^h_{j_8,-\boldsymbol{k}+\boldsymbol{q}}) \\
& \times V^{j_5 j_6 j_7 j_8}_{\boldsymbol{q}} \Bigg[\sum_{j_2,j_3} V^{i_1 j_2 j_3 i_4}_{\boldsymbol{q}} (\delta_{j_6,i_2} - f^h_{j_6 j_2,-\boldsymbol{k}'}) f^h_{j_3 j_7,-\boldsymbol{k}'-\boldsymbol{q}} \\
& + \sum_{i_2,i_3} V^{i_1 i_2 i_3 i_4}_{\boldsymbol{q}} p^*_{j_6 i_2,\boldsymbol{k}'} p_{j_7 i_3,\boldsymbol{k}'+\boldsymbol{q}} \Bigg] p_{j_8 i_4,\boldsymbol{k}-\boldsymbol{q}} \\
& + \frac{\pi}{\hbar} \sum_{\substack{i_4,i_6,i_7,j_8\\ \boldsymbol{k}',\boldsymbol{q}}} \delta(-\epsilon^h_{j_5,-\boldsymbol{k}} - \epsilon^e_{i_7,\boldsymbol{k}'+\boldsymbol{q}} + \epsilon^e_{i_6,\boldsymbol{k}'} + \epsilon^h_{j_8,-\boldsymbol{k}+\boldsymbol{q}}) \\
& \times V^{j_5 i_6 i_7 j_8}_{\boldsymbol{q}} \Bigg[\sum_{i_2,i_3} V^{i_1 i_3 i_2 i_4}_{\boldsymbol{q}} (\delta_{i_3,i_7} - f^e_{i_3 i_7,\boldsymbol{k}'+\boldsymbol{q}}) f^e_{i_6 i_2,\boldsymbol{k}'} \\
& + \sum_{j_2,j_3} V^{i_1 j_2 j_3 i_4}_{\boldsymbol{q}} p_{j_2 i_6,\boldsymbol{k}'} p^*_{j_3 i_7,\boldsymbol{k}'+\boldsymbol{q}} \Bigg] p_{j_8 i_4,\boldsymbol{k}-\boldsymbol{q}}
\end{aligned}
\tag{6.C.1f}
$$

$$
\begin{aligned}
\Gamma^{he,out(cc,2)}_{j_1 i_5,\boldsymbol{k}} =& \frac{\pi}{\hbar} \sum_{\substack{j_4,i_6,i_7,i_8\\ \boldsymbol{k}',\boldsymbol{q}}} \delta(-\epsilon^e_{i_5,\boldsymbol{k}} - \epsilon^e_{i_6,\boldsymbol{k}'} + \epsilon^e_{i_7,\boldsymbol{k}'+\boldsymbol{q}} + \epsilon^e_{i_8,\boldsymbol{k}-\boldsymbol{q}}) \\
& \times V^{i_5 i_6 i_7 i_8}_{\boldsymbol{q}} \Bigg[\sum_{i_2,i_3} V^{j_1 i_2 i_3 j_4}_{\boldsymbol{q}} f^e_{i_6 i_2,\boldsymbol{k}'} (\delta_{i_7,i_3} - f^e_{i_7 i_3,\boldsymbol{k}'+\boldsymbol{q}}) \\
& + \sum_{j_2,j_3} V^{j_1 j_2 j_3 j_4}_{\boldsymbol{q}} p^*_{j_2 i_6,\boldsymbol{k}'} p_{j_3 i_7,\boldsymbol{k}'+\boldsymbol{q}} \Bigg] p_{j_4 i_8,\boldsymbol{k}-\boldsymbol{q}} \\
& + \frac{\pi}{\hbar} \sum_{\substack{j_4,j_6,j_7,i_8\\ \boldsymbol{k}',\boldsymbol{q}}} (-\epsilon^e_{i_5,\boldsymbol{k}} - \epsilon^h_{j_7,-\boldsymbol{k}'-\boldsymbol{q}} + \epsilon^h_{j_6,-\boldsymbol{k}'} + \epsilon^e_{i_8,\boldsymbol{k}-\boldsymbol{q}}) \\
& \times V^{i_5 j_7 j_6 i_8}_{\boldsymbol{q}} \Bigg[\sum_{j_2,j_3} V^{j_1 j_3 j_2 j_4}_{\boldsymbol{q}} (\delta_{j_7,j_3} - f^h_{j_7 j_3,-\boldsymbol{k}'-\boldsymbol{q}}) f^h_{j_2 j_6,-\boldsymbol{k}'} \\
& + \sum_{i_2,i_3} V^{j_1 i_3 i_2 j_4}_{\boldsymbol{q}} p_{j_3 i_7,\boldsymbol{k}'+\boldsymbol{q}} p^*_{j_6 i_2,\boldsymbol{k}'} \Bigg] p_{j_4 i_8,\boldsymbol{k}-\boldsymbol{q}}
\end{aligned}
\tag{6.C.1g}
$$

$$\Gamma^{eh,in(cc,2)}_{i_1 j_5,\boldsymbol{k}} = \frac{\pi}{\hbar} \sum_{\substack{i_4,j_6,j_7,j_8\\ \boldsymbol{k}',\boldsymbol{q}}} \delta(-\epsilon^h_{j_5,x-\boldsymbol{k}} - \epsilon^h_{j_6,-\boldsymbol{k}'} + \epsilon^h_{j_7,-\boldsymbol{k}'-\boldsymbol{q}} + \epsilon^h_{j_8,-\boldsymbol{k}+\boldsymbol{q}})$$

$$\times V^{j_5 j_6 j_7 j_8}_{\boldsymbol{q}} \Bigg[\sum_{j_2,j_3} V^{i_1 j_2 j_3 i_4}_{\boldsymbol{q}} (\delta_{j_6,j_2} - f^h_{j_6 j_2,-\boldsymbol{k}'}) f^h_{j_3 j_7,-\boldsymbol{k}'-\boldsymbol{q}}$$

$$+ \sum_{i_2,i_3} V^{i_1 i_2 i_3 i_4}_{\boldsymbol{q}} p^*_{j_6 i_2,\boldsymbol{k}'} p_{j_7 i_3,\boldsymbol{k}'+\boldsymbol{q}} \Bigg] p_{j_8 i_4,\boldsymbol{k}-\boldsymbol{q}}$$

$$+ \frac{\pi}{\hbar} \sum_{\substack{i_4,i_6,i_7,j_8\\ \boldsymbol{k}',\boldsymbol{q}}} \delta(-\epsilon^h_{j_5,-\boldsymbol{k}} - \epsilon^e_{i_7,\boldsymbol{k}'+\boldsymbol{q}} + \epsilon^e_{i_6,\boldsymbol{k}'} + \epsilon^h_{j_8,-\boldsymbol{k}-\boldsymbol{q}})$$

$$\times V^{j_5 i_7 i_6 j_8}_{\boldsymbol{q}} \Bigg[\sum_{i_2,i_3} V^{i_1 i_3 i_2 i_4}_{\boldsymbol{q}} (\delta_{i_6,i_2} - f^e_{i_6 i_2,\boldsymbol{k}'}) f^e_{i_3 i_7,\boldsymbol{k}'+\boldsymbol{q}}$$

$$+ \sum_{j_2,j_3} V^{j_1 j_3 j_2 i_4}_{\boldsymbol{q}} p^*_{j_2 i_6,\boldsymbol{k}'} p_{j_3 i_7,\boldsymbol{k}'+\boldsymbol{q}} \Bigg] p_{j_8 i_4,\boldsymbol{k}-\boldsymbol{q}} \qquad (6.C.1h)$$

$$\mathcal{U}^{e(cc,2)}_{i_1 j_5,\boldsymbol{k}} = \mathrm{i}\pi \sum_{\substack{i_4,j_6,j_7,j_8\\ \boldsymbol{k}',\boldsymbol{q}}} \delta(-\epsilon^h_{j_5,-\boldsymbol{k}} = \epsilon^e_{i_7,\boldsymbol{k}'+\boldsymbol{q}} + \epsilon^e_{i_6,\boldsymbol{k}'} + \epsilon^h_{j_8,-\boldsymbol{k}+\boldsymbol{q}})$$

$$\times V^{j_5 i_7 i_6 j_8}_{\boldsymbol{q}} \sum_{i_2,i_3} V^{i_1 i_3 i_2 i_4}_{\boldsymbol{q}} \Bigg[f^e_{i_7 i_3,\boldsymbol{k}'+\boldsymbol{q}} \delta_{i_2,i_6}$$

$$- f^e_{i_2 i_6,\boldsymbol{k}'} \delta_{i_3,i_7} \Bigg] p_{j_8 i_4,\boldsymbol{k}-\boldsymbol{q}}$$

$$+ \mathrm{i}\pi \sum_{\substack{i_4,j_6,j_7,j_8\\ \boldsymbol{k}',\boldsymbol{q}}} \delta(-\epsilon^h_{j_5,-\boldsymbol{k}} - \epsilon^h_{j_6,-\boldsymbol{k}'} + \epsilon^h_{j_7,-\boldsymbol{k}'-\boldsymbol{q}} + \epsilon^h_{j_8,-\boldsymbol{k}+\boldsymbol{q}})$$

$$\times V^{j_5 j_6 j_7 j_8}_{\boldsymbol{q}} \sum_{j_2,j_3} V^{i_1 j_2 j_3 i_4}_{\boldsymbol{q}} \Bigg[f^h_{j_3 j_7,-\boldsymbol{k}'-\boldsymbol{q}} \delta_{j_2,j_6}$$

$$- f^h_{j_6 j_2,-\boldsymbol{k}'} \delta_{j_3,j_7} \Bigg] p_{j_8 i_4,\boldsymbol{k}-\boldsymbol{q}} \qquad (6.C.1i)$$

$$
\begin{aligned}
\mathcal{U}^{h(cc,2)}_{j_1 i_5,\boldsymbol{k}} =& \mathrm{i}\pi \sum_{\substack{j_4,j_6,j_7,i_8\\ \boldsymbol{k}',\boldsymbol{q}}} \delta(-\epsilon^e_{i_5,\boldsymbol{k}} - \epsilon^h_{j_7,-\boldsymbol{k}'-\boldsymbol{q}} + \epsilon^h_{j_6,-\boldsymbol{k}'} + \epsilon^e_{i_8,\boldsymbol{k}-\boldsymbol{q}}) \\
& \times V^{i_5 j_7 j_6 i_8}_{\boldsymbol{q}} \sum_{j_2,j_3} V^{j_1 j_3 j_2 j_4}_{\boldsymbol{q}} [f^h_{j_3 j_7,-\boldsymbol{k}'-\boldsymbol{q}} \delta_{j_2,j_6} \\
& - f^h_{j_6 j_2,-\boldsymbol{k}'} \delta_{j_3,j_7}] p_{j_4 i_8,\boldsymbol{k}-\boldsymbol{q}} \\
& + \mathrm{i}\pi \sum_{\substack{i_4,i_6,i_7,i_8\\ \boldsymbol{k}',\boldsymbol{q}}} \delta(-\epsilon^e_{i_5,\boldsymbol{k}} - \epsilon^e_{i_6,\boldsymbol{k}'} + \epsilon^e_{i_7,\boldsymbol{k}'+\boldsymbol{q}} + \epsilon^e_{i_8,\boldsymbol{k}-\boldsymbol{q}}) \\
& \times V^{i_5 i_6 i_7 i_8}_{\boldsymbol{q}} \sum_{i_2,i_3} V^{j_1 i_2 i_3 j_4}_{\boldsymbol{q}} [f^e_{i_7 i_3,\boldsymbol{k}'+\boldsymbol{q}} \delta_{i_2,i_6} \\
& - f^e_{i_2 i_6,\boldsymbol{k}'} \delta_{i_3,i_7}] p_{j_4 i_8,\boldsymbol{k}-\boldsymbol{q}}. \qquad (6.C.1j)
\end{aligned}
$$

REFERENCES

[1] Kadanoff, L. P. and Baym, G. (1962) *Quantum Statistical Mechanics*, Benjamin, New York.

[2] Keldysh, L. V. (1965) *Sov. Phys. JETP* **20**, 1018.

[3] Haug, H. (ed.) (1988) *Optical Nonlinearities and Instabilities in Semiconductors*, Academic, San Diego, p. 53.

[4] Schäfer, W. (1988) *Optical Nonlinearities and Instabilities in Semiconductors* (ed. H. Haug), Academic, San Diego, p. 133.

[5] Schmitt-Rink, S. and Chemla, D. S. (1986) *Phys. Rev. Lett.*, **57**, 2752.

[6] Stahl, A. (1988) *Z. Phys.* B **72**, 371.

[7] Lindberg, M. and Koch, S. W. (1988) *Phys. Rev.* B **38**, 3342.

[8] Kuhn, T. and Rossi, F. (1992) *Phys. Rev.* B, **46**, 7496.

[9] Kubo, R., Toda, M. and Hashitsume, N. (1991) *Statistical Physics II*, Springer, Berlin.

[10] Schultheis, L., Kuhl, J., Honold, A. and Tu, C. W. (1986) *Phys. Rev. Lett.* **57**, 1797.

[11] Honold, A., Schultheis, L., Kuhl, J. and Tu, C. W. (1988) *Appl. Phys. Lett.* **52**, 2105.

[12] Kuhl, J., Honold, A., Schultheis, L. and Tu, C. W. (1989) *Adv. in Solid State Phys.*, **29**, 157.

[13] Lohner, A., Rick, K. and Leisching, P. (1993) *Phys. Rev. Lett.* **71**, 77.

[14] Göbel, E. O., Leo, K., Damen., T. C., Shah, J., Schmitt-Rink, S., Schäffer, W., Müller, J. F. and Köhler, K. (1990) *Phys. Rev. Lett.* **64**, 1801.

[15] Fröhlich, D., Kulik, A., Uebbing, B., Mysyrowicz, A., Langer, V., Stolz, H. and von der Oster, W. (1991) *Phys. Rev. Lett.* **67**, 2343.

[16] Leo, K., Damen., T. C., Shah, J., Göbel, O. and Köhler, K. (1990) *Appl. Phys. Lett.* **57**, 19.

[17] Feldmann, J., Meier, T. and von Plessen, G. (1993) *Phys. Rev. Lett.* **70**, 3027.

[18] Roskos, H. G., Nuss, M. C., Shah, J., Leo, K., Miller, D. A. B., Fox, A. M., Schmitt-Rink, S. and Köhler, K. (1992) *Phys. Rev. Lett.* **68**, 2216.

[19] Waschke, C., Roskos, H. G., Schwedler, R., Leo, K., Kurz, H. and Köhler, K. (1993) *Phys. Rev. Lett.* **70**, 3319.

[20] Planken, P. C. M., Nuss, M. C., Brener, I., Goossen, K. W., Luo, M. S. C. Chuang, S. L. and Pfeiffer, L. (1992) *Phys. Rev. Lett.* **69**, 3800.
[21] Zimmermann, R. (1990) *Phys. Stat. Sol.* (b) **159**, 317.
[22] Tran Thoai, D. B. and Haug, H. (1993) *Phys. Rev.* B **47**, 3574.
[23] Zimmermann, R. and Wauer, J. (1994) *J. Lumin.* **58**, 271.
[24] Schilp, J., Kuhn, T. and Mahler, G. (1994) *Phys. Rev.* B **50**, 5435.
[25] Meden, V., Wöhler, C., Fricke, J. and Schönhammer, K. (1995) *Phys. Rev.* B **52**, 5624.
[26] Wigner, E. (1932) *Phys. Rev.* **40**, 749.
[27] Frensley, W. R. (1987) *Phys. Rev.* B **36**, 1570.
[28] Kluksdahl, N. C., Kriman, A. M. and Ferry, D. K. (1989) *Phys. Rev.* B **39**, 7720.
[29] Leitenstorfer, A., Lohner, A., Elsaesser, T., Haas, S., Rossi, F., Kuhn, T., Klein, W., Boehm, G., Traenkle, G. and Weimann, G. (1994) *Phys. Rev. Lett.* **73**, 1687.
[30] Leitenstorfer, A., Elsaesser, T., Rossi, F., Kuhn, T., Klein, W., Boehm, G., Traenkle, G. and Weimann, G. (1996) *Phys. Rev.* B **53**, 9876.
[31] Lee, I., Goodnick, S. M., Gulia, M., Molinari, E. and Lugli, P. (1995) *Phys. Rev.* B **51**, 7046.
[32] Molinari, E., Baroni, S., Gianozzi, P. and de Gironcoli, S. (1990) *Proc. 20th ICPS, Thessaloniki, Greece* (eds E. M. Anastassakis and J. D. Joannopoulos), World Scientific, Singapore, p. 1429.
[33] Kuhn, T. (1994) *Ladungsträgerdynamik in Halbleitersystemen fern vom Gleichgewicht: Elektronisches Rauschen und Kohärente Prozesse*, Shaker, Aachen.
[34] Hess, O. and Kuhn, T. (1996) *Progress in Quantum Electronics* **20**, 85.
[35] Hess, O. and Kuhn, T. (1996) *Phys. Rev.* A **54**, 3347.
[36] Haug, H. and Koch, S. W. (1993) *Quantum Theory of the Optical and Electronic Properties of Semiconductors*, World Scientific, Singapore.
[37] Haken, H. (1976) *Quantum Field Theory of Solids, An Introduction*, North Holland, New York.
[38] Axt, V. M. and Stahl, A. (1994) *Z. Phys.* B **93**, 195.
[39] Victor, K., Axt, V. M. and Stahl, A. (1995) *Phys. Rev.* B **51**, 14 164.
[40] Axt, V. M., Victor, K. and Stahl, A. (1996) *Phys. Rev.* B **53**, 7244.
[41] Bogoliubov, N. N. (1967) *Lectures on Quantum Statistics* Vol. 1, Gordon and Breach, New York.
[42] McQuarrie, D. A. (1976) *Statistical Mechanics*, Harper and Row, New York.
[43] Scholz, R., Pfeifer, T. and Kurz, H. (1993) *Phys. Rev.* B **47**, 16 229.
[44] Kuznetsov, A. V. and Stanton, C. J. (1994) *Phys. Rev. Lett.* **73**, 3243.
[45] Schilp, J., Kuhn, T. and Mahler, G. (1995) *Phys. Stat. Sol.* (b) **188**, 417.
[46] Quade, W., Schöll, E., Rossi, F. and Jacoboni, C. (1994), *Phys. Rev.* B, **50**, 7398.
[47] Quade, W. (1993) *Konzepte der Stoßionisation in der Halbleiter-Transporttheorie*, Wissenschaft und Technik, Berlin.
[48] Carruthers, P. and Zachariasen, F. (1983) *Rev. Mod. Phys.* **55** 245.
[49] Schröder, H., Buss, H., Kuhn, T. and Schöll, E. (1996) *Hot Carriers in Semiconductors* (eds K. Hess, J.-P. Leburton and U. Ravaioli), Plenum, New York.
[50] Schröder, H., Schöll, E. and Kuhn, T. (1996) *Proc. 23nd ICPS, Berlin, Germany* (eds M. Scheffler and R. Zimmermann), World Scientific, Singapore, p. 1157.
[51] Wyld, H. W. and Fried, B. D. (1963) *Ann. Phys (NY)* **23**, 374.
[52] El Sayed, K., Bànyai, L. and Haug, H. (1994) *Phys. Rev.* **50**, 1541.

[53] Rossi, F., Haas, S. and Kuhn, T. (1994) *Phys. Rev. Lett.* **72**, 152.
[54] Haas, S., Rossi, F. and Kuhn, T. (1996) *Phys. Rev.* B **53**, 12 855.
[55] Kuznetsov, A. V. and Stanton, C. J. (1993) *Phys. Rev.* B **48**, 10 828.
[56] Jauho, A. P. and Johnsen, K. (1996) *Phys. Rev. Lett.* **76**, 4576.
[57] Bastard, G., Mendez, E. E., Chang, L. L. and Esaki, L. (1983) *Phys. Rev.* B **28**, 3241.
[58] Brum, J. A. and Bastard, G. (1985) *Phys. Rev.* B **31**, 3893.
[59] Comte, C. and Mahler, G. (1986) *Phys. Rev.* B **34**, 7164.
[60] Bertoncini, R., Kriman, A. M. and Ferry, D. K. (1990) *Phys. Rev.* B **41**, 1390.
[61] Rossi, F. and Jacoboni, C. (1992) *Semicond. Sci. Technol.* **7**, B383.
[62] Young, J. F. and Kelly, P. J. (1993) *Phys. Rev.* B **47**, 6316.
[63] Kuhn, T. and Rossi, F. (1992) *Phys. Rev. Lett.* **69**, 977.
[64] Binder, E., Kuhn, T. and Mahler, G. (1994) *Phys. Rev.* B **50**, 18 319.
[65] Leo, K., Shah, J., Göbel, E. O., Damen, T. C., Köhler, K. and Ganser, P. (1990) *Appl. Phys. Lett.* **56**, 2031.
[66] Cho, G. C., Kütt, W. and Kurz, H. (1990) *Phys. Rev. Lett.* **65**, 764.
[67] Pfeifer, T., Kütt, W., Kurz, H. and Scholz, R. (1992) *Phys. Rev. Lett.* **69**, 3248.
[68] Dekorsy, T., Kim, A. M. T., Cho, G. C., Kurz, H., Kuznetsov, A. V. and Förster, A. (1996) *Phys. Rev.* B **53** 1531.
[69] Dekorsy, T., Auer, H., Bakker, H. J., Roskos, H. G. and Kurz, H. (1996) *Phys. Rev.* B **53**, 4005.
[70] Zimmermann, R. (1992) *J. Lumin.* **53**, 187.
[71] Haug, H. (1992) *Phys. Stat. Sol.* (b) **173**, 139.
[72] Bányai, L., Tran Thoai, D. B., Reitsamer, E., Haug, H., Steinbach, D., Wehner, M. U., Wegener, M., Marschner, T. and Stolz, W. (1995) *Phys. Rev. Lett.* **75**, 2188.

CHAPTER 7

Dynamic and nonlinear transport in mesoscopic structures

M. Büttiker and T. Christen*

University of Geneva, Department of theoretical physics, 24, Quai E. Ansermet, CH-1211 Geneva, Switzerland

7.1 INTRODUCTION

In this chapter we discuss the capacitance, the linear low-frequency admittance and the weakly nonlinear DC transport of low-dimensional and phase-coherent conductors. Of special interest are quantized Hall samples, the quantum point contact, and the resonant tunnelling barrier. Dynamic and nonlinear transport is a challenging subject since a consistent treatment requires knowledge of the nonequilibrium state. In contrast, the linear static response (DC conductance) of a mesoscopic sample is determined by the equilibrium state only. Once the equilibrium potential is known the transmission and reflection probabilities of electrons in this potential can be calculated. The transmission behaviour of a sample completely specifies its DC conductance. Naturally from such a discussion we learn nothing about the charge density and current distribution inside the mesoscopic conductor. However, the dynamic response and the nonlinear response of a conductor depend on the charge and current distributions established away from equilibrium. Dynamic and nonlinear transport can thus be viewed as a probe of the charge response of the conductor. The dynamic response and the nonlinear response reveal information which is not accessible by linear DC measurements.

An introduction to the theory of mesoscopic conductors and, in particular, to the scattering approach to conduction, is provided by the monograph of Imry (1986), by the reviews of Beenakker and van Houten (1991) and Buot (1993) and by the recent book of Datta (1995). The advantage of the scattering approach lies in the conceptually simple prescription of how to model an open system, i.e. how to couple the sample to external contacts which act as reservoirs of charge carriers and provide the source of irreversibility (Landauer, 1970). Our discus-

*Permanent address: ABB Corporate Research Ltd, CH-5405 Baden-Dättmil, Switzerland

sion expands the scattering approach to electrical conduction in order to permit an investigation of AC transport and weakly nonlinear transport (Büttiker, 1993; Büttiker, Thomas and Prêtre, 1993a; Büttiker, Prêtre and Thomas 1993b, 1994). The success of the transmission approach in explaining many DC transport phenomena is well known. However, the extension of the scattering approach to systems which are not close to equilibrium turns out to be a formidable problem. Whatever approach is eventually used to solve the problem, there are always certain general requirements based on deeper symmetry properties of the underlying system. Below, we emphasize charge and current conservation, and gauge invariance (Büttiker, 1993; Büttiker, Thomas and Prêtre, 1994; Christen and Büttiker, 1996c). Furthermore, microreversibility combined with the thermodynamic properties of the reservoirs leads to the Onsager–Casimir relations of the linear frequency-dependent response coefficients. The solution of a problem in which interactions play an important role necessarily depends on the approximative method used to treat the many-particle problem. A detailed description of the potential landscape can be given in terms of a Hartree approach (Levinson, 1989; Büttiker, 1993). Since density functional theory has the form of an effective Hartree theory, such a discussion can be generalized to include a self-consistent effective potential (Büttiker, 1994). The importance of the self-consistent potential in the context of transport has been pointed out by Landauer (1957, 1987a). However, the consideration of the detailed potential landscape makes analytical progress difficult if not impossible. Both analytical and computational work have to rely on discretization procedures of the potential landscape. A simple discrete potential model which attributes only one self-consistent potential to the entire conductor was investigated by Büttiker *et al.* (1993b) and Prêtre, Thomas and Büttiker (1996). For many cases of interest, in quantum point contacts and quantum Hall samples, the charge distribution is essentially dipolar or of a more complex multipolar form. To treat such conductors we emphasize here an extension of the discrete potential model which includes a finite set of potentials in each conductor (Christen and Büttiker, 1996a, 1996b). This generalized discrete potential model aims at capturing the important physical aspects of the charge distribution and leads to analytical results which can hopefully be used in the laboratory.

The reason why we treat low-frequency transport and weakly nonlinear transport together lies in the fact that both are determined by the **linear quasistatic** electrical response which is due to the piled-up charge in the biased sample. The potential distribution across a quantum point contact was addressed by Levinson (1989). Within a discrete description, the nonequilibrium charge is related to the applied voltages at the contacts via an electrochemical capacitance. The total capacitance is a series contribution of a geometrical capacitance and a quantum capacitance. The quantum capacitance is determined by the density of states in the sample region over which the electrical field can penetrate into the conductor. Such quantum corrections were essential for the early measurement of the density of states of the two-dimensional electron gas (Smith, Wang and Stiles,

1986). The capacitance of small capacitors has been investigated by Fomin (1967), Luryi (1988) and Büttiker, Thomas and Prêtre (1993a). Recently, Field *et al.* (1996) used a quantum dot coupled capacitively to a two-dimensional electron gas to investigate the density of states at the metal-insulator transition. If tunnelling comes into play, the capacitance defined via the static charge distribution is no longer equal to the capacitance defined via the displacement current (Christen and Büttiker, 1996b). The dynamically defined capacitance—which is usually measured in a capacitance experiment—is called **emittance** (Büttiker, 1993). It turns out that the emittance between the condenser plates can have a negative sign relative to the capacitance. Such inductive behaviour occurs whenever transmission of charges between these 'plates' becomes large (Büttiker, 1993; Büttiker, Thomas and Prêtre, 1994; Mikhailov and Volkov, 1995; Christen and Büttiker, 1996b). The low-frequency response for a one-dimensional two-terminal sample was also discussed by Pastawski (1992) and Fu and Dudley (1993), who also obtained an inductive behaviour of the current response. However, the Coulomb interaction was not treated at all and the results were, therefore, in general neither gauge invariant nor current conserving (Jacoboni and Price, 1993).

Nonlinear effects of interest are asymmetric current–voltage characteristics and rectification (Taboryski *et al.* 1994), the evolution of half-integer conductance plateaus (Glazman and Khaetskii, 1989; Patel *et al.*, 1990), the breakdown of conductance quantization (Kouwenhoven *et al.*, 1989), and negative differential conductance and hysteresis (Kluksdahl *et al.*, 1989). Nonlinearities in tunnel contacts have already been addressed in an early paper by Frenkel (1930) without considering self-consistent effects. Landauer (1987b) emphasized the necessity to consider the self-consistent potential. Nevertheless, many works on nonlinear transport consider the noninteracting case or consider interactions which are not gauge invariant (Glazman and Khaetskii, 1989; Patel *et al.*, 1990; Al'tshuler and Khmelnitskii, 1985; Castaño and Kirczenow, 1990; Aronov, Zagoskin, and Jonson, 1995; Zagoskin and Shekhter, 1994). An exception is a self-consistent numerical treatment of a tunnelling barrier by Kluksdahl *et al.* (1989). In general, in the nonlinear regime inelastic processes which destroy the phase coherence can play an important role; below, however, we emphasize phase-coherent transport (Büttiker, 1993; Christen and Büttiker, 1996c).

The emphasis on the role of the long-range Coulomb interaction in this work should be contrasted with the recent discussions of mesoscopic transport in the framework of the Luttinger approach. The Luttinger approach treats the short-range interactions only. However, since a one-dimensional wire or an edge channel cannot screen charges completely, a more realistic treatment should include the long-range Coulomb effect. Without long-range Coulomb interaction the results based only on Luttinger models are not charge and current conserving nor gauge invariant. In this context we note that the theory of the Coulomb blockade has been very successful in describing the important experimental aspects. The Coulomb blockade is a consequence of the long-range Coulomb interaction and not of short-range interactions. In any case the Hartree-like discussion given here is useful as a comparison with other theories which include interactions. To distinguish non-Fermi liquid behaviour from those of a Fermi liquid it will be essential to compare experimental data with a reasonable Fermi-liquid theory.

While a number of experiments and theoretical investigations have been carried out for effects which are both time dependent and nonlinear, the purely linear frequency-dependent admittance of mesoscopic conductors has received little attention. The low-frequency AC transport discussed here should be accessible experimentally with much the same techniques that are currently used for DC measurements. Lock-in techniques used to measure the in-phase response should be equally applicable to measure the out-of-phase response. As an example we cite here only the experiment of Chen *et al.* (1994). Measurements which extend to higher frequencies are more difficult and require special techniques to couple a high-frequency signal to a mesoscopic conductor (Pieper and Price, 1994; Kouwenhoven *et al.*, 1994; Reznikov *et al.*, 1995; Hofbeck *et al.*, 1996). Already the low-frequency admittance of mesoscopic conductors is a very interesting area of research with considerable room for further experimental and theoretical work.

7.2 THEORY

In this section we briefly recall the theory of AC transport and weakly nonlinear transport which has been developed by Büttiker (1993), Büttiker, Prêtre and Thomas (1993b), Büttiker, Thomas and Prêtre (1993a, 1994) and Christen and Büttiker (1996b, 1996c). A review of these works is provided by Büttiker and Christen (1996a).

7.2.1 Mesoscopic conductors

Consider a conducting region which is connected via leads and contacts $\alpha = 1, \dots, N$ to N reservoirs of carriers with charge e as shown in Fig. 7.1(a). In order to include into the formalism the presence of nearby gates (capacitors), some parts of the sample are allowed to be macroscopically large and disconnected from other parts. In principle, a magnetic field B may be present. It is assumed that the leads of the conducting part are so small that carrier motion occurs via one-dimensional subbands (channels) $m = 1, \dots, M_\alpha$ for each contact. Moreover, the distance between these contacts is assumed to be short compared with the inelastic scattering length and the phase-breaking length, such that transmission of charge carriers from one contact to another can be considered to be elastic and phase coherent. Phase-breaking processes are briefly mentioned in section 7.2.7 in the context of voltage probes. A reservoir is at thermodynamic equilibrium and can thus be characterized by its temperature T_α and by the electrochemical potential μ_α of the carriers. While we assume that all reservoirs are at the same temperature T, we consider differences of the electrochemical

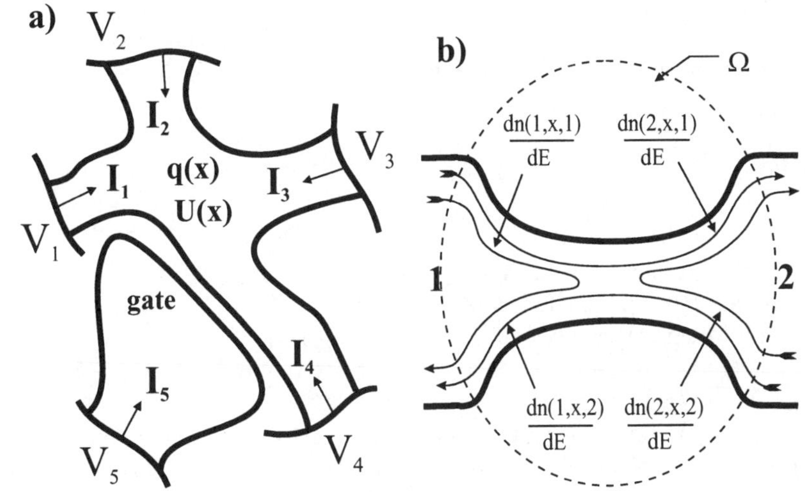

Figure 7.1 *(a) Multiterminal sample including a gate; the voltages V_α induce charges $q(x)$ which influence the potential $U(x)$, and drive currents I_β. (b) Decomposition of the local DOS in a two-terminal sample into local partial densities of states (PDOS).*

potentials which is usually achieved by applying voltages V_α to the reservoirs. We fix the voltage scales such that $V_\alpha \equiv 0$ corresponds to the equilibrium state where all electrochemical potentials are equal to each other, $\mu_\alpha \equiv \mu^{(eq)}$. The problem to be solved is now to find the current $I_\alpha(t)$ entering the sample at contact α in response to a (generally time-dependent) voltage $V_\beta(t)$ at contact β. The general time-dependent nonlinear transport problem is highly nontrivial, and we restrict the discussion to weakly nonlinear DC transport and to linear low-frequency transport.

7.2.2 The scattering approach

Within the scattering approach, the sample is described by a unitary scattering-matrix **S** (Landauer, 1970; Büttiker *et al.*, 1985; Imry, 1986; Büttiker, 1986a). Unitarity together with microreversibility implies that under a reversal of the magnetic field the scattering matrix has the symmetry $\mathbf{S}^T(B) = \mathbf{S}(-B)$ (Büttiker, 1986a, 1988b). Furthermore, the scattering matrix **S** for the conductor can be arranged such that it is composed of submatrices $\mathbf{s}_{\alpha\beta}(E, [U(x)])$ with elements $s_{\alpha\beta nm}(E, [U(x)])$ which relate the out-going current amplitude in channel n at contact α to the incident current amplitude in channel m at contact β. The scattering matrix is a function of the energy E of the carrier and is a functional of the electric potential $U(x)$ in the conductor (Büttiker, 1993). All the physical information which is needed for the DC transport properties is contained in the scattering matrix.

First, the **transmission function** $T_{\alpha\beta}$ is the sum of all transmission probabilities for carriers at energy E incident in contact β to exit the sample at contact α. The reflection function, $R_{\alpha\alpha}$, concerning the reflection probabilities is defined analogously. They are given by

$$T_{\alpha\beta} = \mathrm{Tr}(\mathbf{s}_{\alpha\beta}^{\dagger}\mathbf{s}_{\alpha\beta}), \qquad (\text{if } \alpha \neq \beta) \tag{7.2.1}$$

$$R_{\alpha\alpha} = \mathrm{Tr}(\mathbf{s}_{\alpha\alpha}^{\dagger}\mathbf{s}_{\alpha\alpha}). \tag{7.2.2}$$

The trace corresponds to the sum over all channels in these contacts. It turns out that the transmission functions $T_{\alpha\beta}$ determine the DC conductance (section 7.2.4). As a consequence of the unitary properties of the scattering matrix, the number of quantum channels at contact α is given by $N_\alpha = R_{\alpha\alpha} + \sum_\beta T_{\alpha\beta}$ $= R_{\alpha\alpha} + \sum_\beta T_{\beta\alpha}$.

A second important quantity is the local **density of states** (DOS) of the sample. It can be obtained from the scattering matrix by a functional derivative with respect to the potential (Büttiker, 1993)

$$\frac{\mathrm{d}n(x)}{\mathrm{d}E} = -\frac{1}{4\pi \mathrm{i}}\sum_{\alpha\beta}\mathrm{Tr}\left[\mathbf{s}_{\alpha\beta}^{\dagger}\frac{\delta \mathbf{s}_{\alpha\beta}}{e\delta U(x)} - \frac{\delta \mathbf{s}_{\alpha\beta}^{\dagger}}{e\delta U(x)}\mathbf{s}_{\alpha\beta}\right] \equiv \sum_{\alpha\beta}\frac{\mathrm{d}n(\alpha,x,\beta)}{\mathrm{d}E}. \tag{7.2.3}$$

The global DOS $\mathrm{d}N/\mathrm{d}E$ follows by a spatial integration over the sample region Ω. Note that formula (7.2.3) counts only scattering states. Pure bound states, for example at an impurity, are not included. As indicated by the last equality in equation (7.2.3), the local DOS can be understood as a sum of local **partial densities of states** (PDOS) $\mathrm{d}n(\alpha,x,\beta)/\mathrm{d}E$ (Gasparian, Christen and Büttiker, 1996). The sum is over all injecting contacts β and all emitting contacts α. The meaning of a local PDOS $\mathrm{d}n(\alpha,x,\beta)/\mathrm{d}E$ is then obvious: it is the local DOS associated with carriers which are incident from contact β, reach x and finally are scattered into contact α. The local PDOS are illustrated in Fig. 7.1(b). Integration over the whole sample leads to the global PDOS $\mathrm{d}N_{\alpha\beta}/\mathrm{d}E$. Clearly, it holds $\mathrm{d}N/\mathrm{d}E = \sum_{\alpha\beta}\mathrm{d}N_{\alpha\beta}/\mathrm{d}E$.

7.2.3 Piled-up charge and self-consistent potential

In general, the applied voltages δV_α polarize the sample and drive the conductor into a nonequilibrium state characterized by a piled-up charge density $\delta q(x)$ and a nonequilibrium potential variation $\delta U(x)$. While the linear DC transport depends only on the equilibrium state, a treatment of transport properties at finite frequencies or in the nonlinear regime requires the knowledge of the nonequilibrium state. In the sequel we self-consistently discuss the linear quasistatic response of the conductor to the voltage variation δV_α. The associated response coefficients $u_\alpha(x)$ are called **characteristic potentials** (Büttiker, 1993)

$$\delta U(x) = \sum_\alpha u_\alpha(x)\delta V_\alpha. \tag{7.2.4}$$

The nonequilibrium potential landscape of a two-terminal conductor is illustrated in Fig. 7.2. In order to find the characteristic potentials $u_\alpha(x)$, we first calculate the bare charge density injected by the external voltage shifts at fixed electric potential. Secondly, we determine the screening charge-density induced by the variation of the electric potential at fixed voltages in the reservoirs. The self-consistent loop is then closed by the Poisson equation which relates the total charge density to the electric potential. The bare charge density can be obtained with the help of the PDOS. The part of the injected charge density at point x due to a voltage variation δV_β which eventually leaves the sample through contact α is equal to $e^2(\mathrm{d}n(\alpha, x, \beta)/\mathrm{d}E)\delta V_\beta$. To obtain the total bare charge density, one has to sum over all emitting contacts α. This motivates to define the **injectivity**

$$\frac{\mathrm{d}n(x, \beta)}{\mathrm{d}E} = \sum_\alpha \frac{\mathrm{d}n(\alpha, x, \beta)}{\mathrm{d}E} \tag{7.2.5}$$

which is the local DOS associated with particles injected at contact β. For later use, we also define the **emissivity**

$$\frac{\mathrm{d}n(\alpha, x)}{\mathrm{d}E} = \sum_\beta \frac{\mathrm{d}n(\alpha, x, \beta)}{\mathrm{d}E} \tag{7.2.6}$$

being the local DOS associated with particles emitted at contact α. With the help of the injectivity, the bare charge density can be expressed by

$$\delta q^{(b)}(x) = e^2 \sum_\beta \frac{\mathrm{d}n(x, \beta)}{\mathrm{d}E}\delta V_\beta. \tag{7.2.7}$$

The screening charge density is connected to the electrostatic potential via the Lindhard polarization function $\Pi(x, x')$ (Levinson, 1989)

$$\delta q^{(s)}(x) = -e^2 \int \mathrm{d}^3x'\, \Pi(x, x')\delta U(x'). \tag{7.2.8}$$

The Lindhard function can be expressed in terms of the scattering states. The total charge density $\delta q = \delta q^{(b)} + \delta q^{(s)}$ now acts as the source of the nonequilibrium electric potential in the Poisson equation which reads in terms of the characteristic potentials (Büttiker, 1993)

$$-\Delta u_\alpha(x) + 4\pi e^2 \int \mathrm{d}^3x'\, \Pi(x, x')u_\alpha(x') = 4\pi e^2 \frac{\mathrm{d}n(x, \alpha)}{\mathrm{d}E}. \tag{7.2.9}$$

For simplicity, we assumed that the relative dielectric constant is equal to unity. The boundary conditions for $u_\alpha(x)$ follow from the requirement that the reservoirs are independent of each other and that the fields in the reservoirs are fully

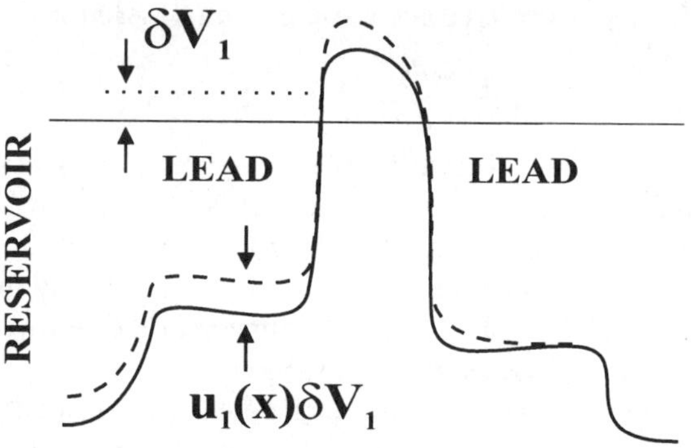

Figure 7.2 *Change (dashed) of the equilibrium potential (solid) in a two-terminal sample due to a voltage variation δV_1 (dotted) in the left reservoir.*

screened. This implies that $u_\alpha(x) = 1$ if x is in reservoir α and that $u_\beta(x)$ $(\beta \neq \alpha)$ vanishes in reservoir α. One can formally solve the inhomogeneous partial differential equation (7.2.9) with the help of Green's function $\mathcal{G}(x, x_0)$. This function is the solution of equation (7.2.9) if the source term $e\,\mathrm{d}n(x,\alpha)/\mathrm{d}E$ on the right-hand side is replaced by a localized test charge $e\delta(x - x_0)$ at point x_0. The characteristic potential $u_\alpha(x)$ can then be written in the form

$$u_\alpha(x) = \int \mathrm{d}^3x'\,\mathcal{G}(x, x')\frac{\mathrm{d}n(x', \alpha)}{\mathrm{d}E}. \qquad (7.2.10)$$

From equations (7.2.4), (7.2.7), and (7.2.8) the total piled-up charge density and the electric potential variation follow immediately. In general, for a typical mesoscopic sample, the Green's function $\mathcal{G}(x, x')$ is nonlocal and, therefore, difficult to evaluate. Below we use two approaches to proceed further. For samples (semiconductors) in which the Green's function is nonlocal we use a capacitive model (discretized Poisson equation). For metals, we will use the Thomas–Fermi approximation: inside a metal the response is (nearly) local and is given by $\Pi(x, x') = \delta(x - x')\mathrm{d}n(x)/\mathrm{d}E$. Assuming additionally full screening, we have $\Delta u_\alpha(x) \equiv 0$ which implies (Büttiker, Thomas, and Prêtre, 1994)

$$\mathcal{G}(x, x') = \delta(x - x')\left(\frac{\mathrm{d}n(x)}{\mathrm{d}E}\right)^{-1}. \qquad (7.2.11)$$

A similarly simple but self-consistent approach for semiconductors is discussed below.

7.2.4 Linear and weakly nonlinear DC transport

Since the electric potential $U(x, \{V_\gamma\})$ is a function of the voltages V_γ in the reservoirs, the scattering matrix depends explicitly on these voltages and the energy of the scattered carriers, $\mathbf{s}_{\alpha\beta}(E, \{V_\gamma\})$. (Whenever we discuss nonlinear effects, we use the notation I_α for the currents and V_α for the voltages rather than δI_α and δV_α.) With the help of the scattering matrix, we can find the current $dI_\alpha(E)$ in contact α due to carriers incident at contact β in the energy interval $(E, E + dE)$. It is convenient to introduce a **spectral conductance**

$$g_{\alpha\beta}(E) = \frac{e^2}{h}\mathrm{Tr}[\mathbf{1}_\alpha(E,\{V_\gamma\})\delta_{\alpha\beta} - \mathbf{s}^\dagger_{\alpha\beta}(E,\{V_\gamma\})\mathbf{s}_{\alpha\beta}(E,\{V_\gamma\})], \qquad (7.2.12)$$

such that the current at energy E becomes $dI_\alpha(E) = g_{\alpha\beta}(E)(dE/e)$. The unity matrix $\mathbf{1}_\alpha$ lives in the space of the quantum channels in lead α with thresholds below the electrochemical potential. Note that this matrix is also a (discontinuous) function of energy and the potential. It changes its dimension by one whenever the band bottom of a new subband passes the electrochemical potential. The current through contact α is the sum of all spectral currents weighted by the Fermi functions $f(z) = (1 + \exp(z/k_B T))^{-1}$ of the reservoirs (Büttiker, 1992b) at temperature T

$$I_\alpha = \sum_{\beta=1}^{N} \int (dE/e) f(E - E_F - eV_\beta) g_{\alpha\beta}(E,\{V_\gamma\}). \qquad (7.2.13)$$

To discuss weakly nonlinear transport, we expand the current according to

$$I_\alpha = \sum_\beta G^{(0)}_{\alpha\beta} V_\beta + \sum_{\beta\gamma} G^{(0)}_{\alpha\beta\gamma} V_\beta V_\gamma + \mathcal{O}(V^3). \qquad (7.2.14)$$

The coefficients $G^{(0)}_{\alpha\beta}$ and $G^{(0)}_{\alpha\beta\gamma}$ are obtained from an expansion of equation (7.2.13) with respect to the voltages V_α. One obtains for the linear conductance

$$G^{(0)}_{\alpha\beta} = \int dE\,(-\partial_E f) g_{\alpha\beta}(E), \qquad (7.2.15)$$

where the $g_{\alpha\beta}$ are taken at $V_1 = \cdots = V_N = 0$. An expansion of equation (7.2.13) up to $\mathcal{O}(V^2)$ yields $G^{(0)}_{\alpha\beta\gamma} = \frac{1}{2}\int dE\,(-\partial_E f)(2\partial_{V_\gamma} g_{\alpha\beta} + e\partial_E g_{\alpha\beta}\delta_{\beta\gamma})$. Writing $\partial_{V_\gamma} g_{\alpha\beta}$ in terms of the functional derivative $\delta g_{\alpha\beta}/\delta U(x)$ and the characteristic potentials yields

$$G_{\alpha\beta\gamma} = \frac{1}{2}\int dE\,(-\partial_E f)\int d^3x\, \frac{\delta g_{\alpha\beta}}{\delta U(x)} \left(2u_\gamma(x) - \delta_{\beta\gamma}\right). \qquad (7.2.16)$$

In section 7.3.3 we will apply this result to the resonant tunnelling barrier.

7.2.5 Linear low-frequency transport

We are interested in the admittance $G_{\alpha\beta}(\omega)$ which determines the Fourier amplitudes of the current, δI_α, at a contact α in response to an oscillating voltage $\delta V_\beta \exp(-i\omega t)$ at contact β

$$\delta I_\alpha = \sum_\beta G_{\alpha\beta}(\omega)\delta V_\beta. \qquad (7.2.17)$$

A restriction to the low-frequency limit means that we expand the admittance in powers of frequency

$$G_{\alpha\beta}(\omega) = G^{(0)}_{\alpha\beta} - \mathrm{i}\omega E_{\alpha\beta} + \omega^2 K_{\alpha\beta} + \mathcal{O}(\omega^3). \tag{7.2.18}$$

The DC conductance $G^{(0)}_{\alpha\beta}$ has already been derived in equation (7.2.15). The first-order term is determined by the emittance matrix $E_{\alpha\beta} \equiv \mathrm{i}(\mathrm{d}G_{\alpha\beta}/\mathrm{d}\omega)_{\omega=0}$. To get some feeling for the emittance and the second-order term $K_{\alpha\beta}$, consider a macroscopic capacitor C in series with a resistor R. For this purely capacitive structure with vanishing DC conductance, $G^{(0)} = 0$, one finds that the emittance is the capacitance, $E = C$, and that $K = R^{(q)}C^2$ contains the $R^{(q)}C$ time with the charge relaxation resistance $R^{(q)} = R$. For any N-terminal capacitor, the emittance matrix is just the capacitance matrix. These simple results must be modified for mesoscopic conductors and conductors which connect different reservoirs. First, it is not the geometrical capacitance but rather the electrochemical capacitance depending on the DOS which relates charges on mesoscopic conductors to voltage variations in the reservoirs (Büttiker, Thomas and Prêtre, 1993a). Secondly, conductors which connect different reservoirs allow a transmission of charge which leads to inductance-like contributions to the emittance (Büttiker, 1993; Büttiker, Prêtre and Thomas, 1993b; Christen and Büttiker, 1996b). Thirdly, the charge relaxation resistance cannot, in general, be calculated like a DC resistance (Büttiker, Thomas and Prêtre, 1993a; Büttiker and Christen, 1996b).

To find a general expression for the emittance, we first notice that for the purely capacitive case, the current (displacement current) at contact α is the time derivative of the total charge δQ_α on the capacitor plates connected to this contact. Hence, $\delta I_\alpha = -\mathrm{i}\omega\delta Q_\alpha$ and the emittance is given by $E_{\alpha\beta} = \partial Q_\alpha/\partial V_\beta$. If transmission between different reservoirs is allowed, the charge δQ_α scattered through a contact can no longer be attributed to a unique spatial region. Indeed, charge at a given location was injected from various contacts and will be ejected at various contacts. However, the charge emitted at a contact can still be calculated within the scattering approach. We decompose it in a bare and a screening part $\delta Q_\alpha = \delta Q^{(b)}_\alpha + \delta Q^{(s)}_\alpha$. The bare part of the charge corresponds to the charge which is scattered through the contact α for fixed electric potential and is thus given by

$$\delta Q^{(b)}_\alpha = e^2 \sum_\beta \int \mathrm{d}^3x \frac{\mathrm{d}n(\alpha, x, \beta)}{\mathrm{d}E} \delta V_\beta \equiv e^2 \sum_\beta \frac{\mathrm{d}N_{\alpha\beta}}{\mathrm{d}E} \delta V_\beta. \tag{7.2.19}$$

The screening part, on the other hand, is associated with the charge which is scattered through contact α for a variation in the electric potential. Since a shift of the band bottom contributes with a negative sign and since the potential is in general spatially varying, $\delta Q_\alpha^{(s)}$ is connected to the local PDOS by

$$\delta Q_\alpha^{(s)} = -e^2 \sum_\beta \int \mathrm{d}^3x \frac{\mathrm{d}n(\alpha, x, \beta)}{\mathrm{d}E} \delta U(x). \tag{7.2.20}$$

Recalling equations (7.2.4) and (7.2.10), one then finds the emittance (Büttiker, 1993)

$$E_{\alpha\beta} = e^2 \frac{\mathrm{d}N_{\alpha\beta}}{\mathrm{d}E} - e^2 \int \mathrm{d}^3x \int \mathrm{d}^3x' \frac{\mathrm{d}n(\alpha, x)}{\mathrm{d}E} \mathcal{G}(x, x') \frac{\mathrm{d}n(x', \beta)}{\mathrm{d}E}. \tag{7.2.21}$$

Importantly, the emittance is in general not capacitance-like, i.e. the diagonal and the off-diagonal emittance elements are not restricted to positive and negative values, respectively. Whenever the transmission of carriers between two contacts predominates the reflection, the associated emittance element changes sign and behaves inductance-like (Büttiker, 1993; Büttiker, Thomas and Prêtre, 1994; Christen and Büttiker, 1996a, 1996b).

We do not present here the general theory for the second-order coefficient $K_{\alpha\beta}$. This involves a rather cumbersome discussion of the low-frequency DOS and the frequency-dependent scattering probabilities (Prêtre, Thomas, and Büttiker, 1996). Below, however, we calculate $K_{\alpha\beta}$ for a set of simple examples and in the framework of the discrete potential approximation.

7.2.6 Charge conservation, gauge invariance and Onsager–Casimir relations

Since the system under consideration includes all conductors and nearby gates, the theory must satisfy charge (and current) conservation and gauge invariance. Due to microreversibility, the linear response matrices must additionally satisfy the Onsager–Casimir symmetry relations.

Let us first discuss charge conservation, which states that the total charge in the sample remains constant under a bias. This also implies current conservation, $\sum_\alpha I_\alpha = 0$. Imagine a volume Ω which encloses the entire conductor including a portion of the reservoirs which is so large that at the place where the surface of Ω intersects the reservoir all the characteristic potentials are either zero or unity. According to the law of Gauss, one concludes that the total charge remains constant. Application of a bias voltage results only in a redistribution of the charge. If the conductor is poor, i.e. nearly an insulator, the contacts act like plates of capacitors. In this case long-range fields exist which run from one reservoir to the other and from a reservoir to a portion of the conductor. But if we choose the volume Ω to be large enough then all field lines stay within this volume. Current conservation implies for the response coefficients the sum rules

$$\sum_\alpha G_{\alpha\beta}^{(0)} = \sum_\alpha E_{\alpha\beta} = \sum_\alpha K_{\alpha\beta} = \sum_\alpha G_{\alpha\beta\gamma}^{(0)} = 0. \tag{7.2.22}$$

Gauge invariance means that measurable quantities are invariant under a global voltage shift δV in the reservoirs, $\delta V_\alpha \mapsto \delta V_\alpha + \delta V$ (which corresponds, of course, to only a change of the global energy scale). The characteristic potentials thus satisfy $\sum_\alpha u_\alpha(x) \equiv 1$ (Büttiker, 1993). For the response coefficients the following sum rules must hold

$$\sum_\beta G^{(0)}_{\alpha\beta} = \sum_\beta E_{\alpha\beta} = \sum_\beta K_{\alpha\beta} = \sum_\beta (G^{(0)}_{\alpha\beta\gamma} + G^{(0)}_{\alpha\gamma\beta}) = 0. \qquad (7.2.23)$$

Further implications of charge conservation and gauge invariance can be written down, for example for the Green's function $\mathcal{G}(x, x')$ or for the screened spectral conductances $g_{\alpha\beta}(E)$. A more extensive discussion of charge and current conservation, and of gauge invariance is provided by Büttiker (1993) and Christen and Büttiker (1996c).

The admittance matrix must also satisfy the Onsager–Casimir reciprocity relations which read in the presence of a magnetic field

$$G_{\alpha\beta}(\omega)|_B = G_{\beta\alpha}(\omega)|_{-B}. \qquad (7.2.24)$$

They are a consequence of time-reversal symmetry of the microscopic equations and the fact that we consider transmission from one reservoir to another (see Büttiker, 1986a, 1988b, 1993). The Onsager–Casimir relations are only valid for the linear response coefficients, i.e. close to equilibrium. Of course, the symmetry relations (7.2.24) hold individually for $G^{(0)}_{\alpha\beta}$, $E_{\alpha\beta}$, and $K_{\alpha\beta}$.

7.2.7 Voltage probes and dissipation

It is interesting to consider not only phase coherent samples but also conductors that are much longer than an inelastic scattering length. An initial investigation on dissipative effects and AC conductance was presented by Anantram and Datta (1995) without consideration of self-consistent potential effects. Here we follow the discussion of Christen and Büttiker (1996a) who introduced phase randomizing, inelastic scattering with the help of voltage probes. Even though the net current at a voltage probe vanishes, carriers can escape from the conductor into the reservoir of the voltage probe and are replaced by carriers from the reservoir (Büttiker, 1986b). Energy and phase of emitted and re-injected carriers are uncorrelated. Thus, the voltage probe is both dissipative and phase randomizing. The 'voltage probe' exhibits no dissipation and acts purely as a dephaser if instead of the total current, the current in each narrow energy interval is taken to be zero (de Jong and Beenakker, 1996). Many voltage probes, instead of only one, can be used to describe spatially uniform dephasing processes (Brouwer and Beenakker, 1997).

The admittance elements $G_{\alpha\beta}(\omega)$ introduced in equation (7.2.17) yield the currents passing the N contacts of the conductor, if all voltages in the reservoirs are prescribed. A voltage probe corresponds to a contact which connects the sample to a highly resistive voltmeter. For the ideal voltage probe at contact N it holds $\delta I_N = 0$. With the help of the condition $\delta I_N = 0$, the voltage δV_N can be determined. An elimination of this voltage in the remaining equations (7.2.17) leads to a reduced $(N-1)$-terminal conductance

$$\tilde{G}_{\alpha\beta}(\omega) = G_{\alpha\beta}(\omega) - \frac{G_{\alpha N}(\omega) G_{N\beta}(\omega)}{G_{NN}(\omega)} \tag{7.2.25}$$

for the sample without contact N. With the help of equation (7.2.18) one can derive the reduced-response coefficients $\tilde{G}^{(0)}_{\alpha\beta}$, $\tilde{E}_{\alpha\beta}$, and $\tilde{K}_{\alpha\beta}$ which determine the $N-1$ conductance matrix. We will not display the rather lengthy expressions here but discuss the important special case of the capacitance.

Consider a sample with $N \geq 4$ contacts with two of them, say γ and δ, being purely capacitive contacts (gates). Thus, an element of the DC conductance matrix $G^{(0)}_{\alpha\beta}$ vanishes whenever one of the indices takes the value γ or δ. There are at least two remaining contacts which are assumed to permit transmission and which we label $N-1$ and N. It can be shown that the emittance $E_{\gamma\delta}$ of the two capacitive contacts is just the static capacitance between these contacts and is thus always symmetric. This follows from the fact that the displacement charges δQ_γ and δQ_δ are equal to the quasistatic charges $\delta q_{k=\gamma}$ and $\delta q_{k=\delta}$, respectively.

We now eliminate contact N and show that $\tilde{K}_{\alpha\beta}$ can in general be asymmetric. An expansion of the reduced admittance with respect to frequency yields then the reduced coefficients

$$\tilde{G}_{\gamma\delta} = 0, \tag{7.2.26}$$

$$\tilde{E}_{\gamma\delta} = E_{\gamma\delta}, \tag{7.2.27}$$

$$\tilde{K}_{\gamma\delta} = K_{\gamma\delta} + \frac{E_{\gamma N} E_{N\delta}}{G^{(0)}_{NN}}. \tag{7.2.28}$$

Note that contact N is connected to at least one other contact, say $N-1$, such that $G^{(0)}_{NN} \neq 0$. Since contact N is not purely capacitive, the emittance elements $E_{\gamma N}$ and $E_{N\delta}$ are in general asymmetric (Büttiker, 1993; Chen *et al.*, 1994; Christen and Büttiker, 1996a). Thus $\tilde{K}_{\gamma\delta}$ is in general also asymmetric, even for a symmetric $K_{\gamma\delta}$.

Our discussion demonstrates that dissipation can give rise to an asymmetric dynamic capacitance. An overview of the magnetic field symmetry properties of capacitive arrangements of conductors is presented in section 7.3.1.

7.2.8 The discrete potential approximation

The considerations of the previous sections concern the fully space-dependent problem of screening. In particular, solving equation (7.2.9) is very complicated. We can obtain analytical results by using a convenient coarse-graining model for the conductors (Prêtre, Thomas, and Büttiker, 1996; Christen and Büttiker, 1996a, 1996b, 1996c). By decomposing the conductor into M regions we discretize the electrostatic potential and the charge density. Space-dependent quantities are written as M-dimensional vectors or $M \times M$-matrices.

For example, the electrostatic potential variation $\delta U(x)$ becomes a vector $\delta \boldsymbol{U} = (\delta U_1, \ldots, \delta U_M)^t$, where δU_k denotes an average electrostatic potential shift in region k. The piled-up charge in the sample is denoted by $\delta \boldsymbol{q} = (\delta q_1, \ldots, \delta q_M)^t$, where δq_k is the charge increment in region k. We write the local PDOS as vectors $\boldsymbol{D}_{\alpha\beta} = e^2(\mathrm{d}N_{\alpha 1\beta}/\mathrm{d}E, \ldots, \mathrm{d}N_{\alpha M\beta}/\mathrm{d}E)^t$. The injectivity and emissivity are then obtained by the sum over one of the indices, $\boldsymbol{D}_\beta^{(i)} = \sum_\alpha \boldsymbol{D}_{\alpha\beta}$ and $\boldsymbol{D}_\alpha^{(e)} = \sum_\beta \boldsymbol{D}_{\alpha\beta}$, respectively. The functional derivative with respect to the potential $U(x)$ is replaced by the gradient $\nabla_U = (\mathrm{d}/\mathrm{d}U_1, \ldots, \mathrm{d}/\mathrm{d}U_M)^t$. The polarization function is a matrix $\boldsymbol{\Pi}$. In the case of locality, for example in the semiclassical limit, the polarization matrix is diagonal with elements equal to the local DOS, $\Pi_{kl} = D_k \delta_{kl}$. Note that here we included the factor e^2 already in the definition of $\boldsymbol{\Pi}$. The relation between the charges and the electric potentials which are mediated by the Poisson equation can now be expressed with a geometric capacitance matrix $\boldsymbol{C}$. The nonequilibrium charge distribution is

$$\delta \boldsymbol{q} = \boldsymbol{C} \delta \boldsymbol{U} = \sum_\beta \boldsymbol{D}_\beta^{(i)} \delta V_\beta - \boldsymbol{\Pi} \delta \boldsymbol{U} \tag{7.2.29}$$

which yields for the characteristic potentials

$$\boldsymbol{u}_\beta = (\boldsymbol{\Pi} + \boldsymbol{C})^{-1} \boldsymbol{D}_\beta^{(i)}. \tag{7.2.30}$$

These relations determine the linear nonequilibrium state. The charge δq_k in region k can be written in terms of an electrochemical capacitance matrix (Christen and Büttiker, 1996a, 1996b, 1996c)

$$\delta q_k = \sum_\beta C_{\mu,k\beta} \delta V_\beta, \tag{7.2.31}$$

where

$$C_{\mu,k\beta} = \sum_l C_{kl} u_{l\beta}. \tag{7.2.32}$$

The discrete version of the emittance (7.2.21) reads then

$$E_{\alpha\beta} = D_{\alpha\beta} - \sum_k D_{\alpha k}^{(e)} u_{k\beta}, \tag{7.2.33}$$

with the global PDOS $D_{\alpha\beta} = \sum_k D_{\alpha k\beta} = e^2\, \mathrm{d}N_{\alpha\beta}/\mathrm{d}E$. We like to mention that the nonequilibrium potential distribution of phase coherent mesoscopic conductors can be derived in a very general manner from a generalized thermodynamic approach. As pointed out by Christen (1997), there exists a grand canonical potential such that its second variation with regard to the electrochemical potentials determines the electrochemical capacitance matrix equation (7.2.32).

To conclude this paragraph we present the discrete rectification coefficients (7.2.16). They are given by

$$G_{\alpha\beta\gamma}^{(0)} = \tfrac{1}{2} \int \mathrm{d}E\, (-\partial_E f)(2(\nabla_U g_{\alpha\beta})^t (\boldsymbol{\Pi} + \boldsymbol{C})^{-1} \boldsymbol{D}_\gamma^{(i)} + e\delta_{\beta\gamma} \partial_E g_{\alpha\beta}). \tag{7.2.34}$$

It is straightforward to show that these results satisfy the requirements of section 7.2.6.

7.3 EXAMPLES

We discuss four examples. Throughout we assume zero temperature where the energy derivative $(-\partial_E f)$ of the Fermi function reduces to a delta function centred at the Fermi level. First, we apply the theory of the admittance to quantized Hall conductors within the edge-channel picture (Christen and Büttiker, 1996a). In this case the theory simplifies considerably due to a metallic-like screening and a unidirectional carrier motion (chirality) in the edge states. Secondly, we study the capacitance and low-frequency admittance of a quantum point contact formed with the help of gates (Christen and Büttiker, 1996b). It turns out that steps in the capacitance and emittance occur in synchronism with the well-known conductance steps. A third example is the resonant tunnelling barrier close to a resonance, for which we calculate the nonlinear current–voltage characteristic, and the emittance (Christen and Büttiker, 1996c). While the first three examples are approximated by the discrete potential model, this is different for our last example. There, we investigate the low-frequency conductance of a metallic diffusive conductor with a linear voltage drop (Blanter and Büttiker, 1997). For all examples, we use explicit expressions for the Lindhard function (matrix) $\mathbf{\Pi}$ given within the local Thomas–Fermi approximation, where $\mathbf{\Pi}$ is diagonal with elements determined by the local DOS.

7.3.1 Quantized Hall systems

We consider two-dimensional conductors subject to a perpendicular magnetic field which establishes an integer quantized Hall state (von Klitzing, Dorda, and Pepper, 1980). In the quantum Hall regime the states at the Fermi energy which contribute to conduction are the **edge states** (Halperin, 1982; Büttiker, 1992a; Komiyama and Hirai, 1996). In a classical picture, edge states correspond to the skipping orbits along the sample boundaries. In a quantum-mechanical single-particle picture, edge states correspond to the intersection of the electrochemical potential with the Landau levels. Going beyond the simple single-particle picture, one finds that due to interaction the electron gas decomposes into spatial regions of compressible and incompressible electron fluids (Chklovskii, Shklovskii, and Glazman, 1992; Lier and Gerhardts, 1994; Cooper and Chalker, 1993). Compressible regions are associated with transversally extended edge channels whereas the incompressible regions act like a dielectric medium. Importantly, only the compressible regions can be charged and contribute to screening (McDonald, Rice, and Brinkman, 1983; Yahel, Orgad and Palevski, 1996). We model thus the quantized Hall sample by an arrangement of metal-like edge channels connected to contacts and embedded in a dielectric medium. We assume spin-split Landau levels and neglect spatio-temporal effects inside edge channels (Talyanskii *et al.*, 1995). Since dynamical screening can be treated in a simple way, we will even derive an expression for the second-order term $K_{\alpha\beta}$ in equation (7.2.18). The AC-conductance of quantized Hall systems is of particular interest since there are now efforts to extend the DC resistance standard into the kHz regime (Delahaye, 1994).

(a) Transmission properties of edge channels

An important property of a single edge channel is its unidirectional transparency and the absence of backscattering (Büttiker, 1988a). We will restrict our discussion to the case when the only states at the Fermi energy which connect contacts are edge states. Furthermore, we also consider the case of ideal contacts (Büttiker, 1988a) at which electrons approaching the contact from the interior of the conductor are either transmitted with probability 1 or reflected with probability 1. The part of the scattering matrix associated with edge channel j can be written in the form $s_{\alpha\beta}^{(j)} = \Delta_{\alpha j} t_j \Delta_{j\beta}$ with a scattering amplitude $t_j = \exp(\mathrm{i}\phi_j(E))$. The energy dependence of the phase $\phi_j(E)$ determines the DOS of the edge channel, $\mathrm{d}N_j/\mathrm{d}E = (1/2\pi)(\mathrm{d}\phi_j/\mathrm{d}E)$ at the Fermi energy. $\Delta_{j\alpha}$ and $\Delta_{\alpha j}$ are the injection and emission probability, respectively, of channel j at contact α:

$$\Delta_{j\alpha} = \begin{cases} 1 & \text{if contact } \alpha \text{ injects into channel } j \\ 0 & \text{otherwise} \end{cases} \tag{7.3.1}$$

$$\Delta_{\alpha j} = \begin{cases} 1 & \text{if channel } j \text{ emits into contact } \alpha \\ 0 & \text{otherwise.} \end{cases} \tag{7.3.2}$$

Since an edge channel is always connected to one contact at each of its ends, one has $\sum_\alpha \Delta_{k\alpha} = \sum_\alpha \Delta_{\alpha k} = 1$. Furthermore, $\sum_k \Delta_{k\alpha}$ $(= \sum_k \Delta_{\alpha k})$ is the number of edge channels which enter (leave) the sample at contact α. Clearly, the transmission probabilities are functions of the magnetic field B and obey the microreversibility conditions $\Delta_{\alpha k}(B) = \Delta_{k\alpha}(-B)$. This follows directly from the fact that an inversion of the magnetic field inverses the arrows attached to the edge states. The transmission function (7.2.1) can easily be expressed with the help of $\Delta_{j\alpha}$ and $\Delta_{\alpha j}$ which yields the well-known quantization of the DC conductance

$$G_{\alpha\beta}^{(0)} = \frac{e^2}{h} \sum_k (\Delta_{k\alpha}\Delta_{k\beta} - \Delta_{\alpha k}\Delta_{k\beta}). \tag{7.3.3}$$

It is only the topology of the edge-channel arrangement which accounts for the DC conductance. Current conservation, gauge invariance, and the Onsager–Casimir relations follow immediately from the properties of $\Delta_{j\alpha}$ and $\Delta_{\alpha j}$.

(b) Dynamical screening

To each edge channel j we can assign a single electrochemical potential δV_j, which is just the electrochemical potential of the injecting reservoir, $\delta V_j = \sum_\beta \Delta_{j\beta}\delta V_\beta$. Note that roman and greek labels indicate edge states and reservoirs, respectively; δV_j is a component of the vector $\delta \boldsymbol{V}$ in the discrete potential approximation. Let $\boldsymbol{C}$ and $\boldsymbol{C}_\mu$ denote the geometric and the electrochemical capacitance matrices of the edge-state arrangement, respectively: $\boldsymbol{\delta q} = \boldsymbol{C\delta U} = \boldsymbol{C}_\mu \delta \boldsymbol{V}$. While $\boldsymbol{C}$ is real and frequency independent, $\boldsymbol{C}_\mu(\omega)$ is a function of frequency. Charge conservation and gauge invariance imply the sum rules $\sum_k C_{kl} = \sum_l C_{kl} = \sum_k C_{\mu,kl} = \sum_l C_{\mu,kl} = 0$. Both matrices are symmetric and even functions of the magnetic field. In the present notation the dynamic DOS can be summarized in a diagonal matrix $\mathbf{D}(\omega)$ with elements (Prêtre, Thomas, and Büttiker, 1996)

$$D_{jj}(\omega) = e^2(\mathrm{d}N_j/\mathrm{d}E)(1 + \mathrm{i}\tau_j\omega/2) + \mathcal{O}(\omega^2), \tag{7.3.4}$$

where $\tau_j = h(\mathrm{d}N_j/\mathrm{d}E)$ is the dwell time of carriers in edge channel j. The specific frequency dependence of the $D_{jj}(\omega)$ can be understood by noticing that the charging of edge channel j is a relaxation process with a time $\tau_j/2$. The dynamic local PDOS are obviously given by

$$D_{\alpha j\beta}(\omega) = \Delta_{\alpha j} D_{jj}(\omega)\Delta_{j\beta}, \tag{7.3.5}$$

from which one obtains the dynamic injectivities, emissivities, and the global PDOS. Using further $\boldsymbol{\delta q} = \mathbf{D}(\boldsymbol{\delta V} - \boldsymbol{\delta U})$, one obtains the dynamic electrochemical capacitance

$$\boldsymbol{C}_\mu(\omega) = \boldsymbol{C}_\mu(0) + \mathrm{i}\omega\frac{h}{2e^2}\boldsymbol{C}_\mu^2(0) + \mathcal{O}(\omega^2), \tag{7.3.6}$$

where $\boldsymbol{C}_\mu(0) = [\mathbf{1} + \mathbf{C}\mathbf{D}^{-1}(0)]^{-1}\boldsymbol{C}$ is the static part of the electrochemical capacitance matrix (Christen and Büttiker, 1996). The dependence of the effective capacitance $\boldsymbol{C}_\mu(0)$ on the DOS is reasonable: small DOSs make it more difficult to charge the conductor. For large DOS, on the other hand, one recovers the geometrical capacitance. The frequency-dependent term of the capacitance can be viewed as being due to a damped plasma mode of the mesoscopic sample. The damping is a consequence of the connection of the sample to reservoirs.

(c) Low-frequency admittance

In order to derive the AC part of the admittance (7.2.18), we combine equations (7.2.33) and (7.3.6), together with the specific form of the local PDOS. The final expression for the admittance then reads

$$G_{\alpha\beta}(\omega) = G_{\alpha\beta}^{(0)} - \mathrm{i}\omega \sum_{k,l=1}^{N} \Delta_{\alpha k} C_{\mu,kl}(\omega)\Delta_{l\beta}. \tag{7.3.7}$$

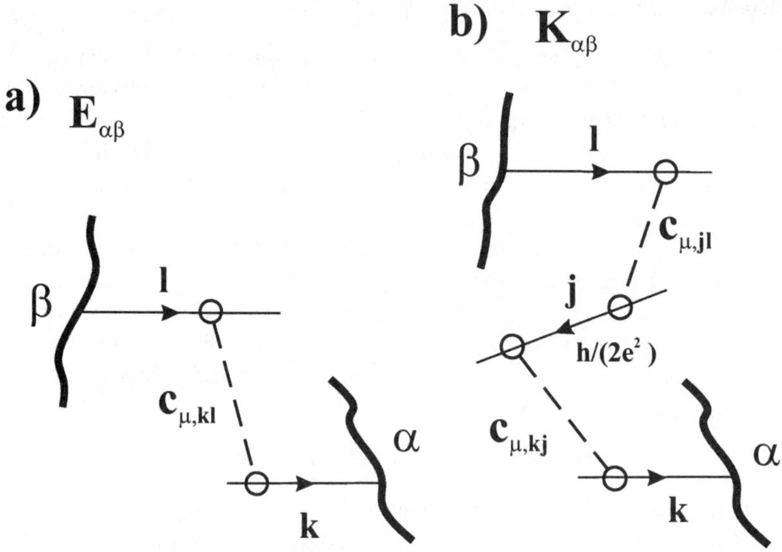

Figure 7.3 *Graphical illustration of the Coulomb interaction processes (dashed lines) and relaxation processes $h/2e^2$ which determine (a) the emittance $E_{\alpha\beta}$ and (b) the charge relaxation $K_{\alpha\beta}$.*

From the microreversibility properties of the $\Delta_{\alpha k}$ and the symmetry of the capacitance matrix it is again easy to prove current conservation, gauge invariance, and the Onsager–Casimir reciprocity relations. Results (7.3.6) and (7.3.7) yield the admittance to order $\mathcal{O}(\omega^2)$. The emittance $E_{\alpha\beta} = \sum_{kl} \Delta_{\alpha k} C_{\mu,kl}(0) \Delta_{l\beta}$ can be understood as the sum of all 'two-body' Coulomb interactions between edge states and/or gates (Christen and Büttiker, 1996a). In a pure capacitive arrangement where $\Delta_{k\alpha} = \Delta_{\alpha k}$, the emittance matrix is a capacitance matrix, i.e. diagonal elements are positive and off-diagonal elements are negative. As we will see below, this can change in the presence of edge channels which connect different reservoirs. The result for the second-order term, $K_{\alpha\beta} = (h/2e^2) \sum_{kjl} \Delta_{\alpha k} C_{\mu,kj}(0)\ C_{\mu,jl}(0) \Delta_{l\beta}$, can be interpreted as a 'three-body' interaction with an intermediate step via edge channel (or gate) j. The interpretations of $E_{\alpha\beta}$ and $K_{\alpha\beta}$ are graphically illustrated in Fig. 7.3.

(d) Two-terminal devices

Let us discuss the Corbino disk and the quantum Hall bar with m channels along each edge, as shown in Figs. 7.4(a) and 7.4(b), respectively, for a magnetic field B such that $m = 2$. The channels at the upper and lower edges are labelled by odd and even numbers, respectively. Our results now permit us to treat the case where different edge states are at different potentials. For the sake of clarity, however, we assume that the capacitances between edge channels at the same edge are very large such that they are at equal potentials. We thus take each set together to form a single m-channel conductor associated with only even or only odd labels. The geometric capacitance between the two sets of channels is denoted by C $(=\sum_{kl} C_{2k2l})$. While for the disk the reservoirs are capacitively coupled, for the bar all edge channels transmit charge between different reservoirs. Hence, the DC conductances of the disk and bar are $G^{(0)} = 0$ and $G^{(0)} = me^2/h$, respectively. The dynamic capacitance between the two sets of edge channels is $C_\mu(\omega) = C_\mu(0) + \mathrm{i}\omega R^{(q)}(C_\mu(0)^2)$ with a static electrochemical capacitance given by $1/C_\mu(0) = 1/C + (\sum_{\mathrm{even}} D_i)^{-1} + (\sum_{\mathrm{odd}} D_i)^{-1}$. The charge relaxation resistance is given by (Büttiker and Christen, 1996b)

$$R^{(q)} = \frac{h}{2e^2}\left(\frac{\sum_{i=\mathrm{odd}} D_i^2}{(\sum_{i=\mathrm{odd}} D_i)^2} + \frac{\sum_{i=\mathrm{even}} D_i^2}{(\sum_{i=\mathrm{even}} D_i)^2}\right). \tag{7.3.8}$$

Note that each quantum channel gives a contribution to the charge relaxation resistance which is proportional to $h/2e^2$ and not h/e^2. This is similar to the spreading resistance $h/2e^2m$ which occurs at the interface of an m channel wire with a reservoir. In our dynamical problem, we find such a Sharvin–Imry contact resistance (Imry, 1986; Landauer, 1987a) only for one-channel and only in the zero-temperature limit. If more than one channel connects to the same reservoir the charge relaxation resistance still scales with m but typically differs from $h/2e^2m$.

For the Corbino disk, the reservoirs are purely capacitively coupled which means that $\Delta_{k\alpha} = \Delta_{\alpha k} = 1$ if $\alpha = 1$ and k odd or $\alpha = 2$ and k even, and zero otherwise. This yields a low-frequency admittance

$$G(\omega) = -\mathrm{i}\omega C_\mu(\omega) = -\mathrm{i}\omega C_\mu(0) + \omega^2 R^{(q)} C_\mu^2(0). \tag{7.3.9}$$

As one expects, this result is equivalent to the low-frequency admittance of a mesoscopic capacitor with m channels on each side (Büttiker, Thomas and Prêtre, 1993a). On the other hand, for the bar $\Delta_{k\alpha}$ remains the same but $\Delta_{\alpha k} = 1$ if $\alpha = 1$ and k even or $\alpha = 2$ and k odd, but $\Delta_{\alpha k} = 0$ otherwise. The low-frequency admittance of the bar is thus given by

$$G(\omega) = m\frac{e^2}{h} + \mathrm{i}\omega C_\mu(\omega) = m\frac{e^2}{h} + \mathrm{i}\omega C_\mu(0) - \omega^2 R^{(q)} C_\mu^2(0). \tag{7.3.10}$$

While the emittance of the disk behaves like a capacitance, the emittance of the bar is inductive-like. Note that if all D_j are of the same order of magnitude, the charge relaxation resistance $R^{(q)}$ scales like $1/m$. Nevertheless, if C_μ scales with m (as is the case on a Hall plateau and for large C) one has $G(\omega) \propto m$. At a transition between two Hall plateaus, however, the DOS of the innermost edge channel becomes very large and one expects the charge relaxation resistance to be of the order of a single resistance quantum. Large oscillations as a function of the magnetic field in the propagation of voltage pulses have been observed by Zhitenev *et al.* (1993).

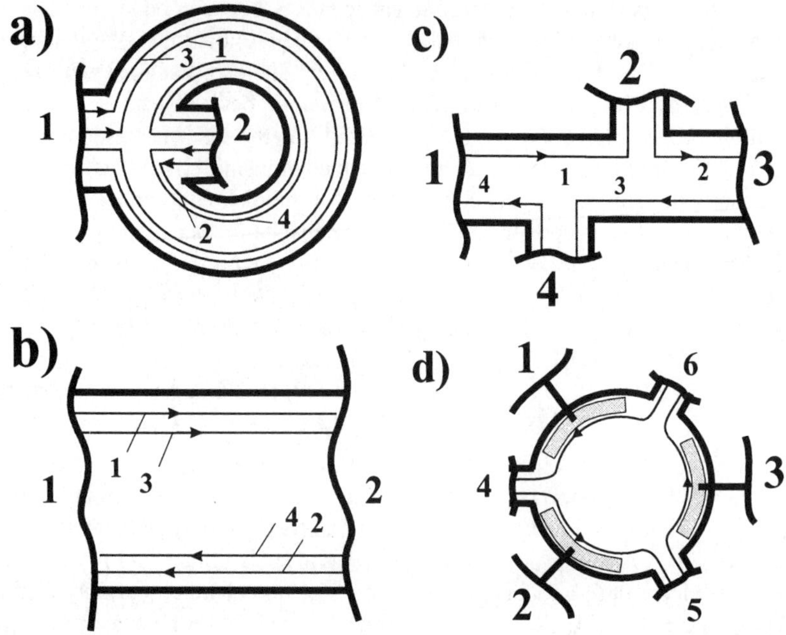

Figure 7.4 *Two-terminal samples: (a) Corbino disk and (b) quantum Hall bar. (c) Four-terminal Hall bar. (d) Disk with three symmetrical gates at the edge and fictitious voltage probes simulating dissipation.*

(e) Four-terminal Hall bar

The integer quantum Hall effect (von Klitzing, Dorda and Pepper, 1980) corresponds to the quantization of the Hall resistance and the vanishing of the longitudinal resistance of the ideal four-probe quantum Hall bar of Fig. 7.4(c) at zero frequency. Two of the contacts serve as voltage probes, whereas the two remaining contacts are used as source and sink for the current. With the help of our theory the DC results can be extended to the low-frequency case. If contacts 3 and 4 are the voltage probes, the longitudinal resistance is defined by $R_L = (\delta V_4 - \delta V_3)/\delta I_1$. The Hall resistance is defined by $R_H = (\delta V_2 - \delta V_4)/\delta I_1$, provided the contacts 2 and 4 are voltage probes. Let us assume that the elements $C_{\mu,kl}$ of the electrochemical capacitance matrix are known. For simplicity we consider a specific geometry (see Fig. 7.4(c)) which is such that $C_{\mu,12} = C_{\mu,24} = C_{\mu,34} = 0$ and calculate the low-frequency admittance (7.3.7). From the result and using equations (7.2.17) we find the longitudinal resistance

$$R_L = \mathrm{i}\omega \frac{h^2}{e^4} C_{\mu,13} + \frac{\omega^2}{2} \frac{h^3}{e^6} C_{\mu,13}(C_{\mu,23} - C_{\mu,14}). \tag{7.3.11}$$

The leading term of the longitudinal resistance is determined by the Coulomb coupling between the current circuit and the voltage circuit which are represented by edge channels 1 and 3, respectively. This is even true for finite $C_{\mu,24}$ (Christen and Büttiker, 1996a). Moreover, for the specific symmetry $C_{\mu,23} = C_{\mu,14}$ the dissipation also vanishes in $\mathcal{O}(\omega^2)$. Note that the second-order term can have either sign depending only on which of the electrochemical capacitances $C_{\mu,23}$ or $C_{\mu,14}$ is dominant. For the Hall resistance we find

$$R_H = \frac{h}{e^2} - \mathrm{i}\omega\frac{h^2}{e^4}C_{\mu,13} + \frac{\omega^2}{2}\frac{h^3}{e^6}C_{\mu,13}(C_{\mu,23} + C_{\mu,14}). \tag{7.3.12}$$

We mention that if we had chosen the symmetry (Christen and Büttiker, 1996a) $C_{\mu,13} = C_{\mu,24}$, the imaginary part of the Hall resistance would have vanished in first order of ω.

(f) Symmetry of dynamic capacitances

The symmetry of the emittance matrix has been discussed earlier by Büttiker (1993) and has been investigated experimentally by Chen *et al.* (1994) who reported on a structure in which some of the capacitance coefficients are not even functions of the magnetic field. They investigated the emittance between a contact of a two-terminal Hall bar and a third contact to a gate close to the edge of the bar. The authors explained their observation in terms of PDOS. It is clear from results (7.2.33) and (7.3.7) that the emittance matrix is in general not symmetric (an even function of the magnetic field) in the presence of transmission between two contacts (Christen and Büttiker, 1996a). However, one expects that the emittance (capacitance) of two separated gates is symmetric. In a recent experiment by Sommerfeld, van der Heijden and Peeters (1996) a striking asymmetry of the dynamic three-terminal capacitance of a Corbino disk was observed, whereas the two-terminal capacitance was symmetric. As an explanation, the authors mentioned the presence of edge magneto-plasmons. Here we show that this result follows from equation (7.2.28), if we allow for a spatially dependent edge potential (Christen and Büttiker, 1997). The approach presented here permits a discussion of a spatially nonuniform edge potential most easily if we introduce fictitious voltage probes along the edge of the sample. As discussed in section 7.2.7 such a sample is dissipative.

Consider a disk of a two-dimensional electron gas in a perpendicular magnetic field as shown in Fig. 7.4(d). Three equivalent gates (grey regions) connected to contacts 1–3 are located at the sample boundary in a symmetric way. We assume that the electron gas is at the first Hall plateau such that one edge channel runs along the sample boundary. Due to the fictitious voltage probes 4–6, this edge channel is divided into three pieces which each couple to the closest gate with a dynamic electrochemical capacitance $C_\mu(\omega) = C_\mu(0) + \mathrm{i}\omega(h/e^2)C_\mu^2(0) + \mathcal{O}(\omega^2)$. Neglecting all capacitances except those between gates and closest edge channel,

Sample \ Symmetry	$E_{12}=E_{21}$	$K_{12}=K_{21}$
	yes	yes
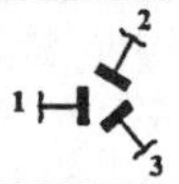	yes	yes
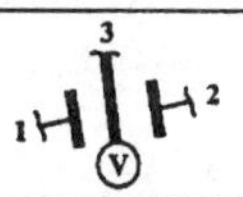	yes	no
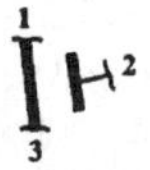	no	no

Table 7.1 *Symmetry properties of the dynamic capacitance of two- and three-terminal samples; the circle (V) indicates a voltage probe. After Christen and Buttiker (1997).*

the current–voltage relations read (Christen and Büttiker, 1997)

$$\begin{aligned}
\delta I_1 &= -\mathrm{i}\omega C_\mu(\omega)(\delta V_1 - \delta V_6)\\
\delta I_2 &= -\mathrm{i}\omega C_\mu(\omega)(\delta V_2 - \delta V_4)\\
\delta I_3 &= -\mathrm{i}\omega C_\mu(\omega)(\delta V_3 - \delta V_5)\\
\delta I_4 &= (e^2/h)(\delta V_4 - \delta V_6) - \mathrm{i}\omega C_\mu(\omega)(\delta V_6 - \delta V_1)\\
\delta I_5 &= (e^2/h)(\delta V_5 - \delta V_4) - \mathrm{i}\omega C_\mu(\omega)(\delta V_4 - \delta V_2)\\
\delta I_6 &= (e^2/h)(\delta V_6 - \delta V_5) - \mathrm{i}\omega C_\mu(\omega)(\delta V_5 - \delta V_3).
\end{aligned} \tag{7.3.13}$$

Using the conditions for voltage probes, $\delta I_4 = \delta I_5 = \delta I_6 = 0$, the elimination of δV_4, δV_5, and δV_6 leads to the reduced admittance matrix $\tilde{G}_{\alpha\beta}$ for contacts 1–3. We find

$$\tilde{G}_{12} = \tfrac{1}{3}(\mathrm{i}\omega C_\mu(0) - (h/e^2)\omega^2 C_\mu^2(0)) \tag{7.3.14}$$

$$\tilde{G}_{21} = \tfrac{1}{3}\mathrm{i}\omega C_\mu(0). \tag{7.3.15}$$

The other elements of $\tilde{G}_{\alpha\beta}$ are determined by the above-mentioned sum rules and the specific symmetry of the sample (i.e. $\tilde{G}_{12} = \tilde{G}_{23} = \tilde{G}_{31}$ and $\tilde{G}_{21} = \tilde{G}_{13} = \tilde{G}_{32}$). There is an obvious asymmetry $(\tilde{G}_{12})_B - (\tilde{G}_{12})_{-B} = -(h/3e^2)\omega^2 C_\mu^2(0)$. For the modulus it holds $|(\tilde{G}_{12})_B| > |(\tilde{G}_{12})_{-B}|$, which is in accordance with the experiment of Sommerfeld, van der Heijden and Peeters (1996). An additional elimination of contact 3 leads to a (scalar and thus symmetric) two-terminal admittance equal to $\frac{3}{2}$ times the admittance $\tilde{G}_{12}$ of equation (7.3.14).

We summarize the main results on the symmetry of capacitances in Table 7.1 for two- and three-terminal samples. To discuss the symmetry properties of the admittance under the reversal of a magnetic field we first have to distinguish the leading-order response (proportional to ω) and the higher-order response (proportional to ω^2 or higher powers of frequency). Secondly, we have to distinguish between different conducting topologies. The dynamic capacitance of two-terminal samples is always symmetric. This is a direct consequence of the Onsager–Casimir relations, gauge invariance, and current conservation. Purely capacitive and phase-coherent multiterminal samples exhibit dynamic capacitance coefficients which are even functions of the magnetic field. In the case of a spatially dependent potential along a conductor (e.g. in the presence of a voltage probe), the emittance of capacitive contacts is symmetric but higher-order terms in frequency are not. In a measurement for which at least one of the conductors is connected to two or more contacts and thus permits transmission from one of its contacts to another one measures an emittance which is not an even function of the magnetic field.

7.3.2 Low-frequency admittance of a quantum point contact

A quantum point contact (QPC) is a small constriction in a two-dimensional electron gas which allows the transmission of only a few conducting channels (van Wees *et al.*, 1988; Wharam *et al.*, 1988). We consider a symmetric QPC with two gates as shown in Fig. 7.5(a). and ask for its capacitance and low-frequency admittance. For simplicity we assume that both gates are at the same potential δV_3 and act thus as a single additional capacitance.

We discretize the QPC by defining to the left and right of the constriction two regions Ω_1 and Ω_2 with sizes of the order of the screening length. A voltage difference across the QPC induces polarization charges δq_1 and δq_2 which reside in Ω_1 and Ω_2, respectively. The charge on the two gates is then $\delta q_3 = -\delta q_1 - \delta q_2$. Each occupied subband j contributes with a transmission probability $T^{(j)}$ to the total transmission function $\mathcal{T}$ $(= T_{11}) = \sum_j T^{(j)}$. The DC conductance (7.2.15) is then given by $G_{11} = G_{22} = -G_{12} = -G_{21} = (2e^2/h)\mathcal{T}$. Clearly, $G_{\alpha\beta}$ vanishes whenever one of the indices corresponds to the gate index 3.

In order to calculate the capacitance and the emittance, we first consider a single channel. It is then straightforward to generalize the result to many channels. We denote by T and $R = 1 - T$ the transmission and reflection probabilities of this single channel. With the help of Fig. 7.1(b), the semiclassical local PDOS $D_{\alpha k\beta}$ can be constructed with simple arguments. For example, D_{211} is given by the transmission probability times the DOS of Ω_1 associated with carriers with positive velocity, hence $D_{211} = TD_1/2$. Since there are no states in Ω_1 associated with scattering from contact 2 back to contact 2 one concludes $D_{212} = 0$. With similar arguments one finds in the semiclassical case for the PDOS

$$D_{\alpha k\beta} = D_k(T/2 + \delta_{\alpha\beta}(R\delta_{\alpha k} - T/2)), \qquad \text{if } \alpha, \beta \neq 3, \tag{7.3.16}$$

where $D_1 = D_2$ holds due to symmetry. The local PDOS of the gate is $D_{\alpha 3\beta} =$

$\delta_{\alpha 3} D_3 \delta_{3\beta}$. Note that D_3 of the macroscopic gate is huge, and we will finally put $1/D_3 \to 0$. The geometric and electrochemical capacitance matrices can be written in the form

$$\mathbf{C}=C_0\begin{pmatrix} 1+\eta & -1 & -\eta \\ -1 & 1+\eta & -\eta \\ -\eta & -\eta & 2\eta \end{pmatrix}, \qquad \mathbf{C}_\mu=\begin{pmatrix} C_\mu & C_g - C_\mu & -C_g \\ C_g - C_\mu & C_\mu & -C_g \\ -C_g & -C_g & 2C_g \end{pmatrix}. \tag{7.3.17}$$

In equation (7.3.17) C_0 is the geometric capacitance between regions Ω_1 and Ω_2 and $C_0\eta$ accounts for the geometric capacitance between these regions and the gates. In the electrochemical capacitance matrix, C_μ is the capacitance if the presence of the gates can be neglected and C_g is the electrochemical gate capacitance. With the help of equation (7.2.32) and the characteristic potentials $u_{k\beta} = (D_{k\beta} - C_{\mu,k\beta})/D_k$ one finds

$$C_g = \frac{1}{(\eta C_0)^{-1} + D_1^{-1}}, \tag{7.3.18}$$

$$C_\mu = \frac{R + \eta(1 - T/2) + C_g D_1^{-1}}{C_0^{-1} + (2+\eta) D_1^{-1}}. \tag{7.3.19}$$

The emittance elements (7.2.33) become

$$\begin{aligned} E_{11} &= E_{22} = RC_\mu - DT^2/4 + C_g T/2 \\ E_{12} &= E_{21} = C_g - E_{11} \\ E_{13} &= E_{31} = E_{23} = E_{32} = -E_{33}/2 = -C_g \end{aligned} \tag{7.3.20}$$

where we introduced the total DOS $D = 2D_1$ of the QPC.

Let us discuss these results for certain limiting cases. First, consider a **three-terminal capacitor**, i.e. the QPC without transmission ($T = 0$). As one expects, the emittance equals the capacitance: $E_{\alpha\beta} = C_{\mu,kl}$ for $k = \alpha$, $l = \beta$. Next, consider the completely open channel, $T = 1$. One finds $C_{\mu,11} = C_{\mu,12} = C_g/2$; the sample then acts like a two-terminal capacitor consisting of the gate and a one-dimensional ballistic channel. The emittance is given by $E_{11} = -DC_g/(4\eta C_0)$ and is inductance-like. Note that the presence of the gate $\eta \neq 0$ diminishes the emittance of the ballistic channel. If the presence of the gates can be neglected, $\eta = 0$ (Christen and Büttiker, 1996b), the charge distribution across the quantum point contact is dipolar. The capacitance is given by

$$C_\mu = \frac{R}{C_0^{-1} + 2D_1^{-1}}. \tag{7.3.21}$$

From (7.3.20) we find that the emittance then has the simple form

$$E_{11} = -E_{12} = RC_\mu - \frac{D}{4}T^2. \tag{7.3.22}$$

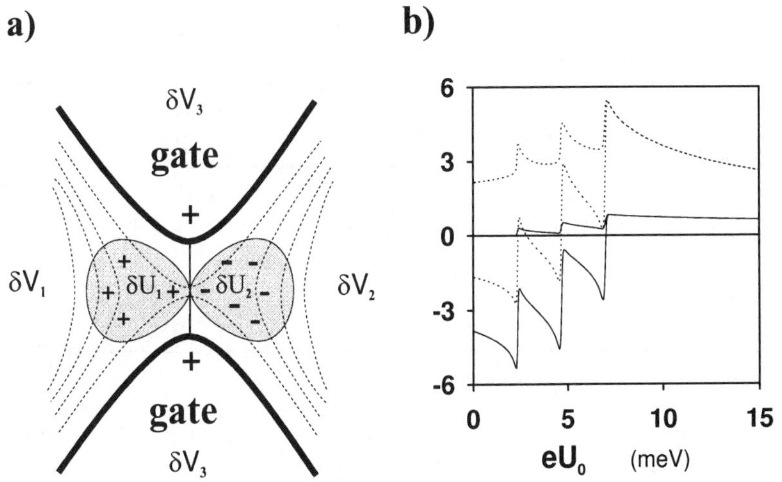

Figure 7.5 *(a) Quantum point contact with gate. The regions $\Omega_{1,2}$ at the left and right of the point contact can be charged. (b) Capacitance C_μ and emittance E_{11} of the QPC. Emittances and capacitances are in units of fF (see text).*

By increasing the transmission from 0 to 1, the capacitance decreases to 0, while the emittance becomes negative.

Let us now consider two additional quantities which can serve to characterize this system. Consider the case where $\delta V_3 = 0$, and denote the voltage difference across the QPC by $\Delta V = \delta V_1 - \delta V_2$. The difference $\Delta q = \delta q_1 - \delta q_2$ of the charges on the QPC must be proportional to the reflection constant. Indeed, we find

$$\frac{\Delta q}{\Delta V} = \frac{R(2+\eta)}{C_0^{-1} + (2+\eta)D_1^{-1}}. \tag{7.3.23}$$

For the same biasing condition the electrostatic voltage drop $\Delta U = \delta U_1 - \delta U_2$ is

$$\frac{\Delta U}{\Delta V} = \frac{D_1 R}{D_1 + (2+\eta)C_0}. \tag{7.3.24}$$

If the capacitive influence of the gates can be neglected ($\eta = 0$) the charge difference is equal to $2C_\mu$ and the electrostatic voltage drop across the quantum point contact is $(C_\mu/C_0)\Delta V$. Since C_μ is proportional to the reflection coefficient R this voltage drop is also proportional to the reflection coefficient. If the coupling to the gate is predominant ($\eta \to \infty$) the charge difference is determined by the total DOS of the wire ($\Delta q/\Delta V = D/2$), and the voltage drop ΔU vanishes. For the case where the capacitance across the quantum point contact can be neglected but the gate capacitance is finite ($C_0 \to 0$, $\eta \to \infty$ such that $\eta C_0 \to C$), the electrochemical gate capacitance is given by $C_g^{-1} = C^{-1} + (D/2)^{-1}$. The

Table 7.2 *Gate capacitance and capacitance, emittance, polarization, and voltage drop of the quantum point contact with a grounded gate for certain limiting cases.*

	$\eta \to 0$ $C_0 =$ constant	$\eta \to \infty$ $C_0 =$ constant	$\eta, C_0^{-1} \to \infty$ $\eta C_0 = C$
C_g	0	D_1	$1/(D_1^{-1} + C^{-1})$
C_μ	$R/(C_0^{-1} + 2D_1^{-1})$	$D_1(1 - T/2)$	$(1 - T/2)C_g$
E_{11}	$RC_\mu - D_1 T^2/2$	RD_1	$RC_g - (C_g/C)D_1 T^2/2$
$\Delta q/\Delta V$	$2C_\mu$	RD_1	RC_g
$\Delta U/\Delta V$	C_μ/C_0	0	RC_g/C

charge difference is determined by RC_g and the electrostatic voltage drop is $\Delta U = R(C_g/C)\Delta V$. If, in addition, the limit $C \to 0$ is taken, the leads are electroneutral: the charge difference vanishes and the potential difference is as in Landauer's (1970) discussion given by $\Delta U = R\Delta V$. To help the reader to gain an overview of these results we have collected them in Table 7.2. The results collected in this table emphasize the profound effect that a gate has not only on the global properties such as the capacitances and emittances but especially on the local properties such as the charge and voltage distributions.

To generalize these results to the many-channel case, we just mention that the local PDOS (7.3.16) $D_k^{(j)}$ of each individual channel simply add. Hence, the local DOS becomes $D_k = \sum_j D_k^{(j)}$ and $T = 1 - R$ is now an average transmission probability defined by $T \equiv T_k = D_k^{-1} \sum_j D_k^{(j)} T^{(j)}$ where ($k = 1, 2$). Note that the average transmission probability T ($\neq \mathcal{T}$) has nothing to do with the DC conductance. As a specific example we consider a quadratic potential $U(x) = U_0(b^2 - x^2)/b^2$ if $|x| \le b$, and $U(x) = 0$ if $b < |x| \le l$. The transmission probabilities $T^{(j)}$ and the PDOS can be calculated analytically from a semiclassical analysis (Miller and Good, 1953). For simplicity, we assume a constant electrostatic capacitance $C_0 = 1 fF$ between Ω_1 and Ω_2 and a fixed number of occupied channels in these regions. The only parameter to be varied is the potential height U_0. In practice, U_0 is tuned with the help of the gates, but we consider it here as an independent control parameter. We assume that no additional channels enter into the regions Ω_k during the variation of U_0. In Fig. 7.5(b) we show the results for a constriction with $b = 500$ nm, $l = 550$ nm, and with three equidistant channels separated by $E_F/3 = \frac{7}{3}$ meV. Whenever a Fermi level of a subband crosses the barrier top, the transmission decreases and a channel gets closed which leads to the well-known DC conductance steps. In synchronism with these steps we also find steps in the capacitance and the emittance. The dotted and solid curves correspond to a gate with coupling $\eta = 0$ and $\eta = 1$, respectively. The upper and lower curves of the same type represent the capacitance C_μ and emittance E_{11}, respectively. For the two-terminal QPC

($\eta = 0$) the capacitance vanishes and the emittance is negative for small U_0 where all channels are open. At each step, the capacitance and the emittance increases and eventually merges when all channels are closed. Due to a weak logarithmic divergence of the WKB DOS at particle energies $E = eU_0$ (where WKB is not appropriate), the emittance shows steep edges between the steps. In the presence of the gate ($\eta \neq 0$), the curves are shifted upwards due to a capacitive contribution of the gate.

7.3.3 Nonlinear conduction and emittance of the resonant tunnelling barrier

Let us now apply our theory to the resonant tunnelling barrier sketched in Fig. 7.6(a). For a comparison of different theoretical approaches we refer the reader to Bruder and Schoeller (1994) and Jauho, Wingreen and Meir (1994). This conductor consists of three discrete potential regions denoted by indices 0, 1 and 2 which correspond to the well, the left side of the barriers, and the right side of the barriers, respectively. We assume fully screened charges. Consider a single resonant level with energy $E_r + eU_0$ which is clearly separated from other resonant levels, from the band bottoms and from the barrier tops. The linear conductance of such a structure is given by $G_{11} = (2e^2/h)T$ where T is the transmission probability of the double-barrier structure. Linear response is only valid if the bias $\Delta V = V_1 - V_2$ is of a much smaller amount than Γ/e, where Γ is the width of the resonant level. To find the nonlinear characteristic $I_1 = I(\Delta V)$, we restrict ourselves to near-resonant conditions where the transmission probability can be approximated by the Breit–Wigner formula $T(E, U_0) = \Gamma_1\Gamma_2/|\Delta|^2$ with $|\Delta|^2 = (E - E_r - eU_0)^2 + \Gamma^2/4$. In this range the nonlinearity is solely due to the resonance. In particular, we assume that Γ_1 and Γ_2, and thus the width $\Gamma = \Gamma_1 + \Gamma_2$ of the resonant level and the asymmetry, $\Delta\Gamma = \Gamma_1 - \Gamma_2$, are constant. The injectivities are (Büttiker, Thomas, and Prêtre, 1994) $D^{(i)}_{0\alpha}(E, U_0) = (e^2/2\pi)\Gamma_\alpha/|\Delta|^2$. The total DOS in the well is $D_0 = D^{(i)}_{01} + D^{(i)}_{02}$. To be specific we investigate the case where the relation between the electrostatic potential shift ΔU_0 and the voltage shifts $V_\alpha (\equiv \Delta U_\alpha)$ is obtained from a local charge-neutrality condition

$$\int_{-\infty}^{\mu_1} D^{(i)}_{01}(E, U_0)\, \mathrm{d}E + \int_{-\infty}^{\mu_2} D^{(i)}_{02}(E, U_0)\, \mathrm{d}E - \int_{-\infty}^{\mu^{eq}} D_0(E, U_0^{(eq)})\, \mathrm{d}E \equiv 0, \tag{7.3.25}$$

which can be integrated analytically due to the simple Breit–Wigner form of the integrands. The result then determines $W = \Delta U_0 - (V_1 + V_2)/2$ which depends on ΔV only. Equation (7.2.13) yields for the current

$$\begin{aligned} I &= \frac{4e}{h}\frac{\Gamma_1\Gamma_2}{\Gamma}\left(\arctan\left[\frac{\Delta E - e(W - \Delta V/2)}{\Gamma/2}\right]\right. \\ &\quad \left. -\arctan\left[\frac{\Delta E - e(W + \Delta V/2)}{\Gamma/2}\right]\right), \end{aligned} \tag{7.3.26}$$

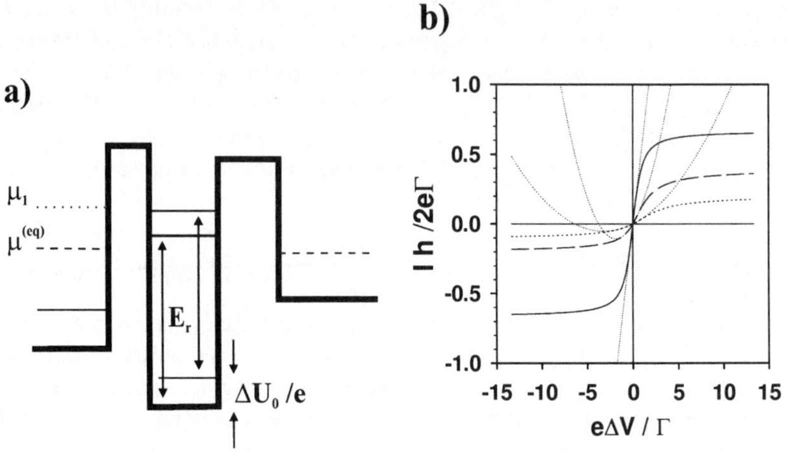

Figure 7.6 *(a) Change of the potential landscape of a resonant tunnelling barrier with a single level due to a voltage* $V_1 = (\mu_1 - \mu_{eq})/e$ *and* $V_2 = 0$. *(b) Nonlinear current–voltage characteristic with* $\Gamma_1 = 0.5$ *meV,* $\Gamma_2 = 1$ *meV,* $\Delta E = 0$ *(solid),* $\Delta E = -\Gamma$ *(dashed),* $\Delta E = -2\Gamma$ *(dotted). The light dotted curves indicate the quadratic approximation.*

where $\Delta E = \mu^{(eq)} - (eU_0^{(eq)} + E_r)$ is the equilibrium distance between the Fermi energy and the resonance. Note that the spin degeneracy is included by a factor of 2 in equation (7.3.26). Without considering the self-consistent shift ΔU_0 of the band bottom one would get a wrong result which is not gauge invariant, i.e. depends not only on the voltage difference. The current given by equation (7.3.26) saturates at a maximum value proportional to $\pi/2 - \arctan(2\Delta E/\Gamma)$. The conduction is optimal for $\Delta E = 0$ and $\Gamma_1 = \Gamma_2$ when $I = 2(e/h)\Gamma\arctan(e\Delta V/\Gamma)$. In Fig. 7.6(b) we have plotted the current–voltage characteristic for an asymmetry $\Delta\Gamma/\Gamma = -\frac{1}{3}$ and for various values of ΔE. Due to the complete screening, the resonant level floats up or down to keep the charge in the well constant. An expansion of the current yields $G_{111} = -(e^3/h)(\Delta\Gamma/\Gamma)\partial_E T$ which is in accordance with (7.2.34) (light dotted curves). The case of incomplete screening can similarly be treated with our approach. At large voltages, the resonance can then eventually fall below the conductance band bottom of the injecting reservoir as is known from the semiconductor double-barrier structures. In general, even an elastically symmetric resonance can be rectifying if the screening is asymmetric. The emittance of the resonant tunnelling barrier has been discussed by Büttiker, Thomas and Prêtre (1994) and Büttiker and Christen (1996a). In a Thomas–Fermi approach the characteristic potentials $u_{0\alpha}$ in the well are determined by $u_{0\alpha} = D_{0\alpha}^{(i)}/D_0 = \Gamma_\alpha/\Gamma$, which leads to an emittance

$$E_{11} = \frac{D_{10}^{(e)} D_{02}^{(i)}}{D_0} \left(\frac{|\Delta|^2 - \Gamma^2/2}{|\Delta|^2} \right). \qquad (7.3.27)$$

Note that the absence of a magnetic field implies the equality of injectivity and emissivity, $D_{0\alpha}^{(i)} = D_{0\alpha}^{(e)}$. For a symmetric resonant tunnelling barrier, equation (7.3.27) simplifies and is given by $E_{11} = (D/4)(R - T)$. At resonance the emittance is negative, reflecting kinetic (inductive) behaviour, it is zero at half width of the resonance, and it is positive (capacitive) if the Fermi level is more than a half width above (below) the resonant energy. Clearly, Thomas–Fermi screening is not very realistic for such a conductor. In semiconductor double barriers screening is not local but long range. Moreover, band bending outside the double barrier might play a decisive role. We refer the reader to Brown, Parker, and Sollner (1989) and Vanbesien *et al.* (1992) for experimental results on the admittance of resonant double barriers. Our considerations indicate the character of the results a more realistic treatment might yield and hopefully stimulate work in that direction. We have also neglected single-charge effects. Such effects will play an important role in quantum dots: a self-consistent discussion on AC transport is given by Bruder and Schöller (1994). A self-consistent treatment of the nonlinear I–V characteristic of a quantum dot including single-charge effects is provided by Stafford (1996). The key point is that a discussion which describes the interaction with a fixed Hubbard U parameter will not lead to a self-consistent result. It is important to remember that the interaction parameter U is a consequence of the long-range Coulomb interaction and in general depends both on the voltages applied to the contacts of the sample as well as on the gate voltages.

7.3.4 The metallic diffusive conductor

Let us consider a mesoscopic metallic diffusive conductor connecting two reservoirs. Metallic diffusive means that although the inelastic mean-free path is still larger than the sample size L, the elastic scattering length l is much smaller, $l \ll L$. The ensemble-averaged conductance $\langle G_0 \rangle$, in the metallic limit, is given by the Drude result and is proportional to $(e^2/h)N(l/L)$, where N is the number of quantum channels. The admittance of a metallic diffusive Aharonov–Bohm ring was investigated experimentally by Pieper and Price (1994) and was discussed theoretically within the discrete potential model by Liu *et al.* (1994). Below we present a discussion by Blanter and Büttiker (1997) of the emittance using a continuous potential. This discussion shows that the emittance of a metallic diffusive conductor vanishes and corrects an earlier result by Büttiker and Christen (1996a).

In order to derive an expression for the average emittance, we mention that in a metallic conductor charge accumulations are screened over a Thomas–Fermi screening length λ_{TF} (apart from miniscule and more subtle Friedellike long-range effects (Levinson, 1989)). We assume that the local Thomas–Fermi approximation holds and calculate the emittance (7.2.21) by using equation (7.2.11) for the case where $\lambda_{TF} \sim \lambda_F \ll l$. There are no electric field lines outside the conductor. First, we determine the ensemble-averaged local PDOS. Suppose that the wire is connected at $x = 0$ and $x = L$ to much wider metallic banks (reservoirs). An increase in the chemical potential in the left reservoir increases (at constant electric potential) the density at the entrance of the wire by $\mathrm{d}n(0) = (\mathrm{d}n/\mathrm{d}E)\mathrm{d}\mu_1$. Here $(\mathrm{d}n/\mathrm{d}E)$ is the local DOS of the metallic wire. Thus, the injectivity of the left reservoir is a solution of the stationary diffusion equation with the boundary conditions $\mathrm{d}n(0) = (\mathrm{d}n/\mathrm{d}E)\mathrm{d}\mu_1$ and $\mathrm{d}n(L) = 0$. The diffusion constant is $D = \frac{1}{3}v_F l_e$ and the solution is $\mathrm{d}n(x, 1)/\mathrm{d}E = (\mathrm{d}n/\mathrm{d}E)(1 - x/L)$. The emissivity is by reciprocity equal to the injectivity $\mathrm{d}n(1, x)/\mathrm{d}E = (\mathrm{d}n/\mathrm{d}E)(1 - x/L)$. The PDOS $\mathrm{d}n(1, x, 1)/\mathrm{d}E$ of carriers that have been injected by the left contact, reach x and are reflected back to contact 1 is equal to the injectivity at x multiplied by the probability of a carrier returning to contact 1. The probability of a carrier returning can be found from the diffusion equation, assuming that a current is injected at x and demanding that the density vanishes at $x = 0$ and $x = L$. The probability is $(1 - x/L)$ and thus $\mathrm{d}n(1, x, 1)/\mathrm{d}E = (\mathrm{d}n/\mathrm{d}E)(1 - x/L)^2$. The probability that a carrier that has reached x will diffuse forward into contact 2 is x/L. Thus, the partial density $\mathrm{d}n(2, x, 1)/\mathrm{d}E = (\mathrm{d}n/\mathrm{d}E)(1 - x/L)(x/L)$. Integrating these results over the volume of the wire determines the global PDOS:

$$\langle \mathrm{d}N_{11}/\mathrm{d}E \rangle = \tfrac{1}{3}\langle \mathrm{d}N/\mathrm{d}E \rangle \tag{7.3.28}$$

and

$$\langle \mathrm{d}N_{21}/\mathrm{d}E \rangle = \tfrac{1}{6}\langle \mathrm{d}N/\mathrm{d}E \rangle. \tag{7.3.29}$$

The corresponding results for carriers injected from the right reservoir are obtained from the above results by permuting indices 1 and 2 and by replacing $1 - x/L$ by x/L and replacing x/L by $(1 - x/L)$. The contribution due to the long-range Coulomb interaction is determined by the local DOS and by the injectivities and emissivities. Using equations (7.2.21) and (7.2.11) we find that the chemical contribution to the emittance is exactly cancelled by the electrically induced contribution. In a diffusive wire the frequency-dependent ensemble-averaged potential drops linearly along the conductor and thus creates no net charge. Thus, the ensemble-averaged emittance of a metallic diffusive conductor vanishes (Blanter and Büttiker, 1997)

$$\langle E \rangle = 0. \tag{7.3.30}$$

It can be shown (Blanter and Büttiker, 1997) that the ensemble-averaged dynamic conductance of a metallic diffusive conductor is independent of frequency over a large range of frequencies. Only when the frequency becomes comparable with the inverse of the Drude relaxation time $\tau = l_e/v_F$ can we expect a significant frequency dependence of the conductance.

7.4 CONCLUSION

Let us allude to a number of additional works which we have not reviewed in this chapter. Computational work might eventually be of importance. The works of Liu *et al.* (1994) and Wang and Guo (1996) mark the beginning. The electrochemical nature of the capacitance with its dependence on the local DOS should also exhibit an Aharonov–Bohm effect if one of the capacitors has the shape of a ring with an Aharonov–Bohm flux through its hole. A general analysis of Aharonov–Bohm effects in capacitances based on a Hartree approach is provided by Büttiker (1994). A more specific prediction of such effects is possible for rings coupled to quantum dots (Büttiker and Stafford, 1996). This requires an extension of the theory of Coulomb blockade to treat the charge transfer quantum coherently. For such systems the capacitance exhibits very narrow flux-sensitive spikes in the capacitance. Much attention in mesoscopic physics has been devoted to characterizing the fluctuations away from the average conductance (in two probe geometries) or the voltage fluctuations away from the average voltage (in multiprobe geometries) as a function of some external parameter (magnetic field, Fermi energy). Similarly the admittance coefficients, capacitances and emittances, will exhibit such fluctuations. Recently, Gopar, Mello and Büttiker (1996) investigated the fluctuations in the capacitance of a chaotic cavity coupled to a single lead and coupled capacitively to a back gate. For the case of a single-channel lead, the distribution functions for the electrochemical capacitance have been derived analytically for Gaussian, unitary and symplectic ensembles. Brouwer and Büttiker (1997) analysed the admittance of a chaotic cavity coupled to two leads with many channels and evaluated the covariance of the conductance and emittance.

Reznikov *et al.* (1995) experimentally investigated the suppression of shot noise in a ballistic quantum point contact. From their results one concludes that the theory of noise in mesoscopic conductors (Büttiker, 1992b) should be generalized to include self-consistently Coulomb interactions.

Schöller (1996) investigated the dynamical capacitance of a one-dimensional wire coupled purely capacitively to the external circuit. In the presence of a gate, Schöller considered the entire capacitance matrix and found an interesting structure due to the one-dimensional plasma modes of the wire. This brief list indicates that there is considerable room for further experimental and theoretical work.

In this chapter we have presented a general theory for the low-frequency admittance and the weakly nonlinear transport of open (i.e. connected to different reservoirs) mesoscopic conductors. We have emphasized the application of this theory to a number of systems of current interest, such as quantized Hall conductors, quantum point contacts, resonant tunnelling barriers and metallic diffusive wires. Our emphasis has been to derive results, which even so they might not be realistic in detail, nevertheless, capture the essential physics. Our results should be useful both for comparison with additional theoretical and experimental work. The work presented here can be extended in a number of directions, for example high-frequency transport, strong nonlinearities, and nonlinear AC transport. The attempt to attack these questions is a quite challenging undertaking but very likely also full of rewards.

ACKNOWLEDGEMENT

This work was supported by the Swiss National Science Foundation under grant no 43966.

REFERENCES

Al'tshuler, B. L. and Khmelnitskii, D. E. (1985) *JETP Lett.* **42**, 359.
Anantram, M. P. and Datta, S. (1995) *Phys. Rev.* B **51**, 7632.
Aronov, I. E., Zagoskin, A. M. and Jonson, M. (1995) *Solid State Commun.* **95**, 647.
Beenakker, C. W. J. and van Houten, H. (1991) *Quantum Transport in Semiconductor Nanostructures* (eds H. Ehrenreich and D. Turnbull), Academic Press, New York.
Blanter, Ya. M. and Büttiker, M. (1997) unpublished.
Brouwer P. W. and Beenakker, C. W. J. (1997) *Phys. Rev.* B **55**, 4695.
Brouwer, P. W. and Büttiker, M. (1997) *Europhys. Lett.* **37**, 441–6.
Brown, E. R., Parker, C. D. and Sollner, T. C. L. G. (1989) *Appl. Phys. Lett.* **54**, 934.
Bruder, C. and Schöller, H. (1994) *Phys. Rev. Lett.* **72**, 1076.
Buot, F. A. (1993) *Phys. Rep.* **234**, 73.
Büttiker, M. (1986a) *Phys. Rev. Lett.* **57**, 1761.
Büttiker, M. (1986b) *Phys. Rev.* B **33**, 3020.
Büttiker, M. (1988a) *Phys. Rev.* B **38**, 9375.
Büttiker, M. (1988b) *IBM J. Res. Develop.* **32**, 317.
Büttiker, M. (1992a) *Semiconductors and Semimetals* (ed. M. Reed), Academic Press, New York, Vol. 35, p. 191.
Büttiker, M. (1992b) *Phys. Rev.* B **46**, 12 485.
Büttiker, M. (1993) *J. Phys.: Condens. Matter* **5**, 9631.
Büttiker, M. (1994) *Physica Scripta*, T **54**, 104–10.
Büttiker, M. and Christen, T. (1996a) *Quantum Transport in Semiconductor Submicron Structures* (ed. B. Kramer), Kluwer Academic Publishers, Dordrecht; NATO ASI Series, Vol. 326, pp. 263–91.
Büttiker, M. and Christen, T. (1996b) Dynamic conductance in quantum Hall systems *The Application of High Magnetic Fields in Semiconductor Physics* (ed. G. Landwehr) (Springer, Heidelberg, 1997) p. 105.
Büttiker, M., Imry, Y., Landauer R. and Pinhas, S. (1985) *Phys. Rev.* B **31**, 6207.
Büttiker, M., Prêtre, A. and Thomas, H. (1993b) *Phys. Rev. Lett.* **70**, 4114.
Büttiker, M. and Stafford, C. A. (1996) *Phys. Rev. Lett.* **76**, 495.
Büttiker, M., Thomas, H. and Prêtre, A. (1993a) *Phys. Lett.* A **180**, 364.
Büttiker, M., Thomas, H. and Prêtre, A. (1994) *Z. Phys.* B **94**, 133.
Castaño, E. and Kirczenow, G. (1990) *Phys. Rev.* B **41**, 3874.
Chen, W., Smith, T. P., Büttiker, M. and Shayegen, M. (1994) *Phys. Rev. Lett.* **73**, 146.
Chklovskii, D. B., Shklovskii, B. I. and Glazman, L. I. (1992) *Phys. Rev.* B **46**, 4026.
Christen, T. (1997) *Phys. Rev.* B **55**, 7606.

Christen, T. and Büttiker, M. (1996a) *Phys. Rev.* B **53**, 2064.

Christen, T. and Büttiker, M. (1996b) *Phys. Rev. Lett.* **76**, 143.

Christen, T. and Büttiker, M. (1996c) *Europhys. Lett.* **35**, 523.

Christen, T. and Büttiker, M. (1997) *Phys. Rev.* B **55**, R1946.

Cooper, N. R. and Chalker, J. T. (1993) *Phys. Rev.* B **48**, 4530.

Datta, S. (1995) *Electronic Transport in Mesoscopic Conductors*, Cambridge University Press, Cambridge.

de Jong, M. J. M. and Beenakker, C. W. J. (1996) *Physica* A **230**, 219.

Delahaye, F. (1994) *Metrologia* **31**, 367.

Field, M., Smith, C. G., Pepper, M., Brown, K. M., Linfield, E. H., Grimshaw, M. P., Ritchie, D. A. and Jones, G. A. C. (1996) *Phys. Rev. Lett.* **77**, 350.

Fomin, N. V. (1967) *Sov. Phys.–Solid State* **9**, 474.

Frenkel, J. (1930) *Phys. Rev.* **36**, 1604.

Fu, Y. and Dudley, S. C. (1993) *Phys. Rev. Lett.* **71**, 466.

Gasparian, V. M., Christen, T. and Büttiker, M. (1996) *Phys. Rev.* A **54**, 4022.

Glazman, L. I. and Khaetskii, A. V. (1989) *Europhys. Lett.* **9**, 263.

Gopar, V. A., Mello, P. A. and Büttiker, M. (1996) *Phys. Rev. Lett.* **77**, 3005.

Halperin, B. I. (1982) *Phys. Rev.* B **25**, 2185.

Hofbeck, K., Genzer, J., Schomburg, E., Ignatov, A. A., Renk, K. F., Pavlev, D. G., Koschurinov, Y., Melzer, B. and Ivanov, S. (1996) *Phys. Lett.* A **218**, 349.

Imry, Y. (1986) *Directions in Condensed Matter Physics* (eds G. Grinstein and G. Mazenko), World Scientific, Singapore, p. 101.

Jacoboni, C. and Price, P. (1993) *Phys. Rev. Lett.* **71**, 464.

Jauho, A. P., Wingreen, N. S. and Meir, Y. (1994) *Phys. Rev.* B **50**, 5528.

Kluksdahl, N. C., Kriman, A. M., Ferry, D. K. and Ringhofer, C. (1989) *Phys. Rev.* B **39**, 7720.

Komiyama, S. and Hirai, H. (1996) *Phys. Rev.* B **54**, 2067.

Kouwenhoven, L. P., van Wees, B. J., Harmans, C. J. P. M., Williamson, J. G., van Houten, H., Beenakker, C. W. J., Foxon, C. T. and Harris, J. J. (1989) *Phys. Rev.* B **39**, 8040.

Kouwenhoven, L. P., Jauhar, S., Orenstein, J., McEuen, P. L., Nagamune, J., Motohisa, J. and Sakaki, H. (1994) *Phys. Rev. Lett.* **73**, 3443.

Landauer, R. (1957) *IBM J. Res. Develop.* **1**, 223.

Landauer, R. (1970) *Philos. Mag.* **21**, 863.

Landauer, R. (1987a) *Z. Phys.* **68**, 217.

Landauer, R. (1987b) in *Nonlinearity in Condensed Matter* (eds A. R. Bishop *et al.*), Springer, Berlin, p. 2.

Levinson, I. B. (1989) *Sov. Phys.–JETP* **68**, 1257.

Lier, K. and Gerhardts, R. R. (1994) *Phys. Rev.* B **50**, 7757.

Liu, D. Z., Hu, B. Y.-K., Stafford, C. A. *et al.* (1994) *Phys. Rev.* B **50**, 5799.

Luryi, S. (1988) *Appl. Phys. Lett.* **52**, 501.

McDonald, A. H., Rice, T. M., Brinkman, W. F. *et al.* (1983) *Phys. Rev.* B **28**, 3648.

Mikhailov, S. A. and Volkov, V. A. (1995) *Pis'ma Zh. Éksp. Teor. Fiz.* **61**, 508 [*JETP Lett.* **61**, 524].

Miller, S. C. and Good, R. M. (1953) *Phys. Rev.* **91**, 174.

Pastawski, H. (1992) *Phys. Rev.* B **46**, 4053

Patel, N. K., Martin-Moreno, L., Pepper, M., Newbury, R., Frost, J. E. F., Ritchie,

D. A., Jones, G. A. C., Janssen, J. T. M. B., Singleton, J. and Perenboom, J. A. A. J. (1990) *J. Phys.: Condens. Matter* **2**, 7247.

Pieper, J. B. and Price, J. C. (1994) *Phys. Rev. Lett.* **72**, 3586.

Prêtre, A., Thomas, H. and Büttiker, M. (1996) *Phys. Rev.* B **54**, 8130.

Reznikov, M., Heiblum, M., Shtrikman, H. and Mahalu, D. (1995) *Phys. Rev. Lett.* **75**, 3340.

Schöller, H. (1996) unpublished correspondence.

Shepard, K. L., Roukes, M. L. and van der Gaag, B. P. (1992) *Phys. Rev. Lett.* **68**, 2660.

Smith III, T. P., Wang W. J. and Stiles, P. J. (1986) *Phys. Rev.* B, **34**, 2995.

Sommerfeld, P. K. H., van der Heijden, R. W. and Peeters, F. M. (1996) *Phys. Rev.* B **53**, 13 250.

Stafford, C. A. (1996) *Phys. Rev. Lett.* **77**, 2770.

Taboryski, R., Geim, A. K., Persson, M. and Lindelof, P. E. (1994) *Phys. Rev.* B **49**, 7813.

Talyanskii, V. I., Mace, D. R., Simmons, M. Y., Pepper, M., Churchill, A. C., Frost, J. E. F., Ritchie, D. A. and Jones, G. A. C. (1995) *J. Phys.: Condens. Matter* **7**, L435.

Vanbesien, O., Sadaune, V., Lippens, D., Vinter, B., Bois, P. and Nagle, J. (1992) *Microwave and Optical Technology Lett.* **5**, 351.

van Wees, B. J., van Houten, H., Beenakker, C. W. J., Williamson, J. G., Kouwenhoven, L. P., van der Marel, D. and Foxon, C. T. (1988) *Phys. Rev. Lett.* **60**, 848.

von Klitzing, K., Dorda, G. and Pepper, M. (1980) *Phys. Rev. Lett.* **45**, 494.

Wang, J. and Guo H. (1996) *Phys. Rev.* B **54**, R11 090.

Wharam, D. A., Thornton, T. J., Newbury, R., Pepper, M., Ahmed, H., Frost, J. E. F., Hasko, D. G., Peacock, D. C., Ritchie, D. A. and Jones, G. A. C. (1988) *J. Phys. C: Solid State Phys.* **21**, L209.

Yahel, E., Orgad, D. and Palevski, A. (1996) *Phys. Rev. Lett.* **76**, 2149.

Zagoskin, A. M. and Shekhter, R. I. (1994) *Phys. Rev.* B **50**, 4909.

Zhitenev, N. B., Haug, R. J., v. Klitzing, K and Eberl, K. (1993) *Phys. Rev. Lett.* **71**, 2292.

CHAPTER 8

Transport in systems with chaotic dynamics: Lateral superlattices

M. Suhrke and P. Rotter

Institut für Theoretische Physik, Universität Regensburg, D-93040 Regensburg, Germany

8.1 INTRODUCTION

The theoretical investigation of electrons subject to both a periodic electrostatic potential and a magnetic field has a long tradition [1]. On the one hand, this situation can be realized in each crystalline solid. On the other hand, it constitutes a highly nontrivial system due to breaking the rotational symmetry of the magnetic field by the (discrete) translational invariance of the periodic potential. A simple version of the problem is obtained in the tight-binding approximation for the electrons or in perturbation theory for very weak periodic potential. In that case Schrödinger's equation reduces to Harper's equation [2], the solution of which has been visualized in the form of the well-known Hofstadter butterfly [3]. The spectrum then critically depends on the number of magnetic flux quanta threading one unit cell of the lattice and becomes a fractal object, a Cantor set, if this number is an irrational one [4]. Experimental studies of such systems promise to provide a confirmation of the theoretical predictions and perhaps new interesting insights into the consequences of the spectral properties on their physical behaviour. However, they turned out to be impossible with conventional crystals because the necessary magnetic fields of several 1000 T for the observation of the peculiarities of the spectrum are still out of the reach of present-day technology.

It has also been noted that classical chaos may play an important role in these systems if one goes beyond Harper's equation [5]. The extended Sinai billiard as a simple realization of a periodic potential (originally without a magnetic field) is one of the prominent examples of classically chaotic systems [6].

The understanding of the quantum-mechanical counterparts of classically chaotic systems has received much interest in the past and significant progress has been achieved during recent decades [7, 8]. Classical chaos manifests itself in correlations in the quantum-mechanical spectrum that can be analysed with statistical methods. Although there is no analytical proof, a variety of numerical investigations demonstrate that the spectral correlations are to a certain extent universal and can be well described by the predictions of random matrix theory (RMT) [9] originally proposed for the description of the spectra of complex nuclei. Up until now most of the investigations have been done for closed systems with discrete energy spectra. For systems with extended states where the classical motion is unrestricted in space, however, the spectrum is not discrete. Particularly, in periodic systems energy bands are formed and the spectrum is absolutely continuous. Nevertheless, they should also exhibit the influence of classical chaos in the spectrum. Some work has been done in this respect by investigating the spectral correlations of a system described by Harper's equation [10]. These calculations have been extended to incorporate classical chaos [11, 12]. Moreover, the predictions of RMT have been found to be fulfilled in the band structure of real crystals such as silicon [13].

The interest in the problem of electrons in a periodic potential and a magnetic field has recently renewed because of the possibility of an experimental realization of systems with parameters allowing access to the interesting regimes mentioned above. The precondition was the realization of two-dimensional (2D) semiconductor heterostructures (usually on the basis of GaAs) with extremely high mobility leading to an electron mean-free path of more than 10 μm at liquid helium temperature. The electrons occupy only the lowest electrical subband at usual densities forming a 2D electron gas (2DEG). The next step was the ability of modern semiconductor technology to transfer highly regular patterns with periods of a few hundred nanometers onto the 2DEG without destroying the high mobility. The electrons then move ballistically in the 2D plane feeling essentially the potential from the periodic pattern. In this situation one only needs magnetic fields below 10 T for a possible observation of Hofstadter's butterfly. These structures are now called lateral superlattices. Figure 8.1 shows some realizations of them together with the Fermi energy, E_F, which in relation to the modulation amplitude allows us to distinguish different situations.

The first lateral superlattices investigated experimentally have been systems with weak one-dimensional (1D) modulation (Fig. 8.1(a)) [14]. They showed unusual oscillations of the magneto-resistance for weak fields well below the onset of Shubnikov–de Haas (SdH) oscillations. Those are now known as Weiss oscillations and have been explained successfully as being due to the commensurability between the potential period and the cyclotron diameter both quantum mechanically [15, 16] and classically [17] within a perturbational approach for small modulation amplitude. This approach is no longer applicable for stronger modulation (Fig. 8.1(c)) if the electron trajectories differ clearly from cyclotron orbits [18, 19]. However, the classical dynamics of the electrons remains regular even in this case.

The dynamics for weak 2D modulation (Fig. 8.1(b)) is also essentially regular

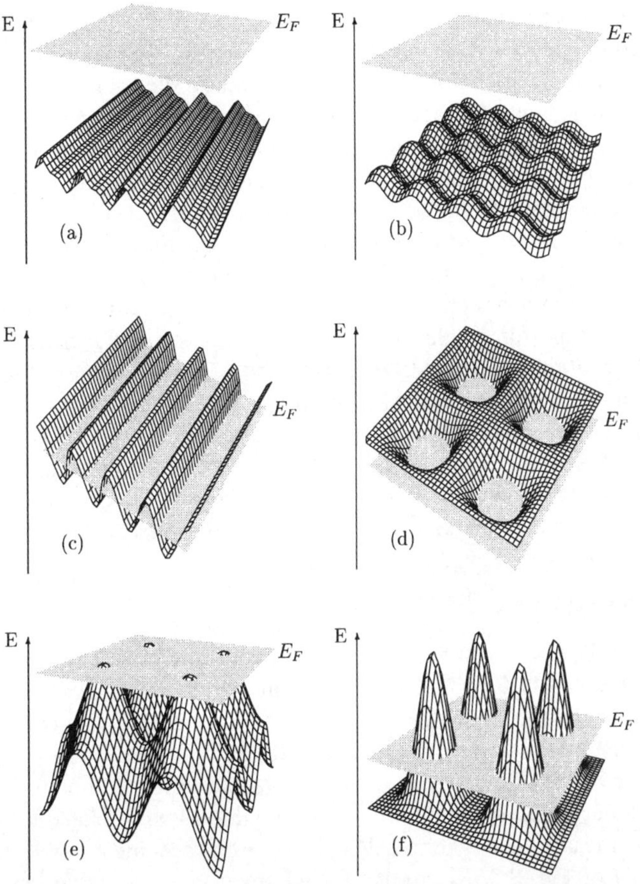

Figure 8.1 *Different realizations of lateral superlattices: (a) weak 1D modulation, (b) weak 2D modulation, (c) strong 1D modulation, (d) quantum dot lattice, (e) strong 2D modulation and (f) antidot lattice. The shaded area indicates the position of the Fermi energy. The magnetic field is usually applied perpendicular to the superlattice plane.*

and the experiments [20] have been described successfully within perturbation theory neglecting the influence of chaos [21, 22].

A particular situation is presented in Fig. 8.1(d) in which the electrons are confined classically (array of quantum dots).

In the following we concentrate on lateral superlattices with strong 2D modulation (Fig. 8.1(e)) and the so-called antidot lattices (Fig. 8.1(f)) characterized by a nonnegligible influence of chaos on the classical dynamics of the electrons. Besides in the classical characterization of chaos and its quantum-mechanical signatures in the spectra of these systems in a perpendicular magnetic field we will especially be interested in their low-field magnetotransport properties which have recently been in the centre of experimental efforts. After an introduction into the consequences of classical dynamics for the magnetoresistivity, we will focus on the implications of a quantum-mechanical description of transport.

Let us finally note that another interesting class of systems exists in which the influence of classical chaos on the transport properties has been studied both experimentally [23] and theoretically [24], namely mesoscopic, ballistic cavities realized on the basis of GaAs heterostructures, too. Depending on the shape of the boundaries, these cavities resemble closed chaotic billiards. The electrons maintain their phase coherence over the whole sample (compare with Chapter 7) in contrast to the situation in lateral superlattices. A negative magnetoresistance differing in shape for regular and chaotic boundaries and aperiodic but reproducible fluctuations of the resistance depending on the magnetic field are found experimentally similar to the observations in disordered systems with diffusive motion of the electrons. The explanation of these observations as quantum interference phenomena is based on semiclassical calculations starting from the Landauer–Büttiker formula for the conductance and on those using RMT as well as on direct numerical simulations.

8.2 EXPERIMENTS

Recently, there has been a number of experiments performed on antidot lattices, these begin with [25–32]. The most important finding was again the influence of the commensurability between lattice period and cyclotron diameter on maxima in the magneto-resistance at weak fields [28, 29].

In the pinball model of Weiss *et al.* [29] the electrons are thought to move on classical trajectories in a quadratic lattice of antidots with relatively hard walls. At the commensurate magnetic fields pinned orbits become possible as sketched in the inset of Fig. 8.2. Electrons moving along these orbits do not contribute to the transport process and the resistance increases. This picture accounts well for the dependence of the positions of the resistance maxima on lattice period and electron density. In the experiments displayed in Fig. 8.2 samples 1 and 3 have a lattice constant of 300 nm and electron densities of 1.4×10^{11} cm^{-2} and 2.8×10^{11} cm^{-2}, respectively, without illumination. The latter (e.g. $1 \rightarrow 1^*$) increases the electron density by a factor of less than 2 (and changes the shape of the antidots). Sample 2 has a lattice constant of 200 nm and an electron density of 2.0×10^{11} cm^{-2} with illumination.

The weak temperature dependence of the commensurability peaks shown in Fig. 8.3 [33] indicates a classical origin as assumed in the above explanation. Phase coherence effects are already destroyed at lower temperatures.

Small oscillations superimposed onto the classical resistance peaks at very low temperatures (Fig. 8.4) which are essentially periodic in the magnetic field [34, 35] have been explained with semiclassical oscillations in the density of states [34]. At weak field the experiments displayed the Altshuler–Aronov–Spivak effect [36, 37].

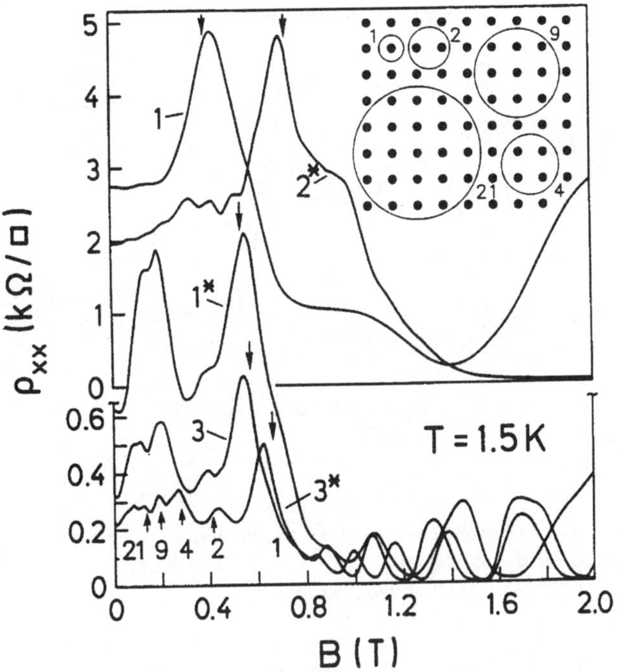

Figure 8.2 *Magnetoresistance of quadratic antidot lattices with different lattice periods and electron densities. The inset shows possible commensurate orbits. The arrows mark the magnetic field positions of the orbits around one antidot and, for sample* 3*, *around several antidots [29].*

Recently, finite antidot lattices have been fabricated where the electrons maintain phase coherence over the whole sample [38, 39]. They allowed the observation of the h/e Aharonov–Bohm effect and, in addition, showed reproducible fluctuations of the magnetoresistance [38] as predicted theoretically [24] and found experimentally [23] in mesoscopic cavities with chaotic electron dynamics.

Another quantum-mechanical effect, namely the direct resolution of the magnetic band structure in samples with strong 2D modulation and relatively small lattice constants, might have been seen in a recent magnetotransport experiment via a splitting of the SdH peaks for certain values of the magnetic flux [40].

Other experiments have been carried out on rectangular [41–45], hexagonal [46, 47, 43] and disordered [48, 45] antidot lattices. Due to the additional independent length scale, experiments on rectangular lattices allow a more detailed investigation of the transport mechanisms. First, they are characterized by an anisotropic behaviour of the diagonal components of the resistance tensor as shown in Fig. 8.5.

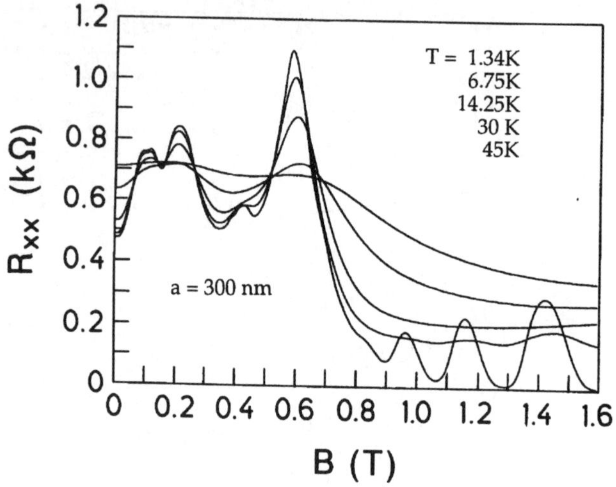

Figure 8.3 *Temperature dependence of the magnetoresistance of sample 3 from Fig. 8.2. The resistance maxima at $B \sim 0.2$ T and $B \sim 0.6$ T correspond to classical trajectories around four antidots and one antidot, respectively [33].*

Besides the anisotropy in the low and especially in the intermediate field regions [45] which can be reproduced classically [49] one recognizes quantum oscillations on top of the main classical resistance peak at $T = 30$ mK which are already destroyed at liquid helium temperature. Their periodicity differs from that of conventional SdH oscillations as indicated by the dashed vertical lines in Fig. 8.5. The simultaneous observation of different periods is unique to rectangular lattices. The Hall resistance is also isotropic in the rectangular lattice as expected from the Onsager relations.

8.3 CLASSICAL CHAOS AND TRANSPORT

The above experiments can be described by classical dynamics of electrons in a 2D periodic potential if the Fermi wavelength is much smaller than the lattice period and if phase coherence effects play no role. One simple example is the pinball model explained above. Otherwise one has to solve the equations of motion and calculate the resistance with the classical Kubo formula [50].

The 2D periodic potential of antidots in the (x, y)-plane with lattice constants a_x and a_y is described by

$$V(x, y) = V_0 \cos\left(\frac{\pi x}{a_x}\right)^{2\alpha} \cos\left(\frac{\pi y}{a_y}\right)^{2\beta} \tag{8.3.1}$$

in the following. This potential (as visualized for a quadratic lattice in Fig. 8.1(f))

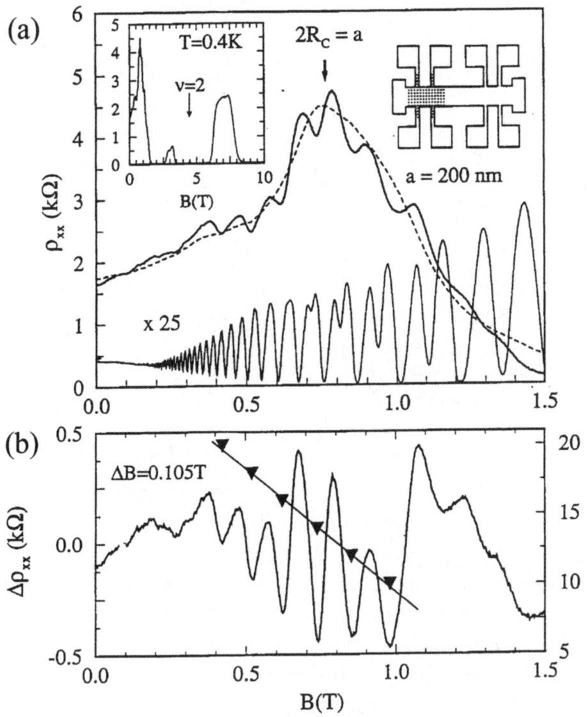

Figure 8.4 *(a) Magnetoresistance ρ_{xx} measured in patterned (top traces) and unpatterned (bottom trace) segments of sample* 2* *from Fig. 8.2 for* $T = 0.4$ K *(solid curves) and* $T = 4.7$ K *(dashed curve). The left inset shows ρ_{xx} up to* $B = 10$ T. *The filling factor* $\nu = 2$ *is marked. The right inset shows the sample layout. (b)* $\rho_{xx}(T = 0.4\text{ K}) - \rho_{xx}(T = 4.7\text{ K})$ *versus B. For* $B < 1$ T *the minima (triangles) are spaced by* $\Delta B = 0.105$ T *[34].*

models the situation in experiment [29] for parameters as for example $V_0 = 25$ meV, $\alpha = \beta = 4$ and $a_x = a_y = 300$ nm together with a typical Fermi energy of about 10 meV.

As an example, typical classical trajectories at the energy $E = 12$ meV are displayed in Fig. 8.6 for a rectangular lattice. Together with a pronounced anisotropy in the intermediate field region one recognizes clear differences in the type of motion for the different magnetic fields. It is essentially chaotic for the lowest field and becomes increasingly regular for larger fields. This can be clarified by looking at the Poincaré surfaces of section. In a regular system the electrons move on invariant tori appearing as closed lines in the Poincaré plot. Due to the KAM theorem [7] the transition to chaos is smooth resulting in a mixed phase space with regular islands in the chaotic sea. From Fig. 8.7 it is clear that the influence of classical chaotic dynamics is not negligible for the systems under consideration. The classical phase space contains at least a considerable

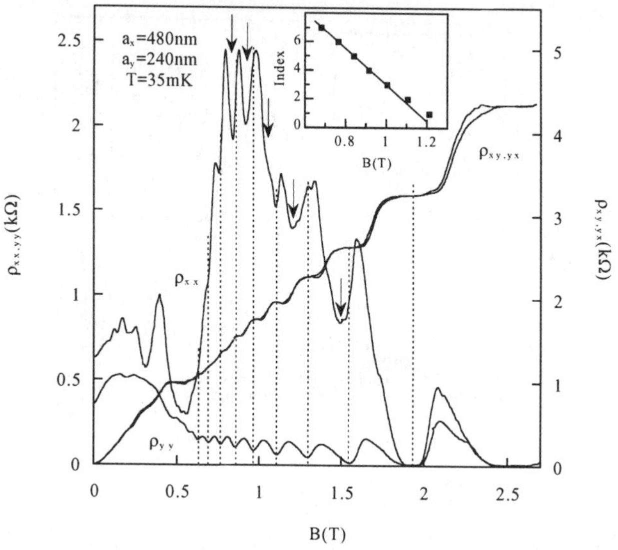

Figure 8.5 *Magnetoresistance and Hall resistance of a rectangular antidot lattice with periods* $a_x : a_y = 480$ nm : 240 nm $= 2 : 1$. *The inset shows the magnetic field periodicity of the additional structure (marked by arrows) on top of the resistance peak in* ρ_{xx} *corresponding to the classical orbit around one antidot [63].*

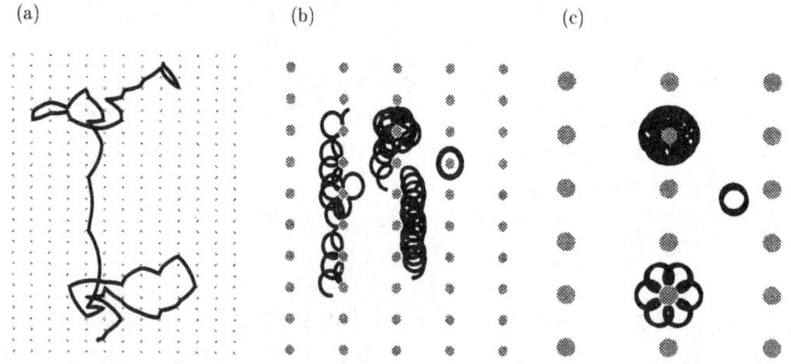

Figure 8.6 *Typical classical trajectories in a rectangular lattice with* $a_x : a_y = 2 : 1$ *for (a)* $2R_c > a_x, a_y$, *(b)* $a_x > 2R_c > a_y$ *and (c)* $a_x, a_y > 2R_c$.

chaotic part at low magnetic fields [51]. Note that the regular island in the middle of the Poincaré plots is especially pronounced at the commensurability condition $2R_c = a$ (Fig. 8.7(b)). It should be mentioned here that the classical phase-space structure depends strongly on the magnetic field as well as on the energy of the particle.

First classical simulations of the transport in antidot lattices revealed the suppression of the diffusion for commensurate magnetic fields also for more realistic smooth antidots [52]. Later it became clear that classical chaos plays

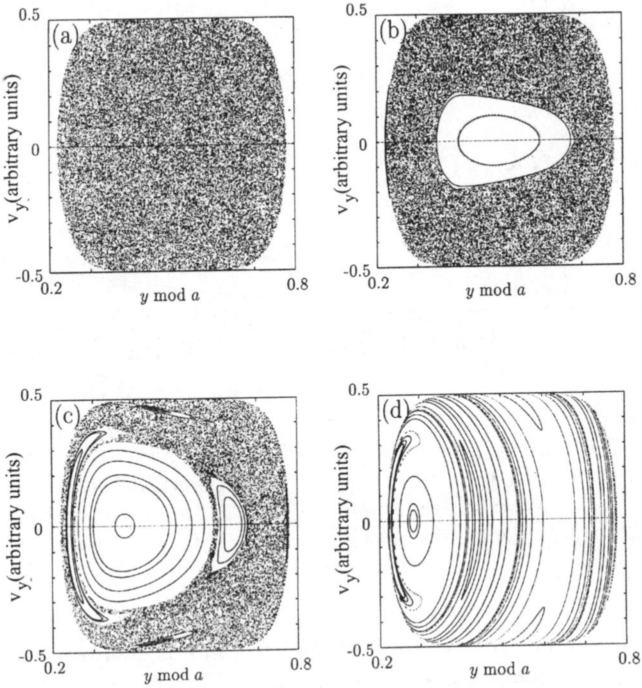

Figure 8.7 *Poincaré surfaces of section at* $E = V_0/2$ *and* $x(\text{mod } a) = 0$ *for the antidot lattice of Fig. 8.1(f) at the magnetic fields (a)* $B = 0.23$ T, *(b)* $B = 0.64$ T, *(c)* $B = 0.92$ T *and (d)* $B = 2.3$ T *[75].*

an important role in a more detailed explanation of the effect [51]. Figure 8.8 taken from [51] shows the magnetoresistance calculated with the classical Kubo formula for different potential shapes. The differences between the steep and soft potentials cannot be understood in the simple pinball model. This especially applies to the peak corresponding to an orbit around two antidots for the steep potential and to peak X for the smooth potential which is shifted to a notably lower field compared with the orbit around four antidots in the pinball model as well as to the shoulder S for the smooth potential.

For the potential parameters of [51] the resistance maxima are caused by chaotic orbits in the vicinity of regular islands on which the electrons are pinned for some time between sections of motion through the lattice. An alternative explanation of the resistance maxima has been given on the basis of runaway trajectories [53, 54]. These trajectories, following rows of antidots for a long time, lead to an increase in conductivity at the commensurate fields which causes, however, again peaks in the resistance due to tensor inversion [55, 56]. They may play a remarkable role for antidot diameters which are small in comparison

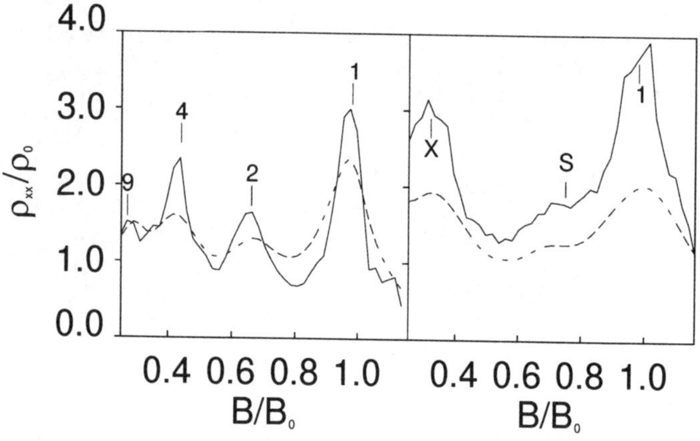

Figure 8.8 *Classical magnetoresistivity normalized to the zero–field resistivity* ρ_0 *for a quadratic antidot lattice with steep (left) and smooth (right) potentials.* B_0 *follows from* $2R_c = a$. *The peaks are marked by the number of antidots enclosed within respective cyclotron-like regular orbits. The dotted curves are experimental traces for samples* 3* *and* 1* *of Fig. 8.2 [51].*

with the lattice constant as shown exemplarily in Fig. 8.6(b). The role of yet another set of trajectories has been discussed in [57].

Finally, the quenched or negative Hall effect in antidot lattices at the lowest magnetic fields has been explained by the existence of channelling trajectories along rows of antidots at these fields which are again located in the vicinity of regular islands in phase space. These trajectories have a large probability to be scattered into the direction opposite to the free $\boldsymbol{E} \times \boldsymbol{B}$ drift [58].

8.4 QUANTUM-MECHANICAL BAND STRUCTURE

In the quantum-mechanical version of the above problem one has to study the eigenstates of

$$H = \frac{1}{2m^*}(\boldsymbol{p} + e\boldsymbol{A})^2 + V(x, y) \tag{8.4.1}$$

with the vector potential $\boldsymbol{A} = B(-y, x, 0)/2$ in symmetric gauge and the potential $V(x, y)$ from equation (8.3.1). The homogeneous system without periodic potential gives the Landau levels (magnetic quantization) whose degeneracy is lifted by the periodic potential. The magnitude of the splitting and, eventually, mixing of the original Landau levels depends on the strength of the lateral superlattice potential for a given magnetic field.

In calculating the eigenstates of H, equation (8.4.1), the translational symmetry of the periodic potential cannot be exploited immediately because due to the vector potential the lattice translations do not commute with the kinetic energy operator. Instead, one has to use magnetotranslations [59]

$$S(\boldsymbol{R}) = \exp\left(-\frac{\mathrm{i}}{\hbar}\boldsymbol{R}\,(\boldsymbol{p} + e\boldsymbol{A})\right) \tag{8.4.2}$$

where $\boldsymbol{R}$ is a lattice vector that commutes with H, equation (8.4.1). The algebra of these operators,

$$[S(\boldsymbol{R}_1), S(\boldsymbol{R}_2)] = -2\mathrm{i}\sin\left(\frac{1}{2l_B^2}(\boldsymbol{R}_1 \times \boldsymbol{R}_2)\hat{z}\right) S(\boldsymbol{R}_1 + \boldsymbol{R}_2) \tag{8.4.3}$$

(where $(\boldsymbol{R}_1 \times \boldsymbol{R}_2)\hat{z}$ is the area spanned by $\mathbf{R}_1$ and $\boldsymbol{R}_2$ and l_B is the magnetic length) shows that $S(\boldsymbol{R}_1)$ and $S(\boldsymbol{R}_2)$ do not commute in general (in contrast to ordinary translations). The phase

$$|\varphi| = \frac{|\boldsymbol{R}_1 \times \boldsymbol{R}_2|}{2l_B^2} = 2\pi\frac{|\boldsymbol{R}_1 \times \boldsymbol{R}_2|}{2h/e}B = \pi\frac{\Phi}{\Phi_0} \tag{8.4.4}$$

is π times the number n_Φ of magnetic flux quanta $\Phi_0 = h/e$ threading the area $(\boldsymbol{R}_1 \times \boldsymbol{R}_2)\hat{z}$. Thus, commutation of $S(\mathbf{R}_1)$ and $S(\boldsymbol{R}_2)$ is achieved for $\Phi = n_\Phi \Phi_0$, $n_\Phi \in N$ and the concepts of Bloch functions and Brillouin zones can be applied as in the magnetic field free case. This allows us to consider magnetic fields for which an integer number of magnetic flux quanta threads the unit cell of the lattice. For rational flux $n_\Phi = p/q$ $(p, q \in N)$ the unit cell can be increased such that one obtains integer flux with p flux quanta per unit cell and uses the same concepts as for $n_\Phi \in N$. The translational symmetry in the presence of the magnetic field is exploited in the following by diagonalization of the Hamiltonian H, equation (8.4.1) in a symmetry adapted basis. It is given by the set of functions Ψ_Θ [60, 61] which are characterized by a magnetic wavevector $\Theta \in [0, 2\pi]^2$ according to

$$S(\boldsymbol{R})\Psi_\Theta = \mathrm{e}^{\mathrm{i}\Theta}\Psi_\Theta. \tag{8.4.5}$$

Results for the band structure along two symmetry lines of the first magnetic Brillouin zone (MBZ), the direction of one Cartesian coordinate and the diagonal (cf. inset in Fig. 8.12), are shown in Fig. 8.9 for the quadratic antidot lattice used in Fig. 8.1(f) and for four values of the magnetic field with the corresponding ladder of Landau levels in between for comparison. For the lowest magnetic field one recognizes a dense system of overlapping magnetic minibands which do not exhibit any similarity with the corresponding Landau levels. Note that one qualitatively obtains the same behaviour for the system without magnetic field [62]. With increasing magnetic field the dispersion gets weaker and the minibands cluster towards the Landau levels if the classical cyclotron diameter at the corresponding energy gets smaller than the lattice constant. It is remarkable that the classical resistance maximum (e.g. at approximately $B = 0.6$ T for sample 3 in Fig. 8.2 [29]) is not reflected in an obvious way in changes in the band structure even at the Fermi energy (Fig. 8.9(b)).

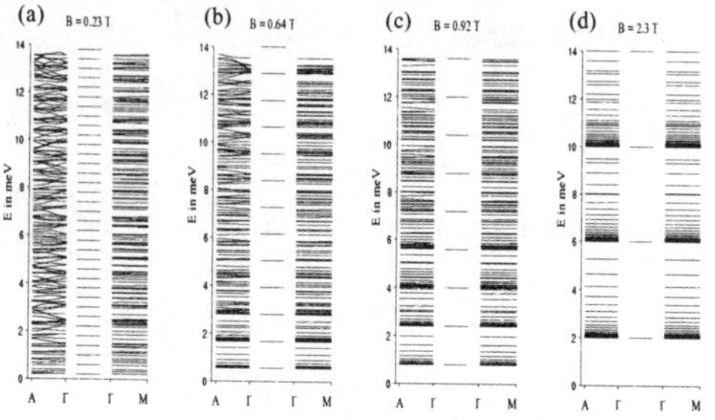

Figure 8.9 *Band structure for the potential parameters* $V_0 = 25$ meV, $a_x = a_y = 300$ nm *and* $\alpha = \beta = 4$ *and for the magnetic fields (a)* $B = 0.23$ T *(*$n_\Phi = 5$*), (b)* $B = 0.64$ T *(*$n_\Phi = 14$*), (c)* $B = 0.92$ T *(*$n_\Phi = 20$*) and (d)* $B = 2.3$ T *(*$n_\Phi = 50$*)[62].*

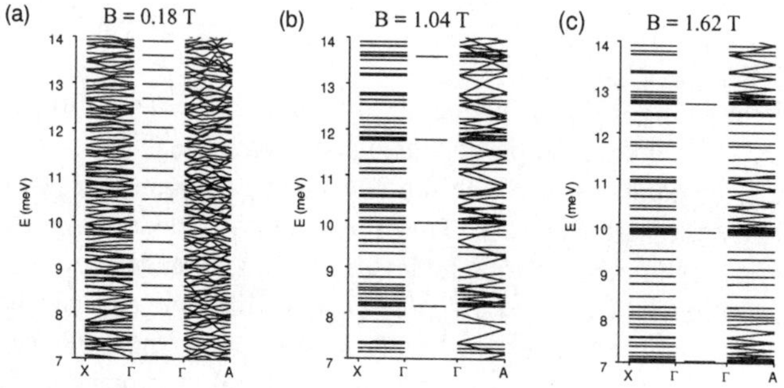

Figure 8.10 *Miniband structure for a rectangular antidot lattice with* $a_x = 480$ nm, $a_y = 240$ nm, $\alpha = 8$, $\beta = 2$, $V_0 = 25$ meV *and for three different magnetic fields (a)* $B = 0.18$ T *or* $2R_c > a_x, a_y$*, (b)* $B = 1.04$ T *or* $a_x > 2R_c > a_y$ *and (c)* $B = 1.62$ T *or* $a_x, a_y > 2R_c$ *[87].*

As a second example, Fig. 8.10 shows the magnetic minibands along the directions of the two Cartesian coordinates $\Gamma - X$ and $\Gamma - A$ (inset in Fig. 8.12) in the MBZ of a rectangular antidot lattice with $a_x : a_y = 2 : 1$ (cf. Fig. 8.5, [63]). The magnetic field values (again visualized by the equidistant Landau levels of the homogeneous 2DEG) are chosen to meet three distinct situations. If the diameter of the cyclotron orbit $2R_c$ at the Fermi energy $E_F = 12.5$ meV is larger than either lattice constant, the minibands again exhibit well-pronounced dispersion in either direction. For $a_x > 2R_c > a_y$, however, the situation becomes strongly anisotropic. Along $\Gamma - X$ corresponding to propagation along the direction with the larger lattice constant the minibands are altogether flat. Along $\Gamma - A$ the flat minibands at the Landau level energies can be ascribed to periodic cyclotron orbits, while strongly dispersive bands in between can be identified with extended skipping orbits along rows of antidots (Fig. 8.6(b)). The anticrossing between both types of minibands will turn

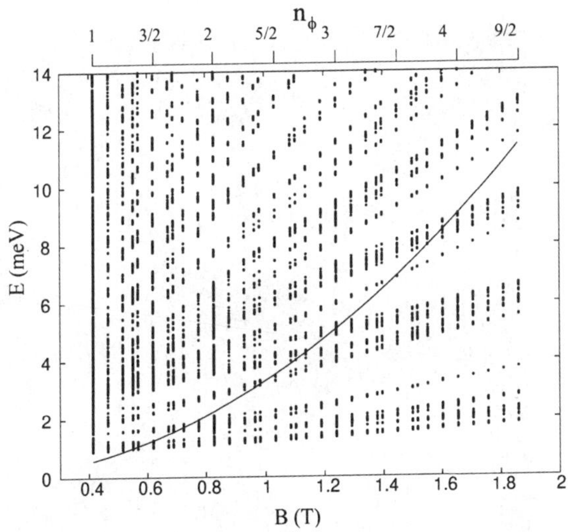

Figure 8.11 *Projection of the miniband structure onto the energy axis of a quadratic antidot lattice with* $a = 100$ nm, $\alpha = 1$, $V_0 = 5$ meV *versus the magnetic field or* n_ϕ. *The solid curve corresponds to* $2R_c = a$. *On the left- and right-hand sides of this curve the band structure are dominated by the potential and by the magnetic field, respectively [92].*

out to be important for the interpretation of the observed quantum oscillations (Fig. 8.5, [63]). Finally, for $a_x, a_y > 2R_c$ the clustering to the Landau levels is more pronounced and most minibands become flat also along $\Gamma - A$.

When projecting the miniband dispersion onto the energy axis for different magnetic fields or rational values of n_ϕ one obtains a picture that corresponds to Hofstadter's butterfly. For a short period quadratic antidot lattice this can be seen in Fig. 8.11. The essential feature here is the dramatic change between broad overlapping bands and small gaps at integer (or half integer) flux numbers and narrow bands separated by larger gaps for rational $n_\phi = p/q$ with larger q. Two regions can be distinguished by the condition $2R_c = a$ (solid curve): to its left-hand side the spectra are dominated by the periodic potential and to its right-hand side the dominance of the magnetic field leads to minibands converging towards Landau levels.

8.5 QUANTUM SIGNATURES OF CHAOS

One of the most frequently investigated quantum manifestations of classical chaos is the statistics of the energy levels [8, 9]. Quantum spectra typically show level repulsion if their classical analogue is chaotic in difference to level clustering for classically regular systems. In the case of level repulsion, the

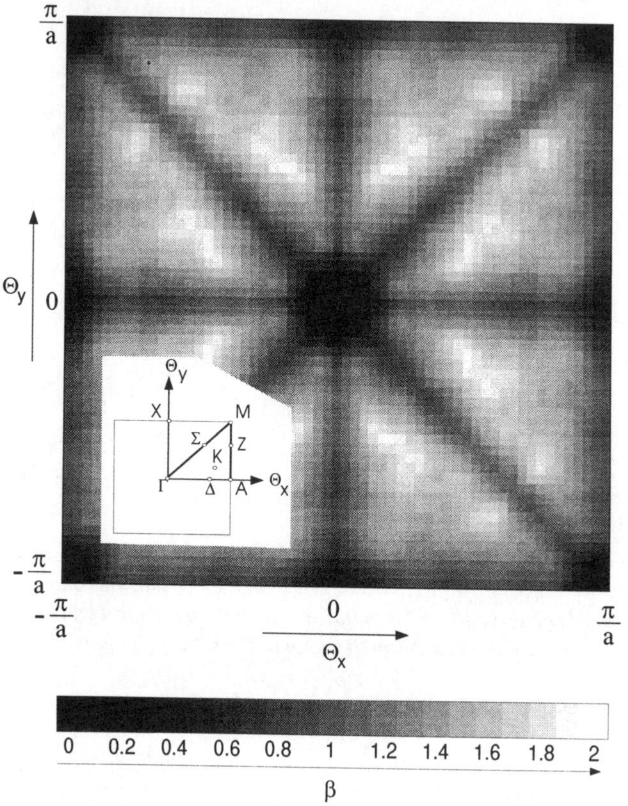

Figure 8.12 *Degree of level repulsion β (obtained from a least square fit of the Izrailev distribution [67] to the numerical data) given in a grey-scale plot in the first MBZ. The scale is given below the plot and ranges from black ($\beta = 0$ corresponding to a Poissonian distribution) over medium grey ($\beta = 1$ corresponding to GOE) to white ($\beta = 2$ corresponding to GUE). The inset gives the designation of inequivalent points in the MBZ used in the text [74].*

distribution of level spacings $P(s)$ vanishes for small spacings with a characteristic power β. This power depends on the symmetry properties of the underlying Hamiltonian which can be divided into three universality classes associated with different ensembles of RMT [8, 9]. It is $\beta = 1$ for the Gaussian orthogonal ensemble (GOE) corresponding to antiunitary symmetry of the Hamiltonian, $\beta = 2$ for the Gaussian unitary ensemble (GUE) for Hamiltonians without antiunitary symmetries and $\beta = 4$ for the Gaussian symplectic ensemble (GSE) for time-reversal-invariant systems with Kramer's degeneracy but without geometrical symmetries, respectively. In contrast, generic regular systems are characterized by a Poisson distribution for $P(s)$. The spectral rigidity $\Delta_3(L)$, as a statistical measure of the spectral correlations on a larger scale, is defined as the mean square deviation of the integrated level density from a linear behaviour over a spectral distance L. It increases logarithmically with L for chaotic systems as opposed by a linear increase for generic regular systems [8, 9]. A transition between different universality classes can be achieved, for example, by switching on a magnetic field, as it breaks time-reversal symmetry. There is continuous interest in the analysis of such transitions and in that from fully chaotic to regular systems which is characterized by a mixed classical phase space [64]. Different proposals have been made to describe the behaviour of the level distribution in this case [65–72].

For statistical analysis spectra corresponding to inequivalent irreducible representations of a geometrical symmetry group have to be considered independently. In the antidot lattice the dominating geometrical symmetry is given by the magnetotranslation group whose irreducible representations are labelled by a magnetic wavevector Θ. Thus, level statistics has to be performed for each wavevector independently. For rational fluxes $n_\Phi = p/q$ one has to start from an MBZ smaller by a factor of q. Intraminiband correlations [73] are not considered in the following.

Numerically, energy spectra have been obtained with sufficient accuracy for fluxes $n_\Phi = p/q$ with small denominators $q \leq 15$. For the statistical analysis only subbands below the potential maxima have been considered where classical dynamics is similar to the Sinai billiard. Depending on the flux denominator q, between 250 and 3000 minibands have been used for statistics. The spectra were unfolded by a standard procedure [8, 9].

The genuine feature for periodic systems is the dependence of the energy spectrum—and consequently of its statistical properties—on the (magnetic) wavevector. To display this dependence the level-spacing distribution has been fitted for a mesh in the MBZ by the Izrailev distribution [67]

$$P_\beta(s) = C_1 \left(\frac{\pi s}{2}\right)^\beta \exp\left[-\frac{1}{16}\beta\pi^2 s^2 - \left(C_2 - \frac{1}{4}\pi\beta\right)s\right] \tag{8.5.1}$$

which allows us to describe the transition between different universality classes using the degree of level repulsion β as a single parameter. With $\beta \in [0, 2]$ one recovers level-spacing distributions for Poisson ($\beta = 0$), Gaussian orthogonal (GOE, $\beta = 1$) and Gaussian unitary ensemble (GUE, $\beta = 2$) of RMT, respectively. C_1 and C_2 are normalization constants.

The values of β obtained for a fixed magnetic field of $B = 0.23$ T on a 59×59 mesh are shown in Fig. 8.12 as a grey-scale plot over the MBZ. Symmetry considerations explain their range from 0 to almost 2. Approximately GUE ($\beta \sim 1.7$) is obtained for the unsymmetric points, since the magnetic field breaks the conventional time-reversal symmetry (TRS). Along the symmetry lines Σ, Δ and Z a generalized TRS can be constructed using the mirror symmetries of the quadratic lattice and leads to the observed GOE behaviour ($\beta \sim 1$). The invariance under discrete rotations finally results in a superposition of two (A)

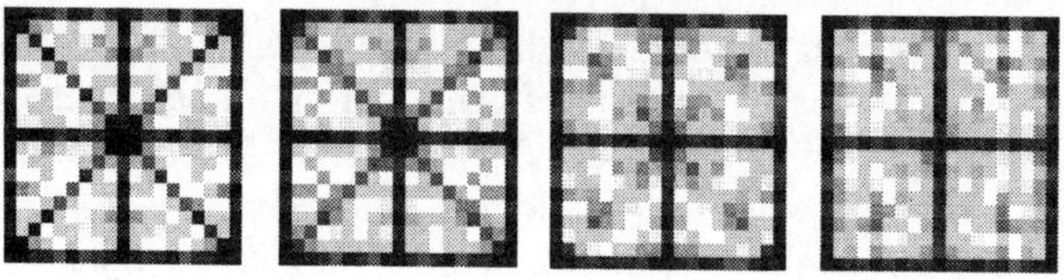

(a) (b) (c) (d)

Figure 8.13 *Degree of level repulsion β given in a grey-scale plot in the first MBZ (Fig. 8.12) for the transition from a quadratic to a rectangular lattice for $B = 0$. The lattice constants are (a) $a_x = 300$ nm and $a_y = 301$ nm, (b) $a_y = 302$ nm, (c) $a_y = 304$ nm and (d) $a_y = 310$ nm.*

or four (Γ and M) subspectra and the corresponding Poisson-like correlations. Thus, in one given magnetic band structure several universality classes and continuous transitions between them are realized [74].

Figure 8.13 shows the parameter β from equation (8.5.1) depending on the position in the MBZ for the transition from a quadratic to a rectangular antidot lattice at $B = 0$. β increases rapidly on the diagonals of the MBZ as the geometrical symmetry of the quadratic lattice is lost. This is a good illustration of the applicability of the above symmetry arguments.

The dependence of the spectral correlations on the magnetic field is demonstrated in Fig. 8.14 for integer values of n_Φ. Figure 8.14(a) shows the dependence of the parameter β on B for the points K and Σ in the MBZ. The increase of β for the smallest n_Φ is due to the breaking of time-reversal invariance by switching on the magnetic field. Afterwards, β decreases monotonously from almost 2 (GUE) for the asymmetric point K and from about 1 (GOE) for the the symmetric point down to $\beta = 0$ (Poisson) for the highest fields. Simultaneously one observes the transition from strongly chaotic behaviour through a mixed phase space to regular behaviour in the classical dynamics of electrons as shown in Fig. 8.7 [75]. Note the remarkable coincidence for both cases for $\beta < 1$. It indicates that the magnetic field rather than symmetry is essential for the transition. Figure 8.14(b) displays the corresponding dependence of the spectral rigidity showing its normalized deviations from the GOE averaged over L [74]. One recognizes an almost identical dependence on the magnetic field as in Fig. 8.14(a) which demonstrates the independence of the observations of the special interpolation formula equation (8.5.1).

For rational flux values $n_\Phi = p/q$ with higher denominators one observes level clustering for magnetic fields where the classical system is widely chaotic [74]. Although the magnetic field values differ only slightly (leading to very similar classical chaotic dynamics) level statistics shows level repulsion for $q = 1$ and (almost) level clustering for $q = 13$. Thus, the universal signatures of quantum chaos decrease with increasing fractality of the spectrum. The spectral rigidity also gives a perfectly linear increase for $n_\Phi = \frac{105}{13}$ as expected for

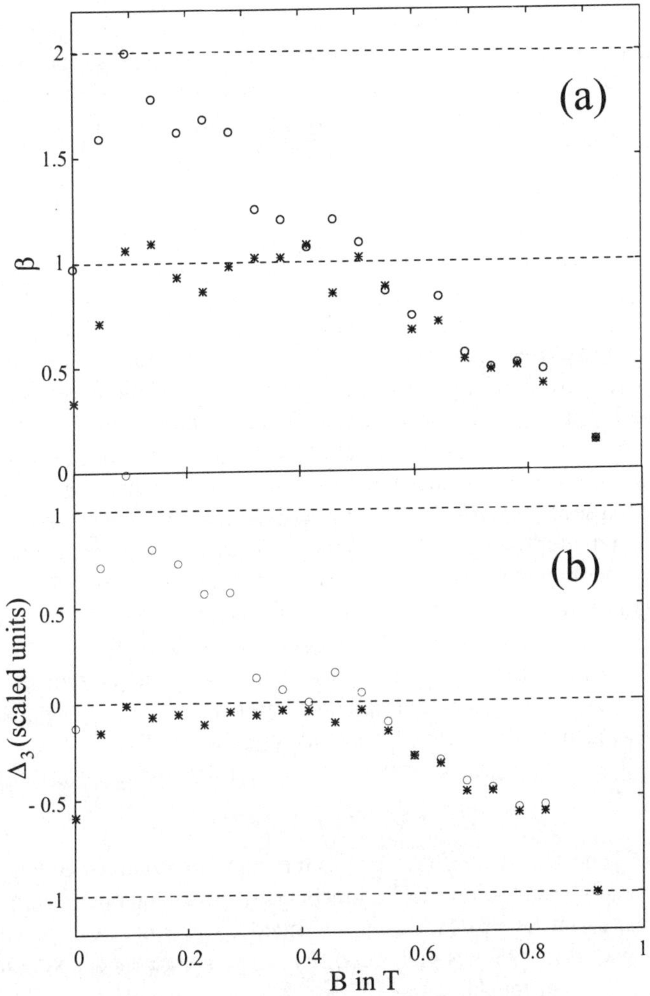

Figure 8.14 *(a) Degree of level repulsion β and (b) spectral rigidity, rescaled and averaged over $2 < L < 10$, versus magnetic field. Only integer flux values are plotted comparing an unsymmetric point K in the MBZ (circles) and a point in the symmetric Σ-direction (stars) [74].*

a Poisson distribution. A similar observation has been made for the (classical regular) Harper model in the high denominator case ($q = 10\,946$) and has been related to the fractal properties of the spectrum [10].

Another type of transition from chaotic to regular classical dynamics occurs if the potential amplitude is reduced at a low magnetic field. Then, one obtains a system with weak 2D modulation as discussed in the introduction. The energy levels become degenerate Landau levels in that case instead of being Poisson distributed, i.e. the regular limit is not generic. Nevertheless, one observes level repulsion with an exponent $\beta = 2$ on a scale which decreases rapidly with the modulation amplitude [76]. The spectral rigidity is well fitted by a simple model with equidistantly split Landau levels as long as the latter are still well separated.

8.6 QUANTUM TRANSPORT

8.6.1 Theory

There are now different approaches to a quantum-mechanical description of the transport properties of antidot lattices. In early calculations the antidots were modelled by simple delta scatterers [77]. In another approach only the band conductivity (for a definition see below) was calculated for realistic potentials [78]. In a third one the conductance for a finite system with antidots was obtained from the transmission probability with the help of the Landauer–Büttiker formula [79–82]. In the following we discuss the method starting from the band structure and the wave functions calculated as described above [75, 83, 84].

Transport properties at low temperatures depend not only on the properties of the pure quantum system, but are also strongly influenced by the scattering of the carriers by impurities. For samples with highly mobile carriers as used in the underlying experiments one may assume short-range impurity scattering which is here included within the well-established self-consistent Born approximation (SCBA). Thus, one must solve the self-consistency equation [22]

$$\Sigma(z) = \gamma^2 \mathrm{tr}(G(z)) \tag{8.6.1}$$

for the self-energy, where $G(z)$ denotes the impurity-averaged Green's function. The self-energy is assumed to be independent of the quantum numbers n and Θ and leads to a level broadening, described by γ, which can be derived in a standard way from the zero-field mobility of the homogeneous system which is known from experimental data.

The static conductivity tensor is calculated with the Kubo formula. The irreducible vertex part of the Bethe–Salpeter equation vanishes if evaluated consistently with ansatz (8.6.1) for the self-energy. The diagonal components of the conductivity tensor (in units of e^2/h) become

$$\sigma_{\mu\mu} = \frac{2\pi}{n_\Phi} \int \mathrm{d}E \left(-\frac{\mathrm{d}f}{\mathrm{d}E}\right) \sigma_{\mu\mu}(E) \tag{8.6.2}$$

where

$$\sigma_{\mu\mu}(E) = \sum_{n_1 n_2} \int \mathrm{d}^2\Theta |\langle n_1\Theta|k_\mu|n_2\Theta\rangle|^2 A_{n_1\Theta}(E) A_{n_2\Theta}(E) \tag{8.6.3}$$

with the Fermi distribution function $f(E)$, the impurity-averaged spectral function $A_{n\Theta}(E)$ and the components k_μ of the kinetic momentum operator. The nondiagonal components are

$$\sigma_{\mu\nu} = \frac{4}{n_\Phi} \int dE\, f(E)\sigma_{\mu\nu}(E) \tag{8.6.4}$$

where

$$\sigma_{\mu\nu}(E) = \sum_{n_1 \neq n_2} \int \Im(\langle n_1\Theta|k_\mu|n_2\Theta\rangle\langle n_2\Theta|k_\nu|n_1\Theta\rangle)\Re\left(\frac{dG_{n_1\Theta}}{dE} A_{n_2\Theta}\right)(E) \tag{8.6.5}$$

with the impurity-averaged retarded Green's function $G_{n\Theta}(E)$. All quantities have been made dimensionless in equations (8.6.2)–(8.6.5) with the cyclotron energy $\hbar\omega_c$ and the magnetic length l_B.

In contrast to weakly modulated 2D electron systems, for which the above formulae could be further simplified by treating the Landau level index as a good quantum number, Landau levels are strongly mixed in antidot lattices, in particular for small magnetic fields. Thus, a completely numerical evaluation of both the longitudinal and Hall conductivities has to be performed from equations (8.6.2)–(8.6.5) without simplifying approximations.

Two contributions have to be distinguished for the diagonal components, equation (8.6.3). The first one (with $n_1 = n_2 = n$) has its origin in the nonvanishing group velocity of carriers in a dispersive miniband. Hereafter, it is called **band conductivity**. In contrast, the other contribution (with $n_1 \neq n_2$), which is also responsible for the conductivity of the homogeneous 2DEG, accounts for scattering between minibands. It is called **scattering conductivity**. Note that the band conductivity decreases with increasing scattering strength whereas this is the other way around for the scattering conductivity. Finally, tensor inversion gives the resistances.

The terms (Landau) band and scattering conductivity, respectively, have also been discussed in samples with weak 2D modulation [22, 85, 86, 40]. They have, however, been used in a slightly different sense because Landau levels are still well separated in these systems.

8.6.2 Classical features

The resistance components R_{xx} and R_{xy} for the potential of Fig. 8.1(f) which models sample 3 of [29] (Fig. 8.2) are shown in Fig. 8.15 in comparison with the experimental data. All features of the experiment are reproduced by the quantum transport calculation: the commensurability peaks in R_{xx}, the non-quantized plateaus in R_{xy} and the quenching of the Hall effect at very low magnetic fields. It is remarkable that one obtains quantitative agreement even on absolute scales for the resistance values with the mobility taken from the experiment and the simple scattering model explained above. The calculated commensurability peaks shift to higher magnetic fields with increasing electron

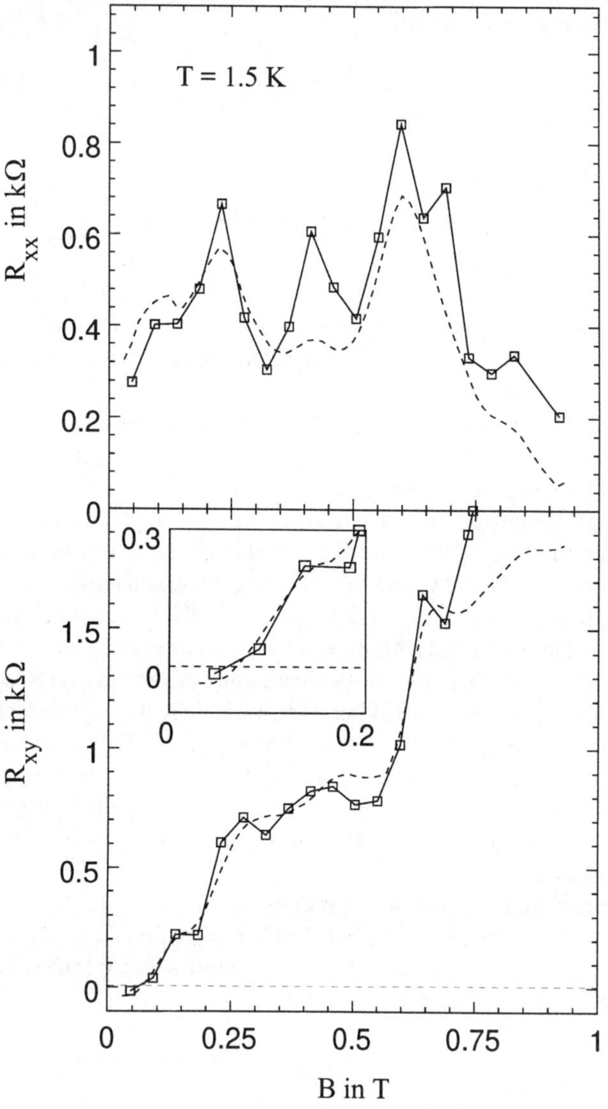

Figure 8.15 *Longitudinal and Hall resistances for the potential of Fig. 8.1(f) (squares) in comparison with experimental data from Fig. 8.2, sample 3 [29] (dashed curves). The solid curves are guides for the eyes. The inset in the lower part of the figure shows the negative Hall resistance at weak magnetic field [83].*

density corresponding to an increased cyclotron radius [75], again in agreement with the experimental result (Fig. 8.2).

The weak temperature dependence in the calculations shown in Fig. 8.16 is

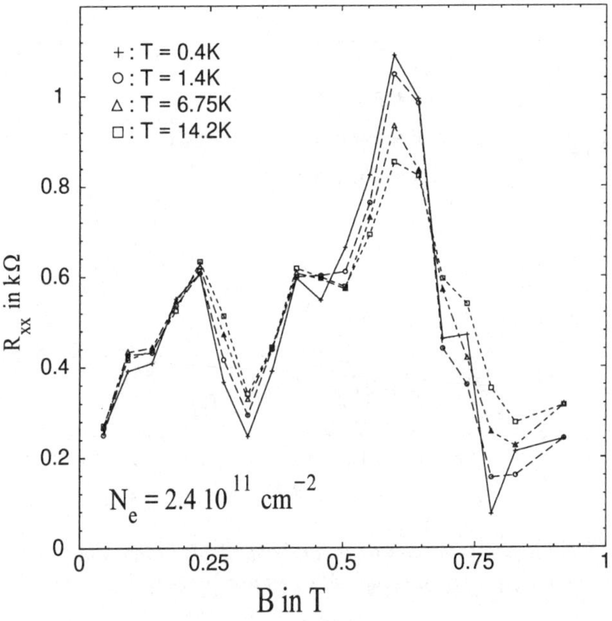

Figure 8.16 *Temperature dependence of the magnetoresistance for the potential of Fig. 8.1(f). The solid curves are guides for the eyes.*

exclusively due to the Fermi function in equations (8.6.2) and (8.6.4). Nevertheless, the experimental observations (Fig. 8.3) are reproduced very well indicating that temperature-dependent scattering mechanisms can be neglected for the classical effects in the range of temperatures considered.

In Fig. 8.17 theoretical results for the rectangular lattice of Fig. 8.5 (solid curves) are compared with the experimental data (dashed curve) for $T = 4.2\,\mathrm{K}$. The inset indicates classical cyclotron orbits encircling one, two and six antidots. Their diameter $2R_c$ is commensurate with the lattice periodicity at magnetic fields indicated with arrows and numbers. The mobility $\mu = 50\ \mathrm{m}^2\ (\mathrm{Vs})^{-1}$ and the electron density $n_s = 3 \times 10^{11}\ \mathrm{cm}^{-2}$ are taken from the experiment. Again, the calculations reproduce even quantitatively all observed features as the anisotropy in ρ_{xx} compared with ρ_{yy}, the commensurability peaks in ρ_{xx} and the SdH-like oscillations in ρ_{yy} [87]. The broad peak at low field in ρ_{yy} can be understood similarly to the boundary scattering peak in quantum wires [88]. The corrugation of the regular antidot potential acts here analogous to the roughness along the wire edges.

Most of these features arise from the competition between band and scattering conductivities. Three magnetic field regimes can be distinguished (Figs. 8.6 and 8.10). For $B < 0.5$ T or $2R_c > a_x, a_y$ the band conductivity contributes significantly to σ_{xx} as well as to σ_{yy} as the minibands show dispersion (though

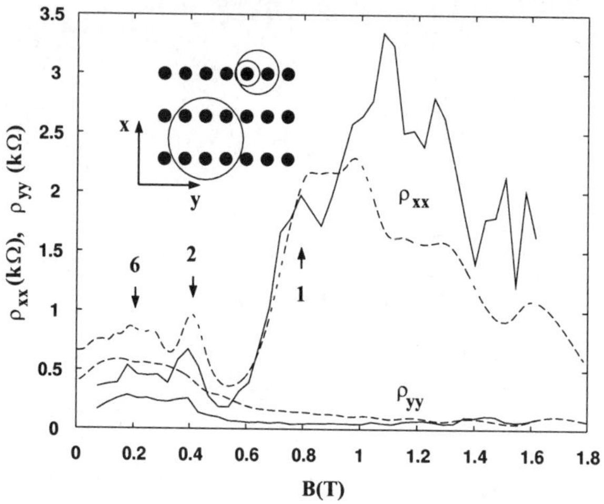

Figure 8.17 *Experimental (dashed lines) and calculated (solid lines) magnetoresistance components* ρ_{xx} *and* ρ_{yy} *for the rectangular antidot lattice of Fig. 8.5 at* $T = 4.2\,\mathrm{K}$ *[87].*

of different strength) in both directions. Beginning at $B \approx 0.5\,\mathrm{T}$ the miniband dispersion in the direction of the large lattice constant and consequently the band conductivity contribution to σ_{xx} becomes very small and σ_{xx} is dominated by scattering conductivity showing SdH oscillations with increasing B. In contrast, the miniband dispersion in the y-direction is still pronounced and the total σ_{yy} is essentially due to band conductivity, where commensurability oscillations show up. Note that in this regime and for higher B $\rho_{\mu\mu} \sim \sigma_{\nu\nu}$. Consequently, the SdH oscillations in ρ_{yy} have their origin in those of σ_{xx}. For $B > 1.5\,\mathrm{T}$ and $a_x, a_y > 2R_c$ the electron motion decouples from the antidot lattice, i.e. the minibands lose their dispersion in either direction and the magnetotransport becomes increasingly similar to that of a homogeneous 2DEG.

Let us finally mention that the large band conductivity in the region of the main resistance peak is due to extended states corresponding to runaway orbits in the classical picture. Thus, the importance of these orbits is shown by the calculations at least for the above potential parameters and for the smoother potential discussed in section 6.3 in connection with the semiclassical oscillations, too (cf. also the discussion in [80]).

8.6.3 Semiclassical features

In section 2 it was noted that the small oscillations on top of the classical resistance peaks at very low temperatures (Figs. 8.4 and 8.5) cannot be explained with classical concepts. Periodic orbit theory, however, represents a well-known

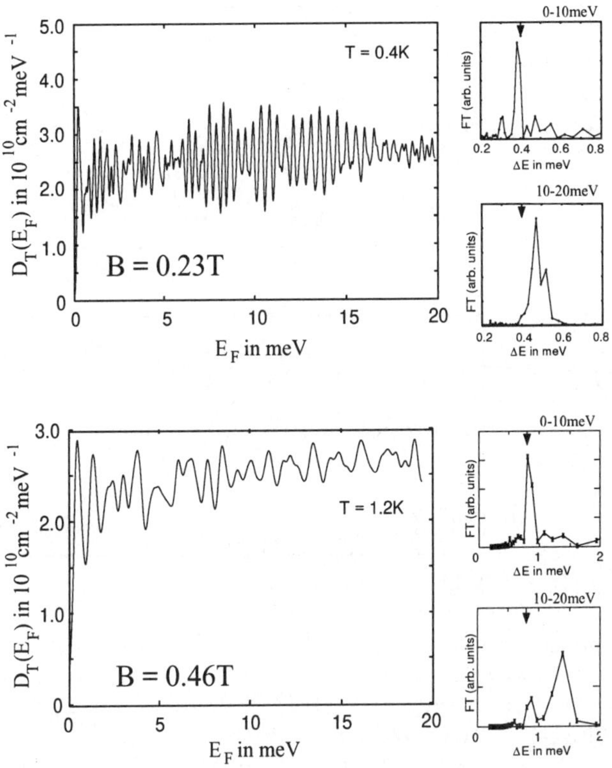

Figure 8.18 *Semiclassical oscillations in the TDOS for the potential of Fig. 8.1(f) at two magnetic fields. The Fourier transforms in the right panel for two energy windows at each field show at higher energies an increased period of the oscillations compared to the cyclotron energy indicated by the arrows [83].*

tool for a semiclassical analysis. According to Gutzwiller's trace formula [7] the oscillatory part of the density of states (DOS) of a classically chaotic system can be written as a sum over contributions $A\,\exp[\mathrm{i}(S/\hbar + \pi\alpha/2)]$ from each classical periodic orbit of the system. Here, $S = \oint \boldsymbol{p}\,\mathrm{d}\boldsymbol{q}$ is the classical action calculated along the periodic orbit, α is the Maslov index and A depends on the stability of the orbit. Due to scattering and finite temperature in experiment, only short periodic orbits are relevant for the smoothed DOS. Their contribution was calculated in [34] and good agreement with experiment has been found for position and periodicity of the oscillations. Especially, the deviations from the $1/B$ periodicity as expected for unperturbed cyclotron orbits have been explained with the influence of the antidot potential leading to a modified B-dependence of the classical action along the orbits.

Figure 8.18 shows these semiclassical oscillations in the thermodynamic DOS

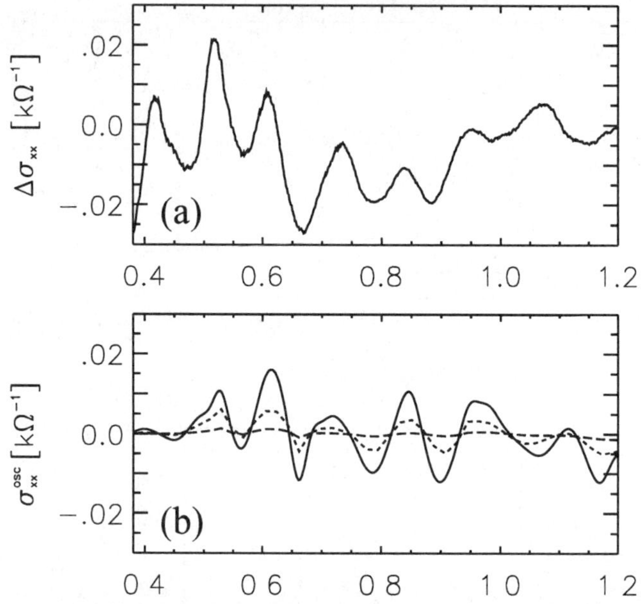

Figure 8.19 *(a) Oscillatory part $\Delta\sigma_{xx}$ (cf. Fig. 8.4) of the experimental longitudinal conductivity in dependence on B. (b) Oscillatory part σ_{xx}^{osc} calculated with the semiclassical Kubo formula for three temperatures* $T = 0.4$ K *(solid curve),* $T = 2.5$ K *(dotted curve) and* $T = 4.7$ K *(dashed curve) [89].*

(TDOS) $D_T(E_F) = \int \mathrm{d}E(-\mathrm{d}f(E)/\mathrm{d}E)D(E)$ obtained from the quantum-mechanical calculation. From the Fourier transforms one recognizes the deviations from the periodicity of unperturbed cyclotron orbits caused by the deformations of orbits at higher energy by the antidot potential.

More recently, a semiclassical expression for the conductivity tensor has been derived from the quantum-mechanical Kubo formula along the lines of Gutzwiller's trace formula [89, 90]. According to this calculation the oscillations in the conductivity are governed by the same contributions from each periodic orbit as those in the DOS discussed above. The main difference to Gutzwiller's trace formula is the appearance of the classical velocity correlation functions along the orbits in the expressions for the oscillatory parts of the conductivity components.

In Fig. 8.19 the semiclassical calculation [89] is compared with the experimental data from [34] for a quadratic antidot lattice. The agreement is very good in view of the simple model for the antidot potential according to equation (8.3.1). A direct comparison between a semiclassical and quantum-mechanical calculation showed, however, that the conductivity components are not equally well approximated by the semiclassical calculation as the DOS [84, 81]. This has been

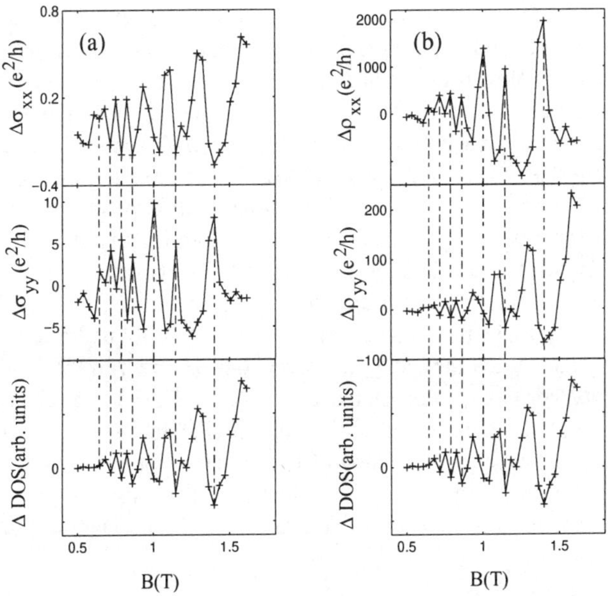

Figure 8.20 *(a) Semiclassical oscillations in the longitudinal conductivities and (b) resistivities, respectively, for the rectangular lattice of Fig. 8.5 in comparison with the oscillations in the TDOS. The values at high temperature* $T = 10$ K *have been subtracted from those at low temperature* $T = 2$ K *in each case. The vertical dashed lines indicate the minima in the TDOS [87].*

attributed to the coupling between different periodic orbits which is formally of higher order in $\hbar$ and is, thus, not included in the semiclassical expression.

In Fig. 8.20 the semiclassical oscillations in the region of the classical commensurability peak for trajectories around one antidot are analysed on the basis of a quantum-mechanical calculation for a 2 : 1 antidot lattice with a steep potential ($V_0 = 25$ meV, $\alpha = 8$, $\beta = 2$). The comparison with the TDOS at the Fermi energy shows that the oscillations in $\Delta\sigma_{xx}$ are SdH-type periodic in $1/B$. In the given range of B, σ_{xx} is dominated by scattering contributions. In the same range, σ_{yy}, dominated by band conductivity, shows the same periodicity but opposite phase of the oscillations. Large (small) scattering conductivity in σ_{xx} is parallel to small (large) band conductivity in σ_{yy}. This behaviour has not been explained by semiclassical concepts [89, 90] so far, presumably because of the neglection of the coupling between periodic orbits and extended states similar to the findings in [84] which results in the anticrossing of bands as discussed in connection with Fig. 8.10. The semiclassical calculations would lead to quantum oscillations in σ_{xx} and σ_{yy} in phase with each other because the dominating periodic orbits in the region of the commensurability peak for classical trajectories around one antidot are expected to be almost circular, thus leading to the same velocity correlation functions in the x- and y-directions. The quantum oscillations can be ascribed, however, to the miniband structure (Fig. 8.10). Closely spaced dispersionless minibands along $\Gamma - X$ (leading to high scattering conductivity) correspond to narrow bands along $\Gamma - A$ (small band conductivity), while minibands with large separations along $\Gamma - X$ (small scattering conductivity) are found at energies where minibands along $\Gamma - A$ have a pronounced dispersion (large band conductivity). These oscillations with opposite phase in $\Delta\sigma_{xx}$ and $\Delta\sigma_{yy}$ transform with $\rho_{\mu\mu} \sim \sigma_{\gamma\gamma}$ into the oscillating quantities $\Delta\rho_{yy}$ and $\Delta\rho_{xx}$, respectively, with small changes in the periodicity caused by the tensor inversion.

For a softer antidot potential the oscillations become periodic in B as expected from the semiclassical considerations and observed in quadratic lattices [34]. In the rectangular lattice they stay, however, out of phase in the both longitudinal conductivity components as stated above [87]. This can again be understood from the band structure which looks very similar to the one for the steep antidots (Fig. 8.10) especially in view of the anticrossing between flat and strongly dispersive bands along the $\Gamma - A$ direction. The only difference is that the positions of the flat minibands shift away from the Landau levels.

8.6.4 Quantum-mechanical features

In the quantum regime when the lattice constant approaches the Fermi wavelength one expects to detect the magnetic band structure directly in transport experiments. A first step in this direction was reported in [40]. Earlier the suppression of the Landau band conductivity was observed in systems with weak 2D modulation which was interpreted as an indirect manifestation of the magnetic band structure [85, 86].

In the following the resistivity is discussed for a strongly modulated 2DEG ($E_F \gtrsim V_0$) in the quantum regime. In addition to classical and semiclassical effects with strong temperature dependence, one finds quantum oscillations which are only weakly affected by temperature. As their origin is the dependence of the band structure on the magnetic field [61, 1–3], the leading period is exactly one flux quantum per unit cell (cf. [78, 84] for the antidot case). They differ from the oscillations with period $1/B$ in the resistivity of a weakly modulated 2DEG ($E_F \gg V_0$) [22, 85, 86] caused by a similar mechanism. In the resistivity they are better observed in a strongly modulated system than in the antidot regime ($E_F < V_0$) as indicated by corresponding calculations. Moreover, an experimental realization of the former seems to be easier to achieve at small lattice constants.

The influence of temperature is taken into account in the calculations only via the Fermi function, i.e. temperature-dependent scattering mechanisms are neglected in the considered temperature range ($T < 15$ K). This is justified because phase breaking by electron–electron and electron–phonon scattering occurs on a length scale larger than the thermal length $L_T = \hbar v_F / kT$ introduced by energy averaging [91].

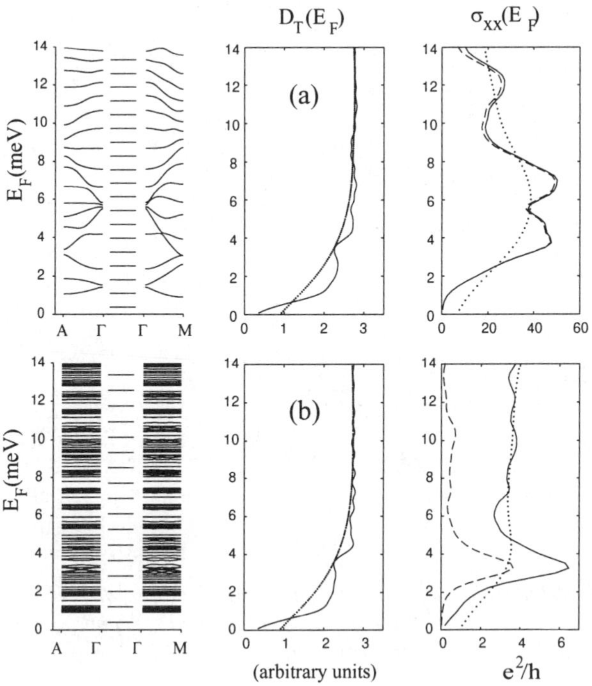

Figure 8.21 *Band structure, TDOS and longitudinal conductivity of a quadratic lattice with strong 2D modulation (*$V_0 = 5$ meV*) and lattice constant* $a = 100$ nm *for two values of the magnetic flux (a)* $n_\Phi = 1$ *and (b)* $n_\Phi = \frac{9}{8}$*. The mobility is* $\mu = 50$ m^2 (Vs)$^{-1}$ *[92].*

The parameters $a = b = 100$ nm for the lattice constant, $V_0 = 5$ meV for the amplitude and $\alpha = \beta = 1$ for the steepness are representative for a strongly modulated system at typical densities $n_s \simeq 1$–3×10^{11} cm^{-2} corresponding to $E_F \simeq 4$–12 meV or $\lambda_F \simeq 75$–40 nm.

The connection between band structure, TDOS and longitudinal conductivity is visualized in Fig. 8.21 for two flux values. In the left panel the energy bands are shown for (1,0) and (1,1) directions in the MBZ, respectively (Fig. 8.12). The positions of the Landau levels are indicated in the middle. The strong dependence of the miniband structure on n_ϕ for similar magnetic fields but different denominators of the rational flux values does not influence the TDOS (middle panel: $T = 4$ K solid line, $T = 15$ K dotted line). This is clearly different for the longitudinal conductivity displayed in the right panel for $T = 4$ K (solid line) and for $T = 15$ K (dotted line) together with the miniband conductivity for $T = 4$ K (dashed line).

For $n_\phi = 1$ (Fig. 8.21(a)) the conductivity is dominated by the miniband contribution due to the strong dispersion with the most dispersive bands corre-

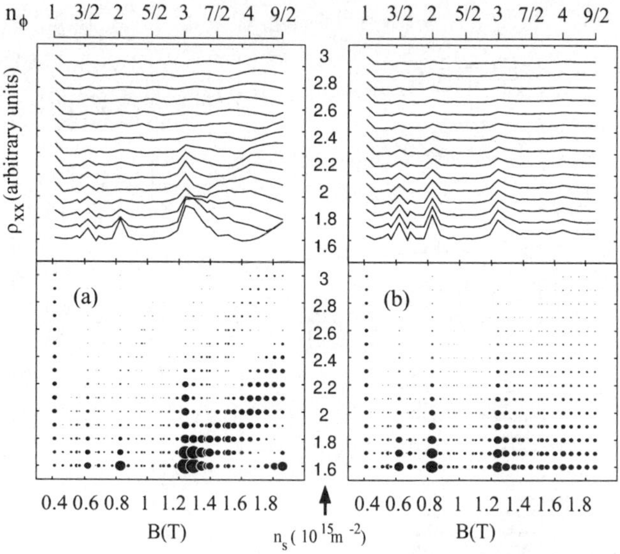

Figure 8.22 *Longitudinal resistivity versus magnetic field for different densities in steps of* 0.1×10^{15} m^{-2}, *vertically offset for clarity, at (a)* $T = 4$ K *and (b)* $T = 15$ K. *In the lower part the dot size indicates the magnitude of* ρ_{xx} *[92].*

sponding to maxima in σ_{xx}. The miniband contribution is strongly suppressed for $n_\phi = \frac{9}{8}$ (Fig. 8.21(b)) and σ_{xx} is dominated by the (slightly enhanced) scattering between closely spaced, almost dispersionless minibands. The comparison of the curves for $T = 4$ K and $T = 15$ K shows that only the fine structure is smeared out by temperature broadening. The suppression of the conductivity by an order of magnitude remains if the flux is changed from an integer to a rational value. Therefore, neither the temperature nor the position of the Fermi energy (i.e. electron density) have considerable influence on this effect in the regime of strong modulation. In the magnetic field dependence of σ_{xx} and, after tensor inversion, of ρ_{xx} and ρ_{xy} this leads to oscillations which are only weakly dependent on temperature and electron density and have their quantum-mechanical origin in the magnetic field-dependent miniband structure.

In Fig. 8.22 the longitudinal resistivity ρ_{xx} for $\mu = 50$ m^2 (Vs)$^{-1}$ is shown depending on the magnetic field for different electron densities n_s corresponding to Fermi energies between 7 meV and 12 meV. Two types of structures can be distinguished. The most pronounced ones for $n_\phi = 1, \frac{3}{2}, 2$ and 3 do not change their position with n_s or T and decrease in amplitude with increasing n_s. SdH-type oscillations (mainly above $B = 1.2$ T) shift to higher magnetic fields with increasing n_s and disappear at higher temperatures (Fig. 8.22(b)). The structure at $B \approx 1.2$ T for the lowest density corresponds to the third Landau level (Fig. 8.11). At higher densities a similar but weaker structure shows up in connection with the fourth Landau level. These structures can also be described by semiclassical concepts [89, 90]. In contrast, the features which do not shift with changing electron density and persist even at higher temperature derive from a magnetic field-dependent miniband structure. They will be called quantum peaks. Note that for $T = 4$ K (Fig. 8.22(a)) these peaks respond differently to changing n_s depending on the position of E_F relative to the gaps in the miniband structure. As for higher temperatures, the gaps loose their relevance and all quantum peaks decrease almost simultaneously with increasing n_s (Fig. 8.22(b)).

Lowering the mobility by a factor of 10 essentially destroys the quantum peak [92]. In that case the scattering contribution dominates the longitudinal conductivity and the reduction of the band part for noninteger fluxes plays only a minor role. Only the semiclassical features survive this increase of scattering strength. Other reasons for a significant suppression of the quantum oscillations not considered in the calculations are fluctuations of lattice constant or antidot shape in experiments on macroscopic samples. Thus, an important experimental prerequisite for an observation of the quantum peaks is either a macroscopic homogeneous sample or a finite small sample if boundary effects can be eliminated. Then a temperature window should exist for an exclusive observation of the quantum peaks when the semiclassical feature is already destroyed by energy averaging but the phase coherence length is still larger than the lattice period.

8.7 SUMMARY AND OUTLOOK

We have discussed classical and quantum-mechanical aspects of transport in lateral superlattices with strong 2D modulation and in antidot lattices. On the one hand, these systems are characterized by a notable influence of classical chaos on the dynamics of the electrons. On the other hand, they have a peculiar magnetic field-dependent band-structure similar to Hofstadter's butterfly.

For the experimental parameters accessible at present, a classical description of the systems has proven to be very successful. It is especially appealing due to the intuitive understanding of classical trajectories. Detailed studies of the classical dynamics have shown, however, that the situation is not as simple as it seemed at the first glance due to the presence of chaos. In view of the dependence of the dominating trajectories on the detailed shape of the electrostatic potential it is rather surprising that the resistivity which is the natural quantity measured in usual experiments always exhibits maxima at the commensurate magnetic fields. The analysis of the quantum spectrum of antidot lattices revealed that the classical chaos also has clear implications on their quantum-mechanical properties.

The illustrative character of classical trajectories has also been exploited in a semiclassical explanation of the main quantum effects in present-day samples. There are, however, some limitations of this plausible approach as it has been shown for the semiclassical oscillations in rectangular lattices with a strong influence of runaway orbits as an example.

The concept of miniband and scattering conductivities used previously for weakly modulated systems has also proven to be very useful in understanding the influence of the quantum-mechanical band structure on the transport properties of antidot lattices. This applies particularly to band-structure effects in the quantum regime which has still to be realized experimentally in the future. Difficulties exist both for the preparation of samples with smaller lattice constants (new methods of patterning) and of such with small electron densities (poor screening of unavoidable impurities by the electrons).

There are several aspects of interest for further investigations of lateral superlattices. The transition from an antidot to a dot lattice corresponds to an artificial metal-insulator transition [93, 94] which implies the possibility of a better experimental control of the system properties. Systems with magnetic modulation represent another realization of a nontrivial band structure which shows differences to the electrostatic modulation already on a classical level [95–100]. In order to access such a band structure more directly than in transport measurements, optical and tunnelling experiments [101] are interesting candidates.

ACKNOWLEDGEMENTS

We gratefully acknowledge the collaboration with R. Neudert, U. Rössler, H. Silberbauer and O. Steffens and helpful discussions with K. Ensslin, D. Weiss, R. Schuster, T. Schlösser, J. P. Kotthaus, K. Richter, R. Gerhardts, W. Schirmacher, R. Fleischmann, R. Ketzmerick and F. von Oppen. Research was supported by the Deutsche Forschungsgemeinschaft within SFB 348. Part of the calculations was done on the Cray Y–MP supercomputers at the Leibnitz–Rechenzentrum München and at the HLRZ Jülich.

REFERENCES

[1] Peierls, R. (1933) *Z. Phys.* **80**, 763.
Fischbeck, H. J. (1970) *Phys. Stat. Sol.* **38**, 11.
Zak, J. (1972) *Solid State Physics* (eds H. Ehrenreich, F. Seitz and W. Turnbull), Academic Press, New York, Vol. 27, p. 1.
[2] Harper, P. G. (1955) *Proc. R. Soc. Lond.* A **68**, 874.
Azbel, M. Ya. (1964) *Sov. Phys. JETP* **19**, 634.
[3] Hofstadter, D. R. (1976) *Phys. Rev.* B **14**, 2239.
[4] Bellissard, J. and Simon, B. J. (1982) *Funct. Anal.* **48**, 408.
[5] Wagenhuber, J., Geisel, T., Niebauer, P. *et al.* (1992) *Phys. Rev.* B **45**, 4372.
[6] Sinai, Y. G. (1959) *Dokl. Akad. Nauk.* **25**, 768.
[7] Gutzwiller, M. C. (1990) *Chaos in Classical and Quantum Mechanics*, Springer, New York.
[8] Haake, F. (1991) *Quantum Signatures of Chaos*, Springer, New York.
[9] Mehta, M. L. (1991) *Random Matrices*, Academic Press, Boston.

[10] Geisel, T., Ketzmerick, R. and Petschel, G. (1991) *Phys. Rev. Lett.* **66**, 1651.
[11] Geisel, T., Ketzmerick, R. and Petschel, G. (1991) *Phys. Rev. Lett.* **67**, 3635.
[12] Petschel, G. and Geisel, T. (1993) *Phys. Rev. Lett.* **71**, 239.
[13] Mucciolo, E. R., Capaz, R. B., Altshuler, B. L. *et al.* (1994) *Phys. Rev.* B **50**, 8245.
[14] Weiss, D., von Klitzing, K., Ploog, K. *et al.* (1989) *Europhys. Lett.* **8**, 179.
[15] Gerhardts, R. R., Weiss, D. and von Klitzing, K. (1989) *Phys. Rev. Lett.* **62**, 1173.
[16] Winkler, R. W., Kotthaus, J. P. and Ploog, K. (1989) *Phys. Rev. Lett.* **62**, 1177.
[17] Beenakker, C. W. J. (1989) *Phys. Rev. Lett.* **62**, 2020.
[18] Beton, P. H., Alves, E. S., Main, P. C. *et al.* (1990) *Phys. Rev.* B **42**, 9229.
[19] Müller, G., Streda, P., Weiss, D. *et al.* (1994) *Phys. Rev.* B **50**, 8938.
[20] Gerhardts, R. R., Weiss, D. and Wulf, U. (1991) *Phys. Rev.* B **43**, 5192.
[21] Gerhardts, R. R. (1992) *Phys. Rev.* B **45**, 3449.
[22] Pfannkuche, D. and Gerhardts, R. R. (1992) *Phys. Rev.* B **46**, 12 606.
[23] Marcus, C. M., Rimberg, A. J., Westervelt, R. M. *et al.* (1992) *Phys. Rev. Lett.* **69**, 506.
Marcus, C. M., Westervelt, R. M., Hopkins, P. F. *et al.* (1993) *Phys. Rev.* B **48**, 2460.
Marcus, C. M., Westervelt, R. M., Hopkins, P. F. *et al.* (1993) *Chaos* **3**, 643.
Marcus, C. M., Westervelt, R. M., Hopkins, P. F. *et al.* (1994) *Surf. Sci.* **305**, 480.
Berry, M. J., Katine, J. A., Westervelt, R. M. *et al.* (1994) *Surf. Sci.* **305**, 495.
Keller, M. W., Millo, O., Mittal, A. *et al.* (1994) *Surf. Sci.* **305**, 501.
Berry, M. J., Baskey, J. H., Westervelt, R. M. *et al.* (1994) *Phys. Rev.* B **50**, 8857.
Chang, A. M., Baranger, H. U., Pfeiffer, L. N. *et al.* (1994) *Phys. Rev. Lett.* **73**, 2111.
Chan, I. H., Clarke, R. M., Marcus, C. M. *et al.* (1995) *Phys. Rev. Lett.* **74**, 3876.
Keller, M. W., Mittal, A., Sleight, J. W. *et al.* (1996) *Phys. Rev.* B **53**, R1693.
[24] Jalabert, R. A., Baranger, H. U. and Stone, A. D. (1990) *Phys. Rev. Lett.* **65**, 2442.
Baranger, H. U., Jalabert, R. A. and Stone, A. D. (1993) *Phys. Rev. Lett.* **70**, 3876.
Baranger, H. U., Jalabert, R. A. and Stone, A. D. (1993) *Chaos* **3**, 665.
Baranger, H. U. and Mello, P. A. (1994) *Phys. Rev. Lett.* **73**, 142.
Baranger, H. U. and Mello, P. A. (1995) *Phys. Rev.* B **51**, 4703.
Baranger, H. U. and Mello, P. A. (1996) *Europhys. Lett.* **33**, 465.
Pluhar, Z., Weidenmüller, H. A., Zuk, J. A. *et al.* (1994) *Phys. Rev. Lett.* **73**, 2115.
Baranger, R. A., Pichard, J.–L. and Beenakker, C. W. J. (1994) *Europhys. Lett.* **27**, 255.
Efetov, K. B. (1995) *Phys. Rev. Lett.* **74**, 2299.
[25] Roukes, M. L. and Scherer, A (1989) *A. Bull. Am. Phys. Soc.* **34**, 633.
[26] Fang, H., Zeller, R. and Stiles, P. J. (1989) *Appl. Phys. Lett.* **55**, 1433.
[27] Ensslin, K. and Petroff, P. M. (1990) *Phys. Rev.* B **41**, 12307.
[28] Lorke, A., Kotthaus, J. P. and Ploog, K. (1991) *Superlatt. Microstruct.* **9**, 103.
[29] Weiss, D., Roukes, M. L., Menschig, A. *et al.* (1991) *Phys. Rev. Lett.* **66**, 2790.
[30] Gusev, G. M., Kvon, Z. D., Kudryashov, V. M. *et al.* (1991) *JETP Lett.* **54**, 368.
[31] Berthold, G., Smoliner, J., Rosskopf, V. *et al.* (1992) *Phys. Rev.* B **45**, 11 350.
[32] Schuster, R., Ensslin, K., Kotthaus, J. P. *et al.* (1992) *Superlatt. Microstruct.* **12**, 93.
[33] Weiss, D. (1992) *Habilitationsschrift*, Universität Stuttgart.
[34] Weiss, D., Richter, K., Menschig, A. *et al.* (1993) *Phys. Rev. Lett.* **70**, 4118.
[35] Nihey, F. and Nakamura, K. (1993) *Physica* B **184**, 398.
[36] Gusev, G. M., Kvon, Z. D., Litvin, L. V. *et al.* (1992) *JETP Lett.* **55**, 123.
Gusev, G. M., Kvon, Z. D., Litvin, L. V. *et al.* (1992) *J. Phys.: Condens. Mat.* **4**, L269.

[37] Nihey, F., Hwang, S. W. and Nakamura, K. (1995) *Phys. Rev.* B **51**, 4649.
[38] Schuster, R., Ensslin, K. and Wharam, D. *et al.* (1994) *Phys. Rev.* B **49**, 8510.
[39] Lenssen, K.–M. H., Boonman, M. E. J., Harmans, C. J. P. M. *et al.* (1995) *Phys. Rev. Lett.* **74**, 454.
[40] Schlösser, T., Ensslin, K. and Kotthaus, J. P. (1996) *Europhys. Lett.* **33**, 683.
[41] Ensslin, K., Sasa, S., Deruelle, T. *et al.* (1992) *Surf. Sci.* **263**, 319.
[42] Schuster, R., Ensslin, K., Kotthaus, J. P. *et al.* (1993) *Phys. Rev.* B **47**, 6843.
[43] Weiss, D., Richter, K., Vasiliadou, E. *et al.* (1994) *Surf. Sci.* **305**, 408.
[44] Schuster, R. and Ensslin, K. (1994) *Advances in Solid State Physics* **34**, 195.
[45] Tsukagoshi, K., Wakayama, S., Oto, K. *et al.* (1995) *Phys. Rev.* B **52**, 8344.
[46] Fang, H. and Stiles, P. J. (1990) *Phys. Rev.* B **41**, 10 171.
[47] Yamashiro, T., Takahara, J., Tagagaki, Y. *et al.* (1991) *Solid State Commun.* **79**, 885.
[48] Gusev, G. M., Kvon, Z. D., Litvin, L. V. *et al.* (1992) *JETP Lett.* **56**, 170.
[49] Wenchang Lu, (1996) *Phys. Rev.* B **54**, 8049.
Wenchang Lu and Andersen, O. K. (1996) *The Physics of Semiconductors* (eds M. Scheffler and R. Zimmermann), World Scientific, Singapore, P. 1497.
[50] Kubo, R. (1957) *J. Phys. Soc. Jpn.* **12**, 570.
[51] Fleischmann, R., Geisel, T. and Ketzmerick, R. (1992) *Phys. Rev. Lett.* **68**, 1367.
[52] Lorke, A., Kotthaus, J. P. and Ploog, K. (1991) *Phys. Rev.* B **44**, 3447.
[53] Baskin, E. M., Gusev, G. M., Kvon, Z. D. *et al.* (1992) *JETP Lett.* **55**, 678.
[54] Hofmann, A. (1995) Diplomathesis, Ludwig–Maximilians–Universität München.
[55] Schuster, R., Ernst, G., Ensslin, K. *et al.* (1994) *Phys. Rev.* B **50**, 8090.
[56] Tsukagoshi, K., Haraguchi, M., Takaoka, S. *et al.* (1996) *J. Phys. Soc. Jpn.* **65**, 811.
[57] Aoki, K. (1996) *The Physics of Semiconductors* (eds M. Scheffler and R. Zimmermann), World Scientific, Singapore, p. 1521.
[58] Fleischmann, R., Geisel, T. and Ketzmerick, R. (1994) *Europhys. Lett.* **25**, 219.
[59] Zak, J. (1964) *Phys. Rev.* **134**, A1602.
[60] Ferrari, R. (1990) *Phys. Rev.* B **42**, 4598.
[61] Silberbauer, H. (1992) *J. Phys.: Condens. Mat.* **4**, 7355.
[62] Rotter, P., Rössler, U., Silberbauer, H. *et al.* (1995) *Physica* B **212**, 231.
[63] Schuster, R., Ensslin, K., Kotthaus, J. P. *et al.* (1997) *Phys. Rev.* B. **55**, 2237.
[64] Pandey, A. (1981) *Ann. Phys.* **134**, 110.
French, J. B., Kota, V. K. B., Pandey, A. *et al.* (1988) *Ann. Phys. N. Y.* **181**, 198.
Bohigas, O., Tomsovic, S. and Ullmo, D. (1993) *Phys. Rep.* **223**, 43.
[65] Brody, T. A. (1973) *Lett. Nuovo Cimento* **7**, 482.
[66] Berry, M. V. and Robnik, M. (1984) *J. Phys. A: Math. Gen.* **17**, 2413.
[67] Izrailev, F. M. (1988) *Phys. Lett.* A **134**, 13.
[68] Lenz, G. and Haake, F. (1990) *Phys. Rev. Lett.* **65**, 2325.
Lenz, G. and Haake, F. (1991) *Phys. Rev. Lett.* **67**, 1.
[69] Leyvraz, F. and Seligman, T. H. (1990) *J. Phys. A: Math. Gen.* **23**, 1555.
[70] Caurier, E., Grammaticos, B. and Ramani, A. (1990) *J. Phys. A: Math. Gen.* **23**, 4903.
[71] Leitner, D. M. (1993) *Phys. Rev.* B **48**, 2536.
[72] Prosen, T. and Robnik, M. (1993) *J. Phys. A: Math. Gen.* **26**, 2371.
[73] Simons, B. D. and Altshuler, B. L. (1993) *Phys. Rev. Lett.* **70**, 4063.
Simons, B. D. and Altshuler, B. L. (1993) *Phys. Rev.* B **48**, 5422.

Taniguchi, N. and Altshuler, B. L. (1993) *Phys. Rev. Lett.* **71**, 4031.
[74] Silberbauer, H., Rotter, P., Rössler, U. *et al.* (1995) *Europhys. Lett.* **31**, 393.
[75] Silberbauer, H., Rotter, P., Suhrke, M. *et al.* (1994) *Semicond. Sci. & Technol.* **9**, 1906.
[76] Steffens, O., Rotter, P. and Suhrke, F.M. (1997) *Phys. Rev.* B **55**, 4486.
[77] Huang, D. and Gumbs, G. (1993) *Phys. Rev.* B **48**, 2835.
Huang, D., Gumbs, G. and MacDonald, A. H. (1993) *Phys. Rev.* B **48**, 2843.
[78] Oakeshott, R. B. S. and. MacKinnon, A (1993) *J. Phys.: Condens. Matt.* **5**, 6991.
Oakeshott, R. B. S. and. MacKinnon, A (1994) *J. Phys.: Condens. Matt.* **6**, 1519.
[79] Hongqi Xu, Zhen-Li Ji and Berggren, K.–F. (1992) *Superlatt. & Microstruct.* **12**, 237.
Hongqi Xu (1994) *Phys. Rev.* B **50**, 8469.
Hongqi Xu (1994) *Phys. Rev.* B **50**, 12 254.
Hongqi Xu (1996) *The Physics of Semicondutors* (eds M. Scheffler and R. Zimmermann), World Scientific, Singapore, p. 1489.
[80] Zozulenko, I. V., Maaø, F. A., Hauge, E. H. (1995) *Phys. Rev.* B **51**, 7058.
Zozulenko, I. V., Maaø, F. A. and Hauge, E. H. (1996) *Phys. Rev.* B **53**, 7975.
Zozulenko, I. V., Maaø, F. A. and Hauge, E. H. (1996) *Phys. Rev.* B **53**, 7987.
Zozulenko, I. V., Maaø, F. A. and Hauge, E. H. (1996) *The Physics of Semicondutors* (eds M. Scheffler and R. Zimmermann), World Scientific, Singapore, p. 1493.
[81] Uryu, S. and Ando, T. (1996) *Phys. Rev.* B **53**, 13 613.
Uryu, S. and Ando, T. (1996) *The Physics of Semicondutors* (eds M. Scheffler and R. Zimmermann), World Scientific, Singapore, p. 1505.
[82] Wenchang Lu and Andersen, O. K. (1996) *The Physics of Semicondutors* (eds M. Scheffler and R. Zimmermann), World Scientific, Singapore, p. 1497.
[83] Silberbauer, H. and Rössler, U. (1994) *Phys. Rev.* B **50**, 11 911.
[84] Ishizaka, S., Nihey, F., Nakamura, K. *et al.* (1995) *Phys. Rev.* B **51**, 9881.
[85] Weiss, D., Menschig, A., von Klitzing, K. *et al.* (1992) *Surf. Sci.* **263**, 314.
[86] Schlösser, T., Ensslin, K., Kotthaus, J. P. *et al.* (1996) *Surf. Sci.* **847**, 361–362.
[87] Neudert, R., Rotter, P., Rössler, U. *et al.* (1997) *Phys. Rev.* B. **55**, 2242.
Rotter, P., Rössler, U., Suhrke, M. *et al.* (1996) *The Physics of Semicondutors* (eds M. Scheffler and R. Zimmermann), World Scientific, Singapore, p. 1517.
[88] Thornton, T. J., Roukes, M. L., Scherer, A. *et al.* (1989) *Phys. Rev. Lett.* **63**, 2128.
[89] Richter, K. (1995) *Europhys. Lett.* **29**, 7.
[90] Hackenbroich, G. and von Oppen, F. (1995) *Europhys. Lett.* **29**, 151.
[91] Yacoby, A., Sivan, U., Umbach, C. P. *et al.* (1991) *Phys. Rev. Lett.* **66**, 1938.
Yacoby, A., Umansky, V., Shtrikman, H. *et al.* (1994) *Phys. Rev. Lett.* **73**, 3149.
Kurdak, C., Chang, A. M., Chin, A. *et al.* (1992) *Phys. Rev.* B **46**, 6846.
[92] Rotter, P., Suhrke, M. and Rössler, U. (1996) *Phys. Rev.* B **54**, 4452.
[93] Lütjering, G., Weiss, D., Tank, R. *et al.* (1996) *Surf. Sci.* **925**, 361–362.
[94] Nihey, F., Kastner, M. A. and Nakamura, K. (1996) *The Physics of Semiconductors* (eds M. Scheffler and R. Zimmermann), World Scientific, Singapore, p. 1525.
[95] Izawa, S., Katsumoto, S., Endo, A. *et al.* (1995) *J. Phys. Soc. Jpn.* **64**, 706.
Carmona, H. A., Geim, A. K., Nogaret, A. *et al.* (1994) *Phys. Rev. Lett.* **74**, 3009.
Ye, P. D., Weiss, D., Gerhardts, R. R. *et al.* (1994) *Phys. Rev. Lett.* **74**, 3013.
[96] Ye, P. D., Weiss, D., von Klitzing, K. *et al.* (1995) *Appl. Phys. Lett.* **67**, 1441.
[97] Vasilopoulos, P. and Peeters, F. M. (1990) *Superlatt. & Microstruct.* **7**, 393.

Xue, D. P. and Xiao, G. (1992) *Phys. Rev.* B **45**, 5986.
Peeters, F. M. and Vasilopoulos, P. (1993) *Phys. Rev.* B **47**, 1466.
Oakeshott, R. B. S. and MacKinnon, A. (1993) *J. Phys.: Condens. Matt.* **5**, 9355.
[98] Schmidt, G. J. O. (1993) *Phys. Rev.* B **47**, 13 007.
Schmelcher, P. and Shepelyansky, D. L. (1994) *Phys. Rev.* B **49**, 7418.
Gerhardts, R. R. (1996) *Phys. Rev.* B **53**, 11 064.
[99] Menne, R. and Gerhardts, R. R. (1997) *Proc. 12th Int. Conf. Application of High Magnetic Fields in Semiconductor Physics* (eds G. Landwehr and W. Ossau), World Scientific, Singapore, p. 323.
[100] Chang, M. C. and Niu, Q. (1994) *Phys. Rev.* B **50**, 10 843.
Gerhardts, R. R., Pfannkuche, D. and Gudmundsson, V. (1996) *Phys. Rev.* B **53**, 9591.
[101] Kardynał, B., Barnes, C. H. W. and Linfield, E. H. (1996) *Phys. Rev. Lett.* **76**, 3802.

CHAPTER 9

Bloch oscillations and Wannier–Stark localization in semiconductor superlattices

Fausto Rossi

Istituto Nazionale per la Fisica della Materia, Dipartimento di Fisica, Università di Modena, Via G. Campi 213/A, I-41100 Modena, Italy

9.1 INTRODUCTION

Ever since the initial applications of quantum mechanics to the dynamics of electrons in solids, the analysis of Bloch electrons moving in a homogeneous electric field has been of central importance.

In 1928 Bloch [1] demonstrated, by employing semiclassical arguments, that a wave packet given by a superposition of single-band states peaked about some quasimomentum, $\hbar\mathbf{k}$, moves with a group velocity given by the gradient of the energy-band function with respect to the quasimomentum and that the rate of change of the quasimomentum is proportional to the applied field $\mathbf{F}$. This is often referred to as the 'acceleration theorem':

$$\hbar\dot{\mathbf{k}} = e\mathbf{F}. \qquad (9.1.1)$$

Thus, in the absence of interband tunnelling and scattering processes, the quasimomentum of a Bloch electron in a homogeneous and static electric field will be uniformly accelerated into the next Brillouin zone in a repeated-zone scheme (or equivalently undergoes an Umklapp process back in to the first zone). The corresponding motion of the Bloch electron through the periodic energy-band structure, shown in Fig. 9.1, is called 'Bloch oscillation'. It is characterized by an oscillation period $\tau_B = h/eFd$, where d denotes the lattice periodicity in the field direction.

There are two mechanisms impeding a fully periodic motion: interband tunnelling and scattering processes. Interband tunnelling is an intricate problem and is still at the centre of a continuing debate. Early calculations of the tunnelling

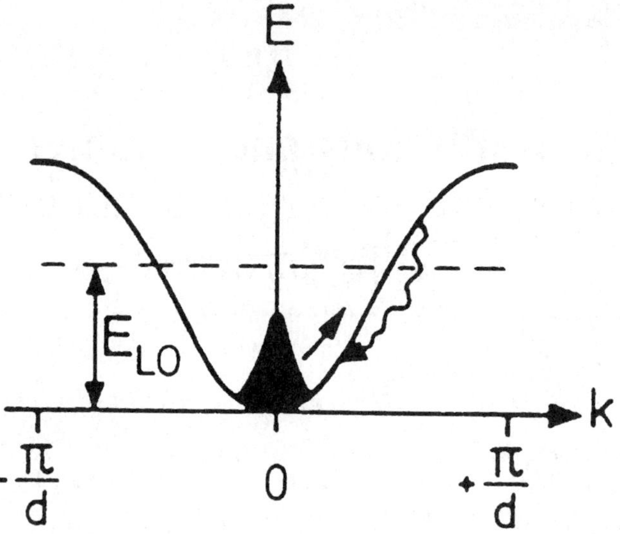

Figure 9.1 *Schematic illustration of the field-induced coherent motion of an electronic wave packet initially created at the bottom of a miniband. Here, the width of the miniband exceeds the LO-phonon energy E_{LO}, so that LO-phonon scattering is possible. After [40].*

probability into other bands in which the electric field is represented by a time-independent scalar potential were made by Zener [2] using a Wentzel–Kramers–Brillouin generalization of Bloch functions, by Houston [3] using accelerated Bloch states (Houston states) and subsequently by Kane [4] and Argyres [5] who employed the crystal-momentum representation. Their calculations lead to the conclusion that the tunnelling rate per Bloch period is much less than unity for electric fields up to 10^6 V cm^{-1} for typical band parameters corresponding to elemental or compound semiconductors.

Despite the apparent agreement among these calculations, the validity of employing the crystal-momentum representation or Houston functions to describe electrons moving in a nonperiodic (crystal plus external field) potential has been disputed. The starting point of the controversy was the original paper by Wannier [6]. He pointed out that, due to the translational symmetry of the crystal potential, if $\phi(\boldsymbol{r})$ is an eigenfunction of the scalar-potential Hamiltonian (corresponding to the perfect crystal plus the external field) with eigenvalue ϵ, then any $\phi(\boldsymbol{r} + n\boldsymbol{d})$ is also an eigenfunction with eigenvalue $\epsilon + n\Delta\epsilon$, where $\Delta\epsilon = eFd$ is the so-called Wannier–Stark splitting ($\boldsymbol{d}$ being the primitive lattice vector along the field direction). He concluded that the translational symmetry of the crystal gives rise to a discrete energy spectrum, the so-called Wannier–Stark ladder. The states corresponding to these equidistantly spaced levels are localized states, as schematically shown in Fig. 9.2 for the case of a semicon-

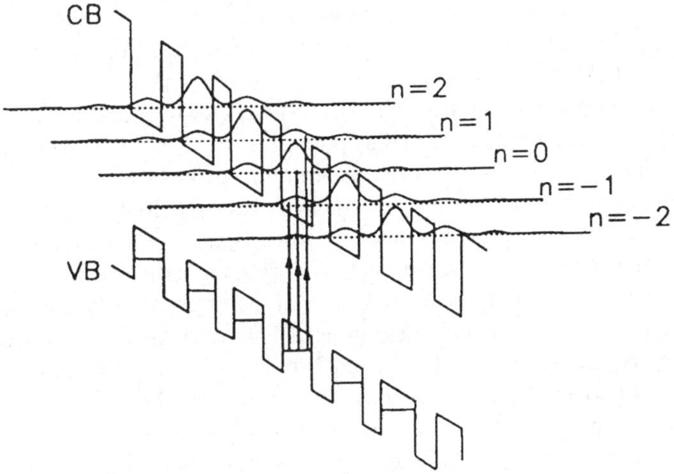

Figure 9.2 *Schematic representation of the transitions from the valence to the conduction band of a superlattice in the Wannier–Stark localization regime. After [22].*

ductor superlattice. The degree of this Wannier–Stark localization depends on the strength of the applied field.

The existence of such energy quantization was disputed by Zak [7], who pointed out that for the case of an infinite crystal the scalar potential $-\boldsymbol{F} \cdot \boldsymbol{r}$ is not bounded, which implies a continuous energy spectrum. Thus, the main point of the controversy was related to the existence (or absence) of Wannier–Stark ladders. More precisely, the point was to decide if interband tunnelling (neglected in the original calculation by Wannier [6]) is strong enough to destroy the Wannier–Stark energy quantization (and the corresponding Bloch oscillations) or not. We will give a brief historical account of this long-standing controversy in the following section.

It is only during the last decade that this controversy has come to an end. From a theoretical point of view, most of the formal problems related to the non-periodic nature of the scalar potential (superimposed onto the periodic crystal potential) were finally removed by using a vector-potential representation of the applied field [8, 9]. Within such vector-potential picture, upper boundaries for the interband tunnelling probability have been established at a rigorous level, which shows that an electron may execute a number of Bloch oscillations before tunnelling out of the band [9–11], in qualitatively good agreement with the earlier predictions of Zener and Kane [2, 4].

The second mechanism impeding a fully periodic motion is scattering by phonons, impurities, etc. (Fig. 9.1). This results in shorter lifetimes than the Bloch period τ_B for all reasonable values of the electric field, so that Bloch oscillations should not be observable in conventional solids.

In superlattices, however, the situation is much more favourable because of the smaller Bloch period τ_B resulting from the small width of the mini-Brillouin zone in the field direction [12].

Indeed, the existence of Wannier–Stark ladders as well as Bloch oscillations in superlattices has been confirmed by a number of recent experiments [13]. The photoluminescence and photocurrent measurements of the biased GaAs/GaAlAs superlattices performed by Mendez, Agullo-Rueda and Hong [14], together with the electroluminescence experiments by Voisin *et al.* [15], provided the earliest experimental evidence of the field-induced Wannier–Stark ladders in superlattices. A few years later, Feldmann *et al.* [16] were able to measure Bloch oscillations in the time domain through a four-wave-mixing (FWM) experiment originally suggested by von Plessen and Thomas [17]. A detailed analysis of the Bloch oscillations in the FWM signal (which reflects the interband dynamics) has also been performed by Leo and coworkers [18, 19].

In addition to the above interband-polarization analysis, Bloch oscillations have also been detected by monitoring the intraband polarization which, in turn, is reflected by anisotropic changes in the refractive index [13]. Measurements based on transmittive electro-optic sampling (TEOS) have been performed by Dekorsy and coworkers [20, 21]. Finally Bloch oscillations were recently measured through a direct detection of the terahertz (THz) radiation in semiconductor superlattices [22, 23].

The aim of this chapter is to present a general approach to the study of the ultrafast carrier dynamics in semiconductor superlattices. Our theoretical description, based on the density matrix formalism discussed in Chapter 6, is presented in section 9.3. It allows us to derive a set of kinetic equations which accounts for interband tunnelling as well as scattering processes and it is valid in any quantum-mechanical representation.

In section 9.4 the freedom of choice of the basis states in our kinetic formulation will be used to introduce the two typical pictures commonly used for the description of semiconductor superlattices, namely, the Bloch-oscillation and Wannier–Stark pictures. In particular, we will see that they correspond to the two equivalent vector- and scalar-potential representations of the applied field. This will implicitly state the total equivalence of the Bloch-oscillation and Wannier–Stark representations, which in turn shows that the so-called 'semiclassical Bloch picture' is in contrast, a rigorous quantum-mechanical result.

Finally, in section 9.5 we will review and discuss some simulated experiments.

9.2 HISTORICAL BACKGROUND

In this section we give a brief historical account concerning the controversy on the existence of the above-mentioned Wannier–Stark ladders. Some of the main criticisms to the pioneering works on Bloch oscillations may be summarized as follows.

1. The eigenvalues of the time-independent Schrödinger equation are not quantized but they form a continuous spectrum.

2. Since the Hamiltonian within the scalar-potential representation is not periodic, it is not clear whether one can employ the periodic Bloch states or Houston functions: a superposition of Bloch functions will automatically yield a periodic function while the solution of the time-dependent Schrödinger equation is, in general, not periodic.

3. The position operator (entering the scalar potential) is ill-defined within the crystal-momentum representation.

In a series of papers, Wannier [6, 24] and coworkers [25, 26] argued that in the presence of a homogeneous electric field, one can modify the Bloch states in such a way that there is no interband coupling and an electron in a crystal will move within one band with its $\boldsymbol{k}$ changing in time according to the acceleration theorem in equation (9.1.1). Furthermore, if $\boldsymbol{k}(t = 0)$ is in the direction of a reciprocal-lattice vector, the periodic motion in $\boldsymbol{k}$-space gives rise to an energy quantization with $\Delta\epsilon = eFd$ ($d = 2\pi/G$ being the lattice constant along the field direction), the so-called Wannier–Stark ladders. The basis for this idea is that energy bands arise from the translational symmetry of the crystalline field and this symmetry is not removed physically by the presence of the applied field [24], i.e. the field is still periodic with the lattice period.

These arguments have been refuted by Zak [7], who showed that, although it immediately follows from the one-dimensional time-independent Schrödinger equation for an infinite crystal with lattice constant d, that if ϵ is an eigenvalue, $\epsilon + n\Delta\epsilon$ is also an eigenvalue, the spectrum of ϵ is continuous with $-\infty < \epsilon < +\infty$, so the Wannier–Stark ladders do not exist.

Wannier [27] argued that Zak's criticisms of his proof are not valid, but concedes that the Stark ladders may be metastable resonant states limited by interband tunnelling, as for the case of the hydrogen atom in the presence of a static field. However, Wannier's arguments were immediately rejected by Zak [28], who claimed that Wannier's original equation was incorrect.

A few years later, Rabinovitch and Zak [29] extended Zak's [7] earlier arguments to the question of Bloch oscillations. They argued that since neglecting the interband coupling terms in the time-independent Schrödinger equation leads incorrectly to energy quantization, then the interband terms cannot be neglected in the lowest approximation because they are the same order as the terms retained. By applying the same reasoning to the time-dependent equation, they conclude (without offering a proof) that neglecting the interband terms as a first approximation, as done by Houston [3], is incorrect for times equal to or longer than the Bloch-oscillation period τ_B. From their conclusions it follows that the typical diagrams which are commonly used to portray trajectories of $\boldsymbol{k}(t)$ superimposed upon the energy-band structure (Fig. 9.1) are incorrect and misleading.

Nevertheless, shortly before these latter arguments appeared, experimental results were obtained by Koss and Lambert [30], which were interpreted as supporting the existence of Wannier–Stark levels. They found that the observed low-temperature optical absorption of GaAs in a strong electric field ($F = 10^5$ V cm^{-1}) closely followed the theoretical predictions of Callaway [31], which, in turn, were based on employing Kane's wavefunctions and Wannier–Stark quantized energy levels.

9.3 THEORETICAL ANALYSIS

In this section we will try to review and discuss in a systematic way the basic ideas used in the theoretical analysis of semiconductor superlattices. As already pointed out in section 9.1, the phenomena under investigation, i.e. Bloch oscillations and Wannier–Stark localization, are peculiar of any lattice structure. Therefore, even if most of the results discussed in this chapter refer to semiconductor superlattices, the general formulation presented in this section applies to any crystalline structure.

9.3.1 Physical System

In order to study the optical and transport properties of semiconductor superlattices, let us consider a gas of carriers in a crystal under the action of an applied electromagnetic field. The carriers will experience their mutual interaction as well as the interaction with the phonon modes of the crystal. Such a physical system can be described by the following Hamiltonian:

$$\boldsymbol{H} = \boldsymbol{H}_c + \boldsymbol{H}_p + \boldsymbol{H}_{cc} + \boldsymbol{H}_{cp} + \boldsymbol{H}_{pp}. \tag{9.3.1}$$

The first term describes the noninteracting carrier system in the presence of the external electromagnetic field while the second one refers to the free-phonon system. The last three terms describe many-body contributions: they refer, respectively, to carrier–carrier, carrier–phonon and phonon–phonon interactions.

In order to discuss their explicit form, let us introduce the usual second-quantization field operators $\boldsymbol{\Psi}^\dagger(\boldsymbol{r})$ and $\boldsymbol{\Psi}(\boldsymbol{r})$. They describe, respectively, the creation and annihilation of a carrier in $\boldsymbol{r}$. In terms of the above field operators the carrier Hamiltonian $\boldsymbol{H}_c$ can be written as

$$\boldsymbol{H}_c = \int \mathrm{d}\boldsymbol{r}\, \boldsymbol{\Psi}^\dagger(\boldsymbol{r}) \left[\frac{(-\mathrm{i}\hbar\nabla_{\boldsymbol{r}} - \frac{e}{c}\boldsymbol{A}(\boldsymbol{r},t))^2}{2m_0} + e\varphi(\boldsymbol{r},t) + V^l(\boldsymbol{r}) \right] \boldsymbol{\Psi}(\boldsymbol{r}). \tag{9.3.2}$$

Here, $V^l(\boldsymbol{r})$ denotes the periodic potential due to the perfect crystal while $\boldsymbol{A}(\boldsymbol{r},t)$ and $\varphi(\boldsymbol{r},t)$ denote, respectively, the vector and scalar potentials corresponding to the external electromagnetic field. Since we are interested in the electro-optical properties as well as in the ultrafast dynamics of photo-excited carriers, the electromagnetic field acting on the crystal—and the corresponding electromagnetic potentials—will be the sum of two different contributions: the high-frequency laser field responsible for the ultrafast optical excitation and the additional electromagnetic field acting on the photo-excited carriers on a longer timescale. More specifically, by denoting with the labels 1 and 2 these two contributions, we can write

$$\boldsymbol{A}(\boldsymbol{r},t)=\boldsymbol{A}_1(\boldsymbol{r},t)+\boldsymbol{A}_2(\boldsymbol{r},t), \qquad \varphi(\boldsymbol{r},t)=\varphi_1(\boldsymbol{r},t)+\varphi_2(\boldsymbol{r},t) \tag{9.3.3}$$

and recalling that

$$\boldsymbol{E}(\boldsymbol{r},t)=-\frac{1}{c}\frac{\partial}{\partial t}\boldsymbol{A}(\boldsymbol{r},t)-\nabla_{\boldsymbol{r}}\varphi(\boldsymbol{r},t), \qquad \boldsymbol{B}(\boldsymbol{r},t)=\nabla_{\boldsymbol{r}}\times\boldsymbol{A}(\boldsymbol{r},t) \tag{9.3.4}$$

we have

$$\boldsymbol{E}(\boldsymbol{r},t)=\boldsymbol{E}_1(\boldsymbol{r},t)+\boldsymbol{E}_2(\boldsymbol{r},t), \qquad \boldsymbol{B}(\boldsymbol{r},t)=\boldsymbol{B}_1(\boldsymbol{r},t)+\boldsymbol{B}_2(\boldsymbol{r},t). \tag{9.3.5}$$

Equation (9.3.4), which gives the electromagnetic fields in terms of the corresponding vector and scalar potentials, reflects the well-known gauge freedom: there is an infinite number of possible combinations of $\boldsymbol{A}$ and φ which give rise to the same electromagnetic field $\{\boldsymbol{E},\boldsymbol{B}\}$. We will use such freedom of choice for the laser field (term 1): we assume a homogeneous (space-independent) laser field $\boldsymbol{E}_1(t)$ fully described by the scalar potential

$$\varphi_1(\boldsymbol{r},t)=-\boldsymbol{E}_1(t)\cdot\boldsymbol{r}. \tag{9.3.6}$$

This assumption, which corresponds to the well-known dipole approximation, is well justified as long as the space-scale of interest is small compared with the light wavelength. The explicit form of the laser field considered in this chapter is

$$E_1(t)=E^+(t)+E^-(t)=E_0(t)\mathrm{e}^{\mathrm{i}\omega_L t}+E_0^*(t)\mathrm{e}^{-\mathrm{i}\omega_L t}, \tag{9.3.7}$$

where $E_0(t)$ is the amplitude of the light field and ω_L denotes its central frequency.

With this particular choice of the electromagnetic potentials describing the laser field, the Hamiltonian in equation (9.3.2) can be rewritten as

$$\boldsymbol{H}_c=\boldsymbol{H}_c^0+\boldsymbol{H}_{cl}, \tag{9.3.8}$$

where

$$\boldsymbol{H}_c^0=\int \mathrm{d}\boldsymbol{r}\,\boldsymbol{\Psi}^\dagger(\boldsymbol{r})\left[\frac{\left(-\mathrm{i}\hbar\nabla_{\boldsymbol{r}}-\frac{e}{c}\boldsymbol{A}_2(\boldsymbol{r},t)\right)^2}{2m_0}+e\varphi_2(\boldsymbol{r},t)+V^l(\boldsymbol{r})\right]\boldsymbol{\Psi}(\boldsymbol{r}) \tag{9.3.9}$$

describes the carrier system in the crystal under the action of the electromagnetic field 2 only, while

$$\boldsymbol{H}_{cl}=e\int \mathrm{d}\boldsymbol{r}\,\boldsymbol{\Psi}^\dagger(\boldsymbol{r})\varphi_1(\boldsymbol{r},t)\boldsymbol{\Psi}(\boldsymbol{r}) \tag{9.3.10}$$

describes the carrier–light (cl) interaction due to the laser photo-excitation.

In analogy with the carrier system, by denoting with $b^\dagger_{q,\lambda}$ and $b_{q,\lambda}$ the creation and annihilation operators for a phonon of mode λ and wave vector $\boldsymbol{q}$, the free-phonon Hamiltonian takes the form

$$\boldsymbol{H}_p = \sum_{q\lambda} \hbar\omega_{q\lambda} b^\dagger_{q\lambda} b_{q\lambda}, \tag{9.3.11}$$

where $\omega_{q\lambda}$ is the dispersion relation for the phonon mode λ.

Let us now discuss the explicit form of the many-body contributions. The carrier–carrier interaction is described by the two-body Hamiltonian

$$\boldsymbol{H}_{cc} = \tfrac{1}{2} \int \mathrm{d}\boldsymbol{r} \int \mathrm{d}\boldsymbol{r}'\, \Psi^\dagger(\boldsymbol{r})\Psi^\dagger(\boldsymbol{r}') V_{cc}(\boldsymbol{r}-\boldsymbol{r}')\Psi(\boldsymbol{r}')\Psi(\boldsymbol{r}), \tag{9.3.12}$$

where V_{cc} denotes the Coulomb potential.

Let us now introduce the carrier–phonon interaction Hamiltonian

$$\boldsymbol{H}_{cp} = \int \mathrm{d}\boldsymbol{r}\, \Psi^\dagger(\boldsymbol{r}) V_{cp}(\boldsymbol{r})\Psi(\boldsymbol{r}), \tag{9.3.13}$$

where

$$V_{cp} = \sum_{q\lambda} \left[\tilde{g}_{q\lambda} b_{q\lambda} \mathrm{e}^{\mathrm{i}\boldsymbol{q}\cdot\boldsymbol{r}} + \tilde{g}^*_{q\lambda} b^\dagger_{q\lambda} \mathrm{e}^{-\mathrm{i}\boldsymbol{q}\cdot\boldsymbol{r}} \right] \tag{9.3.14}$$

is the electrostatic phonon potential induced by the lattice vibrations. Here, the explicit form of the coupling function $\tilde{g}_{q\lambda}$ depends on the particular phonon mode λ (acoustical, optical, etc.) as well as on the coupling mechanism considered (deformation potential, polar coupling, etc.).

Let us finally discuss the phonon–phonon contribution $\boldsymbol{H}_{pp}$. The free-phonon Hamiltonian $\boldsymbol{H}_p$ introduced in equation (9.3.11), which describes a system of noninteracting phonons, by definition accounts only for the harmonic part of the lattice potential. However, nonharmonic contributions of the interatomic potential can play an important role in determining the lattice dynamics in highly excited systems [32], since they are responsible for the decay of optical phonons into phonons of lower frequency. In our second-quantization picture, these nonharmonic contributions can be described in terms of a phonon–phonon interaction which induces, in general, transitions between free-phonon states. Here, we will not discuss the explicit form of the phonon–phonon Hamiltonian $\boldsymbol{H}_{pp}$ responsible for such a decay. We will simply assume that such a phonon–phonon interaction is efficient enough to maintain the phonon system in thermal equilibrium. This corresponds to neglecting hot-phonon effects [33].

It is well known that the coordinate representation used so far is not the most appropriate one at describing the electron dynamics within a periodic crystal. In contrast, it is in general more convenient to employ the representation given by the eigenstates of the noninteracting carrier Hamiltonian—or a part of it—since it automatically accounts for some of the symmetries of the system. For the moment we will simply consider an orthonormal basis set $\{\phi_n(\boldsymbol{r})\}$ without specifying which part of the Hamiltonian is diagonal in such a representation.

This will allow us to write down equations valid in any quantum-mechanical representation. Since the noninteracting carrier Hamiltonian is, in general, a function of time, also the basis functions ϕ_n may be time dependent. Here, the label n denotes, in general, a set of discrete and/or continuous quantum numbers. In the absence of electromagnetic field, the above wave functions will correspond to the well-known Bloch states of the crystal and the index n will reduce to the wave vector $\boldsymbol{k}$ plus the band index ν. In the presence of a homogeneous magnetic field the eigenfunctions ϕ_n may instead correspond to Landau states. Finally, for the case of a constant and homogeneous electric field, two equivalent representations exist: the accelerated Bloch states and the Wannier–Stark picture. Such equivalence results to be of crucial importance in understanding the relationship between Bloch oscillations and Wannier–Stark localization and, for this reason, it will be discussed in more detail in section 9.4.

Let us now reconsider the system Hamiltonian introduced so far in terms of such ϕ_n representation. As a starting point, we may expand the second-quantization field operators in terms of the new wave functions:

$$\Psi(\boldsymbol{r}) = \sum_n \phi_n(\boldsymbol{r})a_n, \qquad \Psi^\dagger(\boldsymbol{r}) = \sum_n \phi_n^*(\boldsymbol{r})a_n^\dagger. \tag{9.3.15}$$

The above expansion defines the new set of second-quantization operators $a_n^\dagger$ and a_n; they describe, respectively, the creation and annihilation of a carrier in state n.

For the case of a semiconductor crystal (which will be the only one considered in this chapter), the energy spectrum of the noninteracting carrier Hamiltonian in equation (9.3.9)—or a part of it—is always characterized by two well-separated energy regions: the valence and conduction band. Also in the presence of an applied electromagnetic field, the periodic lattice potential V^l gives rise to a large energy gap. Therefore, we deal with two energetically well-separated regions, which suggests the introduction of the so-called electron–hole picture. This corresponds to a separation of the set of states $\{\phi_n\}$ into conduction states $\{\phi_i^e\}$ and valence states $\{\phi_j^h\}$. Thus, also the creation (annihilation) operators $a_n^\dagger$ (a_n) introduced in equation (9.3.15) will be divided into creation (annihilation) electron and hole operators: $c_i^\dagger$ (c_i) and $d_j^\dagger$ (d_j). In terms of the new electron–hole picture, the expansion in equation (9.3.15) is given by:

$$\begin{aligned}\Psi(\boldsymbol{r}) &= \sum_i \phi_i^e(\boldsymbol{r})c_i + \sum_j \phi_j^{h*}(\boldsymbol{r})d_j^\dagger \\ \Psi^\dagger(\boldsymbol{r}) &= \sum_i \phi_i^{e*}(\boldsymbol{r})c_i^\dagger + \sum_j \phi_j^h(\boldsymbol{r})d_j.\end{aligned} \tag{9.3.16}$$

If we now insert the above expansion into equation (9.3.9), the noninteracting carrier Hamiltonian takes the form

$$H_c^0 = \sum_{ii'} \epsilon_{ii'}^e c_i^\dagger c_{i'} + \sum_{jj'} \epsilon_{jj'}^h d_j^\dagger d_{j'} = H_e^0 + H_h^0, \tag{9.3.17}$$

where

$$\epsilon_{ll'}^{e/h} = \pm \int \mathrm{d}\boldsymbol{r}\, \phi_l^{e/h*}(\boldsymbol{r}) \left[\frac{\left(-\mathrm{i}\hbar\nabla_{\boldsymbol{r}} - \frac{e}{c}A_2\right)^2}{2m_0} + e\varphi_2 + V^l - \epsilon_0 \right] \phi_{l'}^{e/h}(\boldsymbol{r}) \tag{9.3.18}$$

are just the matrix elements of the Hamiltonian in the ϕ-representation. The $\pm$ sign refers, respectively, to electrons and holes, while ϵ_0 denotes the conduction-band edge. Here, we neglect any valence-to-conduction band coupling due to the external electromagnetic field and vice versa. This is well fulfilled for the systems and field regimes we are going to discuss in this chapter. As already pointed out, the above Hamiltonian may be time dependent. We will discuss this aspect in the following section, where we will derive our set of kinetic equations.

Let us now write in terms of our electron–hole representation the carrier–light interaction Hamiltonian introduced in equation (9.3.10):

$$H_{cl} = -\sum_{i,j} \left[\mu_{ij}^{eh} E^-(t) c_i^\dagger d_j^\dagger + \mu_{ij}^{eh*} E^+(t) d_j c_i \right]. \tag{9.3.19}$$

The above expression has been obtained within the well-known rotating-wave approximation by neglecting intraband transitions, absent for the case of optical excitations. Here, μ_{ij}^{eh} denotes the optical dipole matrix element between states ϕ_i^e and ϕ_j^h.

Similarly, the carrier–carrier Hamiltonian (9.3.12) can be rewritten as:

$$\begin{aligned} H_{cc} \;=\; & \tfrac{1}{2} \sum_{i_1 i_2 i_3 i_4} V_{i_1 i_2 i_3 i_4}^{cc} c_{i_1}^\dagger c_{i_2}^\dagger c_{i_3} c_{i_4} \\ & + \tfrac{1}{2} \sum_{j_1 j_2 j_3 j_4} V_{j_1 j_2 j_3 j_4}^{cc} d_{j_1}^\dagger d_{j_2}^\dagger d_{j_3} d_{j_4} \\ & - \sum_{i_1 i_2 j_1 j_2} V_{i_1 j_1 j_2 i_2}^{cc} c_{i_1}^\dagger d_{j_1}^\dagger d_{j_2} c_{i_2}, \end{aligned} \tag{9.3.20}$$

where

$$V_{l_1 l_2 l_3 l_4}^{cc} = \int \mathrm{d}\boldsymbol{r} \int \mathrm{d}\boldsymbol{r}'\, \phi_{l_1}^*(\boldsymbol{r}) \phi_{l_2}^*(\boldsymbol{r}') V^{cc}(\boldsymbol{r} - \boldsymbol{r}') \phi_{l_3}(\boldsymbol{r}') \phi_{l_4}(\boldsymbol{r}) \tag{9.3.21}$$

are the Coulomb matrix elements within our ϕ-representation. The first two terms describe the repulsive electron–electron and hole–hole interaction while the last one describes the attractive electron–hole interaction. Here, we neglect terms that do not conserve the number of electron–hole pairs, i.e. impact-ionization and Auger-recombination processes [34], as well as the interband exchange interaction. This monopole–monopole approximation is justified as long as the exciton binding energy (which in semiconductors is less than 20 meV) is small compared with the energy gap (which is more than 1 eV).

Finally, let us rewrite the carrier–phonon interaction Hamiltonian introduced in equation (9.3.13):

$$\boldsymbol{H}_{cp} = \sum_{ii',\boldsymbol{q}\lambda} [g^{e}_{ii',\boldsymbol{q}\lambda} c^{\dagger}_{i} b_{\boldsymbol{q}\lambda} c_{i'} + g^{e*}_{ii',\boldsymbol{q}\lambda} c^{\dagger}_{i'} b^{\dagger}_{\boldsymbol{q}\lambda} c_{i}] - \sum_{jj',\boldsymbol{q}\lambda} [g^{h}_{jj',\boldsymbol{q}\lambda} d^{\dagger}_{j} b_{\boldsymbol{q}\lambda} d_{j'} + g^{h*}_{jj',\boldsymbol{q}\lambda} d^{\dagger}_{j'} b^{\dagger}_{\boldsymbol{q}\lambda} d_{j}] \quad (9.3.22)$$

with

$$g^{e/h}_{ll',\boldsymbol{q}\lambda} = \tilde{g}_{\boldsymbol{q}\lambda} \int \mathrm{d}\boldsymbol{r}\, \phi^{e/h*}_{l}(\boldsymbol{r}) \mathrm{e}^{\mathrm{i}\boldsymbol{q}\cdot\boldsymbol{r}} \phi^{e/h}_{l'}(\boldsymbol{r}). \quad (9.3.23)$$

In equation (9.3.22) we can clearly recognize four different contributions corresponding to electron and hole phonon absorption and emission.

9.3.2 KINETIC DESCRIPTION

Our kinetic description of the ultrafast carrier dynamics in semiconductor superlattices, presented in this section, is based on the density-matrix formalism. Since this approach has already been reviewed and discussed in Chapter 6, here we will simply recall in our notation the kinetic equations relevant for the analysis of carrier dynamics in superlattices, generalizing the approach of Chapter 6 to the case of a time-dependent quantum-mechanical representation.

The set of kinetic variables is the same as those considered in Chapter 6. Given our electron–hole representation $\{\phi^{e}_{i}\}$, $\{\phi^{h}_{j}\}$, we will consider the intraband electron and hole single-particle density matrices

$$f^{e}_{ii'} = \langle c^{\dagger}_{i} c_{i'} \rangle, \qquad f^{h}_{jj'} = \langle d^{\dagger}_{j} d_{j'} \rangle \quad (9.3.24)$$

as well as the corresponding interband density matrix

$$p_{ji} = \langle d_{j} c_{i} \rangle. \quad (9.3.25)$$

Here, the diagonal elements f^{e}_{ii} and f^{h}_{jj} correspond to the electron and hole distribution functions of the Boltzmann theory while the nondiagonal terms describe intraband polarizations. In contrast, the interband density-matrix elements p_{ji} describe interband (or optical) polarizations.

In order to derive the set of kinetic equations, i.e. the equations of motion for the above kinetic variables, the standard procedure starts by deriving the equations of motion for the electron and hole operators introduced in equation (9.3.16):

$$c_{i} = \int \mathrm{d}\boldsymbol{r}\, \phi^{e*}_{i}(\boldsymbol{r}) \Psi(\boldsymbol{r}), \qquad d_{j} = \int \mathrm{d}\boldsymbol{r}\, \phi^{h*}_{j}(\boldsymbol{r}) \Psi^{\dagger}(\boldsymbol{r}). \quad (9.3.26)$$

By applying the Heisenberg equation of motion for the field operator $\boldsymbol{\Psi}$, i.e.

$$\frac{\mathrm{d}}{\mathrm{d}t}\boldsymbol{\Psi} = \frac{1}{\mathrm{i}\hbar}[\boldsymbol{\Psi}, \boldsymbol{H}], \tag{9.3.27}$$

it is easy to obtain the following equations of motion:

$$\begin{aligned}\frac{\mathrm{d}}{\mathrm{d}t}c_i &= \frac{1}{\mathrm{i}\hbar}[c_i, \boldsymbol{H}] + \frac{1}{\mathrm{i}\hbar}\sum_{i'} Z^e_{ii'}c_{i'} \\ \frac{\mathrm{d}}{\mathrm{d}t}d_j &= \frac{1}{\mathrm{i}\hbar}[d_j, \boldsymbol{H}] + \frac{1}{\mathrm{i}\hbar}\sum_{j'} Z^h_{jj'}d_{j'}\end{aligned} \tag{9.3.28}$$

with

$$Z^{e/h}_{ll'} = \mathrm{i}\hbar \int \mathrm{d}\boldsymbol{r} \left(\frac{\mathrm{d}}{\mathrm{d}t}\phi^{e/h*}_l(\boldsymbol{r})\right)\phi^{e/h}_{l'}(\boldsymbol{r}). \tag{9.3.29}$$

As for the case of equation (9.3.17), here we neglect again valence-to-conduction band coupling and vice versa. Compared with the more conventional Heisenberg equations of motion, the above equations contain an extra term, the last one. It accounts for the possible time dependence of our ϕ-representation which will induce transitions between different states according to the matrix elements $Z_{ll'}$.

By combining the above equations of motion with the definitions of the kinetic variables in equations (9.3.24) and (9.3.25), our set of kinetic equations can be schematically written as:

$$\begin{aligned}\frac{\mathrm{d}}{\mathrm{d}t}f^e_{i_1i_2} &= \left.\frac{\mathrm{d}}{\mathrm{d}t}f^e_{i_1i_2}\right|_{\boldsymbol{H}} + \left.\frac{\mathrm{d}}{\mathrm{d}t}f^e_{i_1i_2}\right|_{\phi} \\ \frac{\mathrm{d}}{\mathrm{d}t}f^h_{j_1j_2} &= \left.\frac{\mathrm{d}}{\mathrm{d}t}f^h_{j_1j_2}\right|_{\boldsymbol{H}} + \left.\frac{\mathrm{d}}{\mathrm{d}t}f^h_{j_1j_2}\right|_{\phi} \\ \frac{\mathrm{d}}{\mathrm{d}t}p_{j_1i_1} &= \left.\frac{\mathrm{d}}{\mathrm{d}t}p_{j_1i_1}\right|_{\boldsymbol{H}} + \left.\frac{\mathrm{d}}{\mathrm{d}t}p_{j_1i_1}\right|_{\phi}.\end{aligned} \tag{9.3.30}$$

They exhibit the same structure of the equations of motion (9.3.28) for the electron and hole creation and annihilation operators: a first term induced by the system Hamiltonian $\boldsymbol{H}$ (which does not account for the time variation of the basis states) and a second one induced by the time dependence of the basis functions ϕ.

Let us start discussing this second term, whose explicit form is given by:

$$\begin{aligned}
\left.\frac{\mathrm{d}}{\mathrm{d}t} f^e_{i_1 i_2}\right|_\phi &= \frac{1}{\mathrm{i}\hbar} \sum_{i_3 i_4} [Z^e_{i_2 i_4} \delta_{i_1 i_3} - Z^e_{i_3 i_1} \delta_{i_2 i_4}] f^e_{i_3 i_4} \\
\left.\frac{\mathrm{d}}{\mathrm{d}t} f^h_{j_1 j_2}\right|_\phi &= \frac{1}{\mathrm{i}\hbar} \sum_{j_3 j_4} [Z^h_{j_2 j_4} \delta_{j_1 j_3} - Z^h_{j_3 j_1} \delta_{j_2 j_4}] f^h_{j_3 j_4} \\
\left.\frac{\mathrm{d}}{\mathrm{d}t} p_{j_1 i_1}\right|_\phi &= \frac{1}{\mathrm{i}\hbar} \sum_{i_2 j_2} [Z^h_{j_1 j_2} \delta_{i_1 i_2} + Z^e_{i_1 i_2} \delta_{j_1 j_2}] p_{j_2 i_2}.
\end{aligned} \tag{9.3.31}$$

Such terms were not considered in Chapter 6, where a time-independent representation has been used. As we will see in section 9.4, they will play a central role in the description of Bloch oscillations within the vector-potential representation.

Let us now come to the first term. This, in turn, is the sum of different contributions, corresponding to the various parts of the Hamiltonian. In particular, the total Hamiltonian can be regarded as the sum of two terms, a single-particle contribution plus a many-body one:

$$\boldsymbol{H} = \boldsymbol{H}_{sp} + \boldsymbol{H}_{mb} = (\boldsymbol{H}^0_c + \boldsymbol{H}_{cl} + \boldsymbol{H}_p) + (\boldsymbol{H}_{cc} + \boldsymbol{H}_{cp} + \boldsymbol{H}_{pp}). \tag{9.3.32}$$

The explicit form of the time evolution due to the single-particle Hamiltonian $\boldsymbol{H}_{sp}$ (noninteracting carriers plus carrier–light interaction plus free phonons) is given by:

$$\begin{aligned}
\left.\frac{\mathrm{d}}{\mathrm{d}t} f^e_{i_1 i_2}\right|_{sp} &= \frac{1}{\mathrm{i}\hbar} \Bigg\{ \sum_{i_3 i_4} [\epsilon^e_{i_2 i_4} \delta_{i_1 i_3} - \epsilon^e_{i_3 i_1} \delta_{i_2 i_4}] f^e_{i_3 i_4} \\
&\quad + \sum_{j_1} [U_{i_2 j_1} p^*_{j_1 i_1} - U^*_{i_1 j_1} p_{j_1 i_2}] \Bigg\} \\
\left.\frac{\mathrm{d}}{\mathrm{d}t} f^h_{j_1 j_2}\right|_{sp} &= \frac{1}{\mathrm{i}\hbar} \Bigg\{ \sum_{j_3 j_4} [\epsilon^h_{j_2 j_4} \delta_{j_1 j_3} - \epsilon^h_{j_3 j_1} \delta_{j_2 j_4}] f^h_{j_3 j_4} \\
&\quad + \sum_{i_1} [U_{i_1 j_2} p^*_{j_1 i_1} - U^*_{i_1 j_1} p_{j_2 i_1}] \Bigg\} \\
\left.\frac{\mathrm{d}}{\mathrm{d}t} p_{j_1 i_1}\right|_{sp} &= \frac{1}{\mathrm{i}\hbar} \Bigg\{ \sum_{i_2 j_2} [\epsilon^h_{j_1 j_2} \delta_{i_1 i_2} + \epsilon^e_{i_1 i_2} \delta_{j_1 j_2}] p_{j_2 i_2} \\
&\quad + \sum_{i_2 j_2} U_{i_2 j_2} [\delta_{i_1 i_2} \delta_{j_1 j_2} - f^e_{i_2 i_1} \delta_{j_1 j_2} - f^h_{j_2 j_1} \delta_{i_1 i_2}] \Bigg\}
\end{aligned} \tag{9.3.33}$$

with $U_{i_1 j_1} = -\mu^{eh}_{i_1 j_1} E^-(t)$.

This is a closed set of equations, which is a consequence of the single-particle nature of $\boldsymbol{H}_{sp}$. In addition, we stress that the structure of the two contributions entering equation (9.3.30) is very similar: one can include the contribution (9.3.31) into equation (9.3.33) by replacing ϵ with $\epsilon + Z$.

Let us finally discuss the contributions due to the many-body part of the Hamiltonian: carrier–carrier and carrier–phonon interactions (the phonon–phonon one is not explicitly considered here). As discussed in Chapter 6, for both interaction mechanisms one can derive a hierarchy of equations involving higher-order density matrices and, in order to close such equations with respect to our set of kinetic variables, approximations are needed. The lowest-order contributions to our equations of motion are given by first-order terms in the many-body Hamiltonian: Hartree–Fock level. Since we will neglect coherent-phonon states, the only Hartree–Fock contributions will come from the carrier–carrier interaction. They simply result in a renormalization

$$\Delta\epsilon^{e/h}_{l_1 l_2} = -\sum_{l_3 l_4} V^{cc}_{l_1 l_3 l_2 l_4} f^{e/h}_{l_3 l_4} \tag{9.3.34}$$

of the single-particle energy matrices $\epsilon^{e/h}$ as well as in a renormalization

$$\Delta U_{i_1 j_1} = -\sum_{i_2 j_2} V^{cc}_{i_1 j_1 j_2 i_2} p_{j_2 i_2} \tag{9.3.35}$$

of the external field U. (The explicit form of the renormalization terms considered in this chapter accounts for the Fock contributions only, i.e. no Hartree terms. The general structure of Hartree–Fock contributions, relevant for the case of a strongly nonhomogeneous system, is discussed in Chapter 6.)

We stress that the Hartree–Fock approximation, which consists of factorizing average values of four-point operators into products of two density matrices, is independent from the quantum-mechanical picture. This is a general property: any mean-field approximation gives the same result in different representations. The reason is that the mean-field operation commutes with any unitary transformation connecting different basis states. It is then clear that the above kinetic equations are valid in any quantum-mechanical representation.

All the contributions to the system dynamics discussed so far describe a fully coherent dynamics, i.e. no scattering processes. In order to treat incoherent phenomena, for example energy relaxation and dephasing, one has to go one step further in the perturbation expansion taking into account also second-order contributions (in the perturbation Hamiltonian $\boldsymbol{H}_{mb}$). The derivation of these higher-order contributions, discussed in Chapter 6, will not be repeated here. Again, as for the first-order contributions (Hartree–Fock terms), in order to obtain a closed set of equations (with respect to our set of kinetic variables (9.3.24) and (9.3.25)) additional approximations are needed, namely the mean-field and the Markov approximation. As for the Hartree–Fock case, the mean-field approximation allows us to write the various higher-order density matrices as products of single-particle ones. The Markov approximation allows us to eliminate the additional higher-order kinetic variables, for example phonon-assisted density matrices, providing a closed set of equations still local in time, i.e. no memory effects [35–38] This last approximation is not performed in the quantum-kinetic theory discussed in Chapter 6 where, in addition to our single-particle variables, one considers two-particle and phonon-assisted density matrices [34, 38].

While the mean-field approximation is representation independent, this is unfortunately not the case for the Markov limit. This clearly implies that the validity of the Markov approximation is strictly related to the quantum-mechanical representation considered. We will come back to this point in the following section where the two different pictures used for the study of the carrier dynamics in superlattices are discussed.

The above kinetic description, based on intra- and interband density matrices, allows us to evaluate any single-particle quantity. In particular, for the analysis of semiconductor superlattices two physical quantities play a central role: the intra- and interband total (or macroscopic) polarizations:

$$P^{e/h}(t) = \sum_{ll'} M_{ll'}^{e/h} f_{l'l}^{e/h}(t), \qquad P^{eh} = \sum_{ij} \mu_{ij}^{eh} p_{ji}(t), \tag{9.3.36}$$

where $M^{e/h}$ and μ^{eh} denote, respectively, the intra- and interband dipole matrix elements in our ϕ-representation. The time derivative of the intraband polarization $P^{e/h}$ describes the radiation field induced by the Bloch-oscillation dynamics (which for a superlattice structure is in the terahertz range) while the Fourier transform of the interband (or optical) polarization P^{eh} provides the optical-absorption spectrum.

9.4 TWO EQUIVALENT PICTURES

In this section we will apply the theoretical approach presented so far to the case of a semiconductor superlattice in the presence of a uniform (space-independent) electric field. The noninteracting carriers within the superlattice crystal will then be described by the Hamiltonian $\boldsymbol{H}_c^0$ in equation (9.3.9), where now the electrodynamic potentials A_2 and φ_2 (in the following simply denoted with $\boldsymbol{A}$ and φ) correspond to a homogeneous electric field $\boldsymbol{E}_2(\boldsymbol{r}, t) = \boldsymbol{F}(t)$.

As pointed out in section 9.3.1, the natural quantum-mechanical representation is given by the eigenstates of this Hamiltonian:

$$\left[\frac{\left(-i\hbar\nabla_{\boldsymbol{r}} - \frac{e}{c}\boldsymbol{A}(\boldsymbol{r}, t)\right)^2}{2m_0} + e\varphi(\boldsymbol{r}, t) + V^l(\boldsymbol{r})\right]\phi_n(\boldsymbol{r}) = \epsilon_n\phi_n(\boldsymbol{r}). \tag{9.4.1}$$

However, due to the gauge freedom discussed in section 9.3.1, there is an infinite number of possible combinations of $\boldsymbol{A}$ and φ—and therefore of possible Hamiltonians—which describe the same homogeneous electric field $\boldsymbol{F}(t)$. In particular, one can identify two independent choices: the vector-potential gauge

$$\boldsymbol{A}(\boldsymbol{r}, t) = -c \int_{t_0}^{t} \boldsymbol{F}(t')\,\mathrm{d}t', \qquad \varphi(\boldsymbol{r}, t) = 0 \tag{9.4.2}$$

and the scalar-potential gauge

$$\boldsymbol{A}(\boldsymbol{r}, t) = 0, \qquad \varphi(\boldsymbol{r}, t) = -\boldsymbol{F}(t) \cdot \boldsymbol{r} \tag{9.4.3}$$

(previously employed for the description of the laser photo-excitation in equation (9.3.6)).

As we will see, the two independent choices correspond, respectively, to the well-known Bloch-oscillation and Wannier–Stark pictures. They simply reflect two equivalent quantum-mechanical representations and, therefore, any physical phenomenon can be described in both pictures.

9.4.1 The Bloch-oscillation picture

The vector-potential approach presented in this section, originally proposed by Kittel [8], is discussed in [9].

Within the vector-potential gauge (9.4.2), the above eigenvalue equation (9.4.1) reduces to:

$$\left[\frac{\left(-\mathrm{i}\hbar\nabla_{\boldsymbol{r}} - \frac{e}{c}\boldsymbol{A}(t)\right)^2}{2m_0} + V^l(\boldsymbol{r})\right]\phi_n(\boldsymbol{r}) = \epsilon_n\phi_n(\boldsymbol{r}). \tag{9.4.4}$$

In this gauge the vector potential is space-independent but, even for the case of a static field ($\boldsymbol{F}(t) = \boldsymbol{F}_0$), it is always time dependent. Therefore, the above Hamiltonian (together with its eigenvalues and eigenfunctions) will be time dependent as well. However, for any time t we can consider its 'instantaneous' eigenstates $\phi_n(\boldsymbol{r}, t)$. They can be easily evaluated by means of the following transformation:

$$\phi_n(\boldsymbol{r}, t) = \phi_n^0(\boldsymbol{r}, t)\mathrm{e}^{\frac{\mathrm{i}e}{\hbar c}\chi(\boldsymbol{r}, t)} \tag{9.4.5}$$

with

$$\chi(\boldsymbol{r}, t) = \boldsymbol{A}(t) \cdot \boldsymbol{r}. \tag{9.4.6}$$

By applying this transformation to the eigenvalue problem in equation (9.4.4), we obtain [9]:

$$\left[-\frac{\hbar^2\nabla_{\boldsymbol{r}}^2}{2m_0} + V^l(\boldsymbol{r})\right]\phi_n^0(\boldsymbol{r}) = \epsilon_n\phi_n^0(\boldsymbol{r}), \tag{9.4.7}$$

i.e. the wave functions $\phi_n^0(\boldsymbol{r})$ are just the Bloch states $\phi_{\boldsymbol{k}\nu}^0(\boldsymbol{r})$ of our semiconductor, and the energy spectrum ϵ_n coincides with the carrier band structure $\epsilon_{\boldsymbol{k}\nu}$. Therefore, from equation (9.4.5) the desired eigenfunctions result to be of the form:

$$\phi_{k\nu}(\boldsymbol{r}, t) = \phi^0_{k\nu}(\boldsymbol{r}, t)\mathrm{e}^{\frac{ie}{\hbar c}\boldsymbol{A}(t)\cdot\boldsymbol{r}}. \tag{9.4.8}$$

Apart from a phase factor, they coincide with the conventional Bloch states of the crystal. The reason can be understood as follows. Also in the presence of the electric field $\boldsymbol{F}$, the Hamiltonian in equation (9.4.4) is still invariant under a lattice translation corresponding to the crystal potential V^l. Thus, the crystal momentum $\hbar\boldsymbol{k}$ is still a 'good' quantum number and the band dispersion remains the same: $\epsilon = \epsilon_{k\nu}$. Therefore, at each time t our time-dependent eigenstates seem to coincide (apart from the phase factor) with the Bloch states of the crystal and, at a first glance, it is not clear which is the role played by the applied field.

In order to answer this question, let us consider again the general form of our eigenstates in equation (9.4.8). At the initial time t_0 the vector potential $\boldsymbol{A}$ is equal to zero and, therefore, the two basis sets coincide: $\phi_{k_0\nu}(\boldsymbol{r}, t_0) = \phi^0_{k\nu}(\boldsymbol{r})$. (Here, $\boldsymbol{k}_0$ and $\boldsymbol{k}$ denote the carrier wave vectors at times t_0 and t, respectively. They are, in principle, independent quantities, since they correspond to two different eigenvalue problems.) In other words, the Bloch states $\phi^0_{k\nu}$ can also be regarded as the states $\phi_{k_0\nu}$ at the initial time t_0 and vice versa. This allows us to rewrite equation (9.4.8) as

$$\phi_{k\nu}(\boldsymbol{r}, t) = \phi_{k_0\nu}(\boldsymbol{r}, t_0)\mathrm{e}^{\frac{ie}{\hbar c}\boldsymbol{A}(t)\cdot\boldsymbol{r}} \tag{9.4.9}$$

or equivalently

$$\phi^0_{k\nu}(\boldsymbol{r}) = \phi^0_{k_0\nu}(\boldsymbol{r})\mathrm{e}^{-\frac{ie}{\hbar c}\boldsymbol{A}(t)\cdot\boldsymbol{r}}. \tag{9.4.10}$$

Moreover, in view of the translational symmetry of the crystal, at each time t the Bloch states ϕ^0 should obey the Bloch theorem [8]:

$$\phi^0_{k\nu}(\boldsymbol{r} + \boldsymbol{a}) = \phi^0_{k\nu}(\boldsymbol{r})\mathrm{e}^{\mathrm{i}\boldsymbol{k}\cdot\boldsymbol{a}}, \tag{9.4.11}$$

where $\boldsymbol{a}$ denotes any periodicity vector of the crystal. If we now apply the Bloch theorem to both sides of equation (9.4.10), i.e. at times t and t_0, we finally obtain:

$$\boldsymbol{k} = \boldsymbol{k}_0 - \frac{e}{\hbar c}\boldsymbol{A}(t). \tag{9.4.12}$$

This result is quite important: the symmetry properties of the Hamiltonian require a precise relationship between the (formally independent) wave vectors $\boldsymbol{k}_0$ and $\boldsymbol{k}$. More specifically, the carrier wave vector results to be a function of time ($\boldsymbol{k} = \boldsymbol{k}(t)$), i.e. the instantaneous sets of basis states $\{\phi_{k(t)\nu}(\boldsymbol{r}, t)\}$ (corresponding to different times t) are mutually connected through a continuous time evolution of the wave vector $\boldsymbol{k}$. We can then answer the previous question saying that within the vector-potential gauge the application of a homogeneous field $\boldsymbol{F}(t)$ induces a simple 'drift' in $\boldsymbol{k}$-space of the crystal Bloch states, which are therefore not 'distorted' by the presence of the field.

From a physical point of view, the above equation describes the continuous time evolution of a carrier in $\boldsymbol{k}$-space induced by the applied electric field. In particular, taking into account the explicit form of the vector potential $\boldsymbol{A}$ given in equation (9.4.2), we have

$$\boldsymbol{k}(t) = \boldsymbol{k}_0 + \frac{e}{\hbar}\int_{t_0}^{t} \boldsymbol{F}(t')\,\mathrm{d}t', \tag{9.4.13}$$

from which the acceleration theorem in equation (9.1.1) is recovered:

$$\dot{\boldsymbol{k}}(t) = \frac{e}{\hbar}\boldsymbol{F}(t). \tag{9.4.14}$$

As pointed out in the introduction, this is usually regarded as a 'semiclassical' result, i.e. as obtained by applying the laws of classical mechanics to a Bloch electron. However, the above analysis shows that this is a rigorous quantum-mechanical result: the quantum evolution of a carrier within a given band ν under the action of a homogeneous electric field is rigorously described by the acceleration theorem.

We want to stress once again that within the vector-potential gauge discussed so far the acceleration theorem is just a result of the symmetry properties of the crystal and time-dependent eigenstates in equation (9.4.8) describe the quantum analogue of the 'semiclassical motion' of a carrier in $\boldsymbol{k}$-space (Fig. 9.1).

Let us now consider the case of a static field, i.e. $\boldsymbol{F}(t) = \boldsymbol{F}_0$, applied parallel to a symmetry axis of the crystal. In this case, the corresponding vector potential entering equation (9.4.12) is a linear function of time, which induces a uniform drift of the carriers in $\boldsymbol{k}$-space along the field direction:

$$\boldsymbol{k}(t) = \boldsymbol{k}_0 + \dot{\boldsymbol{k}}(t - t_0) = \frac{e\boldsymbol{F}_0}{\hbar}(t - t_0). \tag{9.4.15}$$

Since the carrier energy—given by the eigenvalue in equation (9.4.4)—coincides with the crystal band structure $\epsilon_{\boldsymbol{k}\nu}$, its time evolution is:

$$\epsilon_\nu(t) = \epsilon_\nu(\boldsymbol{k}(t)) = \epsilon_\nu(\boldsymbol{k}_0 + \dot{\boldsymbol{k}}(t - t_0)). \tag{9.4.16}$$

Due to the periodic nature of the band structure as a function of $\boldsymbol{k}$, i.e. $\epsilon_\nu(\boldsymbol{k}) = \epsilon_\nu(\boldsymbol{k}+\boldsymbol{G})$ ($\boldsymbol{G}$ being a reciprocal-lattice vector), the carrier will execute a periodic motion in time with a period

$$\tau_B = \frac{h}{eF_0 d}, \tag{9.4.17}$$

where d is the lattice periodicity (in real space) along the field direction. This coincides with the Bloch period previously introduced (section 9.1). It corresponds to the time needed for the electron to travel from any point k to the energetically equivalent point $k + \frac{2\pi}{d}$.

These periodic oscillations of the carrier over the crystal band structure (Fig. 9.1) are known as Bloch oscillations. As pointed out in section 9.1, they were first introduced by Bloch [1] on the basis of semiclassical arguments.

However, as for the case of the acceleration theorem discussed above, this is a rigorous quantum-mechanical result of the vector-potential picture discussed so far. Such a clear physical interpretation of the quantum-mechanical theory in terms of a semiclassical picture is hard to obtain within the scalar-potential gauge presented in the following section.

Both the acceleration theorem (9.4.14) and the Bloch-oscillation dynamics previously discussed are induced by the noninteracting carrier Hamiltonian in equation (9.4.4) through its time-dependent eigenstates $\phi_{\boldsymbol{k}(t)\nu}$. Therefore, the Bloch-oscillation dynamics considered so far does not account for many-body effects (carrier–carrier and carrier–phonon interactions) as well as for the effects induced by the time variation of our basis states. For a more 'realistic' description of the carrier dynamics within our vector-potential picture we are then forced to employ the general kinetic theory presented in section 9.3.2.

As discussed in Chapter 6, for the case of a homogeneous semiconductor crystal the only relevant terms of the single-particle density matrix in our $\boldsymbol{k}\nu$ representation are those diagonal in $\boldsymbol{k}$. This property, which is again due to the translational symmetry of the Hamiltonian, reduces the set of kinetic variables in equations (9.3.24) and (9.3.25) to the intraband density-matrix elements

$$f^e_{\boldsymbol{k},\alpha\alpha'} = \langle c^\dagger_{\boldsymbol{k}\alpha} c_{\boldsymbol{k}\alpha'} \rangle, \qquad f^h_{-\boldsymbol{k},\beta\beta'} = \langle d^\dagger_{-\boldsymbol{k}\beta} d_{-\boldsymbol{k}\beta'} \rangle \tag{9.4.18}$$

plus the interband density-matrix elements

$$p_{\boldsymbol{k},\beta\alpha} = \langle d_{-\boldsymbol{k}\beta} c_{\boldsymbol{k}\alpha} \rangle \ . \tag{9.4.19}$$

Here, the standard electron–hole picture introduced in section 9.3.1 has been applied to our set of time-dependent eigenstates $\phi_{\boldsymbol{k}(t)\nu}$: the band index ν (which refers to both conduction and valence states) is replaced by two separate band indices α and β for electrons and holes, respectively, while, due to the charge-conjugation symmetry, the hole states are still labelled in terms of the corresponding valence–electron states, i.e. $\boldsymbol{k}^h\beta \equiv -\boldsymbol{k}^e\beta$.

Let us now discuss the explicit form of the kinetic equations (9.3.30) in our vector-potential picture:

$$\begin{aligned}
\frac{\mathrm{d}}{\mathrm{d}t} f^e_{\boldsymbol{k},\alpha_1\alpha_2} &= \left.\frac{\mathrm{d}}{\mathrm{d}t} f^e_{\boldsymbol{k},\alpha_1\alpha_2}\right|_{\boldsymbol{H}} + \left.\frac{\mathrm{d}}{\mathrm{d}t} f^e_{\boldsymbol{k},\alpha_1\alpha_2}\right|_{\phi} \\
\frac{\mathrm{d}}{\mathrm{d}t} f^h_{-\boldsymbol{k},\beta_1\beta_2} &= \left.\frac{\mathrm{d}}{\mathrm{d}t} f^h_{-\boldsymbol{k},\beta_1\beta_2}\right|_{\boldsymbol{H}} + \left.\frac{\mathrm{d}}{\mathrm{d}t} f^h_{-\boldsymbol{k},\beta_1\beta_2}\right|_{\phi} \\
\frac{\mathrm{d}}{\mathrm{d}t} p_{\boldsymbol{k},\beta_1\alpha_1} &= \left.\frac{\mathrm{d}}{\mathrm{d}t} p_{\boldsymbol{k},\beta_1\alpha_1}\right|_{\boldsymbol{H}} + \left.\frac{\mathrm{d}}{\mathrm{d}t} p_{\boldsymbol{k},\beta_1\alpha_1}\right|_{\phi} .
\end{aligned} \tag{9.4.20}$$

Here, the first term is induced by the system Hamiltonian $\boldsymbol{H} = \boldsymbol{H}_{sp} + \boldsymbol{H}_{mb}$ (which does not account for the time variation of the basis states) while the second one is induced by the time evolution of the basis functions ϕ.

The contributions to the carrier dynamics due to the single-particle Hamiltonian $\boldsymbol{H}_{sp}$ are given in equation (9.3.33). They consist of a 'free rotation' plus a term due to the interaction with the external laser field. For the case of an ultrafast laser excitation, after the initial carrier photogeneration the only non-vanishing contributions in equation (9.3.33) are such free-rotation terms. If we now consider that in our vector-potential representation the energy matrix ϵ in equation (9.3.18) is diagonal,

$$\epsilon^{e/h}_{\boldsymbol{k}l,\boldsymbol{k}'l'} = \epsilon^{e/h}_{\boldsymbol{k}l}\delta_{\boldsymbol{k}\boldsymbol{k}'}\delta_{ll'}, \tag{9.4.21}$$

the single-particle contributions after the initial photo-excitation reduce to:

$$\begin{aligned}
\left.\frac{\mathrm{d}}{\mathrm{d}t} f^{e}_{\boldsymbol{k},\alpha_1\alpha_2}\right|_{sp} &= \frac{1}{\mathrm{i}\hbar}\left[\epsilon^{e}_{\boldsymbol{k}\alpha_2} - \epsilon^{e}_{\boldsymbol{k}\alpha_1}\right] f^{e}_{\boldsymbol{k},\alpha_1\alpha_2} \\
\left.\frac{\mathrm{d}}{\mathrm{d}t} f^{h}_{-\boldsymbol{k},\beta_1\beta_2}\right|_{sp} &= \frac{1}{\mathrm{i}\hbar}[\epsilon^{h}_{-\boldsymbol{k}\beta_2} - \epsilon^{h}_{-\boldsymbol{k}\beta_1}] f^{h}_{-\boldsymbol{k},\beta_1\beta_2} \\
\left.\frac{\mathrm{d}}{\mathrm{d}t} p_{\boldsymbol{k},\beta_1\alpha_1}\right|_{sp} &= \frac{1}{\mathrm{i}\hbar}[\epsilon^{h}_{-\boldsymbol{k}\beta_1} + \epsilon^{e}_{\boldsymbol{k}\alpha_1}] p_{\boldsymbol{k},\beta_1\alpha_1}.
\end{aligned} \tag{9.4.22}$$

As we can see, the above equations describe a set of independent many-level systems, i.e. one for each $\boldsymbol{k}$ value. In addition, there is no coupling between different kinetic variables. If, in particular, we assume a diagonal initial condition

$$f^{e/h}_{\boldsymbol{k},ll'} \equiv f^{e/h}_{\boldsymbol{k},l}\delta_{ll'}, \qquad p_{\boldsymbol{k},ll'} \equiv p_{\boldsymbol{k},l}\delta_{ll'} \tag{9.4.23}$$

(which is well fulfilled for the case of a laser photo-excitation of standard bulk semiconductors as well as superlattice structures), the kinetic equations (9.4.22) for the nonzero density-matrix elements, i.e. for the diagonal ones, reduce to:

$$\frac{\mathrm{d}}{\mathrm{d}t} f^{e/h}_{\pm\boldsymbol{k},l} = 0, \qquad \frac{\mathrm{d}}{\mathrm{d}t} p_{\boldsymbol{k},l} = 0, \tag{9.4.24}$$

where the $\pm$ sign refers, respectively, to electrons and holes. If we now remember that in our vector-potential representation the wave vector $\boldsymbol{k}$ is itself a function of time (see equation (9.4.12)), the above kinetic equations can be rewritten as:

$$\frac{\partial}{\partial t} f^{e/h}_{\pm\boldsymbol{k},l} \pm \dot{\boldsymbol{k}} \cdot \nabla_{\boldsymbol{k}} f^{e/h}_{\pm\boldsymbol{k},l} = 0, \qquad \frac{\partial}{\partial t} p_{\boldsymbol{k},l} + \dot{\boldsymbol{k}} \cdot \nabla_{\boldsymbol{k}} p_{\boldsymbol{k},l} = 0. \tag{9.4.25}$$

For both distribution functions f and interband polarizations p we obtain a simple drift equation whose general solution is of the form:

$$y(\boldsymbol{k}(t), t) = y(\boldsymbol{k}_0, t_0), \tag{9.4.26}$$

i.e. the function at time t is obtained through a rigid shift $\Delta\boldsymbol{k} = \boldsymbol{k} - \boldsymbol{k}_0$ of the function at the initial time t_0. Such drift in $\boldsymbol{k}$-space, induced by the external field $\boldsymbol{F}$, is again described by the acceleration theorem in equation (9.4.14). Therefore, as expected, the carrier dynamics described by the above kinetic equations (which accounts for the single-particle Hamiltonian only) is fully equivalent to the

Bloch-oscillation dynamics discussed above. However, such a relatively simple picture of the carrier motion does not account for the time dependence of our basis states as well as for many-body effects, for example carrier–carrier and carrier–phonon interactions.

The contributions to the carrier dynamics induced by the time variation of the basis states are given in equation (9.3.31). In our vector-potential representation (9.4.8) the explicit form of the matrix elements $Z_{ll'}^{e/h}$ introduced in equation (9.3.29) is

$$Z_{\boldsymbol{k}l,\boldsymbol{k}'l'}^{e/h} = Z_{\boldsymbol{k},ll'}^{e/h}\delta_{\boldsymbol{k}\boldsymbol{k}'} \tag{9.4.27}$$

with

$$Z_{\boldsymbol{k},ll'}^{e/h} = \pm e(\delta_{ll'} - 1)\int \mathrm{d}\boldsymbol{r}\,\phi_{\boldsymbol{k}l}^{e/h*}[\boldsymbol{F}(t)\cdot\boldsymbol{r}]\phi_{\boldsymbol{k}l'}^{e/h}. \tag{9.4.28}$$

They result to be strictly related to the matrix elements of the scalar potential in equation (9.4.3), as we will discuss in the following section. The $\boldsymbol{k}l \to \boldsymbol{k}'l'$ transitions induced by the time variation of the basis states are always diagonal in $\boldsymbol{k}$; this reflects the momentum conservation in the carrier–field interaction, i.e. since the momentum $\boldsymbol{q}$ corresponding to a space-independent field is equal to zero, the initial and final carrier wave vectors coincide. Moreover, there are no intraband ($l = l'$) transitions, which confirms that the action of the field within a given band is fully described by the drift terms in equation (9.4.25).

If we now rewrite equation (9.3.31) in our vector-potential representation taking into account the explicit form of the above matrix elements, we finally obtain:

$$\begin{aligned}
\left.\frac{\mathrm{d}}{\mathrm{d}t}f_{\boldsymbol{k},\alpha_1\alpha_2}^{e}\right|_{\phi} &= \frac{1}{\mathrm{i}\hbar}\sum_{\alpha_3\alpha_4}[Z_{\boldsymbol{k},\alpha_2\alpha_4}^{e}\delta_{\alpha_1\alpha_3} - Z_{\boldsymbol{k},\alpha_3\alpha_1}^{e}\delta_{\alpha_2\alpha_4}]f_{\boldsymbol{k},\alpha_3\alpha_4}^{e}\\
\left.\frac{\mathrm{d}}{\mathrm{d}t}f_{-\boldsymbol{k},\beta_1\beta_2}^{h}\right|_{\phi} &= \frac{1}{\mathrm{i}\hbar}\sum_{\beta_3\beta_4}[Z_{-\boldsymbol{k},\beta_2\beta_4}^{h}\delta_{\beta_1\beta_3} - Z_{-\boldsymbol{k},\beta_3\beta_1}^{h}\delta_{\beta_2\beta_4}]f_{-\boldsymbol{k},\beta_3\beta_4}^{h}\\
\left.\frac{\mathrm{d}}{\mathrm{d}t}p_{\boldsymbol{k},\beta_1\alpha_1}\right|_{\phi} &= \frac{1}{\mathrm{i}\hbar}\sum_{\alpha_2\beta_2}[Z_{-\boldsymbol{k},\beta_1\beta_2}^{h}\delta_{\alpha_1\alpha_2} + Z_{\boldsymbol{k}\alpha_1\alpha_2}^{e}\delta_{\beta_1\beta_2}]p_{\boldsymbol{k},\beta_2\alpha_2}.
\end{aligned} \tag{9.4.29}$$

From the above kinetic equations we clearly see that the time variation of our basis states ϕ induces 'vertical' (i.e. $\boldsymbol{k} = \boldsymbol{k}'$) transitions between different bands. Such interband coupling is the well-known Zener tunnelling [2]. This effect is usually described as an interband transition induced by the scalar potential $-\boldsymbol{F}\cdot\boldsymbol{r}$, which is also evident from the explicit form of the Zener matrix elements in equation (9.4.28). However, within our vector-potential picture the Zener tunnelling originates from the time variation of our accelerated Bloch states in equation (9.4.8).

As discussed in [9], for the case of bulk semiconductors this interband coupling results to be very limited even for the case of high applied fields. Therefore, the Bloch-oscillation scenario of the semiclassical theory (Fig. 9.1) is practically unmodified by Zener tunnelling. For the case of interest, i.e. that of semiconductor superlattices, interminiband Zener tunnelling is expected to play a significant role in the high-field regime. However, for relatively low fields (up to 10^4 V cm^{-1}) the effect is again negligible and the Bloch-oscillation regime is fully recovered. In this case, the timescale of Zener-tunnelling processes is much longer than the Bloch-oscillation period τ_B. Therefore, the effect due to the time variation of our basis states is negligible, i.e. the time variation can be regarded as an 'adiabatic transformation'.

Let us finally consider the role played by many-body effects, i.e. carrier–carrier and carrier–phonon interactions. As discussed in section 9.3.2 as well as in Chapter 6, these many-body effects can be divided into coherent and incoherent contributions. With coherent contributions we refer to first-order terms in the many-body Hamiltonian $\boldsymbol{H}_{mb}$. Since we neglect coherent-phonon states, the only nonzero contributions originate from carrier–carrier interaction. The explicit form of these Hartree–Fock terms in our vector-potential representation is obtained from equations (9.3.34) and (9.3.35) by replacing the generic labels i and j with $\boldsymbol{k}\alpha$ and $-\boldsymbol{k}\beta$, respectively.

From a physical point of view, these coherent contributions give rise to excitonic and band-renormalization effects, which result in a modification of the single-particle energy spectrum. In our case, these excitonic effects may lead to modifications of the Bloch-oscillation dynamics, for example small variations of the Bloch period. As we will see, within the Wannier–Stark picture such excitonic effects manifest themselves in a modification of the single-particle Wannier–Stark energy levels, which are no longer equidistantly spaced [39].

Let us now study the role played by incoherent contributions, i.e. second-order contributions in the many-body Hamiltonian. As pointed out in section 9.3.2, these terms are usually treated within the usual Markov approximation. The resulting contributions describe, in general, second-order transitions connecting all possible kinetic variables, i.e. all possible density-matrix elements. The transitions connecting diagonal density-matrix elements, i.e. distribution functions, can be easily described in terms of stochastic scattering processes, i.e. due to the scattering with a partner carrier or phonon, the electron may undergo a transition from an initial state $\boldsymbol{k}\nu$ to a final state $\boldsymbol{k}'\nu'$. In contrast, for the transitions involving nondiagonal terms the second-order coupling is not positive-definite, i.e. it is not a rate, and the intuitive scattering picture cannot be employed.

It is not in the spirit of this chapter to derive and discuss the explicit form of the second-order carrier–carrier and carrier–phonon contributions. From a physical point of view, both carrier–carrier and carrier–phonon scattering processes give rise to energy relaxation and dephasing. It is well known that, due to scattering events, a photogenerated carrier distribution will relax both energy and momentum and, in addition, it will lose its internal degree of coherence, which corresponds to a decay of the interband polarizations. This stochastic dynamics may strongly influence the deterministic Bloch-oscillation regime discussed so far [40–42].

As we will see from some simulated experiments reported in section 9.5, the role played by carrier–carrier and carrier–phonon scattering strongly depends on the physical conditions considered, for example carrier density, excitation energy and lattice temperature. When the scattering rate corresponding to the dominant interaction mechanism is much larger than the Bloch oscillation frequency $\omega_B = \frac{2\pi}{\tau_B}$, the Bloch oscillations are not suppressed, i.e. the carriers perform on average several Bloch oscillations between two scattering events. In contrast, if the scattering rate is larger than the Bloch frequency, the carrier cannot execute a full oscillation without scattering. In this case, the Bloch oscillations are totally suppressed and we deal with a diffusive-transport regime.

As discussed in the introductory part of this chapter, in bulk semiconductors also for very high fields the Bloch-oscillation period is larger than the typical scattering times. In contrast, in semiconductor superlattices the Bloch period is at least one order of magnitude smaller than in bulk systems and, therefore, comparable or even smaller than the typical scattering times.

This allows us to answer the controversial question: Do Bloch oscillations really exist? The analysis presented in this section shows that, in the absence of scattering events, Bloch oscillations exist and, in contrast to the early papers by Rabinovitch and Zak [29], they are not significantly affected by Zener tunnelling, both for bulk and superlattices.

In contrast, due to scattering events, Bloch oscillations are fully suppressed in bulk semiconductors but they still survive in superlattices, as confirmed by several experiments [16, 18, 22, 23, 40].

In the following section we will discuss the so-called Wannier–Stark picture. In contrast to the vector-potential approach discussed so far, this will correspond to a scalar-potential representation. In particular, we will study the link between the two pictures showing that phenomena which are peculiar of one picture can be equally described within the second one.

9.4.2 The Wannier–Stark picture

Within the scalar-potential gauge (9.4.3), the eigenvalue equation (9.4.1) reduces to:

$$H_c^0\phi(\boldsymbol{r}) = \left[-\frac{\hbar^2\nabla_{\boldsymbol{r}}^2}{2m_0} + V^l(\boldsymbol{r}) - e\boldsymbol{F}\cdot\boldsymbol{r}\right]\phi(\boldsymbol{r}) = \epsilon\phi(\boldsymbol{r}). \tag{9.4.30}$$

In this gauge the scalar potential is space dependent but, for the case of a static field considered in this section, it is always time independent.

As originally pointed out by Wannier [6], due to the translational symmetry of the crystal potential $V^l(\boldsymbol{r}) = V^l(\boldsymbol{r}+\boldsymbol{d})$, $\boldsymbol{d}$ being the primitive lattice vector along the field direction, if $\phi(\boldsymbol{r})$ is an eigenfunction of the Hamiltonian in equation (9.4.30) with eigenvalue ϵ, then any $\phi(\boldsymbol{r}+n\boldsymbol{d})$ is also an eigenfunction with eigenvalue $\epsilon + n\Delta\epsilon$, where $\Delta\epsilon = eFd$ is the so-called Wannier–Stark splitting. However, as pointed out by Zak [7], for the case of an infinite crystal the scalar potential $-\boldsymbol{F} \cdot \boldsymbol{r}$ is not bounded, which implies a continuous energy spectrum. This was the starting point of the long-standing controversy on the existence of Wannier–Stark ladders discussed in section 9.2.

Today, after three decades, we know that the problem is by itself ill-defined and eigenstates can be defined only asymptotically [10]. In particular, there exist ladders of metastable Wannier–Stark states weakly coupled (through Zener tunnelling) to a continuous energy spectrum.

We will now study the explicit form of these Wannier–Stark states. As a starting point, let us write the generic eigenstate ϕ as a superposition of conventional Bloch states $\phi^0_{\boldsymbol{k}\nu}$, i.e. let us move to the so-called crystal-momentum representation (CMR):

$$\phi(\boldsymbol{r}) = \sum_{\boldsymbol{k}\nu} s_{\boldsymbol{k}\nu}\phi^0_{\boldsymbol{k}\nu}(\boldsymbol{r}) = \sum_{\boldsymbol{G}\nu} \frac{1}{\Omega} \int^{\Omega} \mathrm{d}\boldsymbol{k}\, s_{\boldsymbol{k}+\boldsymbol{G}\nu}\phi^0_{\boldsymbol{k}+\boldsymbol{G}\nu}(\boldsymbol{r}), \tag{9.4.31}$$

where $\boldsymbol{G}$ is a generic reciprocal-lattice vector while Ω denotes the volume of the first Brillouin zone. Due to the periodicity of the Bloch states in $\boldsymbol{k}$-space ($\phi^0_{\boldsymbol{k}+\boldsymbol{G}\nu} = \phi^0_{\boldsymbol{k}\nu}$), we are allowed to limit the above expansion to the first Brillouin zone, which implies imposing the same periodicity on the coefficients $s_{\boldsymbol{k}\nu}$: $s_{\boldsymbol{k}+\boldsymbol{G}\nu} = s_{\boldsymbol{k}\nu}$. By inserting the above expansion (limited to the first Brillouin zone) into equation (9.4.30), our eigenvalue problem can be rewritten as:

$$\sum_{\nu'} \frac{1}{\Omega} \int^{\Omega} \mathrm{d}\boldsymbol{k}'\, H^0_{\boldsymbol{k}\nu,\boldsymbol{k}'\nu'} s_{\boldsymbol{k}'\nu'} = \epsilon s_{\boldsymbol{k}\nu}, \tag{9.4.32}$$

where

$$H^0_{\boldsymbol{k}\nu,\boldsymbol{k}'\nu'} = \epsilon_{\boldsymbol{k}\nu}\delta_{\boldsymbol{k}\nu,\boldsymbol{k}'\nu'} + H^{\boldsymbol{F}}_{\boldsymbol{k}\nu,\boldsymbol{k}'\nu'} \tag{9.4.33}$$

are the matrix elements of the Hamiltonian H^0_c in the CMR. The first term corresponds to the perfect (field-free) crystal while the second one describes the scalar-potential term:

$$H^{\boldsymbol{F}}_{\boldsymbol{k}\nu,\boldsymbol{k}'\nu'} = -e \int \mathrm{d}\boldsymbol{r}\, \phi^{\circ *}_{\boldsymbol{k}\nu}[\boldsymbol{F} \cdot \boldsymbol{r}]\phi^0_{\boldsymbol{k}'\nu'}. \tag{9.4.34}$$

Following the approach discussed in [34], the above scalar-potential matrix elements can be divided into intraband ($\nu = \nu'$) and interband ($\nu \neq \nu'$) terms. The intraband terms can always be written as a drift operator [34, 43, 44]

$$\frac{1}{\Omega_\parallel} \int^{\Omega_\parallel} \mathrm{d}k'_\parallel\, H^{\boldsymbol{F}}_{\boldsymbol{k}\nu,\boldsymbol{k}'\nu} = -\mathrm{i}eF\delta_{\boldsymbol{k}_\perp \boldsymbol{k}'_\perp} \frac{\partial}{\partial k_\parallel} \tag{9.4.35}$$

($k_\parallel$ and $k_\perp$ being, respectively, the components of the wave vector $\boldsymbol{k}$ parallel and perpendicular to the field) while the nondiagonal terms coincide with the Zener-tunnelling matrix elements Z in equation (9.4.27). As already pointed out in section 9.4.1, the Zener tunnelling, which in the vector-potential representation is induced by the time variation of the basis states, corresponds to interband transitions induced by the scalar potential in equation (9.4.3).

For moderate values of the applied field, Zener tunnelling to other bands can be neglected and, by inserting equations (9.4.33) and (9.4.35) into equation (9.4.32), our eigenvalue equation reduces to

$$\frac{\partial}{\partial k_{\parallel}} s_{\boldsymbol{k}\nu} = -\frac{\mathrm{i}}{eF}(\epsilon_{\boldsymbol{k}\nu} - \epsilon) s_{\boldsymbol{k}\nu}, \tag{9.4.36}$$

whose solution is given by:

$$s_{\boldsymbol{k}\nu} = \delta_{\boldsymbol{k}_\perp \tilde{\boldsymbol{k}}_\perp} \mathrm{e}^{-\frac{\mathrm{i}}{eF}\int_0^{k_\parallel} (\epsilon_{k'_\parallel \tilde{\boldsymbol{k}}_\perp \nu} - \epsilon)\, \mathrm{d}k'_\parallel}. \tag{9.4.37}$$

As expected, the coefficients are diagonal with respect to $\boldsymbol{k}_\perp$, i.e. the linear combination in equation (9.4.31) will only involve Bloch states with the same perpendicular component $\boldsymbol{k}_\perp$. This reflects the translational symmetry of the Hamiltonian with respect to the plane perpendicular to the field. Moreover, the periodicity condition $s_{\boldsymbol{k}+\boldsymbol{G}\nu} = s_{\boldsymbol{k}\nu}$ (applied along the field direction) requires that

$$\frac{1}{eF}\int_{-\frac{\pi}{d}}^{+\frac{\pi}{d}} \left(\epsilon_{k_\parallel \boldsymbol{k}_\perp \nu} - \epsilon\right) \mathrm{d}k_\parallel = 2\pi n \tag{9.4.38}$$

which, in turn, tells us that the only allowed energy values are

$$\epsilon = \epsilon^{\boldsymbol{k}_\perp n\nu} = \epsilon^{\boldsymbol{k}_\perp 0\nu} + neFd \tag{9.4.39}$$

with

$$\epsilon^{\boldsymbol{k}_\perp 0\nu} = \frac{\mathrm{d}}{2\pi}\int_{-\frac{\pi}{\mathrm{d}}}^{+\frac{\pi}{d}} \epsilon_{k_\parallel \boldsymbol{k}_\perp \nu} \mathrm{d}k_\parallel. \tag{9.4.40}$$

What we obtain is the Wannier–Stark ladder mentioned above, whose central ($n = 0$) value is given by the average of the band along the field direction (for a given $\boldsymbol{k}_\perp$).

Let us now discuss the corresponding wave functions. By inserting the explicit form of the coefficients $s_{\boldsymbol{k}\nu}$ given in equation (9.4.37) into expansion (9.4.31) we have:

$$\phi^{\boldsymbol{k}_\perp n\nu}(\boldsymbol{r}) = \frac{\mathrm{d}}{2\pi}\int_{-\frac{\pi}{d}}^{+\frac{\pi}{d}} \mathrm{d}k_\parallel \mathrm{e}^{-\frac{\mathrm{i}}{eF}\int_0^{k_\parallel} (\epsilon_{k'_\parallel \boldsymbol{k}_\perp \nu} - \epsilon^{\boldsymbol{k}_\perp n\nu})\, \mathrm{d}k'_\parallel} \phi^0_{k_\parallel \boldsymbol{k}_\perp \nu}(\boldsymbol{r}). \tag{9.4.41}$$

They are the so-called Wannier–Stark states shown in Fig. 9.2. For each band ν (and for any given $\boldsymbol{k}_\perp$), we have a set of energetically equidistant states $\phi^{\boldsymbol{k}_\perp n\nu}$, obtained from the central ($n = 0$) state $\phi^{\boldsymbol{k}_\perp 0\nu}$ through the spatial translation $\boldsymbol{r} \to \boldsymbol{r} + n\boldsymbol{d}$ discussed above. As schematically depicted in Fig. 9.2, each state is localized around one of the atomic cells of the superlattice and the degree of localization depends on the strength of the applied field F. More precisely, when the Wannier–Stark energy eFd is much smaller than the superlattice miniband width, the wave functions $\phi^{k_\perp n\nu}$ are weakly localized and they extend over several elementary cells. In contrast, when eFd is comparable or larger than the miniband width, the localization increases and the function results to be significantly different from zero only in one cell. Since the miniband width of the holes is smaller than that of the electrons, the hole states exhibit a stronger Wannier–Stark localization (Fig. 9.2).

Due to the neglection of interband Zener tunnelling, each Wannier–Stark state $\phi^{k_\perp n\nu}$ in equation (9.4.41) is obtained as a linear combination of Bloch states belonging to the same band ν. Moreover, we see that all Bloch states have the same weight in the expansion, i.e. the coefficients are just phase factors.

The coefficients s introduced so far can be regarded as the matrix elements of a unitary transformation connecting the Bloch to the Wannier–Stark representation. It is then clear that the inverse transformation allows us to write any Bloch state as a linear combination of Wannier–Stark states:

$$\phi^0_{\boldsymbol{k}\nu} = \sum_n s^{n*}_{\boldsymbol{k}\nu} \phi^{k_\perp n\nu}(\boldsymbol{r}). \tag{9.4.42}$$

The Wannier–Stark states in equation (9.4.41) can now be used as basis states for our kinetic description. The kinetic variables in the Wannier–Stark representation will be formally the same as for the vector-potential picture discussed in section 9.4.1. They are defined according to equations (9.4.18) and (9.4.19) where the three-dimensional wave vector $\boldsymbol{k}$ is replaced by its perpendicular component $k_\perp$ while the band index α/β is replaced by the same band index plus the Wannier–Stark ladder index n.

Provided the above label substitution is performed, the time evolution of the new kinetic variables is again described by equation (9.4.20). However, within our Wannier–Stark representation the basis states are time-independent and, therefore, the ϕ-contributions in equation (9.4.20) are equal to zero. Moreover, in contrast to the vector-potential case, the energy matrix ϵ in equation (9.3.18) is not diagonal. The diagonal terms are now given by the Wannier–Stark ladders in equation (9.4.39) while the nondiagonal terms are once again Zener-tunnelling matrix elements between different Wannier–Stark states.

As already pointed out in section 9.3.2, the coherent contributions entering our kinetic equations are independent from the quantum-mechanical representation considered. This tells us that the Zener-tunnelling contributions to the equations of motion in the scalar- and vector-potential representations should coincide. In fact, despite their different physical interpretations (in the vector-potential picture they are induced by the time variation of the basis states while in the scalar-potential one they are due to interband transitions induced by the field Hamiltonian), their formal structure is exactly the same, as can be seen by comparing equations (9.3.31) and (9.3.33).

In contrast, incoherent contributions, i.e. scattering terms derived within Markov approximation, are expected to be representation dependent. In particular, within the vector-potential picture the scattering terms are usually derived by neglecting the time dependence of the basis states $\phi_{\mathbf{k}(t)\nu}$, which corresponds to neglecting the action of the field, i.e. the time variation of the carrier wave vector $\mathbf{k}(t)$, during the collision. Within the Wannier–Stark picture, the basis states are time independent and the standard Markov limit automatically accounts for the so-called intracollisional field effect [34–36]. However, for moderate values of the applied field incoherent contributions evaluated in the scalar- and vector-potential representations coincide.

Before concluding this section, let us try to better understand the physical link between the accelerated Bloch states of the vector-potential picture and the Wannier–Stark states of the scalar-potential representation. Both basis sets have been introduced as eigenstates of two equivalent Hamiltonians, corresponding to the two different electromagnetic gauges (equations (9.4.4) and (9.4.30)). However, it is well known [45] that the solutions of the corresponding time-dependent Schrödinger equations coincide (apart from a phase factor which is physically irrelevant). More specifically, let us consider the generic time-dependent Schrödinger equation

$$\mathrm{i}\hbar\frac{\mathrm{d}}{\mathrm{d}t}\psi(\mathbf{r},t) = H(t)\psi(\mathbf{r},t) \tag{9.4.43}$$

together with the corresponding eigenvalue problem

$$H(t)\phi_\lambda(\mathbf{r},t) = \epsilon_\lambda(t)\phi_\lambda(\mathbf{r},t). \tag{9.4.44}$$

The generic solution ψ at time t is given by a linear combination of the eigenstates ϕ according to the time evolution induced by the Hamiltonian H:

$$\psi(\mathbf{r},t) = \sum_\lambda S_\lambda \mathrm{e}^{-\frac{\mathrm{i}}{\hbar}\int_{t_0}^{t}\epsilon_\lambda(t')\,\mathrm{d}t'}\phi_\lambda(\mathbf{r},t), \tag{9.4.45}$$

which, for the case of the vector-potential Hamiltonian (9.4.4) and the corresponding eigenstates in equation (9.4.8) reduces to:

$$\psi(\mathbf{r},t) = \mathrm{e}^{\frac{\mathrm{i}e}{\hbar c}\mathbf{A}(t)\cdot\mathbf{r}}\sum_{\mathbf{k}_0\nu} S_{\mathbf{k}_0\nu}\mathrm{e}^{-\frac{\mathrm{i}}{\hbar}\int_{t_0}^{t}\epsilon_{\mathbf{k}(t')\nu}\,\mathrm{d}t'}\phi^0_{\mathbf{k}(t)\nu}(\mathbf{r}), \tag{9.4.46}$$

where $\mathbf{k}_0 = \mathbf{k}(t_0)$ denotes the carrier wave vector at the initial time t_0. As discussed in [9], it is always possible to consider the gauge transformation connecting the above vector-potential picture to the scalar-potential one. The generator of this gauge transformation is the function $-\mathbf{A}(t)\cdot\mathbf{r}$, which tells us that, going from the vector- to the scalar-potential picture, the first phase factor in equation (9.4.46) cancels exactly with the corresponding phase factor of the gauge transformation. Thus, the time-dependent wave function (9.4.46) in the scalar-potential gauge reduces to:

$$\psi(\mathbf{r},t) = \sum_{\mathbf{k}_0\nu} S_{\mathbf{k}_0\nu}\mathrm{e}^{-\frac{\mathrm{i}}{\hbar}\int_{t_0}^{t}\epsilon_{\mathbf{k}(t')\nu}\,\mathrm{d}t'}\phi^0_{\mathbf{k}(t)\nu}(\mathbf{r}). \tag{9.4.47}$$

This is a linear combination of the so-called Houston states [3] originally introduced as time-dependent solutions of the scalar-potential Schrödinger equation.

On the other hand, within the Wannier–Stark representation (9.4.41), the linear combination in equation (9.4.45) reduces to:

$$\psi(\boldsymbol{r}, t) = \sum_{k_\perp n\nu} S^{k_\perp n\nu} \mathrm{e}^{-\frac{\mathrm{i}}{\hbar}\epsilon^{k_\perp n\nu}(t-t_0)} \phi^{k_\perp n\nu}(\boldsymbol{r}). \tag{9.4.48}$$

It is then clear that for a given initial condition $\psi(\boldsymbol{r}, t_0)$ the last two linear combinations must give at any time t the same wave function ψ. Let us consider as an initial condition a single Wannier–Stark state $\phi^{k_\perp n\nu}$. According to equation (9.4.48), at time t the function ψ differs from that at time t_0 only by a phase factor, which implies that this will be a stationary state, i.e. the wave function will remain localized around a given cell and $|\psi|^2$ will not change in time. Therefore, it should be possible to choose the coefficients S entering equation (9.4.47) in such a way that the corresponding expansion in terms of Houston states will provide the same stationary Wannier–Stark state.

In order to determine the explicit form of the coefficients corresponding to a stationary state, let us rewrite equation (9.4.47), replacing at each time t the sum over $\boldsymbol{k}_0$ with an equivalent sum over the instantaneous $\boldsymbol{k} = \boldsymbol{k}_0 + \dot{\boldsymbol{k}}(t - t_0)$ (see equation (9.4.15)):

$$\psi(\boldsymbol{r}, t) = \sum_{\boldsymbol{k}\nu} S_{\boldsymbol{k}-\dot{\boldsymbol{k}}(t-t_0)\nu} \mathrm{e}^{-\frac{\mathrm{i}}{\hbar}\int_{t_0}^{t}\epsilon_{\boldsymbol{k}(t')\nu}\,\mathrm{d}t'} \phi^0_{\boldsymbol{k}\nu}(\boldsymbol{r}). \tag{9.4.49}$$

The stationary-state condition corresponds to imposing that each individual term entering the above expansion will evolve in time according to the constant Wannier–Stark energy $\epsilon^{k_\perp n\nu}$,

$$S_{\boldsymbol{k}-\dot{\boldsymbol{k}}(t-t_0)\nu} \mathrm{e}^{-\frac{\mathrm{i}}{\hbar}\int_{t_0}^{t}\epsilon_{\boldsymbol{k}(t')\nu}\,\mathrm{d}t'} \propto \mathrm{e}^{-\frac{\mathrm{i}}{\hbar}\epsilon^{k_\perp n\nu}(t-t_0)}, \tag{9.4.50}$$

which implies that for each time t

$$S_{\boldsymbol{k}-\dot{\boldsymbol{k}}(t-t_0)\nu} \propto \mathrm{e}^{\frac{\mathrm{i}}{\hbar}\int_{t_0}^{t}(\epsilon_{\boldsymbol{k}(t')\nu}-\epsilon^{k_\perp n\nu})\,\mathrm{d}t'}. \tag{9.4.51}$$

The above time integral over t' can be translated into a corresponding integral over $k' = k_\parallel(t')$:

$$\mathrm{e}^{\frac{\mathrm{i}e}{F}\int_{k_0}^{k}(\epsilon_{k'\nu}-\epsilon^{k_\perp n\nu})\,\mathrm{d}k'} = \mathrm{e}^{\frac{\mathrm{i}e}{F}\int_{0}^{k}(\epsilon_{k'\nu}-\epsilon^{k_\perp n\nu})\,\mathrm{d}k'}\,\mathrm{e}^{-\frac{\mathrm{i}e}{F}\int_{0}^{k_0}(\epsilon_{k'\nu}-\epsilon^{k_\perp n\nu})\,\mathrm{d}k'}. \tag{9.4.52}$$

Since the first phase factor on the right-hand side is time independent, we finally obtain:

$$S_{\boldsymbol{k}_0\nu} \propto \mathrm{e}^{-\frac{\mathrm{i}e}{F}\int_{0}^{k_0}(\epsilon_{k'\nu}-\epsilon^{k_\perp n\nu})\,\mathrm{d}k'}. \tag{9.4.53}$$

As expected, the coefficients S of a stationary state coincide with the coefficients s in equation (9.4.37). Thus, the linear combinations of accelerated Bloch states corresponding to stationary states are just the Wannier–Stark states introduced in equation (9.4.41) as eigenstates of the scalar-potential Hamiltonian.

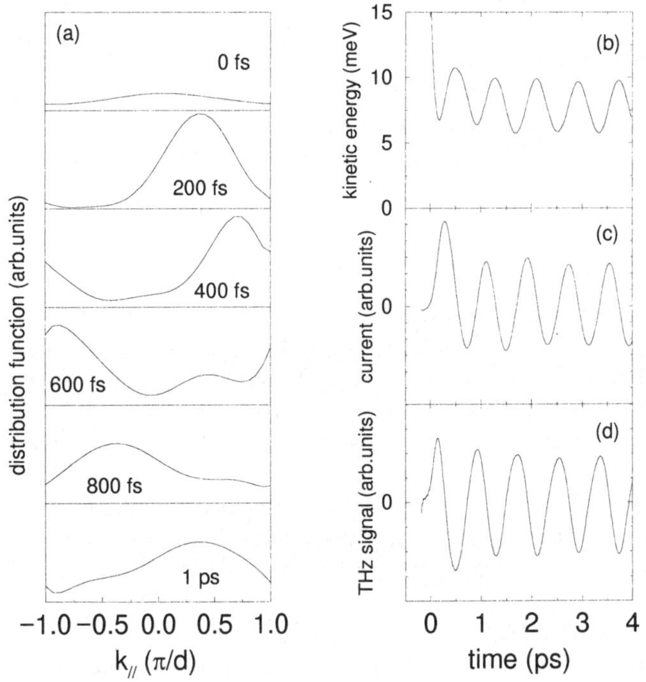

Figure 9.3 *Full Bloch-oscillation dynamics corresponding to a laser photo-excitation resonant with the first-miniband exciton. (a) Time evolution of the electron distribution as a function of* $k_{\parallel}$. *(b) Average kinetic energy, (c) current and (d) THz-signal corresponding to the Bloch oscillations in (a).*

From the above analysis, we see that within a time-dependent approach the scalar- and vector-potential pictures are totally equivalent. According to the initial condition (i.e. depending on the coefficients S), we may have a Bloch oscillation as well as a Wannier–Stark scenario, or any intermediate regime. If we consider, for example, the case of a laser excitation whose energy spectrum is concentrated around a well-defined frequency, this will generate a distribution of photo-excited carriers strongly peaked about a particular k. Each carrier will then be described by a single accelerated Bloch state $\phi_{\boldsymbol{k}\nu}$ and will execute Bloch oscillations. Thus, the overall motion of this packet in $\boldsymbol{k}$-space will resemble the periodic motion of a single electron 'prepared' in a Bloch state (Fig. 9.3). As we will discuss in the following section, such Bloch-oscillation dynamics can be monitored via FWM experiments or THz-signal measurements.

In contrast, if we perform a linear-absorption measurement using as a light source a laser with uniform spectral distribution, we generate a uniform distribution of photo-excited carriers in $\boldsymbol{k}$-space with their proper phase coherence (described by the corresponding interband polarizations). Such distribution is shifted in $\boldsymbol{k}$-space but, being almost constant, there is no macroscopic effect. This resembles the situation corresponding to a single electron prepared in a Wannier–Stark state, which is a uniform superposition of Bloch states. This is confirmed by optical-absorption investigations which clearly show the Wannier–Stark energy quantization (Fig. 9.7).

9.5 SOME SIMULATED EXPERIMENTS

In this section, we will review recent simulated experiments of the ultrafast carrier dynamics in semiconductor superlattices [41, 42, 46–49]. They are based on a generalized Monte Carlo solution [50–53] of the set of kinetic equations (so-called semiconductor Bloch equations) derived in section 9.3.2. This generalized Monte Carlo approach, successfully applied for the interpretation of ultrafast coherent phenomena in bulk semiconductors [54–57] is based on a combined solution of our kinetic equations [53]: the coherent contributions are evaluated by means of a direct numerical integration while the incoherent ones are 'sampled' by means of a conventional Monte Carlo simulation in the three-dimensional $\boldsymbol{k}$-space.

This generalized Monte Carlo method has recently been applied to semiconductor superlattices. As described in [41, 46], the simulation scheme is based on the Bloch-state representation of the vector-potential picture introduced in section 9.4.1.

The following superlattice model has been employed. The energy dispersion and corresponding wave functions along the growth direction ($k_{\|}$) are computed within the well-known Kronig–Penney model [12], while for the in-plane direction ($k_{\perp}$) an effective-mass model has been used. Starting from these three-dimensional wave functions $\phi^0_{\boldsymbol{k}\nu}$, the various carrier–carrier as well as carrier–phonon matrix elements are numerically computed (equations (9.3.21) and (9.3.23)). They are, in general, functions of the various miniband indices and depend separately on $k_{\|}$ and $k_{\perp}$, thus fully reflecting the anisotropic nature of the superlattice structure.

Only coupling to GaAs bulk phonons has been considered. This, of course, is a simplifying approximation which neglects any superlattice effect on the phonon dispersion, such as confinement of optical modes in the wells and in the barriers, and the presence of interface modes [58]. However, while these modifications have important consequences for phonon spectroscopies (such as Raman scattering), they are far less decisive for transport phenomena. Indeed, by now it is well known [58, 59] that the total scattering rates are sufficiently well reproduced if the phonon spectrum is assumed to be bulk-like.

We will start discussing the scattering-induced damping of Bloch oscillations. In particular, we will show that in the low-density limit this damping is mainly determined by optical-phonon scattering [41, 42] while at high densities the main mechanism responsible for the suppression of Bloch oscillations is found to be carrier–carrier scattering [49].

This Bloch-oscillation analysis in the time domain is also confirmed by its counterpart in the frequency domain. As pointed out in section 9.4, the presence of Bloch oscillations, due to a negligible scattering dynamics, should correspond to Wannier–Stark energy quantization. This is confirmed by the simulated optical-absorption spectra, which clearly show the presence of the field-induced Wannier–Stark ladders introduced in section 9.4.2 [48].

9.5.1 Bloch-oscillation analysis

All the simulated experiments presented in this section refer to the superlattice structure considered in [46]: 111 Å GaAs wells and 17 Å $Al_{0.3}Ga_{0.7}As$ barriers. For such a structure there has been experimental evidence for a THz-emission from Bloch oscillations [23].

In the first set of simulated experiments an initial distribution of photo-excited carriers (electron–hole pairs) is generated by a 100 fs Gaussian laser pulse in resonance with the first-miniband exciton ($\hbar\omega_L \approx 1540$ meV). The strength of the applied electric field is assumed to be 4 kV cm^{-1}, which corresponds to a Bloch period $\tau_B = h/eFd$ of about 800 fs.

In the low-density limit (corresponding to a weak laser excitation), incoherent scattering processes do not alter the Bloch-oscillation dynamics. This is due to the following reasons: in agreement with recent experimental [23, 40] and theoretical [41, 42, 46] investigations, for superlattices characterized by a miniband width smaller than the LO-phonon energy—as for the structure considered here—and for laser excitations close to the band gap, at low temperatures carrier–phonon scattering is not permitted. Moreover, in this low-density regime carrier–carrier scattering plays no role. Due to the quasi-elastic nature of Coulomb collisions, in the low-density limit the majority of the scattering processes is characterized by a very small momentum transfer; as a consequence, the momentum relaxation along the growth direction is negligible. As a result, on this picosecond timescale the carrier system exhibits a coherent Bloch-oscillation dynamics, i.e. a negligible scattering-induced dephasing. This can be clearly seen from the time evolution of the carrier distribution as a function of $k_\parallel$ (i.e. averaged over $k_\perp$) shown in Fig. 9.3. During the laser photo-excitation ($t = 0$) the carriers are generated around $k_\parallel = 0$, where the transitions are close to resonance with the laser excitation. According to the acceleration theorem (9.1.1), the electrons are then shifted in $\boldsymbol{k}$-space. When the carriers reach the border of the first Brillouin zone they are Bragg reflected. After about 800 fs, corresponding to the Bloch period τ_B, the carriers have completed one oscillation in $\boldsymbol{k}$-space. As expected, the carriers execute Bloch oscillations without loosing the synchronism of their motion by scattering. This is again shown in Fig. 9.3, where we have plotted: (b) the mean kinetic energy, (c) the current and (d) its time derivative which is proportional to the emitted far field, i.e. the THz-

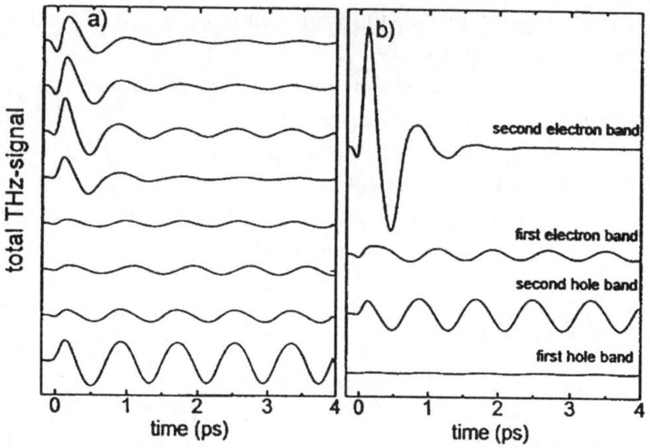

Figure 9.4 *(a) Total THz-signals for eight different spectral positions of the exciting laser pulse:* 1540, 1560, ..., 1680 meV *(from bottom to top). (b) Individual THz-signal of the electrons and holes in the different bands for a central spectral position of the laser pulse of 1640 meV. After [48].*

radiation. (It can be shown that, by neglecting Zener tunnelling, the intraband polarization $P^{e/h}$ in equation (9.3.36) is proportional to the current.) All these three quantities exhibit oscillations characterized by the same Bloch period τ_B. Due to the finite width of the carrier distribution in $\boldsymbol{k}$-space (see Fig. 9.3(a)), the amplitude of the oscillations of the kinetic energy is somewhat smaller than the miniband width. Since for this excitation condition the scattering-induced dephasing is negligible, the oscillations of the current are symmetric around zero, which implies that the time average of the current is equal to zero, i.e. no dissipation.

As already pointed out, this ideal Bloch-oscillation regime is typical of a laser excitation close to gap in the low-density limit. Let us now discuss, still at low densities, the case of a laser photo-excitation high in the band. Figure 9.4(a) shows the THz-signal as obtained from a set of simulated experiments corresponding to different laser excitations [46]. The different traces correspond to the emitted THz-signal for increasing excitation energies. We clearly notice the presence of Bloch oscillations in all cases. However, the oscillation amplitude and decay (effective damping) is excitation dependent.

For the case of a laser excitation resonant with the first-miniband exciton considered above (Fig. 9.3), we have a strong THz-signal. The amplitude of the signal decreases when the excitation energy is increased. Additionally, there are also some small changes in the phase of the oscillations, which are induced by the electron–LO-phonon scattering.

When the laser energy comes into resonance with the transitions between the second electron and hole minibands ($\hbar\omega_L \approx 1625$ meV), the amplitude of the

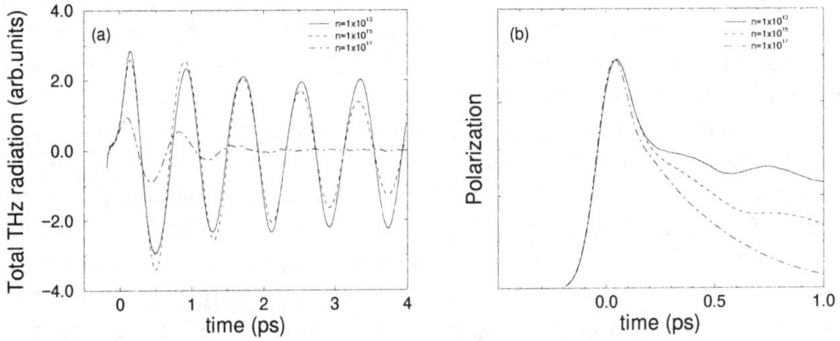

Figure 9.5 *(a) Total THz-radiation as a function of time; (b) incoherently-summed polarization as a function of time. After [49].*

THz-signal increases again. The corresponding THz-transients show an initial part, which is strongly damped and some oscillations for longer times that are much less damped. For a better understanding of these results, in Fig. 9.4(b) we show the individual THz-signals, originating from the two electron and two heavy-hole minibands for the excitation with $\hbar\omega = 1640$ meV. The Bloch oscillations performed by the electrons within the second miniband are strongly damped due to intra- and interminiband LO-phonon scattering processes [41, 46]. Since the width of this second miniband (45 meV) is somewhat larger than the LO-phonon energy, also intraminiband scattering is possible, whenever the electrons are accelerated into the high-energy region of the miniband. The THz-signal, originating from electrons within the first miniband, shows an oscillatory behaviour with a small amplitude and a phase which is determined by the time the electrons need to relax down to the bottom of the band.

At the same time, the holes in both minibands exhibit undamped Bloch oscillations, since the minibands are so close in energy that for these excitation conditions no LO-phonon emission can occur. The analysis shows that at early times the THz-signal is mainly determined by the electrons within the second miniband. At later times the observed signal is due to heavy holes and electrons within the first miniband.

The above theoretical analysis closely resembles experimental observations obtained for a superlattice structure very similar to the one modelled here [23]. In these experiments, evidence for THz-emission from Bloch oscillations has been reported. For some excitation conditions these oscillations are associated with resonant excitation of the second miniband. The general behaviour of the magnitude of the signals, the oscillations and the damping are close to the results shown in Fig. 9.4.

Finally, in order to study the density dependence of the Bloch-oscillation damping, let us go back to the case of laser excitations close to gap. Figure 9.5(a) shows the total (electrons plus holes) THz-radiation as a function of time for three different carrier densities. With increasing carrier density, carrier–carrier scattering becomes more and more important: due to Coulomb screening, the momentum transfer in a carrier–carrier scattering increases (its typical value being comparable with the screening wave vector). This can be seen in Fig. 9.5(a), where for increasing carrier densities we realize an increasing damping of the THz-signal. However, also for the highest carrier density considered here we deal with a damping time of the order of 700 fs, which is much larger than the typical dephasing time, i.e. the decay time of the interband polarization, associated with carrier–carrier scattering. The dephasing time is typically investigated by means of FWM measurements and such multipulse experiments can be simulated as well [54, 55]. From a theoretical point of view, a qualitative estimate of the dephasing time is given by the decay time of the 'incoherently summed' polarization (ISP) [51]. Figure 9.5(b) shows such ISP as a function of time for the same three carrier densities of Fig. 9.5(a). As expected, the decay times are always much smaller than the corresponding damping times of the THz-signals (note the different timescale in Figs. 9.5(a) and (b)). This difference can be understood as follows. The fast decay times of Fig. 9.5(b) reflect the interband dephasing, i.e. the sum of the electron and hole scattering rates. In particular, for the Coulomb interaction this means the sum of electron–electron, electron–hole, and hole–hole scattering. This last contribution is known to dominate and determines the dephasing timescale. On the other hand, the total THz-radiation in Fig. 9.5(a) is the sum of the electron and hole contributions. However, due to the small value of the hole miniband width compared with the electron one, the electron contribution will dominate. This is clearly shown in Fig. 9.6, where (a) the electron and (b) the hole contributions to the THz-radiation are plotted as a function of time (note the different vertical scale). This means that the THz damping in Fig. 9.5(a) mainly reflects the damping of the electron contribution (see Fig. 9.6(a)). This decay, in turn, reflects the intraband dephasing of electrons which is due to electron–electron and electron–hole scattering only, i.e. no hole–hole contributions. This clearly explains the different decay times of Figs. 9.5(a) and (b).

From the above analysis we can conclude that the decay time of the THz-radiation due to carrier–carrier scattering differs considerably from the corresponding dephasing times obtained from a FWM experiment: the first one is a measurement of the intraband dephasing while the second one reflects the interband dephasing.

9.5.2 Optical-absorption analysis

Let us now discuss the frequency-domain counterpart of the Bloch-oscillation picture considered so far. Similar to what happens in the time domain, for sufficiently high electric fields, i.e. when the Bloch period $\tau_B = h/eFd$ becomes smaller than the dephasing time, the optical spectra of the superlattice are ex-

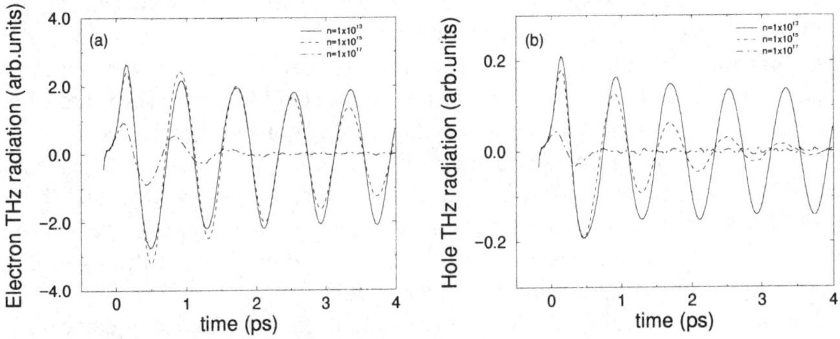

Figure 9.6 *(a) Electron and (b) hole contributions to the total THz-radiation of Fig. 9.5(a). After [49].*

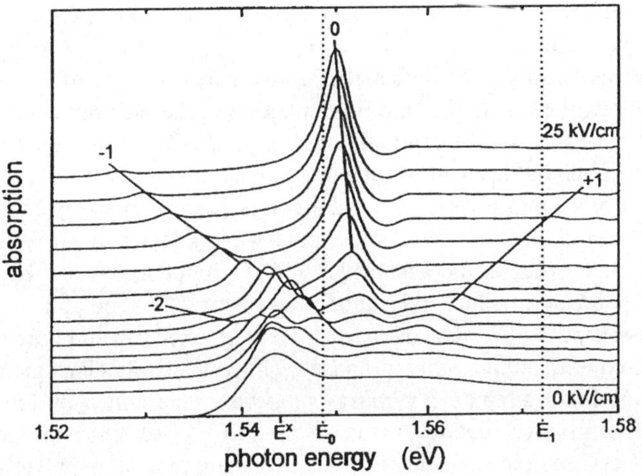

Figure 9.7 *Absorption spectra for various static applied electric fields for a GaAs/Al$_{0.3}$Ga$_{0.7}$As superlattice (well (barrier) width 95 (15) Å). The vertical displacements between any two spectra is proportional to the difference of the corresponding fields. The Wannier–Stark transitions are labelled by numbers, the lower (higher) edge of the combined miniband by E_0 (E_1). After [48].*

pected to exhibit the frequency-domain counterpart of the Bloch oscillations, i.e. the Wannier–Stark energy quantization discussed in section 9.4.2. In the absence of Coulomb interaction, the Wannier–Stark ladder absorption increases as a function of the photon energy in a step-like fashion. These steps are equidistantly spaced. This spacing, named Wannier–Stark splitting, is proportional to the applied electric field (see equation (9.4.39)).

The simulated linear-absorption spectra corresponding to a superlattice structure with 95 Å GaAs wells and 15 Å $Al_{0.3}Ga_{0.7}As$ barriers are shown in Fig. 9.7 [48]. As we can see, the Coulomb interaction gives rise to excitonic peaks in the absorption spectra and introduces couplings between these Wannier–Stark states. Such exciton peaks, which are no longer equidistantly spaced, are often referred to as excitonic Wannier–Stark ladders [39] of the superlattice.

Since for the superlattice structure considered in this simulated experiment [41, 48] the combined miniband width is larger than the typical two- and three-dimensional exciton binding energies, it is possible to investigate the quasithree-dimensional absorption behaviour of the delocalized miniband states as well as localization effects induced by the electric field.

For the free-field case, the electron and hole states are completely delocalized in our three-dimensional $\boldsymbol{k}$-space. The perturbation induced by the application of a low field (here ≈ 5 kV cm^{-1}), couples the states along the field direction and in the spectra the Franz–Keldysh effect, well known from bulk materials, appears: one clearly notices oscillations which increase in amplitude with the field and shift with $F^{2/3}$ from the $n = 0$ and $n = 1$ levels toward the centre of the combined miniband.

For increasing fields the potential drop over the distance of a few quantum wells eventually exceeds the miniband width and the electronic states become more and more localized. Despite the field-induced energy difference $neFd$, the superlattice potential is equal for quantum wells separated by nd. Therefore, the spectra decouple into a series of peaks corresponding to the excitonic ground states of the individual electron–hole Wannier–Stark levels. Each Wannier–Stark transition contributes to the absorption with a pronounced $1 - s$ exciton peak, plus higher bound exciton and continuum states. The oscillator strength of a transition n is proportional to the overlap between electron and hole wave functions centred at quantum wells n' and $n + n'$, respectively. The analysis shows that this oscillator strength is almost exclusively determined by the amplitude of the electron wave function in quantum well $n' + n$ since for fields in the Wannier–Stark regime the hole wave functions are almost completely localized over one quantum well due to their high effective mass (Fig. 9.2). Thus, the oscillator strengths of transitions to higher $|n|$ become smaller with increasing $|n|$ and field.

At high fields (here $> \approx 8$ kV cm^{-1}) the separation between the peaks is almost equal to $neFd$. For example, the peak of the $n = 0$ transition which is shifted by the Wannier–Stark exciton binding energy with respect to the centre of the combined miniband, demonstrates that the increasing localization also increases the exciton binding energy. This increased excitonic binding reflects the gradual transition from a three-dimensional to a two-dimensional behaviour.

For intermediate fields there is an interplay between the Wannier–Stark and Franz–Keldysh effect. Coming from high fields, first the Wannier–Stark peaks are modulated by the Franz–Keldysh oscillations. However, as soon as the separation eFd between neighbouring peaks becomes smaller than their spectral widths, the peaks can no longer be resolved individually so only the Franz–Keldysh structure remains.

ACKNOWLEDGEMENTS

I wish to thank the colleagues of the Marburg group, S.W. Koch, T. Meier and P. Thomas, for their essential contributions to the research activity reviewed in this chapter. I am also grateful to A. Di Carlo, M. Gulia, T. Kuhn, E. Molinari and P.E. Selbmann for stimulating and fruitful discussions.

This work was supported in part by the EC Commission through the Network 'ULTRAFAST'.

REFERENCES

[1] Bloch, F. (1928) *Z. Phys.* **52**, 555.
[2] Zener, C. (1934) *Proc. R. Soc.* **145**, 523.
[3] Houston, W. V. (1940) *Phys. Rev.* **57**, 184.
[4] Kane, E. O. (1959) *J. Phys. Chem. Solids* **12**, 181.
[5] Argyres, P. N. (1962) *Phys. Rev.* **126**, 1386.
[6] Wannier, G. H. (1960) *Phys. Rev.* **117**, 432.
[7] Zak, J. (1968) *Phys. Rev. Lett.* **20**, 1477.
[8] Kittel, C. (1963) *Quantum Theory of Solids*, Wiley, New York, p. 190.
[9] Krieger, J. B. and Iafrate, G. J. (1986) *Phys. Rev.* B **33**, 5494.
[10] Nenciu, G. (1991) *Rev. Mod. Phys.* **63**, 91.
[11] Di Carlo, A., Pötz, W. and Vogl, P. (1994) *Phys. Rev.* B **50**, 8358.
[12] Bastard, G. (1989) *Wave Mechanics of Semiconductor Heterostructures*, Les Editions de Physique, Les Ulis, France.
[13] Shah, J. (1996) *Ultrafast Spectroscopy of Semiconductors and Semiconductor Nanostructures*, Springer, Berlin.
[14] Mendez, E. E., Agullo-Rueda, F. and Hong, J. M. (1988) *Phys. Rev. Lett.* **60**, 2426.
[15] Voisin, P., Bleuse, J., Bouche, C. *et al.* (1988) *Phys. Rev. Lett.* **61**, 1639.
[16] Feldmann, J., Leo, K., Shah, J. *et al.* (1992) *Phys. Rev.* B **46**, 7252.
[17] von Plessen, G. and Thomas, P. (1992). *Phys. Rev.* B **45**, 9185
[18] Leo, K., Bolivar, P. H., Brüggemann, F. *et al.* (1992) *Solid State Commun.* **84**, 943.
[19] Leisching, P., Haring Bolivar, P., Beck, W. *et al.* (1994) *Phys. Rev.* B **50**, 14389.
[20] Dekorsy, T., Leisching, P., Kohler, K. *et al.* (1994) *Phys. Rev.* B **50**, 8106.
[21] Dekorsy, T., Ott, R., Kurz, H. *et al.* (1995) *Phys. Rev.* B **51**, 17275.
[22] Waschke, C., Roskos, H. G., Sscwedler, R. *et al.* (1993) *Phys. Rev. Lett.* **70**, 3319.
[23] Roskos, H. G., Waschke, C., Schwedler, R. *et al.* (1994) *Superlattices and Microstructures* **15**, 281.
[24] Wannier, G. H. (1962) *Rev. Mod. Phys.* **34**, 645.
[25] Wannier, G. H. and Fredkin, D. R. (1962) *Phys. Rev.* **125**, 1910.
[26] Wannier, G. H. and Van Dyke, J. P. (1968) *J. Math. Phys.* **9**, 899.
[27] Wannier, G. H. (1969) *Phys. Rev.* **181**, 1364.
[28] Zak, J. (1969) *Phys. Rev.* **181**, 1366.
[29] Rabinovitch, A. and Zak, J (1972) *J. Phys. Lett.* **40A**, 189.
[30] Koss, R. W. and Lambert, L. M. (1972) *Phys. Rev.* B **5**, 1479.

[31] Callaway, J. (1963) *Phys. Rev.* **130**, 549.

[32] Kash, J. A. and Tsang, J. C. (1989) *Light Scattering in Solids VI* (eds M. Cardona and G. Güntherodt), Springer, Berlin, p. 423.

[33] Pötz, W. and Kocevar, P. (1983) *Phys. Rev.* B **28**, 7040.

[34] Quade, W., Schöll, E., Rossi, F. *et al.* (1994) *Phys. Rev.* B **50**, 7398.

[35] Brunetti, R., Jacoboni, C. and Rossi, F. (1989) *Phys. Rev.* B **39**, 10 781.

[36] Rossi, F. Brunetti, R. and Jacoboni, C. (1992) *Hot Carriers in Semiconductor Nanostructures: Physics and Applications* (ed. J. Shah), Academic Press, Boston, p. 153.

[37] Tran Thoai, D. B. and Haug, H. (1993) *Phys. Rev.* B **47**, 3574.

[38] Schilp, J., Kuhn, T. and Mahler, G. (1994) *Phys. Rev.* B **50**, 5435.

[39] Dignam, M. M. and Sipe, J. E. (1990) *Phys. Rev. Lett.* **64**, 1797.

[40] von Plessen, G., Meier, T., Feldmann, J. *et al.* (1994) *Phys. Rev.* B **49**, 14 058.

[41] Rossi, F. Meier, T., Thomas, P. *et al.* (1995) *Phys. Rev.* B **51**, 16 943.

[42] Rossi, F., Meier, T., Thomas, P. *et al.* (1996) *Hot Carriers in Semiconductors* (eds K. Hess, J.-P. Leburton and U. Ravaioli), Plenum Press, New York, p. 157.

[43] Meier, T., von Plessen, G., Thomas, P. *et al.* (1994) *Phys. Rev. Lett.* **73**, 902.

[44] Meier, T., von Plessen, G., Thomas, P. *et al.* (1995) *Phys. Rev.* B **51**, 14 490.

[45] Schiff, L I (1955) *Quantum Mechanics* 2nd edn., McGraw-Hill, New York, p. 275.

[46] Meier, T., Rossi, F., Thomas, P. *et al.* (1995) *Phys. Rev. Lett.* **75**, 2558.

[47] Je, K-C., Meier, T., Rossi, F. *et al.* (1995) *Appl. Phys. Lett.* **67**, 2978.

[48] Koch, S. W., Meier, T., Stroucken, T. *et al.* (1995) *Microscopic Theory of Semiconductors: Quantum Kinetics, Confinement and Lasers* (ed. S. W. Koch), World Scientific, Singapore, p. 81.

[49] Rossi, F., Gulia, M., Selbmann, P. E. *et al.* (1996) *Proc. 23rd ICPS, Berlin, Germany*, (eds M. Scheffler and R. Zimmermann), World Scientific, Singapore, p. 1775.

[50] Kuhn, T. and Rossi. F. (1992) *Phys. Rev. Lett.* **69**, 977.

[51] Kuhn., T and Rossi, F. (1992) *Phys. Rev.* B **46**, 7496.

[52] Rossi, F., Haas, S. and Kuhn, T. (1994) *Phys. Rev. Lett.* **72**, 152.

[53] Haas, S., Rossi, F. and Kuhn, T. (1996) *Phys. Rev.* B **53**, 12 855.

[54] Lohner, A., Rick, K., Leisching, P. *et al.* (1993) *Phys. Rev. Lett.* **71**, 77.

[55] Leitenstorfer, A., Lohner, A., Rick, K. *et al.* (1994) *Phys. Rev.* B **49**, 16 372.

[56] Leitenstorfer, A., Lohner, A., Elsaesser, T. *et al.* (1994) *Phys. Rev. Lett.* **73**, 1687.

[57] Leitenstorfer, A., Elsaesser, T., Rossi, F. *et al.* (1996) *Phys. Rev.* B **53**, 9876.

[58] Molinari, E., (1994) *Confined Electrons and Photons: New Physics and Applications*, (eds E. Burstein and C. Weisbuch), Plenum, New York.

[59] Rücker, H., Molinari, E. and Lugli, P. (1992) *Phys. Rev.* B **45**, 6747.

CHAPTER 10

Vertical transport and domain formation in multiple quantum wells

Andreas Wacker

Mikroelektronik Centret, Danmarks Tekniske Universitet, 2800 Lyngby, Denmark

now at: Institut für Theoretische Physick, Sekr. PN 7-1, Technische Universität Berlin, Hardenbergstr. 36, 10623 Berlin, Germany

E-mail: wacker@lorenz.physik.tu-berlin.de

10.0 INTRODUCTION

Today's growth techniques allow the construction of semiconductor structures where layers of different semiconductors (exhibiting similar lattice constants) can be grown on each other with the interface being well defined within one atomic monolayer. If two such layers alternate periodically one obtains a periodic structure with an artificial period d in the growth direction (which is defined to be the z-direction). This leads to spatial variations in the conduction and valence band of the material as sketched in Fig. 10.1. Considering only conduction band states in the following, the region with a lower conduction band is called the 'well' and the region with a higher conduction band is called the 'barrier'. An extended discussion of various aspects of such structures can be found in [1]. In this chapter the vertical electronic transport (i.e. in the z-direction) is considered for such structures.

If we consider the electronic properties of such a structure, there are two different approaches. In the first approach one may consider the full structure as an artificial lattice. The energy spectrum can be calculated analogously to the Kronig–Penney model (which is discussed in almost all solid-state physics text-books) resulting in the appearance of energy bands and energy gaps as sketched in Fig. 10.1. Due to this analogy with the atomic lattice of lattice constant a_L, such semiconductor structures are often called superlattices. The corresponding eigenfunctions are the usual Bloch functions which extend over the whole superlattice. The Bloch functions are labelled with the Bloch vector q which is restricted to the Brillouin zone $-\pi/d < q < \pi/d$. This range is much smaller than the Brillouin zone $-\pi/a_L < q < \pi/a_L$ of the atomic lattice as $d \gg a_L$. Therefore these new bands are called minibands. If an electric field is applied,

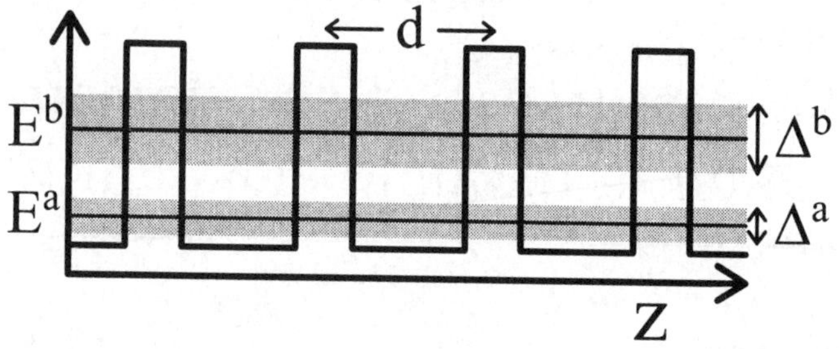

Figure 10.1 *Sketch of the conduction band $E_c(z)$ with minibands $\nu = a, b$.*

the Bloch functions are accelerated and the transport can be treated analogously to the usual transport in bulk systems [2]. Due to the short Brillouin zone many special effects, such as Bloch oscillations can be found here which are not accessible in bulk systems as discussed in Chapter 9.

In the second approach one may consider the wells to be isolated from each other as a first approximation and calculate the eigenstates within each well. This yields a sequence of energy levels and localized wave functions for each well. Such a structure is usually called a multiple quantum well. Of course this approach only makes sense if the coupling between different wells is weak and can be calculated perturbatively. Thinking in terms of Fermi's golden rule, there will be transitions between the levels in different wells yielding sequential tunnelling. Due to energy conservation this tunnelling only takes place if the energy levels align and strong resonances [3] are likely to occur if the relative height of the levels in different wells is varied by a voltage applied in the z-direction.

The width of these resonances is related to scattering processes of the electrons inside the wells. In [4] this feature was used to determine the electron–electron scattering rate from the transport between two weakly coupled quantum wells. This shows that a modelling of scattering processes is necessary for a calculation of the current in such structures and opens the possibility for checking theoretical concepts as well as investigating the significance of various scattering mechanisms by comparison with experimental data.

These resonances yield strong nonlinearities in the local current-field relation. If we now have a long periodic sequence of these coupled quantum wells the full system is a good example of an extended nonlinear system. This leads to the appearance of complicated current–voltage characteristics exhibiting an almost periodic sequence of branches due to the formation of electric field domains as well as to self-sustained oscillations due to moving domains. As the periodicity of the structure is an important feature for the translational invariance, these structures are frequently called superlattices for weak coupling as well. If the translational invariance is broken due to an insufficient control of the growth conditions, the branches lose their periodic structure and information about the actual structure can be obtained from measurements of the current–voltage characteristics.

The first section of this chapter is devoted to the classification of the different regimes depending on the coupling between the wells. In the second section a transport model for weakly coupled quantum wells is presented which allows a microscopic calculation of the current without any fitting parameters. In section 10.3 this model is extended in order to describe domain formation. The fourth section is on the influence of deviations from the periodic structure of the superlattice and the fifth section refers to self-sustained oscillations. Finally, details of the calculations will be given in section 10.6.

10.1 THE DIFFERENT TRANSPORT REGIMES

In this section the precise meaning of weakly and strongly coupled quantum wells is discussed.

Assuming ideal interfaces, the semiconductor structure is translational invariant within the x- and y-direction perpendicular to the growth direction. Therefore the x, y dependence can be separated by the ansatz $e^{i\boldsymbol{k}\cdot\boldsymbol{r}}$ where $\boldsymbol{k}$ and $\boldsymbol{r}$ are vectors within the two-dimensional (x, y) plane. (The effects due to the atomic periodicity of the lattice are treated within the envelope function formalism [5] assuming a parabolic subband.) Within the z-direction we have an artificial period d leading to energy states E_q^ν characterized by the miniband index ν and the quasiwave vectors $-\pi/d < q \le \pi/d$. The energies as well as the wave functions $\varphi_q^\nu(z)$ can be calculated numerically analogously to the well-known Kronig–Penney model. For a given miniband ν the energy E_q^ν has a mean value $E^\nu = d/(2\pi)\int_{-\pi/d}^{\pi/d} \mathrm{d}q\, E_q^\nu$ and varies within the miniband width $\Delta^\nu = \mathrm{Max}_q(E_q^\nu) - \mathrm{Min}_q(E_q^\nu)$ as sketched in Fig. 10.1.

From the Bloch functions $\varphi_q^\nu(z)$ the Wannier functions $\Psi^\nu(z - nd)$ defined by

$$\Psi^\nu(z - nd) = \sqrt{\frac{d}{2\pi}} \int_{-\pi/d}^{\pi/d} \mathrm{d}q \mathrm{e}^{-inqd} \varphi_q^\nu(z) \tag{10.1.1}$$

can be constructed. They are real and localized in well n for an appropriate choice of the complex phase for $\varphi_q^\nu(z)$ [6]. An example is shown in Fig. 10.2.

Using the Fourier expansion

$$E_q^\nu = E^\nu + \sum_{h=1}^{\infty} 2T_h^\nu \cos(hdq) \tag{10.1.2}$$

one obtains the following Hamiltonian in second quantization

$$\hat{H} = \sum_{n,\nu} \left[E^\nu c_n^{\nu\dagger} c_n^\nu + \sum_{h=1}^{\infty} \left(T_h^\nu c_{n+h}^{\nu\dagger} c_n^\nu + T_h^\nu c_{n-h}^{\nu\dagger} c_n^\nu \right) \right] \tag{10.1.3}$$

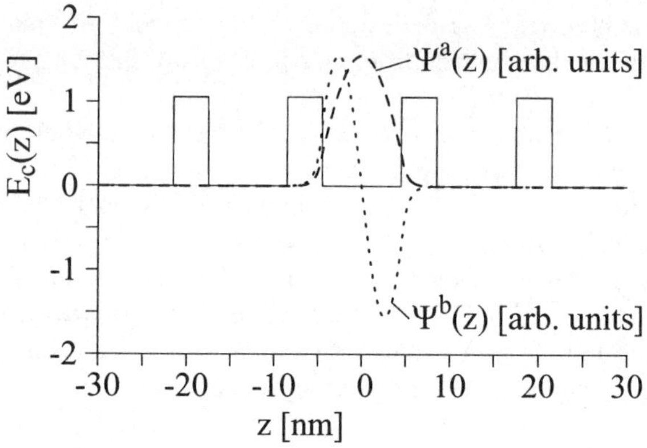

Figure 10.2 *Conduction band $E_C(z)$ together with the Wannier functions calculated for the two lowest subbands.*

for the creation $c_n^{\nu\dagger}$ and annihilation c_n^ν operators of the Wannier functions $\Psi^\nu(z-nd)$. As the Wannier functions are linear combinations of Bloch functions with different energies, there are no stationary states. Neglecting terms with $h > 1$ and setting $E^\nu = 0$ the time evolution of the annihilation operators is given by

$$\mathrm{i}\hbar \frac{\mathrm{d}}{\mathrm{d}t} c_n^\nu = T_1^\nu \left(c_{n+1}^\nu + c_{n-1}^\nu \right). \tag{10.1.4}$$

For the initial condition $c_n^\nu(t=0) = \delta_{n,0} c_0^\nu$ this has the solution

$$c_n^\nu(t) = \mathrm{i}^{-n} J_n \left(\frac{2T_1^\nu}{\hbar} t \right) c_0^\nu \tag{10.1.5}$$

where $J_n(x)$ is the Bessel function of order n. Since $\sum_{n=-\infty}^{\infty} n^2 (J_n(x))^2 = x^2/2$ the average extension $n_a(t)$ as a function of time is found to be

$$n_a(t) = \sqrt{\langle n^2 \rangle} = \frac{\sqrt{2} T_1}{\hbar} t. \tag{10.1.6}$$

This is obviously a coherent process. In a real semiconductor structure there will be scattering (characterized by a scattering time τ_{sc}) destroying the coherent evolution† given by equation (10.1.4). If $n_a(\tau_{\mathrm{sc}}) \ll 1$ holds, the phase coherence is completely lost during the spread and the states of adjacent wells will not maintain a fixed phase relation. In this case the transitions can be described by sequential tunnelling between the Wannier states with loss of phase between

† This scattering can be a phase-breaking process, such as scattering with phonons. But also elastic scattering processes at impurities or interface roughness will contribute here because they destroy the translational invariance of the superlattice, which is both the justification for the existence of the Bloch functions and a WS state.

different tunnelling events. Defining $\Gamma = \hbar/\tau_{sc}$ one may therefore use the picture of

$$\boxed{\text{sequential tunnelling for } T_1 \ll \Gamma/\sqrt{2}.} \tag{10.1.7}$$

If, on the other hand, the extended Bloch functions shall be a reasonable basis set, the phase coherence has to be maintained over a larger number of quantum wells. This gives $n_a(\tau_{sc}) \gg 1$ as a first necessary condition. But a further complication arises if an electric field F is applied in the z-direction. As discussed in Chapter 9 of this book, the eigenvalues of the Hamiltonian take discrete values $E^\nu - neFd$ which are called the Wannier–Stark (WS) ladder. The corresponding eigenfunctions $\phi^\nu_{\text{WS}}(z - nd)$ are localized around well n. Within the one-band limit they are given by

$$\phi^\nu_{\text{WS}}(z) = \sum_n J_n\left(\frac{2T^\nu_1}{eFd}\right)\Psi^\nu(z - nd). \tag{10.1.8}$$

(See, e.g., [7], where the coupling between different bands is also discussed.) Then the spatial extension of the WS states is given by

$$n_{\text{WS}} = \sqrt{\langle n^2\rangle} = \frac{\sqrt{2}T_1}{eFd}. \tag{10.1.9}$$

If now n_{WS} becomes small, the electrons are localized and the use of extended Bloch functions does not make any sense. This provides the second condition $n_{\text{WS}} \gg 1$ for the Bloch functions to be an appropriate basis set of wave functions. Thus it only makes sense to speak about

$$\boxed{\text{minibands for } T_1 \gg \Gamma/\sqrt{2} \quad \text{and} \quad T_1 \gg eFd/\sqrt{2}\,.} \tag{10.1.10}$$

As a third basis set the WS functions $\phi^\nu_{\text{WS}}(z - nd)$ may be used if an electric field is applied. They make sense if coherence is maintained within their extension, i.e. if $n_{\text{WS}} < n_a(\tau_{sc})$. Therefore the transport can be described by

$$\boxed{\text{WS hopping for } eFd > \Gamma\,.} \tag{10.1.11}$$

These different regimes of validity are sketched in Fig. 10.3 where the condition $a \ll b$ has been translated to $a < b/2$ for illustrative purpose.

How can the electric transport be described if an electric field F is applied in the z-direction? For $F = 0$ the system should be in equilibrium and no current I flows. For small fields the current should increase with F. (This may be calculated by linear response which is given by the Kubo formula for a quantum system. See, for example, Chapter 5, where the application of the Kubo formula has been described for the Coulomb drag.) Now it is an important feature of both superlattices and multiple quantum wells that for larger fields there is a range with negative differential conductivity (NDC) $\mathrm{d}I/\mathrm{d}F < 0$. This can be easily understood within the WS picture. As the WS states are orthogonal, transitions between them have to be caused by scattering events.

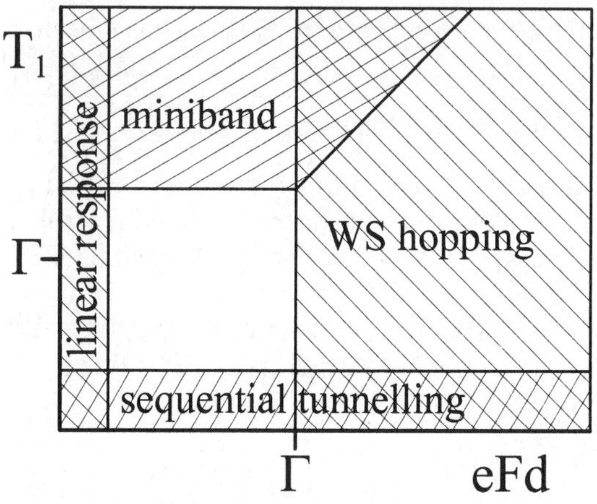

Figure 10.3 *Sketch of the regimes where different approaches are valid.*

Now the matrix element for a given scattering element depends on the spatial overlap of the eigenfunctions. With increasing field the WS states become more and more localized as can be seen from equation (10.1.9). Therefore the matrix element for the scattering is decreased and the hopping rate is diminished [8, 9]. As the WS approach is not justified for low-electric fields (Fig. 10.3) an ohmic behaviour for low fields could not be observed within this model [9].

In the validity range of (10.1.10) the basis of Bloch functions may be used. Due to an electric field the Bloch waves are accelerated according to $\hbar\dot{q} = eF$. As in the standard transport theory for bulk systems a positive conductivity is found for low-electric fields. But for higher fields the Bloch vector q may run through the whole Brillouin zone $2\pi/d$ if $eF\tau_{sc}/\hbar > 2\pi/d$ holds. Then the electrons perform a periodic motion in both q and z space which is called the Bloch oscillation. A quantitative analysis within various approaches [2, 10–12] reveals

$$I \sim \frac{eFd}{(eFd)^2 + \Gamma^2}. \tag{10.1.12}$$

This yields positive differential conductivity for $eFd < \Gamma$ and negative differential conductivity for larger fields which is confirmed by experimental data [13, 14]. For large F there is $I \sim 1/F$. This behaviour is also found from the WS-hopping model for $eFd < \Delta = 4T_1$ [9] which can be easily understood from Fig. 10.3 as this condition includes the region where both approaches are valid. This is an explicit example of the more general equivalence between the WS picture and the Bloch oscillation picture discussed in Chapter 9 of this book.

In the case of weakly coupled multiple quantum wells $T_1 \ll \Gamma/\sqrt{2}$ NDC has also been observed experimentally [15]. Here we can consider the electronic states to be the levels of the single quantum wells which will exhibit a certain width Γ due to scattering. The tunnelling between the wells is caused by the coupling T_1. Due to energy conservation a significant current between adjacent wells may flow if the levels are aligned within accuracy Γ. Therefore NDC is likely to appear if the disalignment eFd becomes larger than Γ [16, 17]. This will be modelled quantitatively in the next section.

For the white region of Fig. 10.3 between these two limits the situation is more complicated. Here Laikhtman and Miller [18] were able to show that equation (10.1.12) holds provided the electron temperature is much larger than T_1 and eFd.

In an extended system the existence of an NDC region typically causes oscillatory behaviour due to travelling field domains [19]. For the miniband regime this was theoretically studied quite a long time ago in [20–22] and recently experimentally [23, 24]. Note that these travelling domain oscillations are self-sustained in [24] while for the usual Bloch oscillations only transient oscillations are found [25]. The same type of oscillations have recently been found in the regime of sequential tunnelling both experimentally [26–28] and by numerical simulations [29, 30]. An analytic treatment of the oscillation mode is given in [31, 32]. Under certain conditions chaotic oscillations also occur [33, 34].

For weakly coupled multiple quantum wells an additional scenario occurs, which has been extensively studied in the last decade. As the electronic states are localized within single wells the domain boundary (which are charge accumulation layers) can be trapped in a single quantum well and a stationary stable domain is formed. As the domain boundary may be located in any well a periodic sequence of branches appears in the current–voltage characteristic. This effect has been observed experimentally by many groups [35–49]. Theoretically such effects can be studied by combining rate equations between the wells and Poisson's equation. Such an approach has already been performed in [50, 51]. Nevertheless, to my knowledge, the full current–voltage characteristic exhibiting the domain structure could first be resolved qualitatively in [52] for slim superlattices exhibiting a few electrons per well and in [53, 54] for two-dimensional wells. A full quantitative calculation has recently been presented [55, 56] for different multiple quantum wells yielding quantitative agreement with experimental data without using any fitting parameters. This model will be presented in the following sections.

10.2 MODELLING OF THE TRANSPORT BETWEEN WEAKLY COUPLED QUANTUM WELLS

This section aims to show how the currents between the wells can be calculated from a microscopic model. Furthermore, some simplifications will be presented in order to obtain estimates for the current. The numerical calculations are performed with the data of the sample used in [26, 40] exhibiting $N = 40$

GaAs wells of width $w = 9$ nm between 41 AlAs barriers of width $b = 4$ nm. The wells are n-doped with a doping density of $N_D = 1.5 \times 10^{11}$ cm^{-2} per well. Details of the specific calculations outlined in this section are given in section 10.6.

As discussed in section 10.1, for weakly coupled quantum wells the products of Wannier functions $\Psi^\nu(z - nd)$, localized in well n and plane waves $e^{i\boldsymbol{k}\cdot\boldsymbol{r}}$ form a reasonable basis set of wave functions. Now we restrict ourselves to the lowest two levels denoted by $\nu = a, b$. The respective annihilation operators are denoted by $a_n = c_n^a$ and $b_n = c_n^b$. Furthermore, we add the contribution $-eFz$ due to a homogeneous electric field F in the Hamiltonian (10.1.3) where $e < 0$ is the charge of the electron. Restricting ourselves to coupling between neighbouring wells ($h = 1$) we obtain the Hamiltonian $\hat{H} = \hat{H}_0 + \hat{H}_1 + \hat{H}_2$:

$$\begin{aligned}
\hat{H}_0^{\rm res} &= \sum_{n,\boldsymbol{k}} [(E^a + E_k - eFdn)a_n^\dagger(\boldsymbol{k})a_n(\boldsymbol{k}) \\
&\quad + (E^b + E_k - eFdn)b_n^\dagger(\boldsymbol{k})b_n(\boldsymbol{k})] \qquad (10.2.1)\\
\hat{H}_1^{\rm res} &= \sum_{n,\boldsymbol{k}} [T_1^a a_{n+1}^\dagger(\boldsymbol{k})a_n(\boldsymbol{k}) + T_1^b b_{n+1}^\dagger(\boldsymbol{k})b_n(\boldsymbol{k}) \\
&\quad - eFR_1^{ab} a_{n+1}^\dagger(\boldsymbol{k})b_n(\boldsymbol{k}) - eFR_1^{ba} b_{n+1}^\dagger(\boldsymbol{k})a_n(\boldsymbol{k}) + \text{h.c.}] \qquad (10.2.2)\\
\hat{H}_2^{\rm res} &= \sum_{n,\boldsymbol{k}} [-eF(R_0^{ab} a_n^\dagger(\boldsymbol{k})b_n(\boldsymbol{k}) + R_0^{ba} b_n^\dagger(\boldsymbol{k})a_n(\boldsymbol{k}))] \qquad (10.2.3)
\end{aligned}$$

with the parabolic dispersion $E_k = \hbar^2 k^2/(2m_w)$ (m_w is the effective mass in the well) and the couplings $R_h^{\nu'\nu} = \int \mathrm{d}z\, \Psi^{\nu'}(z - hd)z\Psi^\nu(z)$. All energies E are given with respect to the bottom of the quantum well. The values of the coefficients are presented in Table 10.1. The term $\hat{H}_2$ can be incorporated into the one-electron states by diagonalizing $\hat{H}_0 + \hat{H}_2$ [16]. This leads to renormalized field-dependent coefficients in $\hat{H}_0$ and $\hat{H}_1$ (which are used in the following) but does not change the structure of the problem for a homogeneous electric field.

$\hat{H}^{\rm res}$ is only considering $\boldsymbol{k}$-conserving processes reflecting an ideal structure with translational invariance in the $\boldsymbol{r}$-plane and neglecting any many-particle processes. Nevertheless, there are also non-$\boldsymbol{k}$-conserving processes which may result from scattering at impurities or interface roughness, for example. These give two further contributions to the Hamiltonian:

$$\begin{aligned}
\hat{H}_0^{\rm scatter} &= \frac{1}{A} \sum_{\boldsymbol{k},\boldsymbol{p}} [U_0^{aa}(\boldsymbol{p})a_n^\dagger(\boldsymbol{k}+\boldsymbol{p})a_n(\boldsymbol{k}) + U_0^{bb}(\boldsymbol{p})b_n^\dagger(\boldsymbol{k}+\boldsymbol{p})b_n(\boldsymbol{k}) \\
&\quad + U_0^{ba}(\boldsymbol{p})b_n^\dagger(\boldsymbol{k}+\boldsymbol{p})a_n(\boldsymbol{k}) + U_0^{ab}(\boldsymbol{p})a_n^\dagger(\boldsymbol{k}+\boldsymbol{p})b_n(\boldsymbol{k})] \qquad (10.2.4)
\end{aligned}$$

Table 10.1 *Calculated level energies and transition elements for equations (10.2.1)–(10.2.3).*

$E^a = 47.1$ meV	$T_1^a = -0.0201$ meV	$R_0^{ba} = -0.149d$
$E^b = 176.6$ meV	$T_1^b = 0.0776$ meV	$R_1^{ba} = 2.66 \times 10^{-4}d$

which is the contribution due to scattering within the well and

$$\hat{H}_1^{\text{scatter}} = \frac{1}{A}\sum_{k,p}[U_1^{aa}(p)a_{n+1}^\dagger(k+p)a_n(k) + U_1^{bb}(p)b_{n+1}^\dagger(k+p)b_n(k) \\ +U_1^{ba}(p)b_{n+1}^\dagger(k+p)a_n(k) + U_1^{ab}(p)a_{n+1}^\dagger(k+p)b_n(k) + \text{h.c.}] \quad (10.2.5)$$

refers to interwell scattering, where the restriction to neighbouring wells is made.

Within the assumptions of local thermal equilibrium in each well and weak coupling between the wells the current from level ν in well n to level μ in well $n+1$ is given by the following expression† (see section 9.3 of [58]):

$$I_{n\to n+1}^{\nu\to\mu} = 2e\sum_{k',k}|H_{(n+1)k',nk}^{\mu,\nu}|^2 \int_{-\infty}^{\infty}\frac{\mathrm{d}E}{2\pi\hbar}A_n^\nu(k,E) \\ \times A_{n+1}^\mu(k',E+eFd)[n_F(E-E_n^F) - n_F(E-E_{n+1}^F+eFd)]. \quad (10.2.6)$$

Here E_n^F is the chemical potential in well n which is measured with respect to the bottom of the quantum well. $n_F(x) = (1+\mathrm{e}^{x/(k_B T_e)})^{-1}$ is the Fermi function and T_e is the electron temperature. $A_n^\nu(k, E)$ denotes the spectral function for the state k of the subband ν in well number n. It represents the weight of the free particle state k contributing to the energy E. Then the total density of states $\rho_n^\nu(E)$ in subband ν is given by

$$\rho_n^\nu(E) = \frac{2}{2\pi A}\sum_k A_n^\nu(k,E) \quad (10.2.7)$$

where the factor of 2 reflects the spin degeneracy and A denotes the sample area. If no scattering is present, the state k has a fixed energy $E^\nu + E_k$ and the spectral function becomes a δ-function $A_n^\nu(k,E) = 2\pi\delta(E - E^\nu - E_k)$. In the continuum limit ($\sum_k \to A/(2\pi)^2 \int \mathrm{d}^2k$) equation (10.2.7) then gives the two-dimensional density of states $\rho_n^\nu(E) = \rho_0\theta(E - E^\nu)$ with $\rho_0 = \frac{m}{\pi\hbar^2}$.

If scattering is present due to $\hat{H}_0^{\text{scatter}}$ the states k are no longer eigenstates

† An important point of the derivation is the assumption of uncorrelated scattering in different wells. For example, this is true if the electrons are dominantly scattered by the impurities localized in the same well. If the scattering occurs at identical impurities for the electrons in well n and in well $n+1$ correlation effects occur which may essentially change the result [57, 16].

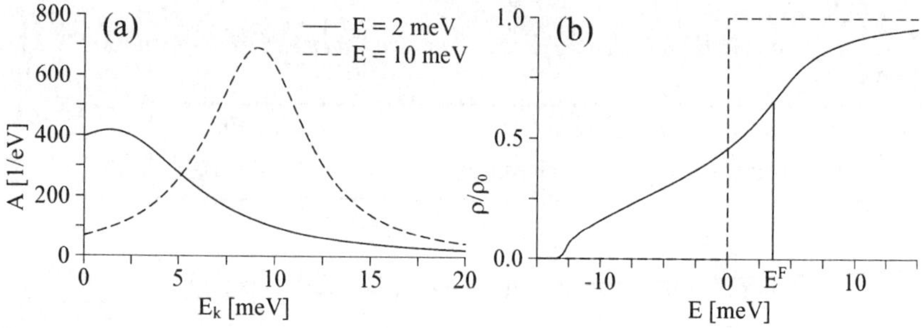

Figure 10.4 *(a) Spectral function $A^a(E_k, E)$ of the lowest level for two different energies E calculated for impurity scattering and interface roughness (section 10.6). (b) Density of states in units of the free-particle density of states (dashed line) where we have also indicated the value of the Fermi energy at zero temperature.*

of the total Hamiltonian. This can be taken into account by calculating the self-energy for the given scattering within standard theory [58]. Assuming equilibrium, the spectral function is then related to the retarded self-energy $\Sigma_n^{\nu\,\mathrm{ret}}(\boldsymbol{k}, E)$ via

$$A_n^\nu(\boldsymbol{k}, E) = \frac{-2\mathrm{Im}\Sigma_n^{\nu\,\mathrm{ret}}(\boldsymbol{k}, E)}{(E - E^\nu - E_k - \mathrm{Re}\Sigma_n^{\nu\,\mathrm{ret}}(\boldsymbol{k}, E))^2 + (\mathrm{Im}\Sigma_n^{\nu\,\mathrm{ret}}(\boldsymbol{k}, E))^2}. \quad (10.2.8)$$

Frequently, the self-energy is taken to be constant for simplicity setting $\Sigma_n^{\nu\,\mathrm{ret}}(\boldsymbol{k}, E) \approx W_n^\nu - \mathrm{i}\Gamma_n^\nu/2$. Then the spectral function is a Lorentzian with a full width at half maximum Γ.

Here the self-energies are calculated from basic scattering processes at impurities and interface roughness without using any fitting parameters. This calculation is presented in section 10.6. The calculated spectral function for different energies E are shown in Fig. 10.4(a) for illustration. One can clearly see that they exhibit a maximum close to $E_k = E$. From the width we may estimate that $\Gamma \approx 10$ meV holds for $E = 2$ meV and $\Gamma \approx 6$ meV for $E = 10$ meV. Fig. 10.4(b) shows the total density of states.

While the full derivation is slightly tedious [58] formula (10.2.6) can be motivated quite easily. In the long-time limit energy has to be conserved during transitions caused by the time-independent interwell couplings $H^{\mu,\nu}_{(n+1)\boldsymbol{k}',n\boldsymbol{k}}$. Therefore we have to consider tunnelling processes for a certain energy E and integrate over E afterwards. The factor $[n_F(E - E_n^F) - n_F(E + eFd - E_{n+1}^F)]$ takes into account the thermal occupation at the given energy in both wells. The free particle state $\boldsymbol{k}$ has a weight $A_n^\nu(\boldsymbol{k}, E)/(2\pi)$ in well n. Its transition probability to the state $\boldsymbol{k}'$ in well $n+1$ is given by $2\pi|H^{\mu,\nu}_{(n+1)\boldsymbol{k}',n\boldsymbol{k}}|^2/\hbar$ (Fermi's golden rule). The final state has a weight $A^\mu_{n+1}(\boldsymbol{k}', E + eFd)/(2\pi)$ at the given energy. Obviously one has to sum over all free particle states $\boldsymbol{k}, \boldsymbol{k}'$. Finally, the factor of 2 is due to the spin degeneracy.

10.2.1 Resonant transitions

Let us first investigate the current due to the transition elements from $\hat{H}_1^{\text{res}}$ in equation (10.2.2). They conserve the momentum $\boldsymbol{k}$ and therefore the kinetic energy $E_{\boldsymbol{k}}$. Using equation (10.2.7) one can rewrite equation (10.2.6) as

$$\begin{aligned} I_{n\to n+1}^{\nu\to\mu,\text{res}} &= A\frac{e|H_1^{\text{res}\,\mu,\nu}|^2}{\hbar}\int_{-\infty}^{\infty} \mathrm{d}E\, \rho_n^\nu(E)\langle A_{n+1}^\mu\rangle(E,eFd) \\ &\quad \times[n_F(E-E_n^F)-n_F(E-E_{n+1}^F+eFd)] \end{aligned} \tag{10.2.9}$$

with the average

$$\langle A_{n+1}^\mu\rangle(E,eFd)=\frac{\sum_{\boldsymbol{k}} A_n^\nu(\boldsymbol{k},E)A_{n+1}^\mu(\boldsymbol{k},E+eFd)}{\sum_{\boldsymbol{k}} A_n^\nu(\boldsymbol{k},E)}. \tag{10.2.10}$$

If the spectral functions are δ-functions, $\langle A_{n+1}^\mu\rangle(E,eFd)=2\pi\delta(E_\nu+eFd-E_\mu)$ holds. Therefore tunnelling only takes place if the levels are exactly aligned. In order to estimate the effect of broadening one may assume constant self-energies $\Sigma_n^{\nu\,\text{ret}}=-\mathrm{i}\Gamma^\nu/2$. Performing the continuum limit and assuming $E-E^\nu\gg\Gamma_n^\nu$ as well as $E+eFd-E^\mu\gg\Gamma^\mu$ one finds

$$\langle A_{n+1}^\mu\rangle(E,eFd)=\frac{\Gamma^\nu+\Gamma^\mu}{(eFd+E^\nu-E^\mu)^2+(\Gamma^\nu+\Gamma^\mu)^2/4}. \tag{10.2.11}$$

This expression has a peak at the resonance $E^\mu=eFd+E^\nu$ and a full width at half maximum of $(\Gamma^\nu+\Gamma^\mu)$. Even if the conditions stated above are not fulfilled, the result is typically similar.

For tunnelling between the lowest levels $a\to a$, equal densities $E_n^F=E_{n+1}^F$, a constant density of states ρ^a and not too high temperatures and fields $k_BT, eFd<E_n^F-E^a$ one finds

$$I_{n\to n+1}^{a\to a}=eA\rho^a\frac{(T_1^a)^2}{\hbar}\frac{2\Gamma^a eFd}{(eFd)^2+(\Gamma^a)^2}. \tag{10.2.12}$$

Therefore the current has a maximum at $eF_{\text{max}}d=\Gamma^a$ where it takes the value $I_{\text{max}}=eA\rho^a(T_1^a)^2/\hbar$. Note that this value depends neither on Γ nor the Fermi level. Estimating $\Gamma\approx 8$ meV from the spectral functions shown in Fig. 10.4 and taking $\rho^a=\rho_0$ yields

$$eF_{\text{max}}d\approx 8\text{ meV}\qquad I_{\text{max}}\approx 0.27\text{ mA}. \tag{10.2.13}$$

for the $a\to a$ transition using $T_1^a=-0.0201$ meV from Table 10.1.

With the translations $2T_1^a\to E_1$, $E^a\to E_0$ and $\hbar/\Gamma^a\to\tau$ equation (10.2.12) is identical to equation (12)† of [10] for the case $E^F>E_0+E_1$. There semiclassical transport in a miniband for the strong coupling limit $T_1^a\gg\Gamma^a$ was regarded while here the limit of weakly coupled quantum wells $T_1^a\ll\Gamma^a$

† Note that there is a factor of τ missing in the numerator due to a misprint.

is considered. Note that in both derivations the case $(E^F - E^a) > \Gamma^a, 2T_1^a$ is considered.

The transitions from the lowest level to the excited level $a \to b$ will become important if $eFd \approx E^b - E^a$. Assuming that $k_B T, E_{n+1}^F \ll E^b - E^a$ we find $n_F(E - E_{n+1}^F + eFd) \approx 0$ in this field range. Using approximation (10.2.11) we find close to the resonance

$$I_{n\to n+1}^{a\to b} \approx eAn_n^a \frac{|H_1^{\mathrm{res}\,a,b}|^2}{\hbar} \frac{\Gamma^a + \Gamma^b}{(eFd + E^a - E^b)^2 + (\Gamma^a + \Gamma^b)^2/4}. \tag{10.2.14}$$

Note that this current is proportional to the density of carriers $n_n^a = \int \mathrm{d}E\, n_F(E - E_n^F)\rho^a(E)$ in well n while this was not the case for the $a \to a$ peak in equation (10.2.12).

10.2.2 Nonresonant current

The transition elements from $\hat{H}_1^{\mathrm{scatter}}$ do not conserve momentum and therefore the kinetic energy E_k is changed during the transition. Thus, these transitions are not as sensitive to the alignment of the levels as the resonant transition discussed above. They yield a background current which may dominate the current between the resonances. Here only nonresonant transitions via interface roughness are considered. The explicit expressions used are given in section 10.6. As the actual shape of the interface is not known and may strongly vary for different wafers it has been parametrized by a reasonable set of parameters in order to show the magnitude of the effect. Therefore the reader has to keep in mind that the nonresonant currents can easily vary by a factor of 3 or even more in the following calculations.

10.2.3 Calculation of the current

In equation (10.2.6) the current depends on the electric field and the Fermi energies in both wells. The Fermi energies E_i^F can easily be calculated from the two-dimensional density of carriers n_i in the well using the relation

$$n_i = \int \mathrm{d}E\, n_F(E - E_i^F) \sum_\mu \rho^\mu(E). \tag{10.2.15}$$

Finally, the total current from well i to well $i+1$ is the sum of the contributions (10.2.6) between the different levels.

$$I_{i\to i+1} = \sum_{\mu,\nu} I_{i\to i+1}^{\nu\to\mu} = I(F, n_i, n_{i+1}). \tag{10.2.16}$$

The result is presented in Fig. 10.5 for $n_i = n_{i+1} = N_D$.

Neglecting the nonresonant transitions (dashed curve) there is a first maximum at $eFd = 9$ meV with a current of 0.277 mA which is in good agreement with estimation (10.2.13) for the $a \to a$ transitions. A second maximum occurs at

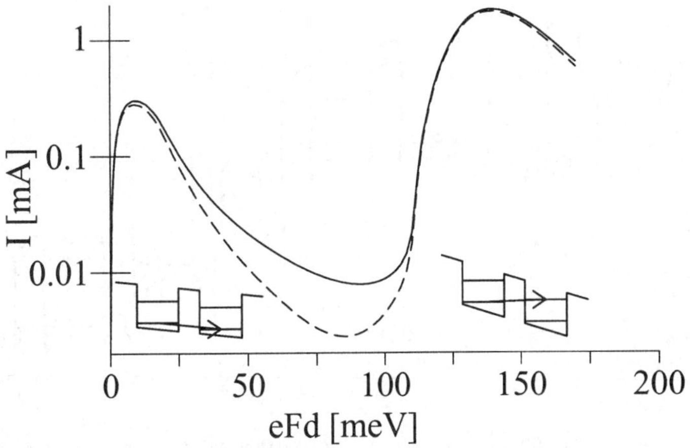

Figure 10.5 *Current* $I_{l\to l+1}$ *for* $n_l = n_{l+1} = N_D$. *The dashed curve gives the current from resonant transition, while the solid curve gives then sum of resonant and nonresonant transitions due to interface roughness.*

$eFd = 140$ meV with $I = 1.75$ mA. In comparison with this, the experimental data (Fig. 6 of [44]) exhibit a first maximum of $I \approx 0.076$ mA and a second maximum of $I \approx 1.45$ mA. While there is a good quantitative agreement for the height of second maximum, the calculated first maximum seems to be too large by a factor of 4. This inconsistency will be resolved in the next chapter where the formation of field domains is considered. In contrast to the currents, the position of the maxima is much more difficult to compare as a part of the voltage may drop outside the superlattice. The nonresonant currents are not very large compared with the currents at the maxima but can dominate the total current between the maxima as can be seen from the solid curve in Fig. 10.5.

Note that the experimental data of [44] indicate that the first excited level is unoccupied close to the onset of the $a \to b$ resonance indicating that the current of approximately 0.1 mA is not carried by the $a \to b$ transitions there. This may be attributed to stronger nonresonant transition in this field range.

An important feature of the model presented here is the fact that only the nominal sample parameters are involved in the calculation of the currents. The resulting currents are in good quantitative agreement with the experimental data of [40, 44]. The same model has also been applied to the samples used in [39] where quantitative agreement could be obtained assuming a smaller barrier width [55]. While these two highly doped samples exhibit a density of states which resembles the free-electron density of states with a smoothed onset (Fig. 10.4), the density of states is much more complicated for low-doped samples due to the presence of impurity bands [59]. Within the single-site approximation for impurity scattering used here (section 10.6) these effects are included in the spectral function. The calculation yields a strong dependence

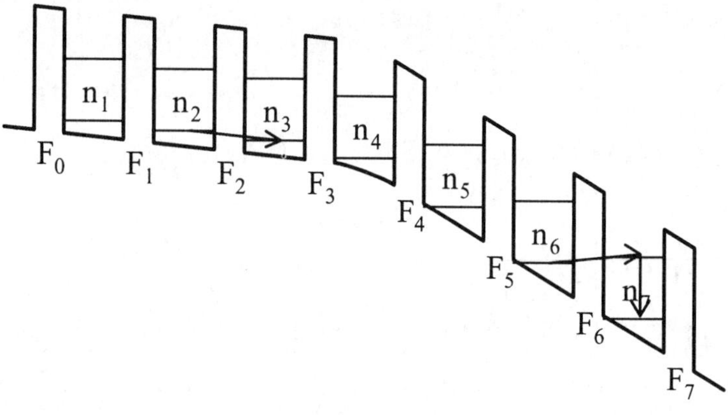

Figure 10.6 *Sketch of the fields and densities for domain formation.*

of the current on temperature [60]. Excellent quantitative agreement has also been found when the sample was irradiated by a strong terahertz field from a free-electron laser source [60, 61]. Therefore we may conclude that the formalism described here allows the quantitative calculation of the currents in weakly coupled multiple quantum wells for a wide range of samples without applying any fitting parameters.

10.3 FORMATION OF FIELD DOMAINS

Now we want to consider the full structure consisting of N wells numbered $i = 1, \ldots, N$ and $N+1$ barriers. Assuming that the electric field is constant over the structure the total voltage is given by $U \approx (N+1)Fd$. Figure 10.5 shows that $I_{i\to i+1}(F, N_D, N_D)$ exhibits a region of NDC between $eF_{\max}d = 9.7$ meV and $eF_{\min}d = 90$ meV. If the electric field is within this NDC region an instability is likely to occur because a spontaneously formed charge accumulation will increase in time instead of decreasing. This is a common situation for samples exhibiting an N-shaped local current-field relation such as the Gunn diode ([19] and references therein). In order to include such effects, charge accumulation has to be allowed for. This means that the carrier density n_i in well i may deviate from the doping N_D per period. Integrating Poisson's equation over one period yields

$$\epsilon_r\epsilon_0(F_i - F_{i-1}) = e(n_i - N_D) \qquad \text{for } i = 1, \ldots, N \qquad (10.3.1)$$

where F_i is the electric field in the middle of the barrier between wells i and $i+1$. Here a constant dielectric permeability $\epsilon_r = 13$ is assumed for simplicity.

The notation as well as a typical potential profile is sketched in Fig. 10.6 for a superlattice with $N = 7$. Within the first wells there is $n_1 = n_2 = n_3 = N_D$ so that $eF_0 = eF_1 = eF_2 = eF_3$. The electric field is low so that the current is dominated by the $a \rightarrow a$ transitions. At well 4 there is charge accumulation $n_4 > N_D$. Therefore we have $eF_4 > eF_3$ and the electric field is large after well 4 so that the current is dominated by the $a \rightarrow b$ resonance. Here we assume that the electrons relax fast from the upper levels into the lower level which is the only level to be populated in thermal equilibrium for level separations larger than the Fermi energy†.

In order to calculate the current between the wells the question arises: Which electric field should be used if an inhomogeneous situation is considered? As the voltage drop between the wells i and $i+1$ can be approximated by $eF_i d$ it is natural to use the electric field F_i in the argument of equation (10.2.16). Nevertheless it has to be stated that this is an approximation done for simplicity and deviations may occur for inhomogeneous field profiles.

Now the temporal evolution of the densities within the wells is given by the continuity equation

$$eA\frac{\mathrm{d}n_i}{\mathrm{d}t} = I_{(i-1)\rightarrow i} - I_{i\rightarrow(i+1)} = I(F_{i-1}, n_{i-1}, n_i) - I(F_i, n_i, n_{i+1}) \quad (10.3.2)$$

for $i = 1, \ldots, N$. In order to obtain a complete set of equations for the temporal evolution of the densities and fields we have to add two more features. At first the voltage condition is now given by

$$U = \int \mathrm{d}z F(z) \approx \sum_{i=0}^{N} \mathrm{d}F_i + U_c \quad (10.3.3)$$

where U_c represents the voltage drop outside the superlattice, which is neglected in the following in order to concentrate on the features of the pure superlattice. At second the currents $I_{0\rightarrow 1}$ and $I_{N\rightarrow(N+1)}$ across the first and last barrier of the structure, respectively, have to be specified. For simplicity one may use expression (10.2.16) with appropriate effective densities

$$n_0 = n_0(F_0, N_D, n_1) \qquad \text{and} \qquad n_{N+1} = n_{N+1}(F_N, N_D, n_N). \quad (10.3.4)$$

In mathematical terms the functions $n_0(F_0, N_D, n_1)$ and $n_{N+1}(F_N, N_D, n_N)$ then represent the boundary conditions of the model. Equations (10.3.1)–(10.3.4) form a complete set of equations for calculating the densities, fields and currents as a function of time for fixed bias voltage U and given initial conditions $n_i(t_0)$ $(i = 1, \ldots, N)$.

Using a different approach for evaluating the current function $I(F_i, n_i, n_{i+1})$ (which is essentially based on the broadening due to the tunnelling time for the $a \rightarrow b$ resonance and miniband conduction for the $a \rightarrow a$ resonance) and the boundary conditions $n_0(F_0, N_D, n_1) = n_1$, $n_{N+1}(F_N, N_D, n_N) = n_N$

† Furthermore, the intersubband relaxation must be fast enough to guarantee thermal equilibrium between the levels. If the level separation is larger than the optical phonon energy of 36 meV, the intersubband relaxation time is of the order of 1 ps [62], which is typically much faster than the tunnelling times. But for wide quantum wells the level separation becomes small and nonequilibrium effects are found [63].

such a model has been used in [53] to calculate the current–voltage characteristics under domain formation. The model of [54] uses the boundary condition $n_0(F_0, N_D, n_1) = N_D$ and the current expression

$$I_{i\to(i+1)} = eAn_i v(F_i) \qquad (10.3.5)$$

where $v(F)$ is a phenomenological tunnelling rate yielding an N-shaped characteristic as shown in Fig. 10.5. This simplification allows the analytical construction of the full current–voltage characteristic [54, 64]. Equation (10.3.5) is motivated by equation (10.2.14) for the $a \to b$ transitions and higher resonances as well. Nevertheless, is seems to be questionable close to the $a \to a$ maximum as the maximum current of equation (10.2.12) is independent of n_i for equal carrier densities in adjacent wells. The reason for this deviation is the existence of backward currents for low fields which are taken into account by the factor $[n_F(E - E_n^F) - n_F(E - E_{n+1}^F + eFd)]$ in equation (10.2.6). If, on the other hand, $eFd \gg E_{n+1}^F - E^a + k_B T_e$ so that $n_F(E - E_{n+1}^F + eFd) \approx 0$ for all relevant energies E, equation (10.3.5) is a reasonable approximation.

10.3.1 Numerical results

In a first simulation the boundary conditions $n_0(F, N_D, n_1) = n_{N+1}(F, N_D, n_N) = 3N_D$ are used. The stationary stable states are determined by simulating equations (10.3.1)–(10.3.4) until a stationary state is reached for given U. Increasing or decreasing U afterwards simulates a sweep-up or sweep-down of the voltage, respectively. The result is shown in Fig. 10.7(a).

We find that the characteristic consists of 40 branches, equal to the number of quantum wells. The maximum current within the branches is about 0.09 mA in good agreement with the experimental situation [40, 42] where around 0.06 mA is observed. This current is significantly lower than the first maximum of $I(F, N_D, N_D)$ (Fig. 10.5) resolving the discrepancy discussed at the end of the last section. Note that another superlattice with the same specification [65] exhibits maximum currents of the branches of 0.14 mA. This indicates that the current is quite sensitive to variations in the sample. The calculation exhibits similar effects. If we ignore the nonresonant transitions, the maxima of the branches are found at 0.06 mA, as shown in Fig. 10.7(b), although the height of the maxima in the homogeneous characteristics (Fig. 10.5) is almost identical. Thus, the extension of the branches are very sensitive to the quality of the interfaces (which causes nonresonant transition via interface roughness). Another uncertainty may be the actual barrier thickness as a variation of one monolayer changes the matrix elements $\hat{H}_1$ by a factor of 1.4 and therefore the current (which is proportional $H_{i,i+1}^2$) by a factor of 2.

The slope of the branches varies between 1.1 and 0.55 mS for low and high voltages in Fig. 10.7 which is significantly larger than the experimental slopes [42] varying between 0.133 and 0.064 mS. This lower slope is responsible for the stronger overlap between the branches yielding pronounced

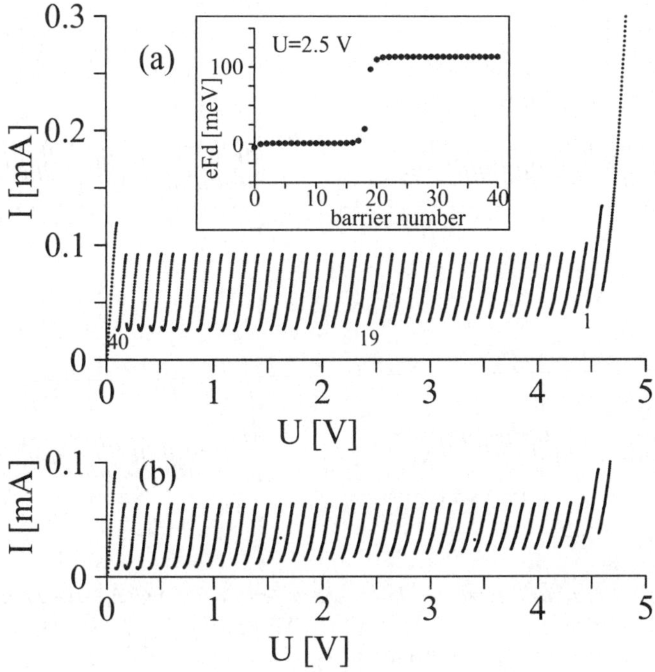

Figure 10.7 *Calculated current–voltage characteristic exhibiting field domains for voltage sweep-up for the boundary conditions* $n_0 = 3N_D$ *and* $n_{N+1} = 3N_D$ *(a) with and (b) without nonresonant transitions due to interface roughness. The inset gives the field distribution at* $U = 2.5$ *V where the domain boundary is located at the 19th well.*

multistability [42]. The discrepancy may be resolved by assuming an appropriate contact voltage U_c in equation (10.3.3) which will depend on I and the fields at the boundaries. Another possibility might be that further nonresonant transitions alter the shape of the homogeneous characteristics of Fig. 10.5.

The field distribution shown in the inset of Fig. 10.7(a) is in good agreement with cathodoluminescence measurements [66] stating that the field distribution consists of one low-field and one high-field domain where the high-field domain is located at the anode. These measurements were performed at the same sample but in the range between the $a \to b$ and the $a \to c$ resonance.

Using different boundary conditions, almost identical domain branches are observed as shown in Fig. 10.8. The main difference occurs at the first and last branch which may be changed significantly. Comparing Figs. 10.8(a–c) with Fig. 10.7(a) shows that the first branch is dominated by the boundary condition $n_{N+1}(F_0, N_D, n_1)$ and the last branch by $n_0(F_N, N_D, n_N)$.

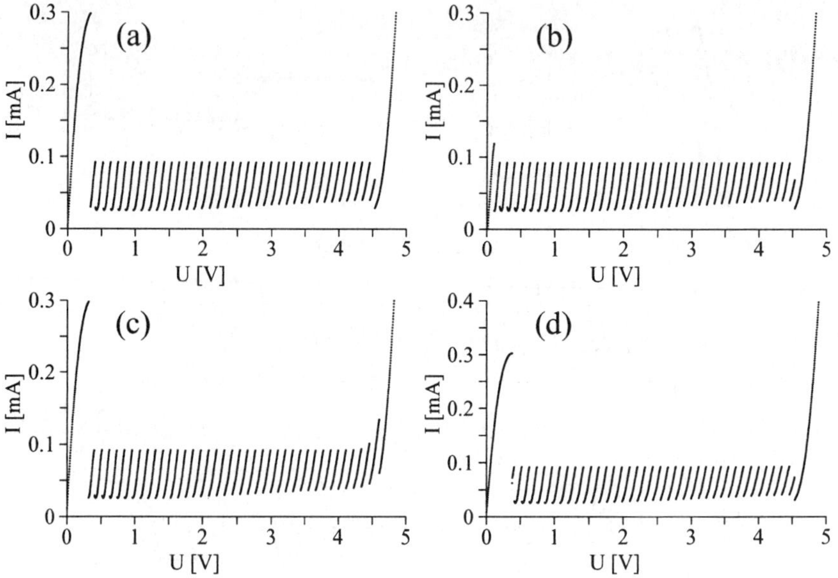

Figure 10.8 *Calculated current–voltage characteristic for voltage sweep-up for the boundary conditions (a)* $n_0 = 1.1N_D$, $n_{N+1} = 1.1N_D$, *(b)* $n_0 = 1.1N_D$, $n_{N+1} = 3N_D$, *(c)* $n_0 = 3N_D$, $n_{N+1} = 1.1N_D$ *and (d)* $n_0 = 1.2N_D$, $n_{N+1} = 0.8N_D$.

10.3.2 General aspects of domain formation

Now a qualitative explanation of the domain formation shall be given. First observe Fig. 10.5. For currents I_0 from the interval $[I_{\min}, I_{\max}]$ where $I_{\max} = 0.30$ mA is the maximum current for low fields and $I_{\min} = 7.8\ \mu$A is the minimum current for medium fields there are three intersections $F_I < F_{II} < F_{III}$ of the $I(eFd)$ curve with I_0. Unlike F_{II}, the fields F_I and F_{III} are in the range of positive differential conductivity (PDC). Therefore charge fluctuations are damped out in spatial regions where the electric field takes the values F_I or F_{III}. The field distribution for $U = 2.5$ V in the inset of Fig. 10.7(a) now shows that the electric field takes the value $F_i \approx F_I$ for $i \leq 17$ and $F_i \approx F_{III}$ for $i \geq 20$. The small deviations for $i = 0, 1$ due to the contact are damped out because the field is in the PDC region there. In between there is a domain boundary where the electric field changes due to a charge accumulation $e(n_i - N_D) = \epsilon_r \epsilon_0 (F_i - F_{i-1})$. As the main jump occurs between F_{18} and F_{19}, the density n_i takes its maximum at $i_D = 19$. If now this charge accumulation layer is shifted by one period to $i_D = 20$, we have almost the identical situation with the same current except for the fact that the voltage is diminished by $(F_{III} - F_I)d$ as a lower fraction of the sample is in the high-field region. This reveals the periodic sequence of branches. If one counts the branches starting from the right-hand side as depicted in Fig. 10.7(a), the maximum of n_i occurs in well i_D for the i_Dth branch. In total one can count 40 branches, which is exactly the number of quantum wells as the domain boundary may be located in each well. If the domain boundary comes close to $i = 1$ the low-field region is not large enough to shield the variation due to the contact. Therefore the first and second branches are strongly dependent on the boundary condition which simulates the contact (compare Fig. 10.7(a) with Fig. 10.8). The field distribution in such stationary domain structures is not arbitrary as the currents across each barrier have to be equal. This provides a condition on the minimal doping density (or minimal carrier generation due to irradiation) as discussed in [54, 67, 64, 68].

Now the question arises as to why these stationary domain states are stable while in other spatially extended NDC systems such as the Gunn diode typically travelling field domains occur. In [64] this question was investigated by using the simplified current relation (10.3.5). There it could be strictly proven that an inhomogeneous field distribution is necessarily stable if all electric fields F_i are in the PDC region of $v(F)$ which coincides with most readers' physical intuition. Therefore a field distribution like that in the inset of Fig. 10.7(a) must be stable if the NDC region $eF_{\text{max}} < eF < eF_{\text{min}}$ is crossed within one jump. Then

$$eF_{i_D-1} \leq eF_{\text{max}} \qquad \text{and} \qquad eF_{i_D} \geq eF_{\text{min}} \tag{10.3.6}$$

hold. In the stationary state the current across each barrier has to be equal to I_0. In particular, there is $I_0 = I(F_{i_D}, n_{i_D}, n_{i_D+1})$. As F_{i_D} is in the high-field region, approximation (10.3.5) may be justified yielding $I(F_{i_D}, n_{i_D}, n_{i_D+1}) \approx I(F_{i_D}, N_D, N_D)n_{i_D}/N_D$. Combining this with Poisson's equation $e(n_{i_D} - N_D) = \epsilon_r\epsilon_0(F_{i_D} - F_{i_D-1})$ exhibits that condition (10.3.6) can be fulfilled if

$$I_0 > I(F_{\text{min}}, N_D, N_D)\left(1 + \frac{\epsilon_r\epsilon_0(F_{\text{min}} - F_{\text{max}})}{eN_D}\right) \tag{10.3.7}$$

holds. It has to be stated that this is only a sufficient condition for stability as there are stable domain states where one field is located within the NDC region. Now I_0 cannot be larger than $I(F_{\text{max}}, N_D, N_D)$ as the low-field domain cannot carry a larger current. Therefore there is a minimum doping density

$$N_{\text{crit}}^{\text{acc}} \sim \frac{\epsilon_r\epsilon_0(F_{\text{min}} - F_{\text{max}})}{e}\frac{I_{\text{min}}}{I_{\text{max}} - I_{\text{min}}} \tag{10.3.8}$$

above which stable domain states with an accumulation layer exist. From the $I(F_i, N_D, N_D)$ relation from Fig. 10.5 $N_{\text{crit}}^{\text{acc}} \approx 1.2 \times 10^{10}$ cm^{-2} is estimated which is much smaller than $N_D = 1.5 \times 10^{11}$ cm^{-2}. Therefore stable domain states are expected in accordance with the experimental and theoretical findings. Note that condition (10.3.8) depends strongly on I_{min} which itself is strongly affected by nonresonant transitions, as shown in Fig. 10.5.

Up until now domain structures have been discussed where the high-field domain is located at the receiving contact. But of course there is the other pos-

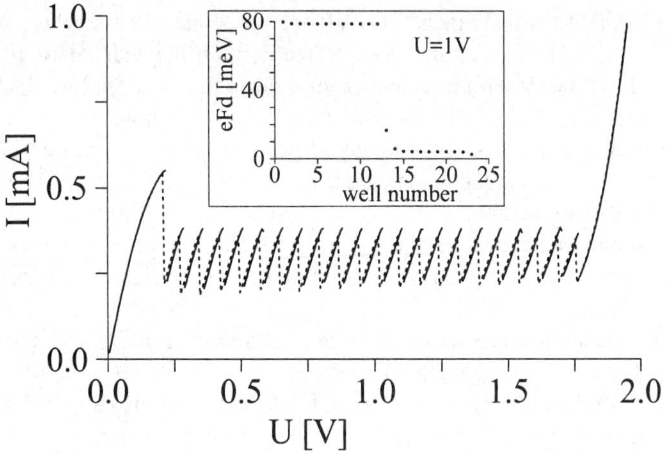

Figure 10.9 *Calculated current–voltage characteristic of the sample from [39] for voltage sweep-up for the boundary conditions* $n_0 = n_{N+1} = 0.95N_D$ *(solid curve),* $n_0 = n_{N+1} = 1.05N_D$ *(dotted curve).*

sibility that the high-field domain is at the injecting contact, i.e. $F_i \approx F_{III}$ for $i < i_D$ and $F_i \approx F_I$ for $i > i_D$. Then Poisson's equation yields a depletion region $n_i < N_D$ at the domain boundary $i \approx i_D$. Such domains have been observed experimentally for highly doped samples [39]. Theoretically, such domains could be both obtained from a simplified model [64] and from the microscopic model [55]. This is shown in Fig. 10.9 using the parameters of the sample from [39]. For the boundary conditions $n_0(F_0, N_D, n_1) = 0.95N_D$, $n_{N+1}(F_N, N_D, n_N) = 0.95N_D$ domain states are found, where the high-field domain is located at the injecting contact as shown in the inset, while for $n_0(F_0, N_D, n_1) = 1.05N_D$, $n_{N+1}(F_N, N_D, n_N) = 1.05N_D$ the high-field domain is located at the receiving contact as in Fig. 10.7. The domain branches themselves look very similar.

Regarding the stability of such domain structures, the argument given above yields a minimum doping

$$N_{\text{crit}}^{\text{dep}} \sim \frac{\epsilon_r \epsilon_0 (F_{\min} - F_{\max})}{e} \frac{I_{\max}}{I_{\max} - I_{\min}} \tag{10.3.9}$$

above which stable domain states with a depletion layer can exist. For the sample of [40] the estimation gives $N_{\text{crit}}^{\text{dep}} \approx 4.4 \times 10^{11}$ cm^{-2} which is three times larger than N_D. Thus, such domain states are not expected to be stable and therefore should not be observed. For comparison the sample used in [39] has $N_D = 8.75 \times 10^{11}$ cm^{-2} and from Fig. 3 of [55] $N_{\text{crit}}^{\text{dep}} \approx 2.5 \times 10^{11}$ cm^{-2} is estimated in good agreement with the observation of stable domain structures with depletion layers.

Note that the proof of stability essentially relies on the discreteness of the system. In a continuous model the NDC region cannot be crossed without having any fields within this region at least for a small spatial interval. This explains the difference to continuous systems such as the Gunn diode where such stable domain states with an arbitrary position of the boundary are not observed.

10.4 IMPERFECT SUPERLATTICES

All the theoretical current–voltage characteristics shown up until now have exhibited an almost regular series of branches whereas in typical experiments the lengths of the branches vary. It is straightforward to assume that this is caused by irregularities in the real superlattice as nothing is perfect. But then the question about the nature of these irregularities arises. At first there are two different possibilities: regarding a wafer as sketched in Fig. 10.10 the irregularity may either be a bad spot localized somewhere in the superlattice as shown on the right-hand side of the wafer or a deviation from periodicity occurring in the whole layer. Such deviations may be a larger or smaller barrier width, or a different doping density in certain wells, for example, which are established due to insufficient control of the growth process. Fortunately, one can distinguish these two cases experimentally. If the local spot would be the essential cause for the irregularities in the characteristics, samples 2 and 3 from the same wafer sketched in Fig. 10.10 should exhibit a different modulation of the branches as their individual spots have different sizes and are located at different positions. But the different samples used in [65] exhibited almost identical modulations of the branches if they originated from the same wafers while there are large differences for samples from different wafers even if the superlattice structure is nominally identical. This would be expected from samples 1 and 2 sketched in Fig. 10.10. The same feature can be observed in [39] where the authors fabricated samples with different numbers of periods originating from one wafer by an etching process. They found very similar sequences of longer and shorter branches which allowed them to conclude that the high-field domain is located at the receiving contact in their samples.

These experimental observations clearly indicate that the dominant deviations from periodicity are not (x, y)-dependent but extend over the whole wafer. This allows us to simulate these irregular superlattices by introducing local fluctuations in the well width, barrier width or doping concentration into equations (10.2.6) and (10.3.1). This was done in [65] within the model of [53] where we found that even small spatial fluctuations (about 7%) of the doping have a significant influence on the length of the branches and can explain the observed behaviour. Furthermore, single fluctuations may be located by just determining the number of the branch which is altered most. This effect is particularly pronounced for fluctuations in the barrier width, where a one-monolayer fluctuation may change the characteristic significantly. The theoretical prediction has been

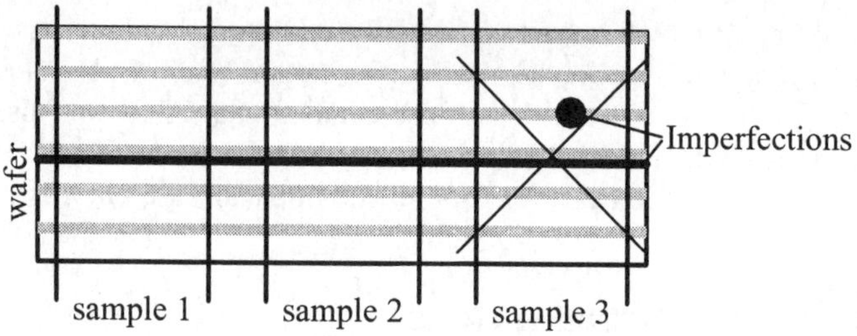

Figure 10.10 *Schematic diagram of a superlattice wafer from which three different samples are obtained. The dot in sample 3 represents a local imperfection while the line represents a modulation of the periodic structure.*

successfully checked experimentally by growing a new sample exhibiting one barrier with a larger thickness [69].

Now doping fluctuations are included into the model discussed before by using a well-dependent doping density N_{Di} in equation (10.3.1). The impact on the impurity scattering is neglected, so that the old function $I(F_i, n_i, n_{i+1})$ is used. The result is shown in Fig. 10.11 and exhibits fluctuating branch heights in accordance with the model used in [65]. Comparing the local fluctuations N_{Di} with the maximum current of the branch, one finds a correlation. Like the findings of [70], the branch $i+1$ counted from the right extends to higher currents if N_{Di} is larger than N_D. Additionally the branch i extends to lower currents as can be seen from Fig. 10.11(b). The deviations in the current seem to be smaller and the effects to neighbouring branches seem to be larger here than observed in [65, 70]. This might be related to the fact that a higher doping density was used in previous works.

10.5 OSCILLATORY BEHAVIOUR

Oscillatory behaviour has been observed in coupled multiple quantum wells both experimentally [26–28] and theoretically [29, 30]. As an example such behaviour is obtained within the model discussed here for the boundary conditions $n_0(F, N_D, n_1) = 0.8N_D$ and $n_{N+1}(F, N_D, n_N) = 1.2N_D$ which is just the reversed sequence used for the calculation of Fig. 10.8(d) where stable domain branches were found. Figure 10.12 shows self-sustained current oscillations between 7.4 and 8.2 μA with a frequency of 0.18 MHz. Similar results are obtained for different biases and different values $n_0(F, N_D, n_1) < N_D$. The oscillations may be described from the field distribution as follows. At $t = 5.71\ \mu$s the field distribution consists of a high- and low-field domain with a depletion layer in between. As for $N_D < N_{\text{crit}}^{\text{dep}}$, such a distribution is not stable the boundary travels to the right thereby decreasing the electric field in the high-field domain because the total voltage has to remain constant ($t = 8.45\ \mu$s). As the high-field domain now enters the region of NDC for $eFd < 90$ meV a new

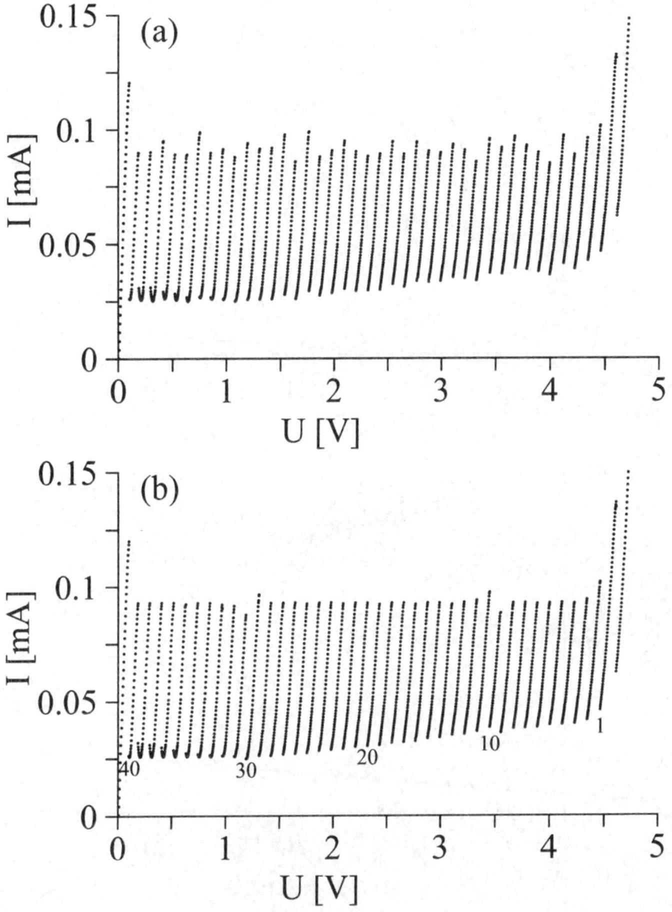

Figure 10.11 *Current-voltage characteristic for doping fluctuations. In (a) a random sequence* N_{Di} *with an average fluctuation of 10% around* N_D *is used. In (b) there is* $N_{D10} = 1.2 N_D$ *and* $N_{D30} = 0.8 N_D$ *while the other densities are not altered.*

positive charge accumulation layer is created there which is slightly visible for $t = 9.21\ \mu$s at $i \approx 10$. This accumulation layer travels to the right and increases in time ($t = 10.16\ \mu$s) until it merges with the old depletion layer. Then the cycle is repeated again. Such an oscillation type has been described in [31] within approximation (10.3.5). The very same behaviour also occurs for charge accumulation layers if domain states with an accumulation layer are unstable. For a full treatment of this oscillation type see [32]. Note that the minimum current in this oscillation cycle roughly coincides with $I(F_{\min}, N_D, N_D)$. This relation seems to hold generally as checked by altering the nonresonant current and thereby $I(F_{\min}, N_D, N_D)$. Therefore the temporal minimum of this

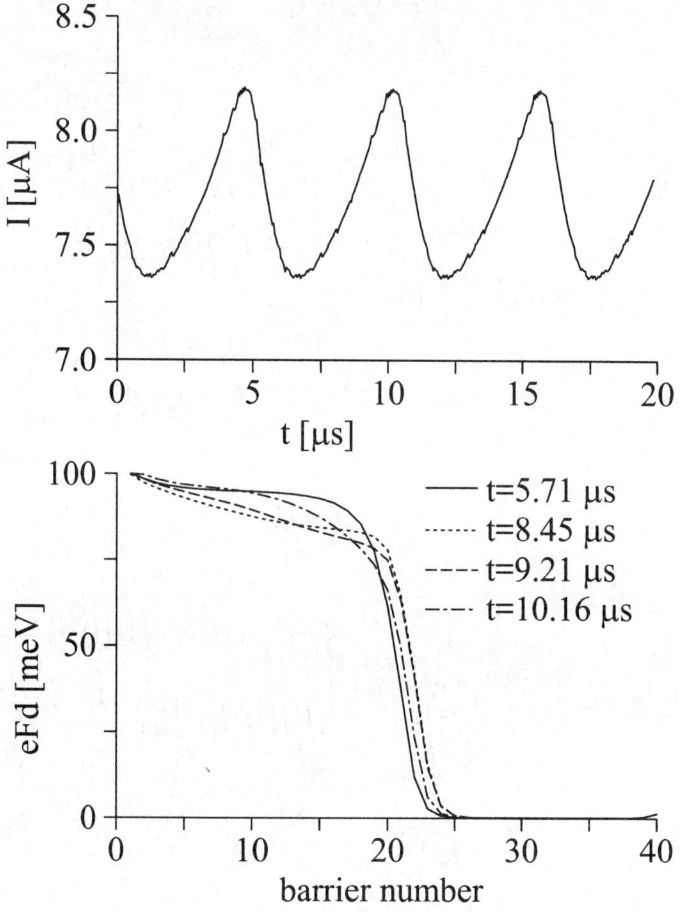

Figure 10.12 *Calculated current oscillations for $U = 2$ V.*

type of current oscillation provides information about $I(F_{\min}, N_D, N_D)$ which is strongly dependent on the nonresonant transitions.

In order to study the influence of the boundary conditions onto the oscillations we use the ohmic contact currents $I_{0\to 1} = \sigma e F_0 d$ and $I_{N\to N+1} = \sigma e F_N d$ in the following. For $\sigma = 0.5$ mA eV^{-1} a completely different oscillation mode is found as shown in Fig. 10.13.

Here one oscillation cycle consists of the nucleation and travelling of a dipole domain which vanishes by leaving the sample at the receiving contact. The scenario is completely analogous to that described in [71] for a continuous system exhibiting NNDC. Additionally, small current spikes appear which are related to the motion of the accumulation layer from one well to the next as

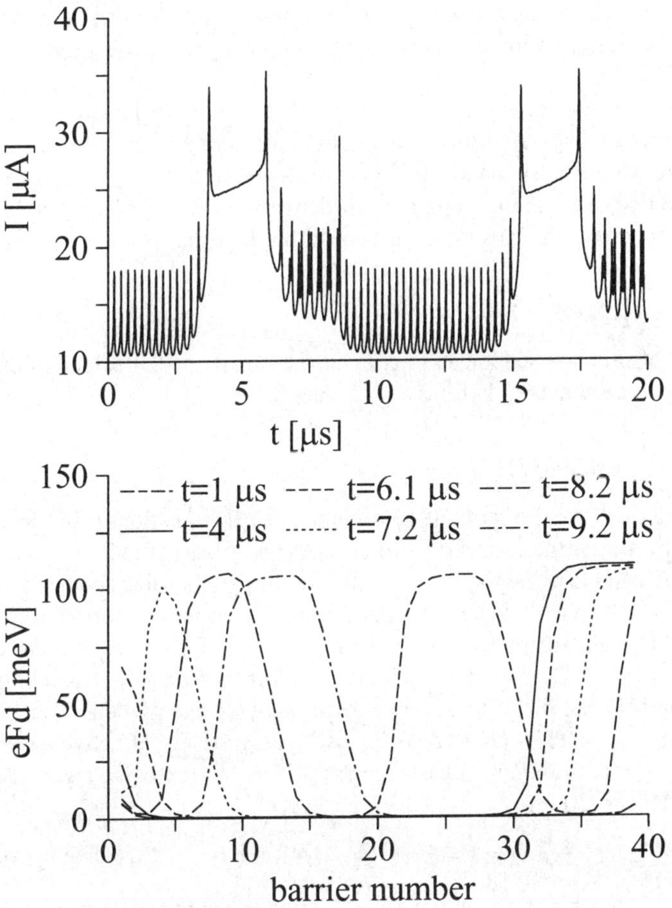

Figure 10.13 *Calculated current oscillations for* $U = 1$ *V using ohmic boundary conditions with* $\sigma = 0.5$ $mA\ eV^{-1}$.

discussed in [72] for the domain formation process. Thus, these spikes reflect the discreteness of the superlattice.

Experimentally, the sample discussed here also exhibits self-sustained current oscillations if a positive bias is applied at the top contact while there are stable domain branches for a negative bias. The experimental data [31] exhibit current oscillations between 20 and 35 μA with a frequency of 0.4 MHz at $U = 0.78$ V. Thus, the theory is in qualitative agreement with the experiment regarding the oscillations as well. As we did not try to model the contact currents microscopically (which strongly influence the type of oscillations) we cannot expect quantitative agreement. Additionally, the shape of the oscillations as well as the frequency depends substantially on the full shape of the current-field relation which is strongly affected by the nonresonant transitions between the resonances. Furthermore, deviations from periodicity may affect the oscillations as well [65, 70, 73, 81].

For larger values of σ (say $\sigma \gtrsim 3$ mA eV^{-1}) the same domain branches like those shown in Fig. 10.8 are found. This indicates that the domain branches themselves are almost identical if their formation is allowed for by the contact conditions and if the doping is sufficiently large. On the other hand, the oscillatory behaviour does strongly depend on the exact contact conditions.

10.6 DETAILS OF THE CALCULATIONS

The aim of this last section is to present the details of the calculations whose results have been shown before.

10.6.1 Calculation of the couplings

In order to calculate the coefficients of Table 10.1 for equations (10.2.1)–(10.2.3) we have to specify the band structure in GaAs and AlAs at first. It is assumed that only the Γ band is of importance. While one may use a parabolic band structure for $Al_xGa_{1-x}As$/GaAs heterostructures for small x this is not appropriate for a GaAs/AlAs heterostructure as the conduction band of GaAs is located far in the band gap of AlAs where the band structure is clearly not parabolic [74]. Following [75], we model the nonparabolicity by an energy-dependent effective mass $m(E) = m_c(1 + (E - E_c)/E_g)$, where m_c is the effective mass at the conduction band minimum of energy E_c and E_g is the energy gap. Then the usual connection rules

$$\varphi(z_0 - \varepsilon) = \varphi(z_0 + \varepsilon) \tag{10.6.1}$$

$$\frac{1}{m(E, z_0 - \varepsilon)} \frac{\mathrm{d}\varphi}{\mathrm{d}z}(z_0 - \varepsilon) = \frac{1}{m(E, z_0 + \varepsilon)} \frac{\mathrm{d}\varphi}{\mathrm{d}z}(z_0 + \varepsilon) \tag{10.6.2}$$

hold for the envelope function $\varphi(z)$ provided that the momentum matrix element $P = \hbar\sqrt{E_g/(2m_c)}$ between the conduction and valence band states is identical in both materials. We use the values [75] $m_c^{\mathrm{GaAs}} = 0.067m_e$, $m_c^{\mathrm{AlAs}} = 0.15m_e$, $E_g^{\mathrm{GaAs}} = 1.52$ eV, $E_g^{\mathrm{AlAs}} = 3.13$ eV and the conduction band discontinuity $\Delta E_c = 1.06$ eV. These parameters yield a relation $E(k) = E_c + \hbar^2k^2/(2m(E))$ which is in excellent agreement with the band structure of AlAs [74] for the energies of interest. It must be stated that for these parameters the value of P is slightly different for the two materials in contrast to the assumption. Nevertheless, these parameters seem to give reasonable agreement with experimental data both in [75] and the calculations presented here.

Imposing the Bloch condition $\varphi_q^\nu(z + d) = \mathrm{e}^{\mathrm{i}qd}\varphi_q^\nu(z)$ the Bloch functions $\varphi_q^\nu(z)$ and eigenvalues E_q^ν are calculated within the Kronig–Penney model. The phase of the Bloch functions is chosen in the following way [6]. Let $z = 0$ be the centre of one quantum well. If $\varphi_0^\nu(0) \neq 0$ we choose the phase in such a way that $\varphi_q^\nu(0)$ is real for each q. For $\varphi_0^\nu(0) = 0$, $\varphi_q^\nu(0)$ is chosen to be purely imaginary. Furthermore, $\varphi_q^\nu(z)$ has to be an analytic function in q for both cases. From equation (10.1.2) the level energies E^ν and couplings T_1^ν are obtained. Equation (10.1.1) provides the Wannier functions which are plotted in Fig. 10.2. Finally, the couplings $R_h^{\nu'\nu}$ are obtained from their definitions $R_h^{\nu'\nu} = \int \mathrm{d}z\, \Psi^{\nu'}(z-hd)z\Psi^\nu(z)$. The calculated values are given in Table 10.1. A complication arises due to the fact that the effective Hamiltonian of the Kronig–Penney model is energy dependent due to the energy dependence of the effective mass. Therefore the envelope functions $\varphi_q^\nu(z)$ for different energies E_q^ν are not strictly orthogonal but exhibit a small overlap which is neglected in the calculation.

10.6.2 Impurity scattering

Here we want to calculate the self-energy† for impurity scattering. The contribution to $\hat{H}_0^{\text{scatter}}$ of equation (10.2.4) for the lowest level a is given by

$$\hat{H}_0^{\text{imp}} = \frac{1}{A} \sum_{\mathbf{k},\mathbf{p},i,n} V_{i,n}^{aa}(\mathbf{p}) a_n^\dagger(\mathbf{k}+\mathbf{p}) a_n(\mathbf{k}). \tag{10.6.3}$$

Here the subscript i denotes the impurity located at the position $(\mathbf{r}_i, z_i)$. The matrix element is calculated with the Wannier functions yielding:

$$\begin{aligned} V_{i,n}^{aa}(\mathbf{p}) &= \int \mathrm{d}^2r\, \mathrm{d}z\, \mathrm{e}^{-\mathrm{i}\mathbf{p}\cdot\mathbf{r}} \Psi_a^*(z)\Psi_a(z) \frac{-e^2}{4\pi\epsilon_s\epsilon_0\sqrt{|\mathbf{r}-\mathbf{r}_i|^2 + (z-z_i+nd)^2}} \\ &= \frac{-e^2}{2\epsilon_s\epsilon_0 p} \int \mathrm{d}z\, \Psi_a^*(z)\Psi_a(z) \mathrm{e}^{-p|z-z_i+nd|} \mathrm{e}^{-\mathrm{i}\mathbf{p}\cdot\mathbf{r}_i}. \end{aligned} \tag{10.6.4}$$

Note that the contribution for $p = 0$ is cancelled by the respective part in the electron–electron interaction as usual if the number of donors is equal to the number of carriers in the whole sample. In order to prevent the divergence of the matrix element for $p \to 0$ screening due to the electron–electron interaction has to be considered. (This is a general problem for the Coulomb interaction, see also Chapters 5 and 6.) The Hamiltonian for the electron–electron interaction reads:

$$\begin{aligned} \hat{H}^{ee} &= \frac{1}{2A} \sum_{\mathbf{k},\mathbf{k}',\mathbf{p},n,h} W_h^{aaaa}(\mathbf{p}) a_n^+(\mathbf{k}+\mathbf{p}) a_{n+h}^+(\mathbf{k}'-\mathbf{p}) a_{n+h}(\mathbf{k}') a_n(\mathbf{k}) \\ &\quad + \text{terms with } b. \end{aligned} \tag{10.6.5}$$

Here we assumed that the overlap of wave functions from different wells is negligible, so that pairs always have to be inside the same well. The matrix

† For readers who are not familiar with the concepts of many-particle physics (such as self-energies, Green functions, etc.) [76] is recommended as a helpful introduction.

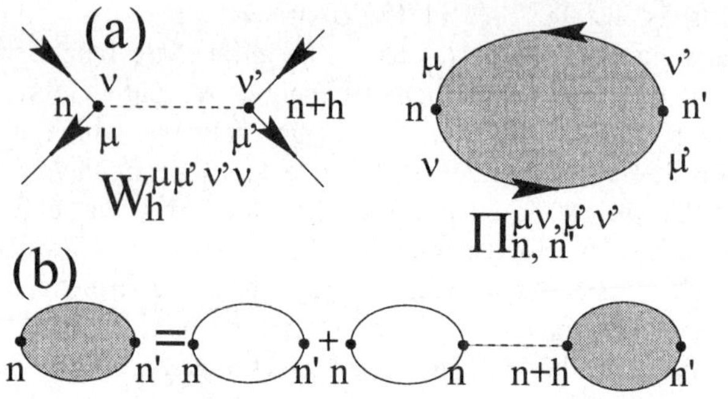

Figure 10.14 *(a) The notation of the matrix element* $W_h^{\mu\mu'\nu'\nu}(\boldsymbol{p})$ *and the polarizability* $\Pi_{n,n'}^{\mu\nu,\mu'\nu'}(\boldsymbol{p},\omega=0)$. *(b) Shows the random-phase approximation diagramatically.*

element reads:

$$W_h^{\mu\mu'\nu'\nu}(\boldsymbol{p}) = \frac{e^2}{2\epsilon_s\epsilon_0 p}\int \mathrm{d}z_1 \int \mathrm{d}z_2\, \Psi^{\mu*}(z_1)\Psi^{\mu'*}(z_2-dh) \times\Psi^{\nu'}(z_2-dh)\Psi^{\nu}(z_1)\mathrm{e}^{-p|z_1-z_2|} \quad (10.6.6)$$

The screening is described by the polarizability $\Pi_{n,n'}^{\mu\nu,\mu'\nu'}(\boldsymbol{p},\omega=0)$, where ν, μ take the values a and b. The notation follows [77], where a similar problem is investigated and is shown in Fig. 10.14. As the $\boldsymbol{p}$ dependence is identical in all parts, we omit it in the notation. Within the random-phase approximation (RPA) we have the Dyson equation [58]:

$$\Pi_{n,n'}^{\mu\nu,\mu'\nu'} = \Pi_n^{0\mu\nu}\left(\delta_{n,n'}\delta_{\mu,\nu'}\delta_{\nu,\mu'} + \sum_{h,\mu'',\nu''} W_h^{\mu\nu''\mu''\nu}\Pi_{n+h,n'}^{\mu''\nu'',\mu'\nu'}\right) \quad (10.6.7)$$

where the vacuum polarizability Π^0 is given by [58]

$$\Pi_n^{0\mu\nu}(\boldsymbol{p},\omega=0) = \frac{2}{A}\sum_{\boldsymbol{k}} \frac{n_F(E_{\boldsymbol{k}+\boldsymbol{p}}+E^\nu-E_n^F)-n_F(E_{\boldsymbol{k}}+E^\mu-E_n^F)}{(E_{\boldsymbol{k}+\boldsymbol{p}}+E^\nu)-(E_{\boldsymbol{k}}+E^\mu)}. \quad (10.6.8)$$

Now we find that $\Pi_n^{0bb}(\boldsymbol{p},\omega=0)=0$ assuming that the upper level is not occupied. $\Pi_n^{0ab}(\boldsymbol{p},\omega=0)$ is quite small as the gap appears in the denominator. It remains the contribution $\Pi_n^{0aa}(\boldsymbol{p},\omega=0)$ which yields for the two-dimensional electron gas at zero temperature [78]:

$$\Pi_n^{0aa}(\boldsymbol{p},\omega=0) = -\frac{m}{\pi\hbar^2}\left(1-\Theta(p-2k_F)\sqrt{1-4\frac{k_F^2}{p^2}}\right) \quad (10.6.9)$$

which is independent of p for $p<2k_F$. As all polarizations only have a-indices we omit these indices in the following. Equation (10.6.7) can be solved by a

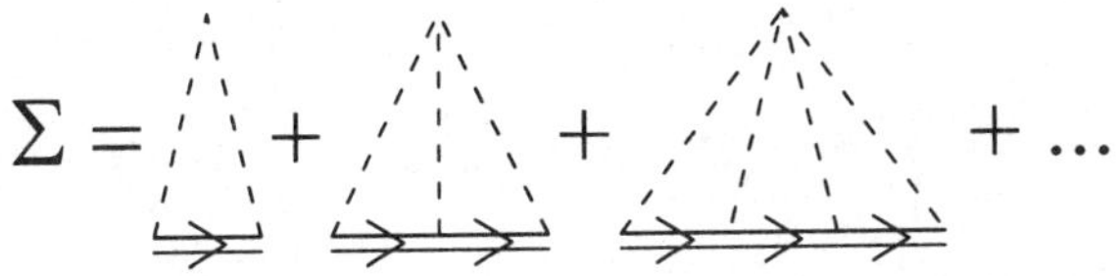

Figure 10.15 *Diagrams contained in the self-consistent single-site approximation.*

Fourier transformation (for an infinite superlattice and assuming $\Pi_n^0 = \Pi^0$ is independent of n). Defining $\tilde{\Pi}_q = \sum_{n'} \Pi_{n,n'} e^{iq(n'-n)}$, $\tilde{W}_q^{aaaa} = \sum_h W_h^{aaaa} e^{iqh}$ we find

$$\tilde{\Pi}_q = \frac{\Pi^0}{1 - \Pi^0 \tilde{W}_q^{aaaa}}. \qquad (10.6.10)$$

Now the screened electron-impurity interaction is given by the bare interaction and a part combined with the polarizability given by

$$V_{i,n}^{aa\,\mathrm{sc}}(\boldsymbol{p}) = V_{i,n}^{aa} + \sum_{h,n'} W_h^{aaaa} \Pi_{n+h,n'} V_{i,n'}^{aa}. \qquad (10.6.11)$$

Defining $\tilde{V}_{i,q}^{aa} = \sum_n V_{i,n} e^{-iqn}$ and using the translational invariance $\Pi_{n+h,n'} = \Pi_{2n+h-n',n}$ we find:

$$V_{i,n}^{aa\,\mathrm{sc}}(\boldsymbol{p}) = \frac{1}{2\pi} \int_{-\pi}^{\pi} dq \frac{\tilde{V}_{i,q}^{aa}}{1 - \Pi^0 \tilde{W}_q^{aaaa}} e^{iqn}. \qquad (10.6.12)$$

Similarly we have:

$$\begin{aligned} V_{i,n}^{bb\,\mathrm{sc}}(\boldsymbol{p}) &= V_{i,n}^{bb} + \sum_{h,n'} W_h^{baab} \Pi_{n+h,n'} V_{i,n'}^{aa} \qquad (10.6.13) \\ &= \frac{1}{2\pi} \int_{-\pi}^{\pi} dq \left[\frac{\tilde{V}_{i,q}^{bb}}{1 - \Pi^0 \tilde{W}_q^{aaaa}} + \frac{(\tilde{W}_q^{baab} \tilde{V}_{i,q}^{aa} - \tilde{W}_q^{aaaa} \tilde{V}_{i,q}^{bb})\Pi^0}{1 - \Pi^0 \tilde{W}_q^{aaaa}} \right] \\ &\quad \times e^{iqn}. \end{aligned}$$

The screened matrix elements from equations (10.6.12) and (10.6.13) are used in the Hamiltonian (10.6.3) in the following.

Now we calculate the self-energy within the single-site approximation which is shown diagrammatically in Fig. 10.15. For the impurity i we find the contribution to the self-energy $\Sigma_n^{a\,\mathrm{ret}}(\boldsymbol{k}, E)$

$$\begin{aligned}\Sigma_n^{a,i}(\boldsymbol{k}, E) \;=\;& \frac{1}{A^2}\sum_{\boldsymbol{k}_1} V_{i,n}^{aa\,\mathrm{sc}}(\boldsymbol{k}-\boldsymbol{k}_1)G(\boldsymbol{k}_1, E)V_{i,n}^{aa\,\mathrm{sc}}(\boldsymbol{k}_1-\boldsymbol{k}) \\ &+\frac{1}{A^3}\sum_{\boldsymbol{k}_1,\boldsymbol{k}_2} V_{i,n}^{aa\,\mathrm{sc}}(\boldsymbol{k}-\boldsymbol{k}_1)G(\boldsymbol{k}_1, E)V_{i,n}^{aa\,\mathrm{sc}}(\boldsymbol{k}_1-\boldsymbol{k}_2) \\ &\times G(\boldsymbol{k}_2, E)V_{i,n}^{aa\,\mathrm{sc}}(\boldsymbol{k}_2-\boldsymbol{k}) \\ &+\frac{1}{A^4}\sum_{\boldsymbol{k}_1,\boldsymbol{k}_2,\boldsymbol{k}_3} V_{i,n}^{aa\,\mathrm{sc}}(\boldsymbol{k}-\boldsymbol{k}_1)G(\boldsymbol{k}_1, E)V_{i,n}^{aa\,\mathrm{sc}}(\boldsymbol{k}_1-\boldsymbol{k}_2)G(\boldsymbol{k}_2, E) \\ &\times V_{i,n}^{aa\,\mathrm{sc}}(\boldsymbol{k}_2-\boldsymbol{k}_3)G(\boldsymbol{k}_3, E)V_{i,n}^{aa\,\mathrm{sc}}(\boldsymbol{k}_3-\boldsymbol{k})+\cdots\end{aligned} \tag{10.6.14}$$

where $G(\boldsymbol{k}, E) = (E - E_k - \Sigma_n^{a\,\mathrm{ret}}(\boldsymbol{k}, E))^{-1}$ is the full retarded Green's function. This sum can be transformed to the self-consistent equation [59]

$$K^{a,i}(\boldsymbol{k}_1, \boldsymbol{k}, E) = V_{i,n}^{aa\,\mathrm{sc}}(\boldsymbol{k}_1-\boldsymbol{k})+\frac{1}{A}\sum_{\boldsymbol{k}_2} V_{i,n}^{aa\,\mathrm{sc}}(\boldsymbol{k}_1-\boldsymbol{k}_2)G(\boldsymbol{k}_2, E)K^{a,i}(\boldsymbol{k}_2, \boldsymbol{k}, E) \tag{10.6.15}$$

which can be solved numerically for a given self-energy function $\Sigma_n^{a\,\mathrm{ret}}(\boldsymbol{k}, E)$ entering $G(\boldsymbol{k}_2, E)$. The contribution to the self-energy is then given by

$$\Sigma_n^{a,i}(\boldsymbol{k}, E) = \frac{1}{A^2}\sum_{\boldsymbol{k}_1} V_{i,n}^{aa\,\mathrm{sc}}(\boldsymbol{k}_1-\boldsymbol{k})G(\boldsymbol{k}_1, E)K^{a,i}(\boldsymbol{k}_1, \boldsymbol{k}, E). \tag{10.6.16}$$

Summing up the contribution from all impurities i and possibly different scattering processes we obtain the self-energy function

$$\Sigma_n^{a\,\mathrm{ret}}(\boldsymbol{k}, E) = \sum_i \Sigma_n^{a,i}(\boldsymbol{k}, E) + \Sigma_n^{a,\mathrm{other\ scattering}}(\boldsymbol{k}, E). \tag{10.6.17}$$

Now a self-consistent solution of equations (10.6.15)–(10.6.17) can be achieved by iteration.

The contribution to $\Sigma_n^{b\,\mathrm{ret}}(\boldsymbol{k}, E)$ is calculated in the same way.

10.6.3 Interface roughness

Interface roughness is modelled as in [56] considering an interface located at $z = z_0$ exhibiting thickness fluctuations $\xi(\boldsymbol{r})$ of the order of $\pm\eta$ (we use $\eta =$ 2.8 Å which is one monolayer of GaAs). We assume the correlations

$$\langle\xi(\boldsymbol{r})\rangle_r = 0 \tag{10.6.18}$$

$$\langle\xi(\boldsymbol{r})\xi(\boldsymbol{r}')\rangle_r = \alpha\eta^2\exp(-|\boldsymbol{r}-\boldsymbol{r}'|/\lambda) \tag{10.6.19}$$

with a correlation length of $\lambda = 7$ nm and an average coverage of $\alpha = 0.5$. Such an exponential correlation function $\langle\xi(\boldsymbol{r})\xi(\boldsymbol{r}')\rangle_r$ seems to be more appropriate than the usual choice of a Gaussian [79] as stated in [80, 56]. As in [79] (where the scattering of Bloch states in a superlattice is regarded) the additional potential† due to the roughness is modelled by a δ-function at the perfect interface

$$U(\boldsymbol{r}, z) = \xi(\boldsymbol{r})\Delta E_c \delta(z - z_0). \tag{10.6.20}$$

This gives the following interface roughness contribution to $\hat{H}^{\text{scatter}}$

$$\begin{aligned}\hat{H}^{\text{rough}} &= \frac{1}{A}\sum_{\boldsymbol{k},\boldsymbol{p},h} U_h^{aa}(\boldsymbol{p}) a_{n+h}^\dagger(\boldsymbol{k}+\boldsymbol{p}) a_n(\boldsymbol{k}) + U_h^{bb}(\boldsymbol{p}) b_{n+h}^\dagger(\boldsymbol{k}+\boldsymbol{p}) b_n(\boldsymbol{k}) \\ &\quad + U_h^{ba}(\boldsymbol{p}) b_{n+h}^\dagger(\boldsymbol{k}+\boldsymbol{p}) a_n(\boldsymbol{k}) + U_h^{ab}(\boldsymbol{p}) a_{n+h}^\dagger(\boldsymbol{k}+\boldsymbol{p}) b_n(\boldsymbol{k})]\end{aligned} \tag{10.6.21}$$

with the matrix elements

$$U_h^{\nu\mu}(\boldsymbol{p}) = \int \mathrm{d}^2 r\, \mathrm{e}^{-\mathrm{i}\boldsymbol{p}\cdot\boldsymbol{r}} \Delta E_c[\xi(\boldsymbol{r})\Psi^{\nu *}(z_0 - hd)\Psi^\mu(z_0)]. \tag{10.6.22}$$

Using the correlation function (10.6.19) we obtain the square of the matrix element

$$|U_h^{\nu\mu}(\boldsymbol{p})|^2 = A\Delta E_c^2 |\Psi_\nu(z_0 - hd)|^2 |\Psi_\mu(z_0)|^2 \frac{2\pi\alpha\eta^2\lambda^2}{(1 + (p\lambda)^2)^{3/2}} \tag{10.6.23}$$

which enters the expressions in the following.

The elements $U_0^{\nu\mu}$ result in scattering within the wells. Their contribution to the self-energy is calculated within the self-consistent Born approximation (which is just the first diagram of the infinite sum in Fig. 10.15)

$$\Sigma_{\text{rough}}^{a\,\text{ret}}(\boldsymbol{k}, E) = \frac{2}{A^2}\sum_{\boldsymbol{k}_1} |U_0^{aa}(\boldsymbol{k} - \boldsymbol{k}_1)|^2 G(\boldsymbol{k}_1, E), \tag{10.6.24}$$

where the factor of 2 takes into account the two interfaces per well. These self-energies contribute to the total self-energy in equation (10.6.17). The calculation for the subband b is performed in the same way.

The elements $U_1^{\nu\mu}$ contribute to the nonresonant current from one well to the next via equation (10.2.6). Here the contributions from all four interfaces of both wells involved are summed up. For weakly coupled wells $U_2^{\nu\mu}, U_3^{\nu\mu}, \ldots$ are small and can be neglected.

10.6.4 *Optical phonons*

In polar materials such as GaAs, the polar interaction with optical phonons provides an important scattering process. As the energy $\hbar\omega_0 = 36$ meV is

† In Chapter 11 of this book the potential is chosen to be the variation of the energy levels due to the well width fluctuation which is only defined for scattering within a given level and a given well. The approach via equation (10.6.20) has the advantage that interwell and interlevel transitions can also be handled. For intrawell and intralevel processes the results are similar to $\mathrm{d}E^\nu/\mathrm{d}w \sim \Delta E_c|\Psi^\nu(z_0)|^2$ where w is the well width and z_0 is the position of the interface.

transferred, this process couples electronic states with different energy in contrast to the two scattering mechanism discussed before. This makes the full calculation of the self-energies much more complicated. But fortunately only a restricted number of processes are allowed at low temperatures where only the emission of phonons takes place. Therefore to any electronic state $\boldsymbol{k}$ in level ν with energy E_k+E^ν affected there must be another state at an energy $E_k+E^\nu-\hbar\omega_0$. For the lowest level $\nu=a$ this means that the condition $E_k>\hbar\omega_0$ must be satisfied in order for phonon scattering to be possible. But for small Fermi levels (5.37 meV for the sample considered) these states are far away from any resonant transition so that the neglection of the phonon contribution to the self-energy hardly affects the currents. The situation is different for the second level. Here the states with $E_k\approx 0$ are in resonance with the occupied states in the ground level if the electric field takes the value $eFd\approx E^b-E^a$. Therefore the actual broadening of these states is crucial for the $a\to b$ resonance. This process is taken into account by calculating the scattering time $\tau_{\rm ph}$ for this process following [62] yielding $\tau_{\rm ph}=0.854$ ps for the structure considered here. For the self-energy contribution the constant value

$$\Sigma^{b\,\rm ret}_{\rm phonon}(\boldsymbol{k},E)=-\mathrm{i}\frac{\hbar}{2\tau_{\rm ph}} \tag{10.6.25}$$

is used which contributes to the total self-energy in equation (10.6.17) for level b.

10.7 CONCLUSIONS

In this chapter the electronic transport in weakly coupled multiple quantum wells has been considered. Within the model of sequential tunnelling the currents have been calculated without any fitting parameters, taking into account the scattering at impurities and interface roughness. The currents are in good quantitative agreement with experimental data stating the physical relevance of the model.

In the NDC region a homogeneous field distribution is unstable. For the actual doping of the sample considered both stable field-domains and self-sustained current oscillations are found theoretically in good agreement with the experimental observation. The sequence of domain branches is almost independent of the contacts as the influence of the boundaries is shielded by the domain regions where the electric field is in the regime of positive differential conductivity. Therefore the branches contain information about the transport in multiple quantum wells itself which is not spoilt by contacts which are often only poorly defined. Furthermore, the domain branches react quite sensitively to local deviations from periodicity which allows for a check of the actual sample quality.

ACKNOWLEDGEMENTS

I would like to thank Luis Bonilla, Holger Grahn, Ben Hu, Anatoli Ignatov, Antti-Pekka Jauho, Kristinn Johnsen, Jörg Kastrup, Miguel Moscoso, Michael

Patra, Frank Prengel, Eckehard Schöll, Georg Schwarz and Stefan Zeuner for fruitful collaboration and helpful discussions. Financial support from the Deutsche Forschungsgemeinschaft is gratefully acknowledged.

REFERENCES

[1] Grahn, H. T. (ed.) (1995) *Semiconductor Superlattices, Growth and Electronic Properties*, World Scientific, Singapore.
[2] Esaki, L. and Tsu, R. (1970) *IBM J. Res. Develop.* **14**, 61.
[3] Capasso, F., Mohammed, K., Cho, A. Y. (1986) *Appl. Phys. Lett.* **48**, 478.
[4] Murphy, S. Q., Eisenstein, J. P., Pfeiffer, L. N. *et al.* (1995) *Phys. Rev.* B **52**, 14 825.
[5] Bastard, G. (1988) *Wave Mechanics Applied to Semiconductor Heterostructures*, Les Editions de Physique, Les Ulis Cedex, France.
[6] Kohn, W. (1959) *Phys. Rev.* **115**, 809.
[7] Emin, D. and Hart, C. F. (1987) *Phys. Rev.* B **36**, 7353.
[8] Tsu, R. and Döhler, G. (1975) *Phys. Rev.* B **12**, 680.
[9] Rott, S., Linder, N. and Döhler, G. H. (1997) *Superlatt. Microstruct.* **21**, 569.
[10] Lebwohl, P. A. and Tsu, R. (1970) *J. Appl. Phys.* **41**, 2664.
[11] Ignatov, A. A., Dodin, E. P. and Shashkin, V. I. (1991) *Mod. Phys. Lett.* B **5**, 1087.
[12] Lei, X. L., Horing, N. J. M. and Cui, H. L. (1991) *Phys. Rev. Lett.* **66**, 3277.
[13] Sibille, A., Palmier, J. F., Wang, H. *et al.* (1990) *Phys. Rev. Lett.* **64**, 52.
[14] Schomburg, E., Ignatov, A. A., Winnerl, S. *et al.* (1996) *Proc. 23rd Int. Conf. Phys. Semicond., Berlin* (eds M. Scheffler and R. Zimmermann), World Scientific, Singapore, vol. 3, pp. 1679–82.
[15] Grahn, H. T., von Klitzing, K., Ploog, K. *et al.* (1991) *Phys. Rev.* B **43**, 12 094.
[16] Kazarinov, R. F. and Suris, R. A. (1972) *Sov. Phys. Semicond.* **6**, 120.
[17] Miller, D. and Laikhtman, B. (1994) *Phys. Rev.* B **50**, 18 426.
[18] Laikhtman, B. and Miller, D. (1993) *Phys. Rev.* B **48**, 5395.
[19] Shaw, M. P., Mitin, V. V., Schöll E. *et al.* (1992) *The Physics of Instabilities in Solid State Electron Devices*, Plenum Press, New York.
[20] Büttiker, M. and Thomas, H. (1977) *Phys. Rev. Lett.* **38**, 78.
[21] Büttiker, M. and Thomas, H. (1979) *Z. Phys.* B **34**, 301.
[22] Ignatov, A. A., Piskarev, V. I. and Shashkin, V. I. (1985) *Sov. Phys. Semicond.* **19**, 1345.
[23] LePerson, H., Minot, C., Boni, L. *et al.* (1992) *Appl. Phys. Lett.* **60**, 2397.
[24] Hofbeck, K., Grenzer, J., Schomburg, E. *et al.* (1996) *Phys. Lett.* A **218**, 349.
[25] Waschke, C., Roskos, H. G., Schwedler, K. *et al.* (1993) *Phys. Rev. Lett.* **70**, 3319.
[26] Kastrup, J., Klann, R., Grahn, H. T. *et al.* (1995) *Phys. Rev.* B **52**, 13 761.
[27] Grahn, H. T., Kastrup, J., Klann, R. *et al.* (1996) *Proc. 23rd Int. Conf. Phys. Semicond., Berlin 1996* (eds M. Scheffler and R. Zimmermann), World Scientific, Singapore, vol. 3, pp. 1671–4.
[28] Ohtani, N., Hosoda, M., Mimura, H. *et al.* (1996) *Proc. 23rd Int. Conf. Phys. Semicond., Berlin* (eds M. Scheffler and R. Zimmermann), World Scientific, Singapore, vol. 3, pp. 1675–8.
[29] Bonilla, L. L. (1995) *Nonlinear Dynamics and Pattern Formation in Semiconductors*, (ed. F. J. Niedernostheide), Springer, Berlin, ch. 1, pp. 1–20.

[30] Wacker, A., Prengel, F. and Schöll, E. (1995) *Proc. 22nd Int. Conf. Phys. Semicond., Vancouver 1994* (ed. D. J. Lockwood), World Scientific, Singapore, vol. 2, p. 1075.
[31] Kastrup, J., Hey, R., Ploog, K. H. *et al.* (1997) *Phys. Rev.* B **55**, 2476.
[32] Bonilla, L. L., Kindelan, M., Moscoso, M. *et al.* (1997) *SIAM J. Appl. Math.* **57**, in print.
[33] Bulashenko, O. M. and Bonilla, L. L. (1995) *Phys. Rev.* B **52**, 7849.
[34] Zhang, Y., Kastrup, J., Klann, R. *et al.* (1996) *Phys. Rev. Lett.* **77**, 3001.
[35] Esaki, L. and Chang, L. L. (1974) *Phys. Rev. Lett.* **33**, 495.
[36] Kawamura, Y., Wakita, K., Asahi, H. *et al.* (1986) *Japan J. Appl. Phys.* **25**, L928.
[37] Choi, K. K., Levine, B. F., Malik, R. J. *et al.* (1987) *Phys. Rev.* B **35**, 4172.
[38] Helm, M., England, P., Colas, E. *et al.* (1989) *Phys. Rev. Lett.* **63**, 74.
[39] Helgesen, P. and Finstad, T. G. (1990) *Proceedings of the 14th Nordic Semiconductor Meeting* (ed. O. Hansen), University of Århus, p. 323.
[40] Grahn, H. T., Haug, R. J., Müller, W. *et al.* (1991) *Phys. Rev. Lett.* **67**, 1618.
[41] Murugkar, S., Kwok, S. H., Ambrazevicius, G. *et al.* (1994) *Phys. Rev.* B **49**, 16 849.
[42] Kastrup, J., Grahn, H. T., Ploog, K. F. *et al.* (1994) *Appl. Phys. Lett.* **65**, 1808.
[43] Zhang, Y., Yang, X., Liu, W. *et al.* (1994) *Appl. Phys. Lett.* **65**, 1148.
[44] Kwok, S. H., Grahn, H. T., Ramsteiner, M. *et al.* (1995) *Phys. Rev.* B **51**, 9943.
[45] Stoklitskii, S. A., Murzin, V. N., Rasulova, G. K. *et al.* (1995) *JETP Lett.* **61**, 405.
[46] Han, Z. Y., Yoon, S. F., Radhakrishnan, K. *et al.* (1995) *Superlatt. Microstruct.* **18**, 83.
[47] Keay, B. J., Allen, S. J., Galán, J. *et al.* (1995) *Phys. Rev. Lett.* **75**, 4098.
[48] Zeuner, S., Keay, B. J., Allen, S. J. *et al.* (1996) *Phys. Rev.* B **53**, 1717.
[49] Mimura, H., Hosoda, M., Ohtani, N. *et al.* (1996) *Phys. Rev.* B **54**, 2323.
[50] Suris, R. A. (1974) *Sov. Phys. Semicond.* **7**, 1030.
[51] Laikhtman, B. (1991) *Phys. Rev.* B **44**, 11 260.
[52] Korotkov, A. N., Averin, D. V. and Likharev, K. K. (1993) *Appl. Phys. Lett.* **62**, 3282.
[53] Prengel, F., Wacker, A. and Schöll, E. (1994) *Phys. Rev.* B **50**, 1705
Prengel, F., Wacker, A. and Schöll, E. (1995) *Phys. Rev.* B **52**, 11 518.
[54] Bonilla, L. L., Galán, J., Cuesta, J. A. *et al.* (1994) *Phys. Rev.* B **50**, 8644.
[55] Wacker, A. and Jauho, A.-P. (1997) *Physica Scripta* **T69**, 321.
[56] Wacker, A. and Jauho, A.-P. *Superlatt. Microstruct.* in print (cond-mat/9610119).
[57] Zheng, L. and MacDonald, A. H. (1993) *Phys. Rev.* B **47**, 10 619.
[58] Mahan, G. D. (1990) *Many-Particle Physics*, Plenum, New York.
[59] Gold, A., Serre, J. and Ghazali, A. (1988) *Phys. Rev.* B **37**, 4589.
[60] Wacker, A., Jauho, A.-P., Zeuner, S. and Allen, S. J. (1997) *Phys. Rev.* B **56**, in print.
[61] Wacker, A. and Jauho, A.-P. *Phys. Status Solidi* (b) in print.
[62] Ferreira, R. and Bastard, G. (1989) *Phys. Rev.* B **40**, 1074.
[63] Mityagin, Y. A., Murzin, V. N., Efimov, Y. A. *et al.* (1997) *Appl. Phys. Lett.* **70**, 3008.
[64] Wacker, A., Moscoso, M., Kindelan, M. *et al.* (1997) *Phys. Rev.* B **55**, 2466.
[65] Wacker, A., Schwarz, G., Prengel, F. *et al.* (1995) *Phys. Rev.* B **52**, 13 788.
[66] Kwok, S. H., Jahn, U., Menniger, J. *et al.* (1995) *Appl. Phys. Lett.* **66**, 2113.
[67] Schwarz, G. and Schöll, E. (1996) *Phys. Status Solidi* (b) **194**, 351.
[68] Mityagin, Y. A. and Murzin, V. N. (1996) *JEPT Lett.* **64**, 155.

[69] Schwarz, G., Prengel, F., Schöll, E. *et al.* (1996) *Appl. Phys. Lett.* **69**, 626.
[70] Patra, M. (1996) *Master's Thesis*, Technische Universität Berlin; Patra, M., Schwarz, G. and Schöll, E. (1998) *Phys. Rev.* B, in print.
[71] Bonilla, L. L., Hernando, P. J., Herrero, M. A. *et al.* (1997) *Physica D*, **108**, 168.
[72] Kastrup, J., Prengel, F., Grahn, H. T. *et al.* (1996) *Phys. Rev.* B **53**, 1502.
[73] Schöll, E., Schwarz, G., Patra, M., *et al.* (1996) *Proc. 9th Int. Conf. on Hot Carriers in Semiconductors, Chicago 1995* (eds K. Hess, J. P. Leburton and U. Ravaioli), Plenum Press, New York, pp. 177–81.
[74] Schulman, J. N. and Chang, Y.-C. (1985) *Phys. Rev.* B **31**, 2056.
[75] Brozak, G., de Andrada e Silva, E. A., Sham, L. J. *et al.* (1990) *Phys. Rev. Lett.* **64**, 471.
[76] Mattuck, R. D. (1992) *A Guide to Feynman Diagrams in the Many Body Problem* (Republication by Dover, New York).
[77] Eliasson, G. L. (1987) *Ph.D. Thesis*, Brown University.
[78] Stern, F. (1967) *Phys. Rev. Lett.* **18**, 546.
[79] Dharssi, I. and Butcher, P. N. (1990) *J. Phys.: Condens. Matter* **2**, 4629.
[80] Goodnick, S. M., Ferry, D. K., Wilmsen, C. W. *et al.* (1985) *Phys. Rev.* B **32**, 8171.
[81] Schöll, E., Schwarz, G. and Wacker, A. (1998) *Physica* B, in print.

CHAPTER 11

Scattering processes in low-dimensional structures

B. K. Ridley

Department of Physics, University of Essex, Colchester, UK

11.1 INTRODUCTION

A fundamental concern of transport theory is the description of how particles are scattered as they move randomly or under the influence of applied fields. The concept of scattering carries with it the implication that the rate of scattering is relatively small and for the interaction that produces scattering to be regarded as a small perturbation on the free motion of the particle, so that the unperturbed states of the particle are well defined. Accepting this implication as basic, scattering theory then proceeds to identify the scattering agents and quantify their interaction strengths. There are three rates that enter the physical description of the system, namely, the scattering rate, the momentum-relaxation rate and the energy-relaxation rate. These are obtained using standard quantum-mechanical time-dependent perturbation theory as embodied in Fermi's golden rule. The scattering rate determines the lifetime of the particle in a given state and is the quantity needed in Monte Carlo simulations. The momentum-relaxation rate determines the current and the energy-relaxation rate determines the average energy of the particles in the presence of fields. In this chapter we describe the principal electron scattering mechanisms that operate in semiconductor heterostructures.

11.2 THE SCATTERING RATE

The scattering rate is given by Fermi's golden rule:

$$W_i = \frac{2\pi}{\hbar} \int |\langle f|H|i\rangle|^2 \delta\{E_f - E_i\} \, \mathrm{d}N_f \qquad (11.2.1)$$

where H is the perturbing energy. The matrix element is

$$\langle f|H|i\rangle = \int \Psi_f^*(\boldsymbol{R}) H(\boldsymbol{R}) \Psi_i(\boldsymbol{R}) \, \mathrm{d}\boldsymbol{R} \qquad (11.2.2)$$

where $\Psi_{f,i}(\boldsymbol{R})$ are the final and initial wave functions of the system. In the case of electron scattering the latter are expressible as the product of a one-electron wave function, $u(\boldsymbol{R})\psi(\boldsymbol{R})$ (where $u(\boldsymbol{R})$ is the cell-periodic wave function and $\psi(\boldsymbol{R})$ is the envelope function) and harmonic oscillator wave functions $\Phi(\boldsymbol{R})$. The Fourier transform of $H(\boldsymbol{R})$ is

$$H(\boldsymbol{R}) = \sum_{\boldsymbol{Q}} H(\boldsymbol{Q})e^{i\boldsymbol{Q}\cdot\boldsymbol{R}}. \tag{11.2.3}$$

The interaction strength, now encapsulated in the operator $H(\boldsymbol{Q})$, depends on the particular scattering mechanism. Common to all processes is the overlap integral of the spatial dependence of the coupling and the electron wave functions:

$$\begin{aligned} I(\boldsymbol{K}', \boldsymbol{K}, \boldsymbol{Q}) &= I(\boldsymbol{K}', \boldsymbol{K})G(\boldsymbol{K}', \boldsymbol{K}, \boldsymbol{Q}) \\ I(\boldsymbol{K}', \boldsymbol{K}) &= \int_{\text{unit cell}} u^*_{\boldsymbol{K}'}(\boldsymbol{R})u_{\boldsymbol{K}}(\boldsymbol{R})\,d\boldsymbol{R} \\ G(\boldsymbol{K}', \boldsymbol{K}, \boldsymbol{Q}) &= \int \psi^*(\boldsymbol{K}', \boldsymbol{R})H(\boldsymbol{Q}, \boldsymbol{R})\psi(\boldsymbol{K}, \boldsymbol{R})\,d\boldsymbol{R}. \end{aligned} \tag{11.2.4}$$

The integral over the unit cell involving the periodic Bloch functions depends on the angular momenta of the initial and final states. In general the latter will consist of mixed states due to nonparabolicity and the effects of quantum confinement. Complications of this sort are often ignored and the standard band-edge bulk results are used. These are

$$\begin{aligned} |I(\boldsymbol{K}', \boldsymbol{K})|^2 &= 1 \qquad J = \tfrac{1}{2} \qquad \text{CB and SO} \\ &= \tfrac{1}{4}(1 + 3\cos^2\beta) \qquad J = \tfrac{3}{2}, \qquad m = \tfrac{1}{2} \qquad \text{LH} \\ &= \tfrac{3}{4}\sin^2\beta \qquad J = \tfrac{3}{2}, \qquad m = \tfrac{3}{2} \qquad \text{HH} \end{aligned} \tag{11.2.5}$$

where J is the total angular-momentum quantum number, m is the component along the z-direction and β is the angle between $\boldsymbol{K}$ and $\boldsymbol{K}'$. (CB = conduction band, SO = split-off band, LH = light-hole band, HH = heavy-hole band.)

The effect of quantum confinement can often be ignored in the case of the unit-cell overlap integral, this is not the case for the overlap integral involving the envelope functions. This integral is always sensitive to electron confinement and when the scattering is by phonons it is also sensitive to the details of phonon confinement. In general, the integral ensures that crystal momentum is conserved in directions along which there is no confinement, but not otherwise.

In the effective-mass approximation electron confinement is described by entailing the continuity of ψ and $(1/m^*)\,d\psi/dz$ (z normal to the interface). (Note that these boundary conditions are not always valid. Comprehensive discussions of the validity of the effective-mass approximation have been given by Burt [1] and Foreman [2].) The form of the wave function is particularly simple for an electron confined between infinitely high potential barriers and in what follows we will illustrate scattering rates using the wave functions of an infinitely deep well. Because confinement introduces mixing of LH and HH states and even

SO states if the spin–orbit splitting is small, hole wave functions are more complex [3]. Moreover, holes are scattered by the deformation potential interaction as well as the polar interaction, in general. These properties make a description of hole scattering rather lengthy, though not intrinsically difficult and such a description cannot be given here for reasons of space. We will therefore limit our attention to the scattering of electrons.

The multilayer nature of low-dimensional structures means that there is mechanical and dielectric mismatch at each interface and this can have a profound effect on the phonon spectrum. Basically, the acoustic modes exhibit folding and the optical modes exhibit confinement. Because the interaction with electrons involves long-wavelength modes (ignoring intervalley scattering for the moment) the effect of zone folding is usually small and it is usual to adopt a bulk-like spectrum for the acoustic modes. Short-wavelength acoustic modes and long-wavelength optical modes are, however, seriously affected by confinement. The long-wavelength optical modes are more important since these tend to dominate electron and hole scattering in direct-cap materials, we will concentrate our attention on their confinement. Space will not allow us to consider the situation in indirect-gap semiconductors such as Ge and Si in which the confinement of short-wavelength modes is important.

11.3 OPTICAL PHONONS IN A QUANTUM WELL

The first theoretical step was taken long ago by Fuchs and Kliewer [4] who described confined long-wavelengths and interface optical modes in a thin ionic slab. They described the ionic slab in terms of a dielectric continuum and the modes they obtained satisfied the usual electric boundary conditions when there are no free charges, i.e. the continuity of tangential field and the normal component of the electric displacement. They assumed that the longitudinally polarized (LO) and transversely polarized (TO) modes were dispersionless, in the sense that in neither case did the frequency depend upon the wavelength. This meant that there were no mechanical effects to worry about. Despite the fact that a microscopic model of an ionic slab by Jones and Fuchs [5] showed a more complex picture, the simplicity of the dielectric-continuum (DC) model, and the observation that the dispersion of LO and TO modes was indeed weak, made this model the first to be applied to the problem of confined optical modes in semiconductor multilayers [6].

In a direct-gap polar semiconductor an electron in the conduction band interacts only with LO modes, via their long-range electric fields and not at all with TO modes, at least to zero order. Satisfying the electric boundary conditions quantizes the component of the wave vector normal to the interface, say q_z and divides the modes into symmetric and antisymmetric forms, assignations that refer to the electric potential, ϕ, relative to the mid-plane of the layer. From the general analysis of Born and Huang [7] it can be deduced that the electric field, $\boldsymbol{E}$, associated with an optical mode is related to the relative displacement of the ions, $\boldsymbol{u}$, as follows

$$\boldsymbol{E} = -\frac{s(\omega)\rho_i \boldsymbol{u}}{\varepsilon_\infty} \tag{11.3.1a}$$

$$s(\omega) = \frac{\omega^2 - \omega_{TO}^2}{\omega_{LO}^2 - \omega_{TO}^2} \tag{11.3.1b}$$

where $s(\omega)$ is the field factor, ρ_i is the ionic charge density and ε_∞ is the high-frequency permittivity. Thus, when $\omega = \omega_{LO}$, $\boldsymbol{E}$ is the LO field and when $\omega = \omega_{TO}$, the field is zero. The electric displacement satisfies

$$\nabla \cdot (\varepsilon(\omega)\boldsymbol{E}) = 0$$
$$\varepsilon(\omega) = \varepsilon_\infty \frac{\omega^2 - \omega_{LO}^2}{\omega^2 - \omega_{TO}^2} \tag{11.3.2}$$

where $\varepsilon(\omega)$ is the permittivity and ε_∞ is the high-frequency limit. For LO modes the permittivity is zero.

In the DC model the allowed optical modes in a polar layer are determined solely by the electric boundary conditions. Four different sorts of mode must be considered: two TO modes, LO modes and interface modes. The TO modes can be classified according to whether the direction of polarization is perpendicular to the plane of incidence on reflection at an interface or whether it lies in the plane of incidence. These are referred to as s-modes and p-modes respectively. In this classification LO modes are always p-modes and so are the interface modes of interest. Very long-wavelength versions of the latter are usually referred to as interface polaritons, which are essentially transverse electromagnetic waves plus ionic polarization. As their wavelength along the surface shortens they move further away from the light line, becoming more and more like polarization waves and although they retain their essentially transverse electromagnetic character, their interaction with electrons can be satisfactorily described in terms of a scalar potential, as though they were, like LO modes, electrostatic waves. (For a discussion on this point, see [8].)

We first look at the Fuchs–Kliewer modes of a single quantum well as these are useful in several ways: they are simple, they are recognizable components of the more complex patterns that occur in reality and they are usually useful for determining scattering rates.

11.3.1 The dielectric-continuum model

In the DC model only the LO and interface modes need to be considered as these are the only waves that have electric fields. Because the permittivity vanishes for LO modes, the boundary conditions entail that the potential vanishes at an interface. The allowed modes in a quantum well formed by two barriers of the same material then have the symmetric form:

$$\begin{aligned} u_x &= q_x A_{\mathrm{LO}} e^{\mathrm{i}(q_x x - \omega t)} \cos(q_z z) \\ u_z &= \mathrm{i} q_z A_{\mathrm{LO}} e^{\mathrm{i}(q_x x - \omega t)} \sin(q_z z) \end{aligned} \tag{11.3.3}$$

$$q_z a = n\pi, \qquad n = 1, 3, 5, \ldots \qquad -\frac{a}{2} \leq z \leq \frac{a}{2}$$

or the antisymmetric form:

$$\begin{aligned} u_x &= \mathrm{i} q_x A_{\mathrm{LO}} e^{\mathrm{i}(q_x x - \omega t)} \sin(q_z z) \\ u_z &= q_z A_{\mathrm{LO}} e^{\mathrm{i}(q_x x - \omega t)} \cos(q_z z) \end{aligned} \tag{11.3.4}$$

$$q_z a = n\pi, \qquad n = 2, 4, 6, \ldots - \frac{a}{2} \leq z \leq \frac{a}{2}.$$

We choose z along the direction perpendicular to the interfaces and x along the direction of travel in the plane. Symmetry is defined with respect to the scalar potential, which is proportional to u_x. The amplitude A_{LO} is obtained by the usual quantization procedure and is given by:

$$\begin{aligned} A_{\mathrm{LO}} &= \left(\frac{\hbar}{N \mu \omega Q^2} \right)^{\frac{1}{2}} \\ Q^2 &= q_x^2 + q_z^2 \end{aligned} \tag{11.3.5}$$

where N is the number of unit cells and μ is the reduced mass.

There are two interface modes associated with the well and two associated with the barrier. In the unretarded limit (the velocity of light regarded as infinite) the symmetric mode associated with the well has the form:

$$\begin{aligned} u_x &= q_x A_I e^{\mathrm{i}(q_x x - \omega t)} \cosh(q_x z) \\ u_z &= -\, \mathrm{i} q_x A_I e^{\mathrm{i}(q_x x - \omega t)} \sinh(q_x z) \end{aligned} \tag{11.3.6}$$

$$\coth(q_x a/2) + r = 0 \qquad r = \frac{\varepsilon(\omega)_{\mathrm{well}}}{\varepsilon(\omega)_{\mathrm{barrier}}}$$

and the antisymmetric form is:

$$\begin{aligned} u_x &= \mathrm{i} q_x A_I e^{\mathrm{i}(q_x x - \omega t)} \sinh(q_x z) \\ u_z &= q_x A_I e^{\mathrm{i}(q_x x - \omega t)} \cosh(q_x z) \\ &\tanh(q_x a/2) + r = 0. \end{aligned} \tag{11.3.7}$$

The antisymmetric mode has the higher frequency which means that it has a more powerful scalar potential and therefore interacts more strongly with electrons. The normalization constant is given by:

$$A_I = \left(\frac{\hbar a}{2N \mu \omega q_x \sinh(q_x a)} \right)^{\frac{1}{2}}. \tag{11.3.8}$$

The interface modes associated with a semi-infinite barrier have amplitudes which fall off exponentially away from the well. This time the higher-frequency mode produces a symmetric potential in the well as given in equation (11.3.6) and obeys the same dispersion relation. The lower-frequency mode has the form of equation (11.3.7). The normalization constant is different from that of the well for the two barrier modes being given by:

$$A_{l\text{barrier}} = 2^{\frac{1}{2}}\left(\frac{\hbar a}{N\mu\omega q_x}\right)^{\frac{1}{2}} \begin{cases} \sinh(q_x a/2) & \text{antisymmetric} \\ \cosh(q_x a/2) & \text{symmetric.} \end{cases} \tag{11.3.9}$$

The DC model has been applied to structures other than a simple quantum well, the main examples of structures being quantum wells containing a monolayer [9, 10], quantum wells with nearby metallic layers [11] and quantum wires [12].

11.3.2 The hybrid-mode model

The optical modes predicted by the DC model are not those observed in Raman scattering experiments [13] nor are they those obtained in computationally intensive lattice-dynamics calculations [14]. It is hardly surprising that this is the case given the total neglect of mechanical boundary conditions in the DC model. Unfortunately, it is not always clear what these conditions should be. It would take us too far astray to go into this problem in depth. Suffice it to say that part of the difficulty is in describing the nature of optical-mode stress in a continuum model, so that even in cases where there are no delta-function-like forces at the interface—such as exist when there is no common ion [15]—and where the acoustic-like boundary conditions—continuity of amplitude and stress—may be expected to apply, the lack of a precise definition of 'optical' stress remains troublesome [16]. Nevertheless, acoustic-like boundary conditions have been applied in some reported cases [17, 18] but without justification. Fortunately, there are many cases in practice where this is not a problem. Where there is a large mechanical mismatch at the interface, i.e. where there is a large disparity of optical-mode frequencies in the adjacent materials, the appropriate mechanical boundary condition is simply:

$$\boldsymbol{u} = \boldsymbol{0} \tag{11.3.10}$$

at, or just beyond, the interface. This condition has been shown to work well in the paradigm case of AlAs/GaAs by comparing the results of continuum theory relating to hybrid dispersion with microscopic calculations [19] and experiment [20]. In what follows we will adopt this simplifying boundary condition.

The condition $\boldsymbol{u} = \boldsymbol{0}$ and the electromagnetic connection rules can be satisfied without hybridization for the s-polarized TO mode and by a linear combination of LO, TO and interface p-polarized modes [21, 22]. For frequencies above ω_{TO} the p-polarized TO mode is evanescent and the solution for p-polarized modes in a single quantum well is given below for the case where the LO dispersion is strong enough to cover the frequency range of the interface polaritons. The antisymmetric solution is (with k instead of q denoting wavevector)

$$u_x(z) = \mathrm{i}k_x A(\sin k_L z - s_T \sinh \alpha_T z - s_p \sinh k_x z) \tag{11.3.11a}$$

$$u_z(z) = k_L A\left(\cos k_L z - \frac{k_x^2}{k_L \alpha_T} s_T \cosh \alpha_T z - \frac{k_x}{k_L} s_p \cosh k_x z\right) \tag{11.3.11b}$$

where

$$s_T = \frac{\sin k_L a/2}{\sinh \alpha_T a/2}(1 - p_a \tanh k_x a/2) \tag{11.3.11c}$$

$$s_p = p_a \frac{\sin k_L a/2}{\cosh k_x a/2} \tag{11.3.11d}$$

$$p_a = \frac{1}{s(\tanh k_x a/2 + r)} \tag{11.3.11e}$$

with

$$\cot k_L a/2 = \frac{k_x}{k_L} p_a + \frac{k_x^2}{\alpha_T k_L}(1 - p_a \tanh k_x a/2) \coth \alpha_T a/2 \tag{11.3.12}$$

and barrier fields

$$E_x(z) = \pm \mathrm{i} E_z(z) = \pm \mathrm{i} k_x A r s \rho_0 p_a \sin(k_L a/2) \mathrm{e}^{\pm k_x (z \pm a/2)} \tag{11.3.13}$$

where the upper sign is for $z \le -a/2$ and the lower sign is for $z \ge a/2$. The symmetric solution is

$$u_x(z) = k_x A[\cos k_L z - c_T \cosh \alpha_T z - c_p \cosh k_x z] \tag{11.3.14a}$$

$$u_z(z) = \mathrm{i} k_L A\left[\sin k_L z + \frac{k_x^2}{k_L \alpha_T} c_T \sinh \alpha_T z + \frac{k_x}{k_L} c_p \sinh k_x z\right] \tag{11.3.14b}$$

where

$$c_T = \frac{\cos k_L a/2}{\cosh \alpha_T a/2}(1 - p_s \coth k_x a/2) \tag{11.3.14c}$$

$$c_p = \frac{p_s \cos k_L a/2}{\sinh k_x a/2} \tag{11.3.14d}$$

$$p_s = \frac{1}{s(\coth k_x a/2 + r)} \tag{11.3.14e}$$

with

$$\tan k_L a/2 = -\frac{k_x}{k_L} p_s - \frac{k_x^2}{\alpha_T k_L}(1 - p_s \coth k_x a/2) \tanh \alpha_T a/2 \tag{11.3.15}$$

and barrier fields

$$E_x(z) = \pm \mathrm{i} E_z(z) = -k_x A r s \rho_0 p_s \cos(k_L a/2) \mathrm{e}^{\pm k_x (z \pm a/2)} \tag{11.3.16}$$

where, once again, the upper sign is for $z \le -a/2$ and the lower sign is for

$z \geq a/2$. In these equations the field, permittivity and charge parameters are, as before,

$$s = \frac{\omega^2 - \omega_{\mathrm{TO}}^2}{\omega_{\mathrm{LO}}^2 - \omega_{\mathrm{TO}}^2}, \qquad r = \frac{\varepsilon_\infty}{\varepsilon_B}\frac{\omega^2 - \omega_{\mathrm{LO}}^2}{\omega^2 - \omega_{\mathrm{TO}}^2},$$
$$\rho_0 = \frac{e^*}{\varepsilon_\infty} = \left(\frac{\mu(\omega_{\mathrm{LO}}^2 - \omega_{\mathrm{TO}}^2)}{\varepsilon_\infty}\right)^{\frac{1}{2}} \tag{11.3.17}$$

where ε_B is the permittivity (assumed to be frequency independent and consistent with the assumption of rigidity) of the barrier and μ is the reduced-mass density. The amplitude A is determined by the usual quantization procedure [22]. Mode patterns are depicted in Fig. 11.1 and the dispersion is depicted in Fig. 11.2.

Modes in the barrier scatter electrons in the well via their associated interface-mode fields [23]. Summing the effects of all barrier hybrids modes turns out to be very close to the DC result for the barrier interface modes, so in this case the simpler DC model can only be used if the total rate is required.

It is straightforward to obtain the hybrid modes for a superlattice [19]. These turn out to be more complicated in that they have mixed symmetry for a general direction of propagation. Usually this complication is ignored in calculations of scattering rates.

The effect of hybridization is to combine LO, TO and interface modes into a coherent whole. Nevertheless, it is possible to distinguish LO-like and interface-like hybrids which scatter rather like the uncoupled modes [24]. Moreover, it is possible to neglect the effect of the TO component, provided the frequency is not too close to ω_{TO}, and hence treat the mode as a double, rather than as a triple, hybrid, which simplifies calculations of scattering rate. We will adopt this approximation in our illustration of scattering rates.

11.3.3 Scattering rates in a quantum well

The scattering rate associated with the polar interaction in a well of width 'a' can be written:

$$W_{ij} = W_0\hbar\left(\frac{2\hbar\omega}{m^*a^2}\right)^{\frac{1}{2}} \sum_{q_L,k}\int\left(n(\omega,q) + \frac{1}{2} \mp \frac{1}{2}\right)$$
$$\times\delta_{\mathbf{k}_{xi}\pm q_x,\mathbf{k}_{xf}}\frac{G_{ij}^2(q_x,q_L)}{Q_\alpha^2(q_x,q_L)}I^2(\mathbf{k}_f,\mathbf{k}_i)\delta(E_f - E_i)q_x\,\mathrm{d}q_x\,\mathrm{d}\theta \tag{11.3.18}$$

where $n(\omega, q)$ is the phonon occupation factor and the upper sign is for absorption, the lower sign is for emission. The strength of the interaction is quantified by $\boldsymbol{W}_0$:

$$W_0 = \frac{e^2}{4\pi\hbar}\left(\frac{1}{\varepsilon_\infty} - \frac{1}{\varepsilon_s}\right)\left(\frac{2m^*\omega}{\hbar}\right)^{\frac{1}{2}} \tag{11.3.19}$$

where ε_∞, ε_s are the high-frequency and static permittivities, m^* is the effective mass of the electron and ω is the angular frequency of the mode. Crystal momentum in the plane is conserved. $G_{ij}(q_x, q_L)$ is the overlap integral involving the envelope function of the potential, with q_x being the in-plane wave vector and q_L being the vector normal to the interfaces restricted by boundary conditions. $I(\boldsymbol{k}_f, \boldsymbol{k}_i)$ is the cell overlap integral (equation (11.2.5)). E_i and E_f are the initial and final energies of the system. $Q_\alpha(q_x, q_L)$ is the energy-normalizing wave vector, with $\alpha = s$ (symmetric mode) or a (antisymmetric mode). The sum is over all allowed values of q_L and the integral is over all values of the in-plane wave vector, with θ being the angle between in-plane phonon and electron wave vectors. Equation (11.3.18) applies for all phonon models provided that for DC interface modes q_L is replaced by q_x.

The scattering rate depends on many factors. Here we will focus on its dependence on well width. This can be illustrated best by the so-called threshold emission rates. For an intrasubband transition this is the rate for an electron with in-plane kinetic energy equal to the phonon energy, which defines the in-plane phonon wavevector to be:

$$q_x = (2m^*\omega/\hbar)^{\frac{1}{2}} = q_0. \tag{11.3.20}$$

As regards intersubband transitions, we look at the rate for a transition from the bottom of subband 2 to a state in subband 1, which defines the operative phonon wave vector to be:

$$q_x = (2m^*(\Delta E - \hbar\omega)/\hbar^2)^{\frac{1}{2}} = q_1 \tag{11.3.21}$$

where ΔE is the subband energy separation. The threshold rates for the AlAs/GaAs system are shown in Fig. 11.3 for the hybrid model and for the DC model.

Despite the different mode patterns, the rates given by the two models agree remarkably well. Several authors have commented on a sum rule for rates which would apply strictly provided that the coupling strength was independent of frequency [25–27]. If this condition held any complete set of modes however chosen it would give the same result. The condition holds approximately for LO modes but not for interface modes. Thus, using a complete set of bulk-like LO modes could not be expected to give valid results. Both DC and hybrid models take proper account of the variation with frequency of the coupling strength of interface components in this particular case and reach approximately the same result. However, it must be stressed that agreement of this sort cannot be relied upon in all structures. Nevertheless, even where the DC and hybrid models predict markedly different spectra—as, for example, in the case of a

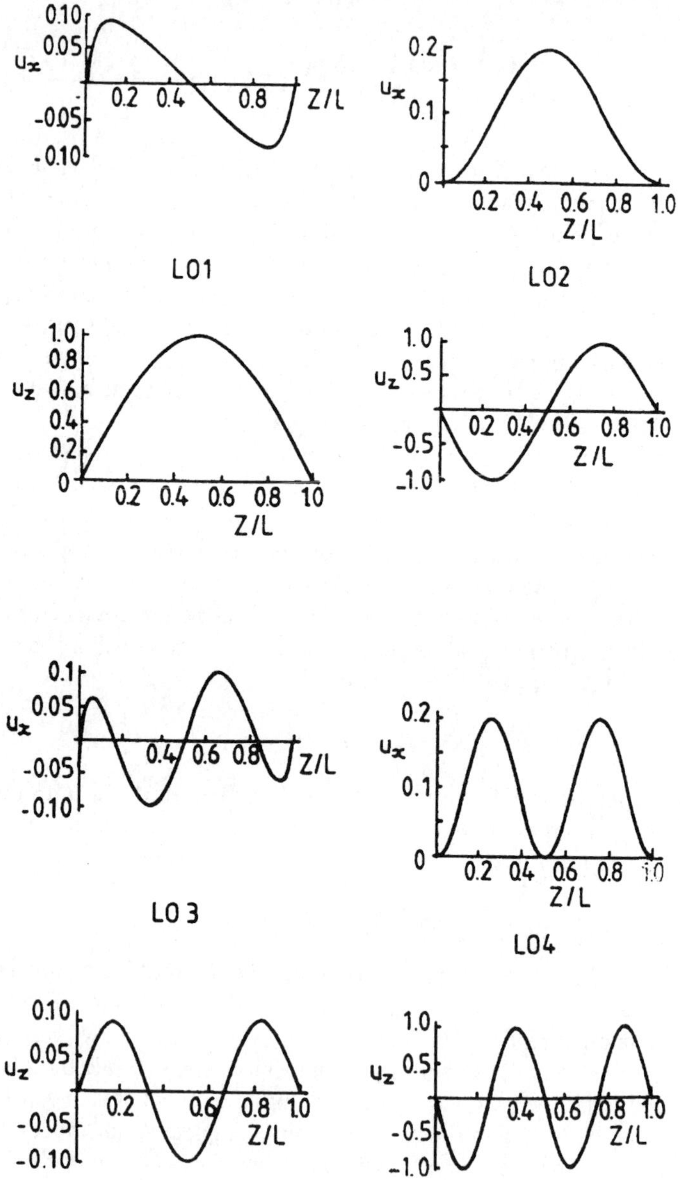

Figure 11.1 *Patterns of optical displacement for hybrid modes in a single quantum well. (a) LO-like modes in a 38 Å well for $k_x a \to 0$.*

well containing a monolayer—the intrasubband and intersubband scattering rates appear to be the same for the two models [10].

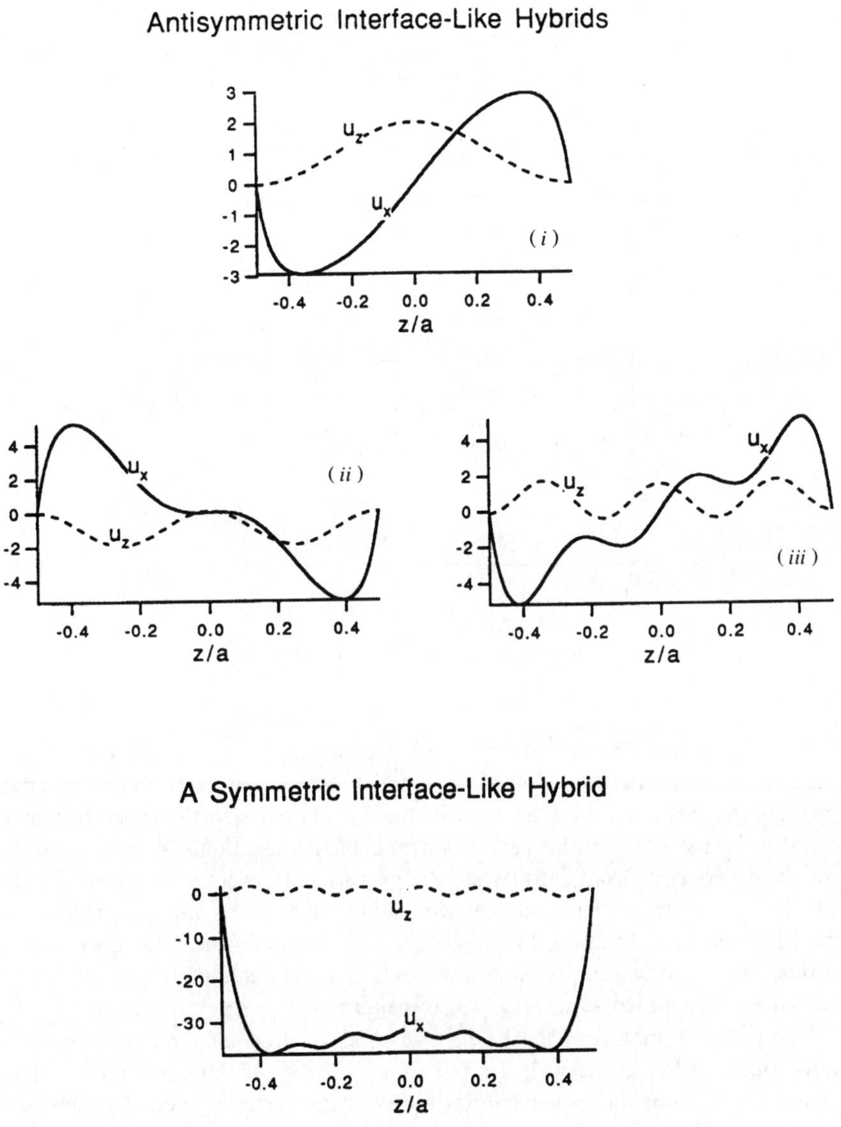

Figure 11.1 *(Continued.) (b) Interface-like modes in a 50 Å well at anticrossings.*

The dependence on well width of the rates associated with the modes at the frequencies of the well material—W modes—illustrate the well-known property of the strength of any long-range polar interaction to fall off with increasing wave vector. In the simplest cases, $q_L = n\pi/a$, where n is an integer, and so the contribution to the intrasubband rate from the LO components vanishes for

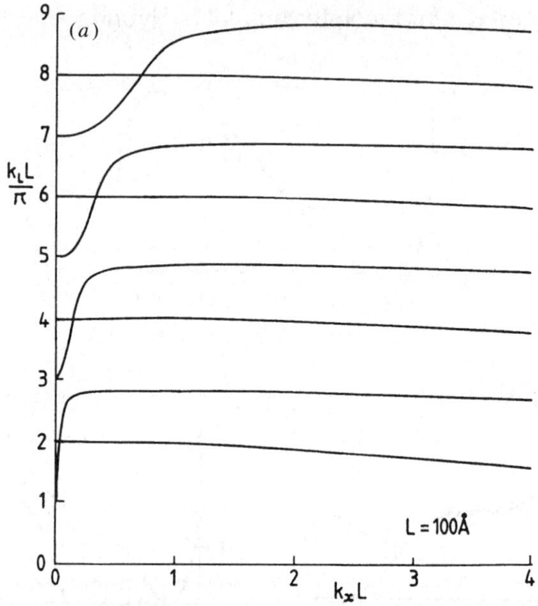

Figure 11.2 *Dispersion (a) in a 100 Å well.*

$a = 0$. The same is true for the interface component, but for a different reason; in this case the frequency of the interactive mode, which is the symmetric one, approaches ω_{TO}, the frequency of the TO phonon and the mode becomes electrically inactive. In the case of intersubband transitions the wave vector involved (q_1) rises from zero (when $\Delta E = \hbar\omega$) and becomes larger and larger as the well narrows. At moderate well widths this effect already makes the intersubband rate significantly smaller than the intersubband rate. Less than a certain width the second subband disappears into the continuum and one has to consider a new process, namely, that of capture.

The bulk LO modes of the barrier—*B* modes—produce symmetric and antisymmetric fields in the well very like those of the DC interface modes. The symmetric component responsible for intrasubband scattering actually increases its strength up to its maximum (that of the LO mode) as the well narrows, so it is this mode which maintains scattering in the limit of a narrow well. The antisymmetric component responsible for intersubband scattering is much weaker.

Schemes to suppress polar–optical-phonon scattering are based on the weakening of the interaction with increasing wave vector. These include narrowing the well [28] or artificially narrowing the well by including one or more monolayers [9] (but also see [10]). In all cases the counteracting effect of the *B* modes remains a problem and this has provoked the idea of incorporating metal layers

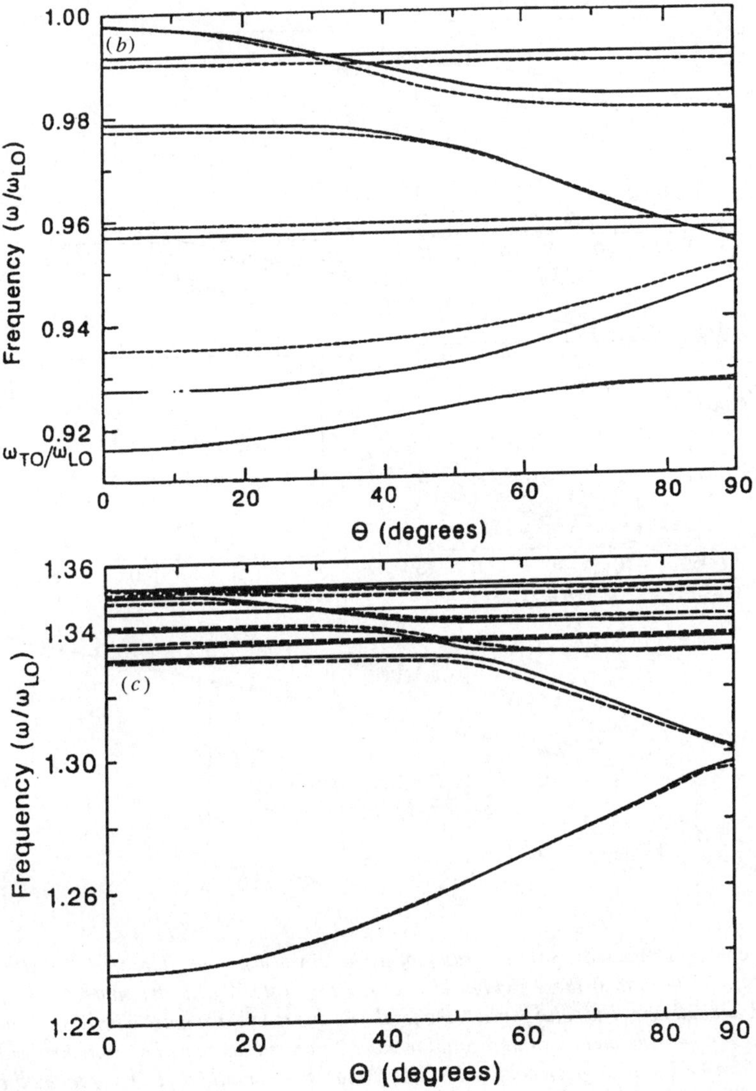

Figure 11.2 *(Continued.) (b) GaAs modes in a GaAs (29.4 Å)/AlAs (32.6 Å) superlattice as a function of the angle between the direction of propagation and the superlattice (growth) axis. Solid curves: microscopic model; dashed curves: hybrid model. (c) AlAs modes in a GaAs (20 Å)/AlAs (20 Å) superlattice. Key as for (b).*

to short out the fields of the B modes [11]. A better approach, though technically difficult, is to replace the polar barriers with nonpolar barriers. Even so, in order to optimize the effect it is necessary to choose a nonpolar barrier with a

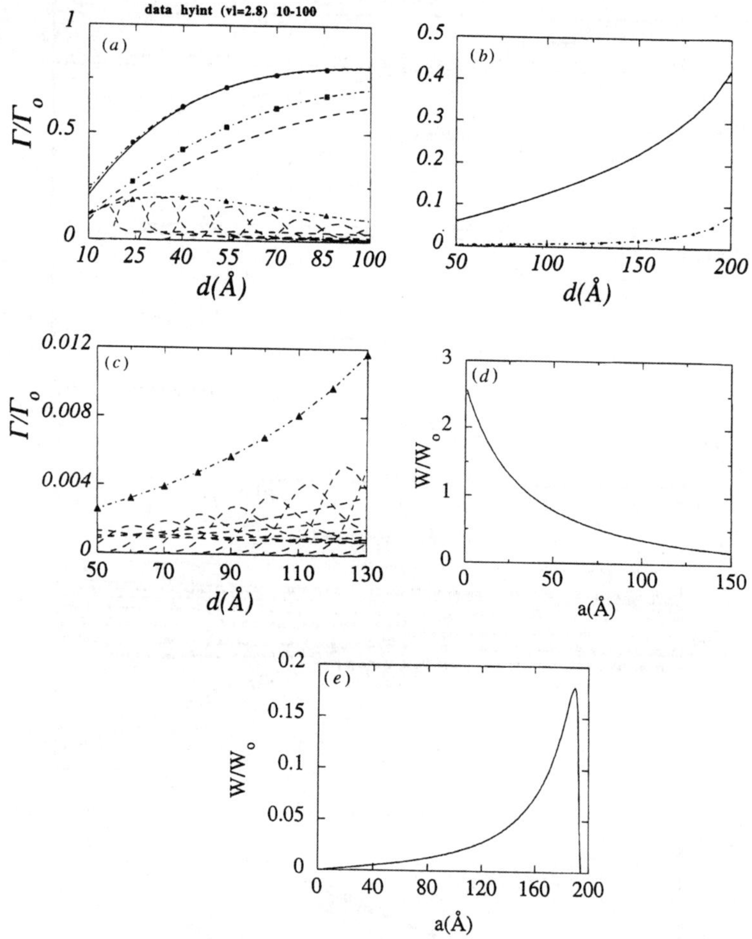

Figure 11.3 *Threshold scattering rates in a GaAs well as a function of well width ((d) or (a)). (Note that in these figures the normalizing rate,* Γ_0*, is the same as* W_0*.) (a) Intrasubband rate. The solid curve is the total rate due to hybrid modes with dashed curves depicting the contribution of individual modes. Chain curves show DC model rates from interface modes (triangles), confined modes (squares) and total (circles). (b) Intersubband rate. The solid curve is the total rate due to hybrid modes; it coincides with the total DC model rate. The chain curve with triangles shows the contribution to the DC rate from the GaAs interface mode. (c) Intersubband rate. The chain curve with triangles is as for (b) and the dashed curves show the contributions from higher-order hybrids which exhibit resonances at the anticrossings where the modes are interface-like. (d) Intrasubband rate due to AlAs barrier modes. Again, DC and hybrid rates agree. (Note that* W_0 *is the GaAs parameter.) (e) Intersubband rate due to AlAs barrier modes.*

permittivity at the W mode frequency as near as possible to the high-frequency permittivity of the well material, as illustrated in Fig. 11.4.

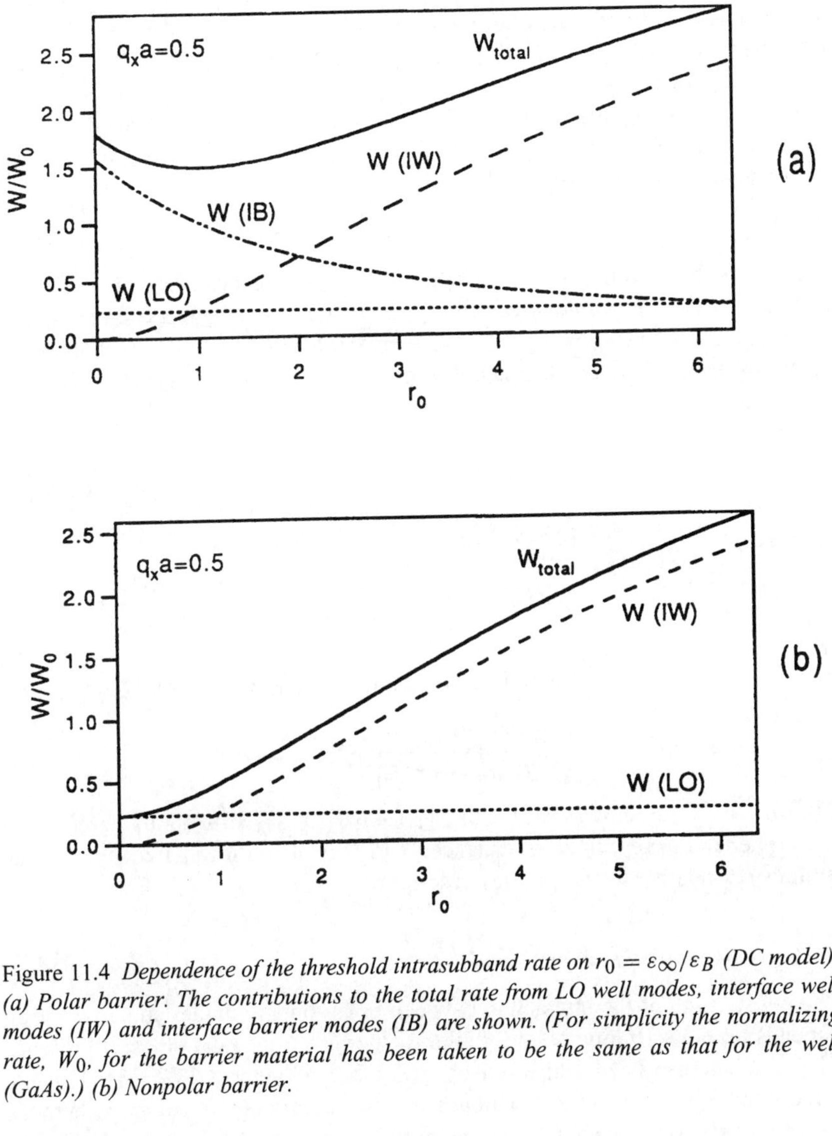

Figure 11.4 *Dependence of the threshold intrasubband rate on* $r_0 = \varepsilon_\infty/\varepsilon_B$ *(DC model). (a) Polar barrier. The contributions to the total rate from LO well modes, interface well modes (IW) and interface barrier modes (IB) are shown. (For simplicity the normalizing rate,* W_0*, for the barrier material has been taken to be the same as that for the well (GaAs).) (b) Nonpolar barrier.*

11.3.4 Scattering in a two-mode alloy

Several heterostructures commonly studied consist of one or more alloys. The interaction of an electron with the optical modes of such a system can be considerably modified. Most strikingly this is true of the situation in Si/Ge alloys where now a polar interaction appears that was entirely absent in Si and Ge on their own. In the case of ternary III–V alloys there are two sets of optical modes to interact with and more sets in quaternaries etc. A number of authors have studied two-mode alloys of the type $A_xB_{1-x}C$ in which there are AC- and BC-like modes [29–31], we will confine our attention to these and give a brief summary of their results.

Usually it is assumed that the TO frequencies remain unchanged but that the different polarizabilities of the two compounds combine to give a mean high-frequency permittivity of the form:

$$\varepsilon_\infty = x\varepsilon_{\infty 1} + (1-x)\varepsilon_{\infty 2} \tag{11.3.22}$$

together with a permittivity ε whose frequency dependence stems from a linear combination of the two polarizabilities. The new LO frequencies are determined from the condition $\varepsilon = 0$. One can then define the strength of coupling to electrons for each mode. The result is that the factor W_0 of equation (11.3.19) is replaced by

$$W_{0i} = \frac{e^2}{4\pi\hbar}\left(\frac{2m^*\omega_{\mathrm{LO}i}}{\hbar}\right)^{\frac{1}{2}}\frac{1}{\varepsilon_i^*} \tag{11.3.23}$$

where:

$$\frac{1}{\varepsilon_i^*} = \left|\frac{(\omega_{\mathrm{LO}i}^2 - \omega_{\mathrm{TO1}}^2)(\omega_{\mathrm{LO}i}^2 - \omega_{\mathrm{TO2}}^2)}{\varepsilon_\infty\omega_{\mathrm{LO}i}^2(\omega_{\mathrm{LO2}}^2 - \omega_{\mathrm{LO1}}^2)}\right| \tag{11.3.24}$$

and $i = 1$ or 2. Note that a simple prescription would have given:

$$\frac{1}{\varepsilon_i^*} = \frac{(\omega_{\mathrm{LO}i}^2 - \omega_{\mathrm{TO}i}^2)}{\varepsilon_\infty\omega_{\mathrm{LO}i}^2}. \tag{11.3.25}$$

The correct result shows that when $\omega_{\mathrm{LO2}}, \omega_{\mathrm{TO2}} > \omega_{\mathrm{LO1}}, \omega_{\mathrm{TO1}}$ the strength of the higher-frequency mode is enhanced and that of the lower-frequency mode diminished relative to the simple prescription.

11.3.5 Screening and coupled-mode effects

The net strength of the interaction between polar–optical modes and an electron depends on the screening effect of all the other electrons (and holes, if present). The problem has been addressed by a number of authors [32, 33]. Dynamic effects and multisubband occupation make the situation quite complex; here we can give only a brief summary of the principal results.

Basically, screening depends on the relation of the plasma frequency of the electrons to the LO frequency and on the efficacy of Landau damping at the wave vectors and temperature involved. In a quasi-two-dimensional gas the square of the plasma frequency rises linearly with electron density and wavevector. As it approaches the LO frequency linearly the coherent oscillations which the electrons make in response to the LO field are out of phase and as a result antiscreening occurs and the interaction with an individual electron is actually enhanced. The size of this effect is severely determined by the degree of Landau damping, i.e. the rate at which the energy of coherent oscillations of the electrons is absorbed by single-electron excitation. When the plasma frequency is much higher than the LO frequency the electron oscillations can respond rapidly to the disturbance created by the LO fields and the latter tend to be screened out. In this case a static-screening approximation is often used, although, once again, the effect of Landau damping must be taken into account. Indeed, it is often considered that in many cases the screening of the polar–optical interaction can be ignored entirely, such being the strength of Landau damping.

11.3.6 The nonpolar interaction

In both polar and nonpolar materials there is an interaction additional to any polar interaction that is associated with the change of energy of the electron caused by the deformation of the lattice, quantified by a deformation-potential constant, D, such that the interaction energy is $\boldsymbol{D} \cdot \boldsymbol{u}$. $\boldsymbol{D}$ is a nonzero vector directed along the $\langle 111 \rangle$ direction for intravalley scattering of electrons in L-valleys and it is also nonzero for the light and heavy holes, but for electrons in a central valley it is zero, at least to zero order. A similar type of interaction is responsible for intervalley scattering. This form of interaction is independent of wavevector and is therefore not affected as markedly as the polar interaction by phonon confinement. The scattering rate will be affected by electron confinement through the general dependence of the scattering rate on the density of final states [34]. The rate for a particle in a parabolic band in an infinitely deep quasi-two-dimensional well is:

$$W = \frac{3D_0^2 m^*}{4\rho\hbar^2\omega a}\langle|I(\boldsymbol{k}', \boldsymbol{k})|^2\rangle \tag{11.3.26}$$

where D_0 is the directionally averaged optical deformation constant and ρ is the mass density. For the case of holes the cell-periodic overlap integral (11.2.5) must be directionally averaged and the degree of confinement must be taken into account.

11.4 ACOUSTIC PHONONS

Acoustic phonons scatter via the deformation potential and, in III–V and II–VI compounds, via the piezoelectric interaction. Like optical–phonon scattering it provides a means whereby the electrons can relax energy, though it is generally a much weaker channel. It becomes important at low temperatures when electrons do not have sufficient energy to emit optical phonons—in GaAs, at temperatures below about 40 K.

Long-wavelength acoustic modes are never confined in quantum wells or superlattices, as optical modes are, because there are always common frequencies of acoustic vibration in well and barrier materials. Mechanical mismatch at the interfaces only affects the dispersion, in that gaps can appear in the frequency spectrum. For a superlattice the dispersion is given by [35]:

$$\cos k_z(a+b) = \cos k_a a \cos k_b b - [(1+\eta^2)/2\eta] \sin k_a a \sin k_b b \qquad (11.4.1)$$

where k_z is the superlattice wave vector and k_a, k_b are the wave vectors in the adjacent layers corresponding to the same frequency, ω. The factor η is given by

$$\eta = \frac{\rho_b v_b}{\rho_a v_a} \qquad (11.4.2)$$

where v_a, v_b are the phase velocities and $\rho_a \rho_b$ are the mass densities. Each velocity is determined by the elastic constant c and density ρ according to the usual equation $v = (c/\rho)^{1/2}$. Superlattice zone boundaries appear at $k_z = n\pi/d$, where $d = (a+b)$. The spectrum is said to be folded. Gaps that appear at the zone boundaries are often ignored in computing the scattering rate and it is usual to take the simplifying approximation that acoustic modes are essentially bulk-like.

The scattering rate can be written:

$$W(\boldsymbol{k}) = \int \sum_{q} W_{ac} f(q)(n(\omega) + \tfrac{1}{2} \pm \tfrac{1}{2})|G(\boldsymbol{k}', \boldsymbol{k}, q)|^2 \times \delta(E(\boldsymbol{k}') - E(\boldsymbol{k}) \pm \hbar\omega(q))\, \mathrm{d}\boldsymbol{k}' \qquad (11.4.3)$$

where

$$W_{ac} = \frac{C_{ac}^2}{8\pi^2 \rho\omega}, \qquad C_{ac} = \Xi \quad \text{or} \quad eK_{av}(c_{av}/\varepsilon_s)^{1/2}. \qquad (11.4.4)$$

Here, ρ is the mass density, Ξ is the deformation potential, K_{av} is the directional average of the electromechanical constant and c_{av} is a similar average of the elastic constant. For the deformation interaction $f(q) = q^2/S(q)$ and for the piezoelectric interaction $f(q) = 1/S(q)$, where $S(q)$ is the screening function.

Acoustic–phonon scattering in semiconductor multilayers is significantly more complicated than in three dimensions, even when folding is ignored. In the three-dimensional case it is usually sufficient to regard the phonon energy as small compared with the energy of the electron, which may allow us to make a quasi-elastic approximation and in most cases to assume equipartition: $n(\omega) = (k_B T/\hbar\omega) \gg 1$. Moreover, the plasma frequency is usually much greater than the phonon frequency, so that the screening can be regarded as essentially static. In the low-dimensional case none of these assumptions can be taken for granted, for the simple reason that crystal momentum is no longer strictly conserved along the confinement direction (which we will continue to take parallel to the z-direction). The overlap integral, shown in Fig. 11.5 for infinitely deep-potential electron wave functions, allows q_z up to, roughly, π/a for intrasubband scattering and $2\pi/a$ for the lowest intersubband scattering, while momentum conservation in the plane restricts q_x to order k_x, the electron wave vector.

It is instructive to analyse the conservation conditions, taking q_z to be an

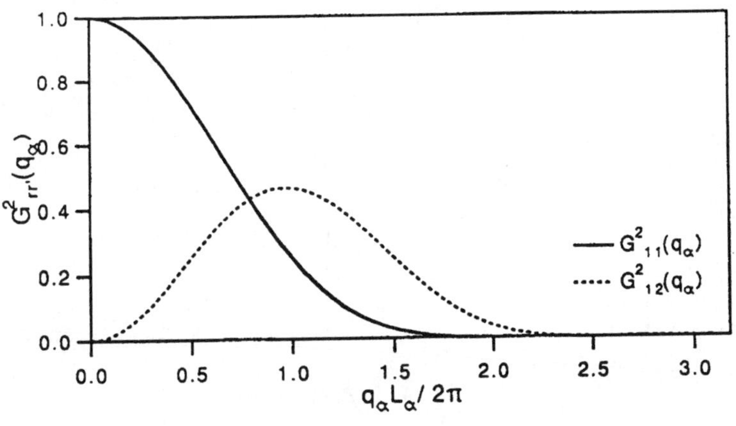

Figure 11.5 *Overlap integrals for intrasubband and intersubband transitions involving bulk phonons in an infinitely deep well.*

unrestricted variable and $\hbar\omega = \hbar s\sqrt{(q_x^2 + q_z^2)}$, where s is the velocity of the acoustic wave [36]. One can easily show that for intrasubband transitions

$$q_z = \frac{\hbar}{2m^*s}|q_x|\sqrt{\left(q_x \pm 2k_x\cos\theta + \frac{2m^*s}{\hbar}\right)\left(q_x \pm 2k_x\cos\theta - \frac{2m^*s}{\hbar}\right)} \tag{11.4.5}$$

where θ is the angle in the plane between q_x and k_x and the upper (lower) sign is for absorption (emission). In one dimension $\cos\theta$ is replaced by unity. Now the wave vector $2m^*s/\hbar$ is a very small quantity so unless $q_x \approx \mp 2k_x\cos\theta$, $q_z \gg q_x$. This will be the case, therefore, for most of the scattering events, which means that if q_z was truly unrestricted, the phonon energy would be large and most of the scattering would be inelastic, quite unlike the situation in three dimensions. The form-factor restriction implies that the condition for inelastic scattering is:

$$E \leq \hbar s\pi/a \tag{11.4.6}$$

where E is the electron energy. For $s = 5 \times 10^5$ cm s^{-1} and $a = 50$ Å the phonon energy is about 2 meV, corresponding to a temperature of about 24 K. For intersubband scattering the energy difference between the subbands is usually sufficient for the phonon energy to be neglected and the transitions to be regarded as quasi-elastic.

The large magnitude of q_z is also a factor in determining phonon occupancy in collisions involving absorption or stimulated emission. The condition for equipartition to hold is:

$$k_BT > \hbar sn\pi/a \tag{11.4.7}$$

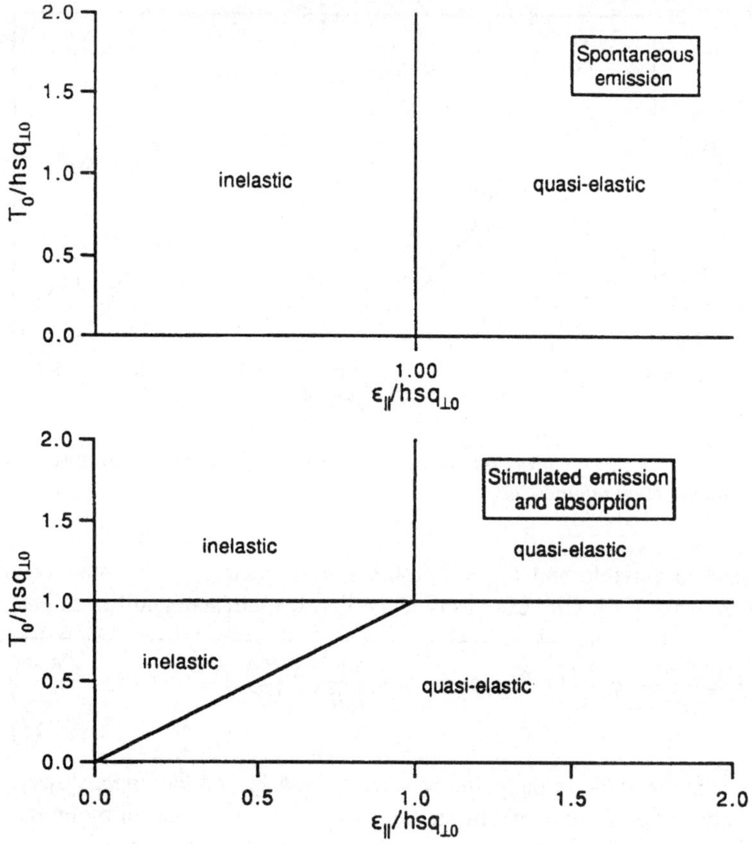

Figure 11.6 *Inelastic and quasi-elastic scattering regimes for acoustic phonons. (Notation:* $T_0 = k_B T$, $q_\perp = q_z$ *and* $\varepsilon_\parallel = E$, *the electron kinetic energy in the plane.)*

where $n = 1$ for intrasubband scattering and $n = 2$ for intersubband scattering. When this condition fails there is an effective restriction on q_z for absorption and stimulated emission given by:

$$q_z \leq k_B T/\hbar s \qquad (11.4.8)$$

which may modify the condition for inelastic scattering. Figure 11.6 summarizes these features.

Screening presents further problems. In one and two dimensions the plasma frequency rises from zero with wave vector and it can be of the same order as the phonon frequency thereby making the screening dynamic. The carrier density is a crucial factor here. In the case of the piezoelectric interaction, which is strongest for small wave vectors, a static screening approximation is often adequate and this is often enough to reduce the piezoelectric scattering rate far below that for the nonpolar interaction. But for the nonpolar interaction, which strengthens towards high wave vectors, each case has to be treated on its merits.

We cannot pursue all these issues here; they are discussed in more detail elsewhere [16]. We will be content to quote the rate for the simplest situation—nonpolar interaction, equipartition, quasi-elastic collisions, no screening, bulk-like modes:

$$W_{ij} = \frac{\Xi^2 m^* k_B T}{\rho s^2 \hbar^3 a}\left(\delta_{ij} + \frac{1}{2}\right). \tag{11.4.9}$$

11.5 CHARGED IMPURITIES

Scattering by charged impurities is, like that by phonons, important at all temperatures irrespective of dimensionality. Unfortunately, the theory of charged-impurity scattering is beset by all sorts of problems. There is, first of all, the classic problem set by the infinite scattering cross section of a single-point charge. Then for real charged centres there are central-cell corrections to be made which depend on the detailed quantum structure of the impurity. There is the question of the applicability of the Born approximation to replace the more complex partial-wave analysis. And last, but by no means least, is the problem of dealing with an array of scattering centres. These questions have been discussed by a number of authors [37, 38]; here we will ignore central-cell corrections and accept the applicability of the Born approximation. The only simple feature in this topic is screening, which we can safely take to be static, although even that is not straightforward when the screening particles occupy a number of subbands.

A characteristic feature of a multilayer system is the scattering by remote, as well as local, impurities. There are two approaches which are commonly made that aim at addressing the problem of scattering by an array. One is the standard model that treats the interaction as being between the electron and each individual impurity independently; the other is the model that treats the interaction as being between the electron and a fluctuation from a regular array. Both models ignore the possibility of coherent scattering from more than one impurity or fluctuation.

The standard model runs the risk of producing infinite cross sections. It avoids this by relying on screening to limit the range of interaction (Brooks–Herring approach [39]), by arbitrarily limiting the range to half the distance apart of the impurities (Conwell–Weisskopf approach [40]) or by restricting the interaction to the nearest impurity (Ridley [41]). The latter approach encompasses the other two so we will quote the momentum-relaxations rates it gives for local and remote impurity scattering in a quantum well:

$$W = W_0 \int \mathrm{d}z N(z) \int_0^{\pi/2} \frac{F(q,z)P(b,z)\sin^2\alpha}{(q+q_s)^2}\,\mathrm{d}\alpha \tag{11.5.1}$$

$$W_0 = \frac{(Ze^2)^2 m^*}{\pi\hbar^3\varepsilon_s^2} \tag{11.5.2}$$

$$F(q,z) = \int_{z_1}^{z_2} \psi^2(z')\psi^2(z)\mathrm{e}^{-q|z'-z|}\,\mathrm{d}z'\,\mathrm{d}z \tag{11.5.3}$$

$$P(b,z) = \mathrm{e}^{-Ndf(b)}. \tag{11.5.4}$$

Here z is the vertical distance measured from the impurity to the electron, $q = 2k\sin\alpha$, where k is the in-plane wave vector of the electron, q_s is the static screening parameter and $N(z)$ is the impurity density. $F(q, z)$ is the intrasubband form factor and z_1, z_2 are the limits of the quantum well. Intersubband transitions are not favoured by the polar nature of the interaction and will not be considered here. $P(b, z)$ is the probability that the scattering impurity is the closest, calculated semiclassically using the impact parameter, b, when the average density is N and the average impurity separation is d. Equation (11.5.1) encompasses local and remote impurity scattering. For local impurity scattering:

$$f(b) = a|b|. \tag{11.5.5}$$

For remote impurity scattering by a population at least z_0 from the centre of the well:

$$f(b) = -Nd(b^2+z^2)\left\{\cos^{-1}\left(\frac{z_0}{\sqrt{b^2+z^2}}\right) - \frac{z_0(b^2+z^2-z_0^2)^{1/2}}{b^2+z^2}\right\}. \tag{11.5.6}$$

The in-plane impact parameter is given by:

$$b = -\frac{W_0}{2v(k)}\int_{\pi/2}^{\alpha} \frac{F(q',z)}{(q'+q_s)^2}\,\mathrm{d}\alpha' \tag{11.5.7}$$

where $v(k)$ is the velocity of the electron and $q' = q(\alpha')$.

Figure 11.7 illustrates a few cases for a GaAs quantum well in an approximation that takes the electron wave function to be a delta function. Clearly, without limiting the interaction to closest encounters the rate is grossly overestimated particularly for low energies and remote-impurity scattering. Good agreement with experiment has been obtained [41]. When the donors can be taken to be delta-function-like in space the probability function can be taken to be:

$$P(b, z_0) = \mathrm{e}^{-2N_s db}. \tag{11.5.8}$$

To a good approximation the in-plane impact parameter can be taken to be equal to the average distance apart from the donors, d. The average distance apart in a

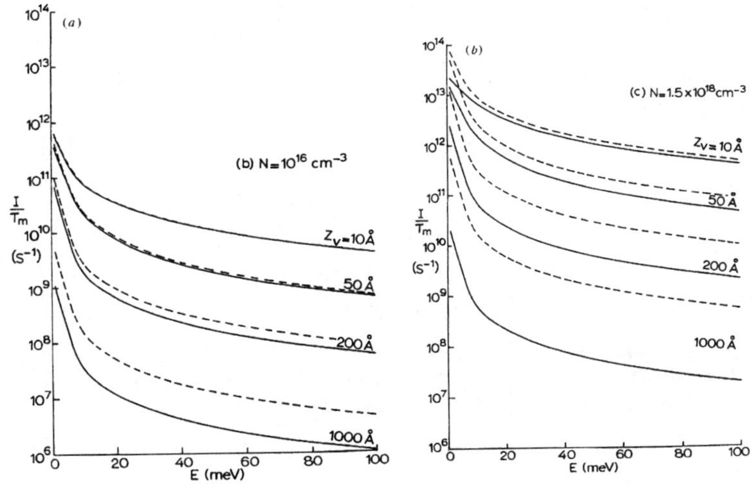

Figure 11.7 *Momentum-relaxation rates associated with statistically screened remote charged-impurity scattering in a GaAs quantum well for the spacer values (measured from the assumed delta-function wave function of the electron) shown on the curves. The dashed curves are the rates without statistical screening. (a)* $N = 10^{16}$ cm^{-3}; *(b)* $N = 1.5 \times 10^{18}$ cm^{-3}.

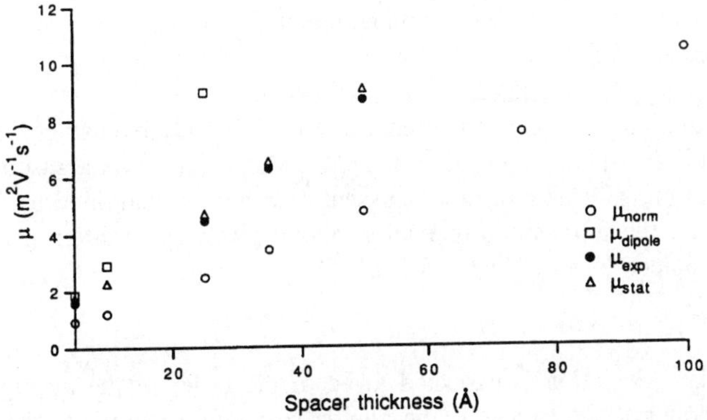

Figure 11.8 *Component of room-temperature mobility in a delta-doped GaInAs FET associated with charged-impurity scattering. Experimental results (Seaford et al. [52]) compared with normal, dipole-fluctuation and statistical-screening models.*

randomly distributed planar population is given by $\langle d^2 \rangle = (\pi N_s)^{-1}$ and hence

$$P(b, z_0) \approx \mathrm{e}^{-2/\pi} = 0.529. \tag{11.5.9}$$

This provides a very simple prescription for the reduction factor to be applied to the usual rate and gives good agreement with experiment (Fig. 11.8) [42].

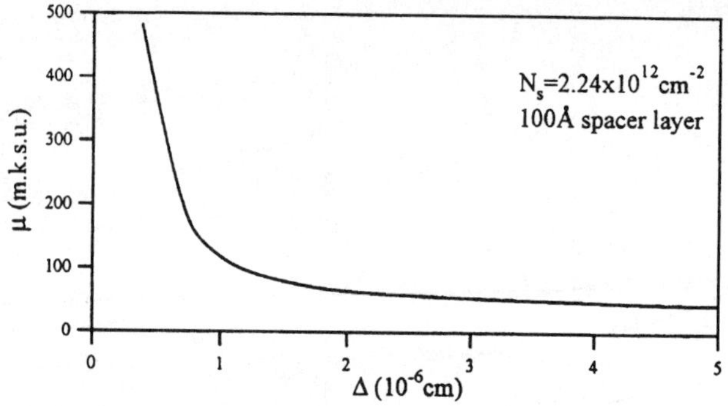

Figure 11.9 *Dependence of mobility in an InGaAs 2DEG at room temperature on the standard deviation describing fluctuations in a delta-doped layer.*

In modulation-doped structures an electron experiences the Coulomb attraction of the donors, partially screened by the other electrons, which is responsible in a simple heterostructure for the quantum confinement. It would therefore be counting the Coulomb interaction twice to sum up the scattering from every donor. Restricting the interaction to closest encounters avoids this. Another approach is based on the observation that a regular array of donors would superimpose onto the average quantizing potential a periodic component which would produce a scarcely noticeable miniband structure but would not otherwise scatter. Scattering then occurs only by the effect of fluctuations from the regular array.

This situation has been analysed by van Hall [43] by using a multipolar expansion in the spatial fluctuation in a plane at $z = z_0$. With the assumption that the fluctuations are random in a Gaussian distribution quantified by a standard deviation Δ, the momentum-relaxation rate is given by equation (11.5.1) with $P(b, z)$ replaced by $G(q)$, where:

$$G(q) = 2(1 - e^{-q^2\Delta^2/2}). \tag{11.5.10}$$

The standard deviation is expected to be of the order of the average spacing between donors. Unfortunately, the rate is extremely sensitive to the choice of Δ, as Fig. 11.9 illustrates. Nevertheless, good agreement with experiment has been reported.

11.6 INTERFACE ROUGHNESS SCATTERING

Despite spectacular progress in crystal growing over recent decades it is not always possible to grow multilayered structures with perfect interfaces or free-of-potential fluctuations. Being confined close to interfaces the electrons are highly sensitive to irregularities, which can be impurities trapped at the interface or monolayer steps. Scattering by charged impurities sitting on the interface can be treated by the methods discussed in the previous section. Here we look at the problem of scattering by geometrical irregularities in the interface, which can become the dominant mechanism at low temperatures and is seldom ignorable.

Let $\Delta(\boldsymbol{r})$ be the deviation in the z-direction (normal to the interface) of the interface from its average position. Any deviation will affect the subband energies and wave functions and thereby affect the motion of the electron in the plane. If $\Delta(\boldsymbol{r})$ is small and the subband separation is large, the wave function will not change appreciably, so we can use the unperturbed solutions in the matrix element quantifying the scattering rate. If the deviation produces a change of energy $\Delta H(\boldsymbol{r}, z)$ the scattering rate will be

$$W = \frac{2\pi}{\hbar} \int |\langle \boldsymbol{k}'|\Delta H(\boldsymbol{r}, z)|\boldsymbol{k}\rangle|^2 \delta(E' - E)\, \mathrm{d}N_f \tag{11.6.1}$$

with

$$\Delta H(\boldsymbol{r}; z) = \Delta T(\boldsymbol{r}, z) + \Delta V(\boldsymbol{r}, z) \tag{11.6.2}$$

where T is the kinetic energy and V is the potential energy.

The case of electrons confined electrostatically at a single heterojunction is complicated by space-charge effects, in general involving image charges. The situation was analysed by Ando [44]. The expression for the rate is somewhat lengthy and we will not reproduce it here. Instead, we look at the simpler case of a quantum well free of space-charge effects. Following Sakaki *et al.* [45] we can express the energy change for an intrasubband process in terms of the dependence of subband energy E on well width L, namely

$$\Delta H(\boldsymbol{r}, z) = \Delta(\boldsymbol{r}) \frac{\mathrm{d}E}{\mathrm{d}L}. \tag{11.6.3}$$

After expanding $\Delta(\boldsymbol{r})$ in a Fourier series, the squared matrix element reduces to

$$|\langle \boldsymbol{k}'|\Delta H(\boldsymbol{r}, z)|\boldsymbol{k}\rangle|^2 = |\Delta(\boldsymbol{q})|^2 \left(\frac{\mathrm{d}E}{\mathrm{d}L}\right)^2, \qquad \boldsymbol{q} = \boldsymbol{k}' - \boldsymbol{k}. \tag{11.6.4}$$

Only the power spectrum of $\Delta(\boldsymbol{r})$ is required. The usual approach is to assume, for mathematical simplicity, that the autocovariance function is isotropic and Gaussian. Thus:

$$\langle \Delta(\boldsymbol{r}')\Delta(\boldsymbol{r}' - \boldsymbol{r})\rangle = \Delta^2 \mathrm{e}^{-r^2/\Lambda^2} \tag{11.6.5}$$

and so

$$|\Delta(\boldsymbol{q})|^2 = \pi \Delta^2 \Lambda^2 \mathrm{e}^{-q^2\Lambda^2/4}. \tag{11.6.6}$$

Including static two-dimensional screening leads, with $x = \sin\theta/2$, θ being the scattering angle, to the momentum-relaxation rate

$$W = 4\left(\frac{\mathrm{d}E}{\mathrm{d}L}\right)^2 \frac{\Delta^2\Lambda^2 m^*}{\hbar^3} \int_0^1 \frac{x^4 \mathrm{e}^{-k^2\Lambda^2 x^2}}{[x + (q_s/2k)]^2(1 - x^2)^{1/2}}\, \mathrm{d}x. \tag{11.6.7}$$

The factors Δ and Λ specify the roughness, Δ being the average deviation and Λ being a length of order of the range of the deviation in the plane of the interface. Mobility measurements suggest that Δ is 1–3 ml (monolayers) (1 ml in GaAs = 2.83 Å) and Λ is 30–70 Å.

The assumption of a Gaussian autocovariance function is quite arbitrary and adopted for mathematical convenience. Experimental evidence from the study of Si/SiO_2 interfaces point rather to a simple exponential form [46], in which case:

$$|\Delta(\boldsymbol{q})|^2 = \pi\Delta^2\Lambda^2[1-(q^2\Lambda^2/2)]^{-3/2}. \tag{11.6.8}$$

In either situation, the rate is very sensitive to the well width. For an infinitely deep well, $W \propto L^{-6}$. A variation with well width of this rapidity was observed [45] in the GaAs/AlAs system.

Irregularities in the width of the well are not the only source of interface-roughness scattering. We have already mentioned space-charge fluctuations in modulation-doped systems (of a larger scale than those responsible for remote impurity scattering, but clearly in the same class). The gathering of impurities at the interface may also be a source of scattering and fluctuations in the barrier height can occur, for example in connection with alloy fluctuations. Indeed, alloy systems are particularly prone to fluctuation scattering of one sort and another. The fact of the matter is that there are a number of mechanisms that come under the title interface-roughness scattering, none of which are very well defined. Many are expected to be determined by the conditions of crystal growth in such a way that a distinct relation is likely to exist between step height and correlation length. This particular feature of interface-roughness scattering is under-researched at the present time.

11.7 ALLOY SCATTERING

The simplest picture of a ternary alloy $A_xB_{1-x}C$ is of a virtual crystal having properties which are some average of the properties of the two binaries AC and BC (a scheme which is obviously extendable to quaternaries). Thus, if V_A and V_B are the potentials experienced by the electron associated with the cations A and B in the corresponding binaries, the virtual-crystal average, in the simplest model, can be taken to be

$$V = V_A(x) + V_B(1-x). \tag{11.7.1}$$

There are bound to be fluctuations of concentration during growth. If in a certain region the concentration of A is x', rather than the average x, then the deviation in potential will be

$$\Delta V(\boldsymbol{r}, z) = (V_A - V_B)(x' - x) \tag{11.7.2}$$

and this produces scattering. The root-mean square (rms) deviation is just that for a binomial distribution

$$\langle \Delta V \rangle = |V_A - V_B| \left(\frac{x(1-x)}{N_c} \right)^{1/2} \tag{11.7.3}$$

where N_c is the number of cation sites.

Expanding $\Delta V(\boldsymbol{r}, z)$ in a Fourier series we can write the quasi-two-dimensional rate as

$$W_{n',n} = \frac{2\pi}{\hbar}(V_A - V_B)^2 x(1-x) N(E) \Omega_0 F_{n',n} \tag{11.7.4}$$

where $N(E) = m^*/2\pi\hbar^2 L$, Ω_0 is the volume of the unit cell and

$$F_{n',n} = \int_{-\infty}^{\infty} \left(\int_0^L \psi_{n'}(z) \mathrm{e}^{\mathrm{i}q_z z} \psi_n(z)\, \mathrm{d}z \right)^2 \frac{L}{2\pi}\, \mathrm{d}q_z. \tag{11.7.5}$$

Here n, n' denote subbands. Alloy scattering is elastic and isotropic and therefore scattering and momentum-relaxation rates are identical.

The potentials V_A and V_B are not firmly known. The difference $|V_A - V_B|$ is obtained from the respective pseudopotentials of the band-structure or from the band-edge discontinuities or from the electron affinities. Values of 0.5 eV for GaInAs and 1.0 eV for AlInAs are often used.

11.8 OTHER SCATTERING

The scattering mechanisms discussed hitherto are for electrons in quasi-two-dimensional structures and they are those that affect the drift and energy relaxation. Another important electron scattering mechanism is electron–electron scattering. Naturally, this mechanism cannot relax the total energy and momentum of the electron gas, but, in bulk and quasi-two-dimensional systems, it is influential in the case of injected fast electrons and it is important for determining the form of distribution function when the electrons are hot. There are a number of subtleties in working out the scattering rate and this may account for differences in published rates up to a factor of 4 [47, 48, 16].

In quasi-one-dimensional systems, restrictions on energy and momentum conservation renders electron–electron scattering ineffective. Besides this striking property, scattering processes in quasi-one dimension are influenced by the characteristic form of the density of states function.

Electron–hole scattering is less complicated in that the particles involved are distinguishable. All the remarks made in the context of charged-impurity scattering apply here. In addition, one must take into account the fact that the mass of the hole is finite and this means replacing the mass of the incident particle by the reduced mass of the electron–hole pair.

Finally, a mention must be made to phonon scattering. In a system disturbed either optically or electrically the response and steady state depend on the scattering of phonons as well as the scattering of electrons. Optical phonons emitted by electrons decay into other phonons at a rate determined by the anharmonic interaction [49–51]. In GaAs, for example, the lifetime of an LO phonon at room temperature is about 4 ps and this is to be compared with the electron–LO-phonon scattering time of about 200 fs. When the electron density is high enough, the phonon population is seriously disturbed and so-called hot-phonon effects, such as the reduction in the net energy-relaxation rate of the electrons, can be observed. Usually the momentum-relaxing processes experienced by the phonon are strong enough to inhibit the phonons from sharing the drift of the electrons in an experiment on transport [51], otherwise an instability would result. This is not the case for acoustic phonons as the well-known acousto-electric effect in bulk material testifies.

REFERENCES

[1] Burt, M. G. (1992) *J. Phys.: Condens. Matter* **4** 6651.

[2] Foreman, B. A. (1995) *Phys. Rev.* B **52** 12241.
Foreman, B. A. (1996) *Phys. Rev.* B **54** 1909.

[3] Foreman, B. A. (1994) *Phys. Rev.* B **49** 1757.

[4] Fuchs, R. and Kliewer, K. L. (1965) *Phys. Rev.* **140A** 2076.

[5] Jones, W. E. and Fuchs, R. (1971) *Phys. Rev.* B **4** 3581.

[6] Lassnig, R. (1984) *Phys. Rev.* B **30** 7132.

[7] Born, M. and Huang, K. (1954) *Dynamical Theory of Crystal Lattices* Clarendon, Oxford.

[8] Babiker, M., Constantinou, N. C. and Ridley, B. K. (1993) *Phys. Rev.* B **48** 2236.

[9] Pozela, J., Juciene, V. and Pozela, K. 1995 *Semicond. Sci. Technol.* **10** 1076.

[10] Ridley, B. K., Babiker, M., Zakhleniuk, N. *et al.* (1996) *Proc. ICPS-23*, Berlin.

[11] Constantinou, N. C. (1993) *Phys. Rev.* B **48** 11 931.

[12] Bennett, C. R., Constantinou, N. C., Babiker, M. *et al.* (1995) *J. Phys.: Condens. Matter* **7** 9819.

[13] Sood, A. K., Menendez, J., Cardona, M. *et al.* (1985) *Phys. Rev. Lett.* **54** 2111.
Sood, A. K., Menendez, J., Cardona, M. *et al.* (1985) *Phys. Rev. Lett.* **54** 2115.

[14] Rucker, H., Molinari, E. and Lugli, P. (1991) *Phys. Rev.* B **44** 3463.
Rucker, H., Molinari, E. and Lugli, P. (1992) *Phys. Rev.* B **45** 6747.

[15] Foreman, B. A. (1995) *Phys. Rev.* B **52** 12260.

[16] Ridley, B. K. (1996) *Electrons and Phonons in Semiconductor Multilayers*, Cambridge University Press, Cambridge.

[17] Perez-Alvarez, R., Garcia-Moliner, F., Velasco, V. R. *et al.* (1993) *J. Phys.: Condens. Matter* **5** 5389.

[18] Ridley, B. K., Al-Dossary, O., Constantinou, N. C. *et al.* (1994) *Phys. Rev.* B **50** 11 701.

[19] Chamberlain, M. P., Cardona, M. and Ridley, B. K. (1993) *Phys. Rev.* B **48** 14 356.

[20] Constantinou, N. C., Al-Dossary, O. and Ridley, B. K. (1993) *Solid State Commun.* **86** 191.
Constantinou, N. C., Al-Dossary, O. and Ridley, B. K. (1993) *Solid State Commun.* P**87** 1087.

[21] Trallero-Giner, C., Garcia-Moliner, F., Velasco, V. R. *et al.* (1992) *Phys. Rev.* B **45** 11 944.
[22] Ridley, B. K. (1992) *Proc. SPIE* **1675** 492.
Ridley, B. K. (1993) *Phys. Rev.* B **47** 4592.
[23] Ridley, B. K. (1994) *Phys. Rev.* B **49** 17 253.
[24] Constantinou, N. C. and Ridley, B. K. (1994) *Phys. Rev.* B **49** 17 065.
[25] Herbert, D. C. (1973) *J. Phys. C: Solid State Phys.* **6** 2788.
[26] Mori, N. and Ando, T. (1989) *Phys. Rev.* B **40** 6175.
[27] Register, L. F. (1992) *Phys. Rev.* B **45** 8756.
[28] Sawaki, N. (1986) *J. Phys. C: Solid State Phys.* **19** 4965.
[29] Chang, I. F. and Mitra, S. S. (1971) *Adv. Phys.* **20** 359.
[30] Kim, O. K. and Spitzer, W. G. J. (1979) *J. Appl. Phys.* **50** 4362.
[31] Nash, K. J., Skolnick, M. S. and Bass, S. J. (1987) *Semicond. Sci. Technol.* **2** 329.
[32] Das Sarma, S. and Mason, B. A. (1985) *Phys. Rev.* B **31** 5536.
[33] Das Sarma, S., Jain, J. and Jalabert, R. (1988) *Phys. Rev.* B **37** 1228.
Das Sarma, S., Jain, J. and Jalabert, R. (1988) *Phys. Rev.* B **37** 4560.
[34] Ridley, B. K. (1982) *J. Phys. C: Solid State Phys.* **15** 5899.
[35] Sytov, S. M. (1956) *Sov. Phys. Acoustics* **2** 67.
[36] Ridley, B. K. and Zakhleniuk, N. (1996) *JETP Lett.* **63** 465.
[37] Moore, E. J. (1967) *Phys. Rev.* **160** 607.
Moore, E. J. (1967) *Phys. Rev.* **160** 618.
[38] Lancefield, D., Adams, A. R. and Fisher, M. A. (1987) *J. Appl. Phys.* **62** 2342.
[39] Brooks, H. and Herring, C. (1951) *Phys. Rev.* **83** 879.
[40] Conwell, E. M. and Weisskopf, V. F. (1950) *Phys. Rev.* **77** 388.
[41] Ridley, B. K. (1977) *J. Phys. C: Solid State Phys.* **10** 1589.
[42] Ridley, B. K. (1996) *Semicond. Sci. Technol.* **11** 1.
[43] van Hall, P. J. (1989) *Superlat. Microstruct.* **6** 213.
[44] Ando, T. (1982) *J. Phys. Soc. Japan* **51** 3900.
[45] Sakaki, H., Noda, T., Hirakawa, K. *et al.* (1987) *Appl. Phys. Lett.* **51** 1934.
[46] Goodnick, S. M., Ferry, D. K., Wilmsen, C. W. *et al.* (1985) *Phys. Rev.* B **32** 8171.
[47] Mosko, M., Moskova, A. and Cambel, V. (1995) *Phys. Rev.* B **51** 16 860.
[48] Goodnick, S. M. and Lugli, P. (1988) *Solid State Elecron.* **31** 463.
Goodnick, S. M. and Lugli, P. (1992) *Hot Carriers in Semiconductor Nanostructures* (ed. J. Shah), Academic Press, New York.
[49] Klemens, P. G. (1966) *Phys. Rev.* **148**, 845.
[50] Ridley, B. K. and Gupta, R. (1991) *Phys. Rev.* **B43**, 4939.
[51] Ridley, B. K. (1996) *J. Phys. Condens. Matter* **8** L511.
[52] Seaford, M. L., Martin, G., Hartzell, D., Massie, S. and Eastman, L. F. (1996) *J. Electron. Mater.* **25** 1551.

Index